Peroxiporins

To produce energy, aerobic organisms transform oxygen molecules into water. This reductive mechanism yields toxic radical intermediates, collectively known as reactive oxygen species (ROS). Paradoxically, these physiological processes entail the production of potentially damaging species. Evolution has turned this apparent disadvantage into an opportunity for transmitting information. As a result, redox signaling within the cell is an efficient, exquisitely organized process. A key element for its regulation is the physical separation of sources and targets into different cell compartments. Peroxiporins, H_2O_2-transporting proteins spanning biological membranes, distribute the signal from emitters to receptors. Thus, these channels are strategically situated in the thin line between life and death, guaranteeing adequate but safe signaling.

Key Features:

- Provides a brief history of the discovery and characterization of peroxiporins
- Reviews key findings on hydrogen peroxide transport across biological membranes
- Discusses the cartography of redox signaling in crowded cells
- Includes lavish illustrations and comprehensive images to facilitate teaching
- Highlights recent findings, outstanding controversies, and open questions

Oxidative Stress and Disease

Series Editors: Helmut Sies, MD Enrique Cadenas, MD, PhD
University of Southern California School of Pharmacy Los Angeles, California

Peroxiporins
Redox Signal Mediators In and Between Cells
edited by Iria Medraño-Fernandez, Gerd Patrick Bienert, and Roberto Sitia

Physiological Consequences of Brain Insulin Action
edited by André Kleinridders

Oxidative Eustress in Exercise Physiology
edited by James N. Cobley and Gareth W. Davison

Redox Regulation of Differentiation and De-differentiation
edited by Carsten Berndt and Horst Lillig

Oxidative Stress in Cancer, AIDS, and Neurodegenerative Diseases
edited by Luc Montagnier, René Olivier, and Catherine Pasquier

Understanding the Process of Aging:
The Roles of Mitochondria, Free Radicals, and Antioxidants
edited by Enrique Cadenas and Lester Packer

Redox Regulation of Cell Signaling and Its Clinical Application
edited by Lester Packer and Junji Yodoi

Antioxidants in Diabetes Management
edited by Lester Packer, Peter Rösen, Hans J. Tritschler, George L. King, and Angelo Azzi

Free Radicals in Brain Pathophysiology
edited by Giuseppe Poli, Enrique Cadenas, and Lester Packer

Nutraceuticals in Health and Disease Prevention
edited by Klaus Krämer, Peter-Paul Hoppe, and Lester Packer

Environmental Stressors in Health and Disease
edited by Jürgen Fuchs and Lester Packer

Handbook of Antioxidants
Second Edition, Revised and Expanded
edited by Enrique Cadenas and Lester Packer

For more information about this series, please visit: www.crcpress.com/Oxidative-Stress-and-Disease/book-series/CRCOXISTRDIS

Peroxiporins

Redox Signal Mediators In and Between Cells

Edited by
Iria Medraño-Fernandez,
Gerd Patrick Bienert,
and Roberto Sitia

CRC Press
Taylor & Francis Group
Boca Raton London New York

CRC Press is an imprint of the
Taylor & Francis Group, an **informa** business

Designed cover image: Iria Medraño-Fernandez

First edition published 2024
by CRC Press
6000 Broken Sound Parkway NW, Suite 300, Boca Raton, FL 33487-2742

and by CRC Press
4 Park Square, Milton Park, Abingdon, Oxon, OX14 4RN

CRC Press is an imprint of Taylor & Francis Group, LLC

Library of Congress Cataloging-in-Publication Data
Names: Medraño-Fernandez, Iria, editor.
Title: Peroxiporins : redox signal mediators in and between cells / edited by Iria Medraño-Fernandez, Gerd Patrick Bienert, and Roberto Sitia.
Description: First edition. | Boca Raton : Taylor and Francis, 2024 | Series: Oxidative stress and disease | Includes bibliographical references and index.
Identifiers: LCCN 2023009972 (print) | LCCN 2023009973 (ebook) | ISBN 9780367745899 (hardback) | ISBN 9780367749866 (paperback) | ISBN 9781003160649 (ebook)
Subjects: LCSH: Aquaporins.
Classification: LCC QP552.A65 P45 2024 (print) | LCC QP552.A65 (ebook) | DDC 572/.696—dc23/eng/20230626
LC record available at https://lccn.loc.gov/2023009972
LC ebook record available at https://lccn.loc.gov/2023009973

ISBN: 978-0-367-74589-9 (hbk)
ISBN: 978-0-367-74986-6 (pbk)
ISBN: 978-1-003-16064-9 (ebk)

DOI: 10.1201/9781003160649

Typeset in Joanna MT
by Apex CoVantage, LLC

CONTENTS

PREFACE

Physicists often say that entropy is the desire of electrons to find peace. From this viewpoint, things got much busier for electrons after oxygen appeared on our planet, and redox signaling invaded virtually all forms of life. Things are complicated also for scientists. Thousands of papers pop up when one searches PubMed with "antioxidants" and "cancer" as query terms. Yet, key information is missing, as clinical trials did not yield, so far, the expected and longed-for fruits. This is because the context in which redox signals are generated, transmitted, and quenched is poorly understood. No doubt remains that hydrogen peroxide (H_2O_2) is a fundamental second messenger, capable of facilitating the growth of living cells or of causing their death. We know little about how H_2O_2 moves from its sources to its targets in and between organelles and cells. The pioneering work of Peter Agre and colleagues told us that, in physiological contexts, the transport of H_2O across membranes needs aquaporin (AQP) channels. Likewise, it has become recently clear that also H_2O_2 needs proteinaceous channels to cross a membrane. This book is dedicated to these molecules, now dubbed peroxiporins (PXPs). AQPs and PXPs belong to the same superfamily of asymmetric channels, which allow the bidirectional passage of generally small, uncharged molecules across various membranes. Not only does facilitated transport satisfy the need to speed up transport through lipid barriers, but, perhaps even more importantly, it allows concentration and regulation of key redox signals, guaranteeing cellular control of the processes. In this vein, most membrane-bound organelles are likely to contain a H_2O_2 conduit, though not necessarily a PXP. Such compartment-specific channels would allow signal targeting and specificity, considering that H_2O_2 transport can be reversibly gated. A thorough dissection of the underlying mechanisms will provide tools to regulate the strength and duration of different signals, and hence to design new treatments for the many diseases in which redox reactions are a target. In agriculture, for instance, a role of PXPs in phytohormone-regulated stomatal closure control and in pathogen-triggered plant immunity responses have just been unveiled. We expect more PXP-mediated processes to be discovered shortly in medicine.

This book summarizes our knowledge of PXPs, from their discovery to their functional roles at the organellar, cellular, and tissue levels in physiology, stress, and disease. As all important discoveries, these findings raised more questions than they solved. Clearly, much remains to be learned about PXPs to target them in beneficial ways. Optimism is justified by the new tools and technologies that are being developed to dynamically measure and map H_2O_2 fluxes *in vivo* and *in vitro*. By hosting diverse contributions from the most active scientists in this arena, the readings we propose should be of value to newcomers and experts and hopefully attract investigators from different fields to redox biology.

IRIA MEDRAÑO-FERNANDEZ,
GERD PATRICK BIENERT, ROBERTO SITIA

ACKNOWLEDGMENTS

Roberto Sitia wants to acknowledge support from grants from the Associazione Italiana Ricerca sul Cancro (IG 2019–23285) and Ministero dell'Istruzione, dell'Università e della Ricerca (MIUR)-PRIN (grant no. 2017XA5J5N). Gerd Patrick Bienert was supported by an Emmy Noether grant from the Deutsche Forschungsgemeinschaft (1668/1-1/2). Iria Medraño-Fernandez acknowledges support from the Madrid Government (Comunidad de Madrid) under the Multiannual Agreement with UC3M in the line of "Research Funds for Beatriz Galindo Fellowships" (REDOXSKIN-CM-UC3M), and in the context of the V PRICIT (Regional Programme of Research and Technological Innovation, and from "Proyectos de I + D + I" (PID2020-114230 GA-I00) funded by MCIN/AEI/10.13039/501100011033/.

All editors warmly acknowledge the great effort exerted by all authors to make this book possible.

EDITORS

Iria Medraño-Fernandez, PhD, is currently Chair of Excellence Beatriz Galindo in the Department of Bioengineering of the University Carlos III de Madrid (UC3M) and head of the core facility CleanRooms for Bioengineering. In 2012 she earned her PhD in biochemistry and molecular biology at the Universidad Complutense of Madrid. Soon afterward, she moved to Milan to deepen her expertise in signal transduction mechanisms by focusing on peroxiporin-controlled signaling pathways. Since 2020 she has led the lab "Redox Signaling in Regenerative Medicine" in the Tissue Engineering in Regenerative Medicine (TERMEG) UC3M unit. Her main research interest is centered on deciphering how cells maintain redoxtasis to achieve desired outcomes, particularly during skin regeneration, using peroxiporins both as subjects of study and as genetically engineered research tools.

Gerd Patrick Bienert, PhD (born in Rothenburg ob der Tauber, Germany, in 1978) is a Professor of Crop Physiology at the Technical University of Munich in Germany since 2020. His main research objectives are to understand metalloid- and water-efficiency mechanisms in crop plants. His research targets the understanding of the physiology of plants down to the structure and function of genes and proteins, as well as relevant mechanisms which allow optimizing biotechnologically and agriculturally important quality and yield traits and the generation of crops, which are beneficial for society in terms of food security and food safety. A second research focus is on the molecular elucidation of whether and how the transport of hydrogen peroxide through plant and mammalian aquaporins is regulated.

Prof. Bienert studied biology at the Universities of Würzburg and Darmstadt, Germany. Following this, he earned his PhD in molecular plant nutrition in Copenhagen, Denmark. Using yeast toxicity growth experiments and direct transport assays, he and colleagues provided first-time functional evidence that specific aquaporins facilitate the transport of hydrogen peroxide across membranes—what are now called "peroxiporins." An assay that is now routinely applied to characterize the transmembrane transport of hydrogen peroxide had been developed. He then moved on to be a postdoctoral researcher at UCLouvain, Belgium. Between 2013 and 2020, he headed an independent research group at the Leibniz Institute of Plant Genetics and Crop Plant Research in Gatersleben dealing with the elucidations of mechanisms regulating metalloid efficiency in crops. In 2021, he was Laureate of the Joseph Schepkens award for Plant Genetics from the Belgian Royal Academy for Science for his work on "Water and Metalloid Transport in Arable Crops."

Roberto Sitia, MD is a Professor in the Division of Genetics and Cell Biology, Università Vita-Salute and San Raffaele Hospital, Milan, Italy. During his training as a hematologist at the University of Genoa, Roberto Sitia became fascinated by the immune system and how our body manages to produce, assemble, and release highly specific antibodies against the most diverse substances it encounters. Always in search of mechanistic explanations to the scientific problem at stake, he worked at the Memorial Sloan Kettering Cancer Center in New York and then at the MRC Laboratory of Molecular Biology in Cambridge. In 1990, he was then recruited at the San Raffaele Hospital in Milan to help in the design, management, and development of a competitive research institute (DiBiT) and a medical school (UniSR), both of which are now considered the best in Italy. For over two decades he has been a full professor of Molecular Biology at UniSR.

Key discoveries include the mechanisms of (1) protein quality control and traffic jams in the secretory pathway; (2) unconventional (leaderless) protein secretion; (3) oxidative folding; (4) de novo ER biogenesis during the differentiation of antibody producing cells; and (5) redox adaptation to increased immunoglobulin secretion and life span control. Recently, his lab pioneered the discovery and characterization of peroxiporins, and their role in inter- and intra-cellular redox signaling and homeostasis. A member-elect of the European Molecular Biology Organization, he has published more than 170 papers, many of which are highly cited breakthroughs. Many investigators trained in his lab now occupy leading positions all over the world. Besides his scientific and managerial activities, he fights for the survival and growth of basic sciences, a tough challenge in a period in which obscurantism is gaining space.

CONTRIBUTORS

KARINA ALLEVA
Universidad de Buenos Aires-CONICET
Buenos Aires, Argentina

MARCO ANTEGHINI
LifeGlimmer GmbH
Berlin, Germany

FERNANDO ANTUNES
Institute of Molecular Sciences
Lisbon, Portugal

EDUARDO AREVALO-NUÑEZ DE ARENAS
University Carlos III of Madrid
Madrid, Spain

EDWARD AVEZOV
University of Cambridge
Cambridge, UK

ERIC BEITZ
Christian-Albrechts-University of Kiel
Kiel, Germany

MILENA BERTOLOTTI
University of Freiburg
Freiburg, Germany

GERD PATRICK BIENERT
Technical University of Munich
Freising, Germany

KATHRIN BLUMENSTEIN
University of Freiburg
Freiburg, Germany

JOAN CERDÀ
Institute of Agrifood Research and Technology (IRTA)
Barcelona, Spain

FRANÇOIS CHAUVIGNÉ
Institute of Marine Sciences
Barcelona, Spain

CLÁUDIO F. COSTA
KU Leuven
Leuven, Belgium

CÉCILE CRAPART
University of Cambridge
Cambridge, UK

LAURA DE CUBAS
Universitat Pompeu Fabra
Barcelona, Spain

BERND H. DREYER
Leibniz Institute of Plant Genetics and Crop Plant Research (IPK)
Seeland, Germany

RODERICK NIGEL FINN
Institute of Agrifood Research and Technology (IRTA)
Barcelona, Spain

MARC FRANSEN
KU Leuven
Leuven, Belgium

STEPHANIE FROHN
Leibniz Institute of Plant Genetics and Crop Plant Research (IPK)
Seeland, Germany

JOSÉ MANUEL GARCÍA-MANTEIGA
IRCCS San Raffaele Scientific Institute
Milan, Italy

TOBIAS HENZLER
Stemmer Imaging AG
Munich, Germany

ELENA HIDALGO
Universitat Pompeu Fabra
Barcelona, Spain

KENICHI ISHIBASHI
Meiji Pharmaceutical University
Tokyo, Japan

TASUKU KONNO
University of Cambridge
Cambridge, UK

HONGLI LI
KU Leuven
Leuven, Belgium

CELIEN LISMONT
KU Leuven
Leuven, Belgium

OLGA LYUBLINSKAYA
Institute of Cytology
St. Petersburg, Russia

IRIA MEDRAÑO-FERNANDEZ
University Carlos III of Madrid
Madrid, Spain

ANGIE KATHERINE MOLINA-OVIEDO
University Carlos III of Madrid
Madrid, Spain

MATÍAS N. MÖLLER
Instituto de Química Biológica
Montevideo, Uruguay

YOSHIYUKI MORISHITA
Jichi Medical University
Saitama, Japan

BARBARA LUDWIG NAVARRO
University of Göttingen
Göttingen, Germany

DANIELA ELISABETH NORDZIEKE
University of Göttingen
Göttingen, Germany

MICHAEL BROBERG PALMGREN
University of Copenhagen
Frederiksberg C, Denmark

RAFAEL RADI
Universidad de la República
Montevideo, Uruguay

IULIA REVENCO
KU Leuven
Leuven, Belgium

ARMINDO SALVADOR
University of Coimbra
Cantanhede, Portugal

JOS H. M. SCHIPPERS
Leibniz Institute of Plant Genetics and Crop Plant Research (IPK)
Seeland, Germany

ROMY R. SCHMIDT-SCHIPPERS
University of Bielefeld
Bielefeld, Germany

HELMUT SIES
Leibniz Research Institute for Environmental Medicine
Düsseldorf, Germany

ROBERTO SITIA
Università Vita-Salute
Milan, Italy

ILARIA SORRENTINO
University Carlos III of Madrid
Madrid, Spain

YASUKO TANAKA-MOCHIZUKI
Meiji Pharmaceutical University
Tokyo, Japan

ARI ZEIDA
Universidad de la República
Montevideo, Uruguay

CHAPTER ONE

Role of the Cellular Hydrogen Peroxide Landscape Profile in Redox Signaling

Helmut Sies

ABBREVIATIONS

AQP	Aquaporin
ER	Endoplasmic reticulum
MAM	Mitochondrial-associated membranes
PTEN	Phosphatase and tensin homolog

1.1 H_2O_2 CELLULAR LANDSCAPE PROFILE

Hydrogen peroxide is a central redox signaling molecule,[1–6] and its physiological steady-state level is in the low nanomolar concentration range.[7–9] Importantly, the intracellular distribution is not uniform, as there is a "landscape profile" of hydrogen peroxide concentrations which is maintained by several mechanisms. As shown in Table 1.1, the various subcellular compartments are characterized by substantially different estimated concentrations of hydrogen peroxide, and the extracellular space (e.g., blood plasma), has a higher steady-state concentration orders of magnitude higher, which is in the low μM range.[10] The question is: what determines this landscape profile, how is the pattern of H_2O_2 concentrations in the various locations generated, and how is it maintained? The answer resides in the control of three components: sources, sinks, and gradient control (i.e., H_2O_2 generation, H_2O_2 reduction, and H_2O_2 transport; Figure 1.1A). The latter is the domain of the peroxiporins, a group of aquaporins capable of transporting H_2O_2, first described by Henzler and Steudle in 2000 in studying algal H_2O_2 transport,[11] and established by Bienert and colleagues in its general importance for facilitating diffusion of H_2O_2 across biological membranes.[12–14] Several aquaporins (AQP3, AQP5, AQP8, AQP9, AQP11) serve as peroxiporins, being located in different cellular membranes (Figure 1.1B; AQP5 is not shown).[4,15] Regulation of these peroxiporins is a field of active research. Work on AQP8 identified a gating mechanism which involves persulfidation (RSSH) of a critical cysteine (C53) in the peroxiporin, opening the possibility for redox-dependent H_2O_2 gradient control.[16,17]

TABLE 1.1
Cellular, Subcellular, and Extracellular Concentrations of Hydrogen Peroxide

	H_2O_2 (nM)	Ref
"Overall cellular"*	1–10	7–9
Cytosol	0.1	33
Mitochondrial matrix	4	34
Endoplasmic reticulum	700	35
Peroxisome	>4	(M. Fransen, pers. comm.)
Lysosome	?	
Nucleus	<<4	36
Extracellular (blood plasma)	1,000–5,000	10
Plasma membrane, localized spots	12,000	9

* For rough orientation.

DOI: 10.1201/9781003160649-1

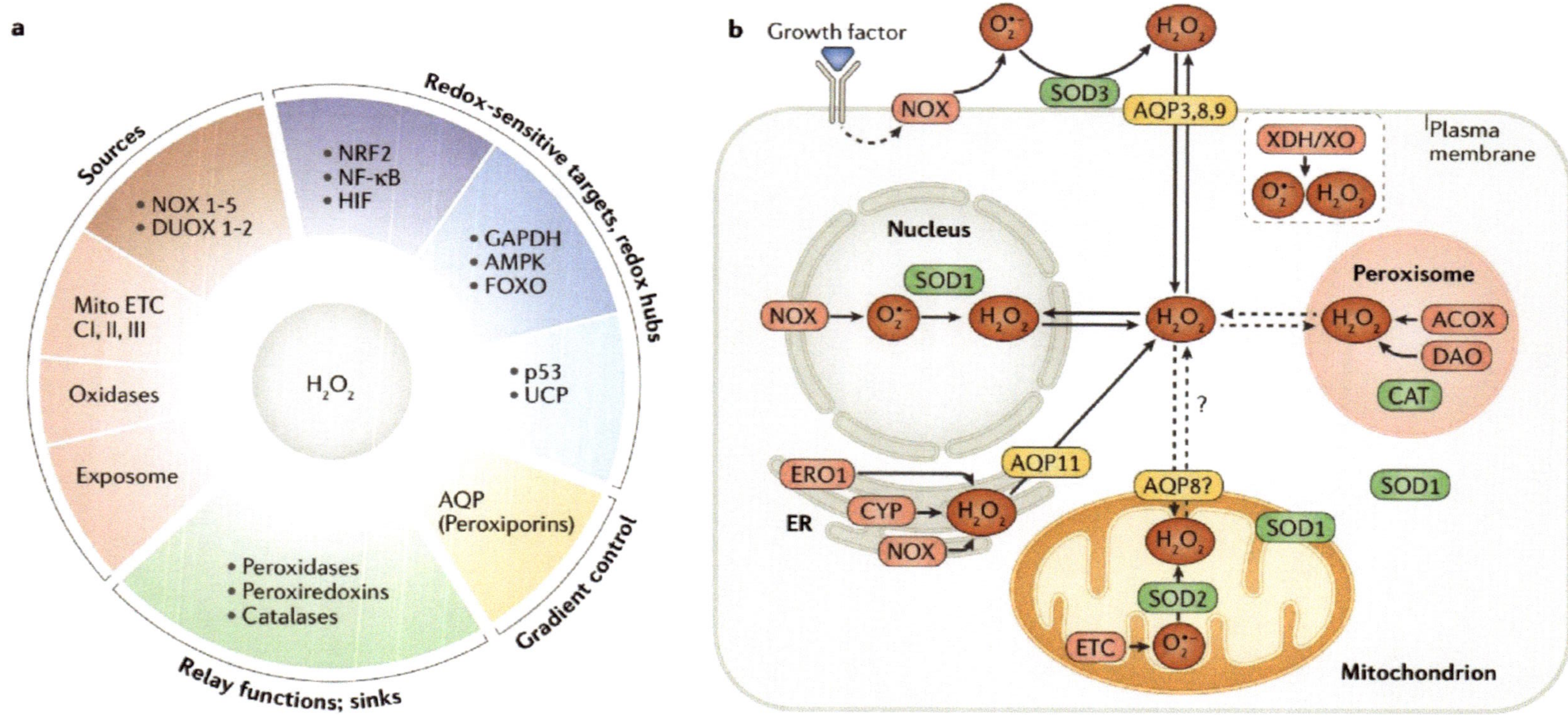

Figure 1.1 Key modulators and targets of H_2O_2. (A) H_2O_2 sources (red), redox-sensitive targets and hubs (blue), aquaporins (AQP/peroxiporins) (yellow), and sinks (green). (B) The major subcellular sites of generation of the superoxide anion radical and of H_2O_2 are shown.

Source: **Originally published in** *Nature Reviews Molecular Cell Biology*. **Sies, H., and Jones, D. P., Reactive oxygen species (ROS) as pleiotropic physiological signaling agents.** *Nat. Revs. Mol. Cell. Biol.* **21, 363–383 (2020).**

AQP11 is localized in the endoplasmic reticulum (ER), accumulating partly in mitochondrial-associated ER membranes (MAM). Its downregulation perturbs the flux of H_2O_2 through the ER, but not through the mitochondrial or plasma membranes. These properties make AQP11 a potential regulator of ER redox homeostasis and signaling.[18]

1.2 CRITICAL APPRAISAL

While recent advances revealed basic information on the landscape profile of H_2O_2 in the cell, it is obvious that the depictions in Figure 1.1 and Table 1.1 provide only a first step for orientation. A full spatiotemporal analysis on cellular H_2O_2 will expose a much more sophisticated picture. It is likely that the cell biology of the peroxiporins will provide more refined insight into the orchestration of cellular redox processes. Major tools for this goal consist of the use of noninvasive technologies, foremost the use of genetically encoded fluorescent probes (e.g., Hyper7[19] or roGFP2-Tsa2deltaCR).[20] While there are many advantages of these probes, a major challenge is the difficulty of calibration in terms of real concentration of H_2O_2 rather than providing color-coded information on relative levels.

The currently available numbers for H_2O_2 concentration as given in Table 1.1 should be taken as a first approximation only. There is room for overestimation or underestimation because of assumptions and calculations involved in obtaining the results. A detailed analysis of the potential pitfalls is beyond the scope of the present chapter.

Another critical aspect concerns the temporal aspect and the cell-type specificity. The numbers given for H_2O_2 concentrations in Table 1.1 relate to resting conditions in a generic cell. Clearly, there are dynamic changes, including circadian rhythm, and there are different fluxes of H_2O_2 in different cell types.

1.3 OUTLOOK

More refined insight into the orchestration of the cellular landscape profile of H_2O_2 will require better understanding of the interplay of sources, sinks, and gradient control of H_2O_2. As to the latter, the fine control of peroxiporins in terms of expression, targeted membrane insertion, and post-translational modifications will continue to be an active field of molecular and cell biological research. The molecular characterization of AQP-facilitated H_2O_2 transport is an ongoing endeavor.[21,22]

Likewise, the role of peroxiporins in biomedical conditions is coming into focus. For example, inhibition of AQP3 in macrophages by a monoclonal antibody prevented experimental liver injury.[23] AQP3 is involved in epidermal growth factor receptor signaling via H_2O_2 transport in cancer cells.[24] AQP3-mediated H_2O_2 uptake inhibits lung adenocarcinoma autophagy by inactivating PTEN.[25] AQP3 and AQP5 are being discussed as potential regulators of redox status in breast cancer.[26] In pancreatic β-cells, interleukin-1β mediates AQP8 overexpression which, in turn, alleviates cytokine-mediated toxicity.[27] AQP8 was suggested to be involved in hepatocyte cholesterologenesis.[28] AQP11 may have a role in obesity-associated inflammation.[29] The roles of AQPs in infectious and non-infectious diseases[30] and in cell signaling and aging[31] have been reviewed, and a wider role of AQPs may reside in functions beyond being channels.[32]

The aforementioned aspects of peroxiporin research are not intended to be comprehensive, they rather exhibit examples of current research on peroxiporins in the biomedical field. It will of interest to see whether peroxiporin AQPs will become useful as targets in redox medicine in the future.

REFERENCES

1 Jones, D. P. & Sies, H. The redox code. *Antioxid Redox Signal* **23**, 734–746 (2015).
2 Stocker, S., Van, L. K., Mijuskovic, A. & Dick, T. P. The conundrum of hydrogen peroxide signaling and the emerging role of peroxiredoxins as redox relay hubs. *Antioxid Redox Signal* **28**, 558–573 (2018).
3 Forman, H. J., Maiorino, M. & Ursini, F. Signaling functions of reactive oxygen species. *Biochemistry* **49**, 835–842 (2010).
4 Sies, H. & Jones, D. P. Reactive oxygen species (ROS) as pleiotropic physiological signalling agents. *Nat Rev Mol Cell Biol* **21**, 363–383 (2020).
5 Sies, H. Hydrogen peroxide as a central redox signaling molecule in physiological oxidative stress: Oxidative eustress. *Redox Biol* **11**, 613–619 (2017).
6 Stone, J. R. & Yang, S. Hydrogen peroxide: a signaling messenger. *Antioxid Redox Signal* **8**, 243–270 (2006).
7 Chance, B., Sies, H. & Boveris, A. Hydroperoxide metabolism in mammalian organs. *Physiol Rev* **59**, 527–605 (1979).
8 Lyublinskaya, O. & Antunes, F. Measuring intracellular concentration of hydrogen peroxide with the use of genetically encoded H(2)O(2) biosensor HyPer. *Redox Biol* **24**, 101200 (2019).

9 Hosogi, S. *et al.* Plasma membrane anchored nanosensor for quantifying endogenous production of H(2)O(2) in living cells. *Biosens Bioelectron* **179**, 113077 (2021).
10 Forman, H. J., Bernardo, A. & Davies, K. J. What is the concentration of hydrogen peroxide in blood and plasma? *Arch Biochem Biophys* **603**, 48–53 (2016).
11 Henzler, T. & Steudle, E. Transport and metabolic degradation of hydrogen peroxide in *Chara corallina*: model calculations and measurements with the pressure probe suggest transport of H(2)O(2) across water channels. *J Exp Bot* **51**, 2053–2066 (2000).
12 Bienert, G. P. *et al.* Specific aquaporins facilitate the diffusion of hydrogen peroxide across membranes. *J Biol Chem* **282**, 1183–1192 (2007).
13 Bienert, G. P., Schjoerring, J. K. & Jahn, T. P. Membrane transport of hydrogen peroxide. *Biochim Biophys Acta* **1758**, 994–1003 (2006).
14 Bienert, G. P. & Chaumont, F. Aquaporin-facilitated transmembrane diffusion of hydrogen peroxide. *Biochim Biophys Acta* **1840**, 1596–1604 (2014).
15 Nordzieke, D. E. & Medraño-Fernandez, I. The plasma membrane: A platform for intra- and intercellular redox signaling. *Antioxidants (Basel)* **7**, 7110168 (2018).
16 Medraño-Fernandez, I. *et al.* Stress regulates aquaporin-8 permeability to impact cell growth and survival. *Antioxid Redox Signal* **24**, 1031–1044 (2016).
17 Bestetti, S. et al. A persulfidation-based mechanism controls aquaporin-8 conductance. *Sci Adv* **4**, eaar5770 (2018).
18 Bestetti, S. et al. Human aquaporin-11 guarantees efficient transport of H(2)O(2) across the endoplasmic reticulum membrane. *Redox Biol* **28**, 101326 (2020).
19 Pak, V. V. et al. Ultrasensitive genetically encoded indicator for hydrogen peroxide identifies roles for the oxidant in cell migration and mitochondrial function. *Cell Metab* **31**, 642–653 (2020).
20 Morgan, B. *et al.* Real-time monitoring of basal H2O2 levels with peroxiredoxin-based probes. *Nat Chem Biol* **12**, 437–443 (2016).
21 Wang, H., Schoebel, S., Schmitz, F., Dong, H. & Hedfalk, K. Characterization of aquaporin-driven hydrogen peroxide transport. *Biochim Biophys Acta Biomembr* **1862**, 183065 (2020).
22 Wragg, D., Leoni, S. & Casini, A. Aquaporin-driven hydrogen peroxide transport: A case of molecular mimicry? *RSC Chem Biol* **1**, 390–394 (2020).
23 Hara-Chikuma, M., Tanaka, M., Verkman, A. S. & Yasui, M. Inhibition of aquaporin-3 in macrophages by a monoclonal antibody as potential therapy for liver injury. *Nat Commun* **11**, 5666 (2020).
24 Hara-Chikuma, M., Watanabe, S. & Satooka, H. Involvement of aquaporin-3 in epidermal growth factor receptor signaling via hydrogen peroxide transport in cancer cells. *Biochem Biophys Res Commun* **471**, 603–609 (2016).
25 Wang, Y. *et al.* AQP3-mediated H(2) O(2) uptake inhibits LUAD autophagy by inactivating PTEN. *Cancer Sci* **112**, 3278–3292 (2021).
26 Milković, L. & Čipak Gašparović, A. AQP3 and AQP5-potential regulators of redox status in breast cancer. *Molecules* **26**, 26092613 (2021).
27 Krüger, C. *et al.* The importance of aquaporin-8 for cytokine-mediated toxicity in rat insulin-producing cells. *Free Radic Biol Med* **174**, 135–143 (2021).
28 Danielli, M., Capiglioni, A. M., Marrone, J. & Marinelli, R. A. Further evidence for the involvement of mitochondrial aquaporin-8 in hepatocyte lipid synthesis. *Biochimie* **188**, 16–19 (2021).
29 Frühbeck, G. et al. Aquaporin-11 contributes to TGF-β1-induced endoplasmic reticulum stress in human visceral adipocytes: Role in obesity-associated inflammation. *Cells* **9**, 9061403 (2020).
30 Azad, A. K. et al. Human aquaporins: Functional diversity and potential roles in infectious and non-infectious diseases. *Front Genet* **12**, 654865 (2021).
31 Tamma, G. et al. Aquaporin membrane channels in oxidative stress, cell signaling, and aging: Recent advances and research trends. *Oxid Med Cell Longev* **2018**, 1501847 (2018).
32 Galli, M., Hameed, A., Żbikowski, A. & Zabielski, P. Aquaporins in insulin resistance and diabetes: More than channels! *Redox Biol* **44**, 102027 (2021).
33 Lim, J. B., Langford, T. F., Huang, B. K., Deen, W. M. & Sikes, H. D. A reaction-diffusion model of cytosolic hydrogen peroxide. *Free Radic Biol Med* **90**, 85–90 (2016).
34 Stein, K. T., Moon, S. J., Nguyen, A. N. & Sikes, H. D. Kinetic modeling of H2O2 dynamics in the mitochondria of HeLa cells. *PLoS Comput Biol* **16**, e1008202 (2020).
35 Gao, C., Tian, Y., Zhang, R., Jing, J. & Zhang, X. Endoplasmic reticulum-directed ratiometric fluorescent probe for quantitive detection of basal H2O2. *Anal Chem* **89**, 12945–12950 (2017).
36 Malinouski, M., Zhou, Y., Belousov, V. V., Hatfield, D. L. & Gladyshev, V. N. Hydrogen peroxide probes directed to different cellular compartments. *PLoS ONE* **6**, e14564 (2011).

CHAPTER TWO

How Gradients and Microdomains Determine H_2O_2 Redox Signaling

Fernando Antunes and Armindo Salvador

2.1 HISTORICAL BACKGROUND

Hydrogen peroxide (H_2O_2) plays a central role in cellular signaling. When compared with other reactive oxygen species, H_2O_2 has a low reactivity. As a consequence, H_2O_2 is an abundant reactive oxygen species and travels relatively long distances to exert its signaling effects away from the production site. The notion that biomembranes constitute a barrier to H_2O_2 diffusion was established in the context of studies on catalase latency in the 1950s. The concept of enzyme latency (i.e., the increase of enzyme activity upon disruption of the compartment where the enzyme is entrapped) was extensively used by de Duve and co-workers to identify and characterize sub-cellular fractions (De Duve 1965). Catalase latency had been identified in *Micrococcus lysodeikticus* and in yeast (Kaplan 1955). In the first case, catalase latency was attributed to low intracellular pH and in the second case to a three-dimensional fold change that activates catalase upon cell disruption. In 1959, Clayton rightly attributed this phenomenon in *Rhodopseudomonas spheroids* to a permeability barrier that limited diffusion of H_2O_2 into the cell, causing a lower catalase activity (Clayton 1959). Later, in erythrocytes a gradient between extracellular and intracellular H_2O_2 concentration of 3 was estimated with a permeability coefficient of 6×10^{-4} s^{-1} (Nicholls 1965). Nevertheless, H_2O_2 did permeate biomembranes, and the notion that H_2O_2 readily diffuses across cellular compartments was established in the literature (Chance, Sies, and Boveris 1979). This notion became more accepted, and the formation of gradients in small compartments was considered as a possibility (Ma and Eaton 1992). Then in 2000, the concept of enzyme latency was revisited and the H_2O_2 gradients across biomembranes were identified in T lymphocytes (Antunes and Cadenas 2000). This was later confirmed by showing H_2O_2 compartmentalization in *Escherichia coli* (Seaver and Imlay 2001). Lipid membranes' biophysical properties are affected by H_2O_2 and thus H_2O_2 can potentially modulate its own permeability coefficient. In 2004, the decrease of the H_2O_2 plasma membrane permeability was identified as a new mechanism of adaptation to H_2O_2 (Branco et al. 2004). This permeability was identified to be under active regulation by changing the lipid and protein composition of the membranes (Matias et al. 2007; Pedroso et al. 2009, 2012). In 2007, aquaporins were identified as mediators of H_2O_2 transport across biomembranes (Bienert, Schjoerring, and Jahn 2006). Subsequently, aquaporins were identified to play an important role in H_2O_2 signaling (Miller, Dickinson, and Chang 2010; Bertolotti et al. 2013), and today aquaporins and their regulation are established members of the archetypal mechanism of redox signaling.

DOI: 10.1201/9781003160649-2

2.2 GRADIENT ESTIMATION—A MATHEMATICAL MODELING APPROACH

The realization that there might be important transmembrane and intracellular H_2O_2 concentration gradients and that these gradients may play key roles in redox signaling prompted attempts to determine or estimate their amplitude and spatial scale. We will first describe the computational approaches as this helps understanding the concepts underlying the experimental ones.

From a theoretical standpoint, the simplest non-trivial case of interest arises in the situation depicted in Figure 2.1. Here, H_2O_2 permeates (with permeability constant κ) between the extracellular medium and the cytosol of a cell, where it is consumed through a first-order process with rate constant k. We will begin with two important simplifying assumptions. First, that the extracellular H_2O_2 concentration (c_0) is constant. This might be either because extracellular processes impose a steady state or because in the experimental setup the volume of medium is far greater than the volume of cells, such that just a small fraction of the supplied H_2O_2 is consumed over the duration of the experiment. Second, that the intracellular consumption is slow enough that H_2O_2 molecules have time to diffuse throughout the cytosol (we will provide a quantitative criterion below). Under these conditions, the cytosolic H_2O_2 concentration (c_i) is approximately uniform throughout the cytosol, and can be estimated from the simple ordinary differential equation

$$\frac{dc_i}{dt} = k_{p,i}(c_o - c_i) - k\,c_i. \qquad (2.1)$$

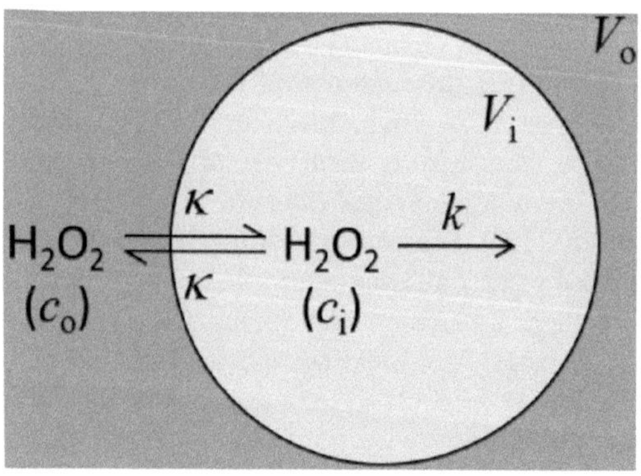

Figure 2.1 Simple model of H_2O_2 permeation and intracellular clearance.

Here, $k_{p,i} = \kappa A/V_i$ stands for the effective permeation (efflux) rate constant, with A and V_i the cell's surface area and volume, respectively. Considering that the cytosolic concentration is initially negligible, this equation has the solution

$$c_i(t) = \frac{c_o}{1 + k/k_{p,i}}\left(1 - e^{-(k+k_{p,i})t}\right). \qquad (2.2)$$

Therefore, the cytosolic concentration will increase until it settles to a steady state $c_i(\infty) = c_0/(1 + k/k_{p,i})$, corresponding to the transmembrane gradient

$$\Gamma = c_o / c_i(\infty) = 1 + k/k_{p,i}. \qquad (2.3)$$

This transmembrane gradient is thus determined by the ratio between the (pseudo-)first-order rate constant for H_2O_2 consumption and the effective permeation rate constant. *That is, a strong transmembrane gradient will ensue if and only if cytosolic H_2O_2 clearance can strongly compete with H_2O_2 efflux.*

The relationships of the transmembrane gradient to enzyme latency and cellular H_2O_2 clearance are discussed in Appendices A and B.

2.3 FORMATION OF H_2O_2 MICRODOMAINS

The situation in real cells is complicated by the fact that the H_2O_2 molecules react with cytosolic peroxiredoxins and peroxidases before penetrating deep into the cytosol. In order to account for this phenomenon, one has to resort to reaction-diffusion models (Poole 1975; Lim et al. 2015, 2016; Travasso et al. 2017). For simplicity, we will consider here the case of a symmetric layer of cells of width L, such that H_2O_2 can permeate equally from both sides (Figure 2.2A). As in the first model, we will assume that the extracellular H_2O_2 concentration is constant. The distribution of the intracellular H_2O_2 concentration at steady state ($c_i(\infty, x)$) as function of the distance (x) from the membrane at the left-hand side is given by the model

$$\begin{aligned} D\frac{\partial^2 c_i}{\partial x^2} &= k\,c_i(\infty, x) \\ \left.\frac{dc_i(\infty)}{dx}\right|_{x=0} &= -\frac{\kappa}{D}(c_o - c_i(\infty, 0)), \\ \left.\frac{dc_i(\infty)}{dx}\right|_{x=\frac{L}{2}} &= 0, \end{aligned} \qquad (2.4)$$

where D stands for the diffusion constant for H_2O_2 in the cytosol. The solution is

$$c_i(\infty,x) = c_0 \frac{1+e^{-2(x-L/2)/\lambda}}{(1-\frac{D}{\kappa\lambda})e^{-x/\lambda}+(1+\frac{D}{\kappa\lambda})e^{(L-x)/\lambda}}, \quad (2.5)$$

where $\lambda = \sqrt{D/k}$. For $\lambda << L$, $x << L/2$, this expression is approximated as

$$c_i(\infty,x) \approx \frac{c_0}{1+\frac{D}{\kappa\lambda}} e^{-x/\lambda}. \quad (2.6)$$

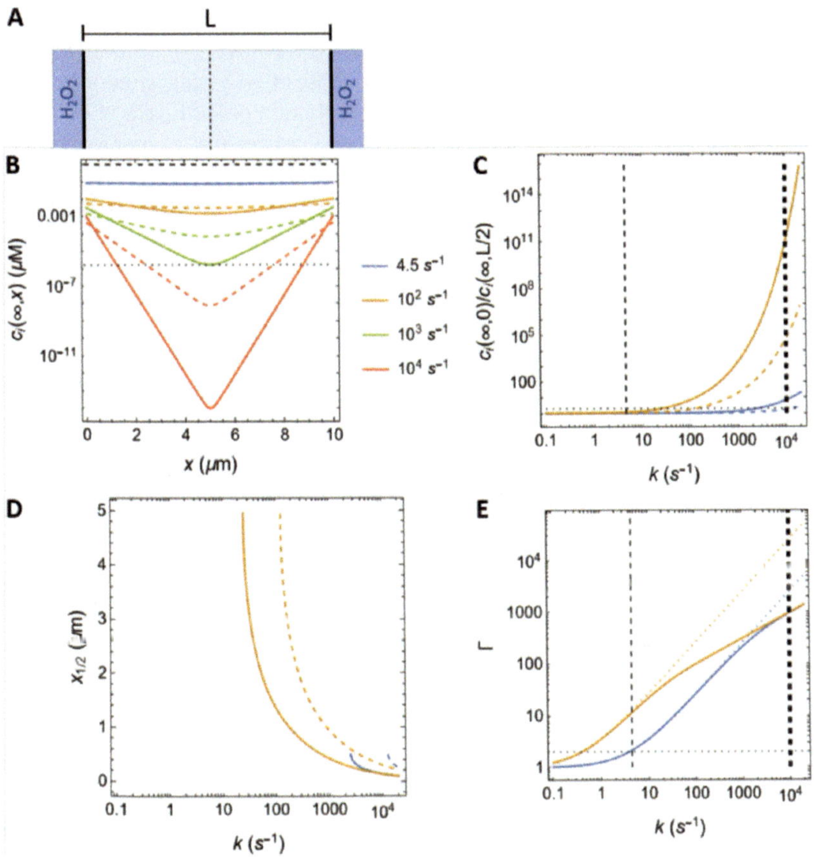

Figure 2.2 H_2O_2 transmembrane and cytosolic gradients. (A) The geometrical setup for the reaction-diffusion model. A cell layer of width L is considered. The extracellular H_2O_2 concentration is considered constant, and it is assumed that H_2O_2 permeates uniformly across the membrane surfaces at each side of the layer, and that the clearance rate constant is uniform throughout the cytosol. Therefore, the system can be treated as one-dimensional. The vertical dashed line indicates the axis of symmetry at the center of the cell layer. (B) Steady-state distribution of the cytosolic H_2O_2 concentration for a 10 μm-wide cell, considering either $D = 1830\ \mu m^2 s^{-1}$ (dashed lines) or $D = 370\ \mu m^2 s^{-1}$ (solid lines), $\kappa = 2\ \mu m\ s^{-1}$, and various clearance rate constants. The black dashed and dotted lines indicate the considered extracellular concentration and the "concentration" corresponding to a single H_2O_2 molecule in a 1000-μm^3 cell, respectively. Note the logarithmic concentration scale. (C) Amplitude of the cytosolic H_2O_2 concentration gradients as function of the clearance rate constant for 1-μm-wide (blue) or 10-μm-wide (yellow) cell layers, considering either $D = 1830\ \mu m^2 s^{-1}$ (dashed lines) or $D = 370\ \mu m^2 s^{-1}$ (solid lines). The vertical thin and thick dashed lines mark $k = 4\ s^{-1}$ and $k = 10^4\ s^{-1}$, respectively. (D) Depth at which the H_2O_2 concentration decreases to half the value obtained by the inner face of the membrane, as function of the clearance rate constant. Same dashing and color code as for panel C. (E) Transmembrane gradient computed in terms of the cytosolic concentration by the inner face of the membrane (solid lines) or of the mean cytosolic concentration (dotted lines), for 1-μm-wide (blue) or 10-μm-wide (yellow) cell layers. The permeation constant $\kappa = 2\ \mu m\ s^{-1}$ and diffusion constant $D = 370\ \mu m^2 s^{-1}$ were assumed. See main text for additional explanations.

The cytosolic gradient, computed as the ratio between the maximal and minimal H_2O_2 concentrations, is given by

$$\Lambda = c_i(\infty,0)/c_i(\infty,L/2) = \left(e^{L/(2\lambda)} + e^{-L/(2\lambda)}\right)/2, \quad (2.7)$$

which for high clearance rate constants ($k >> D/L^2$) is approximately

$$\Lambda \approx e^{L/(2\lambda)}/2 \propto e^{\frac{L}{2\sqrt{D}}\sqrt{k}}. \quad (2.8)$$

Although these precise solutions are only strictly valid for the considered geometry, where H_2O_2 permeates uniformly across the membranes, the results discussed below apply qualitatively for other geometries as well, such as discrete H_2O_2 sources within the cytosol. Figure 2.2B shows a graphical representation of equation (2.5) for a 10-μm-wide cell, considering either $D = 1830\ \mu m^2 s^{-1}$ as determined for buffer (dashed lines) or $D = 370\ \mu m^2 s^{-1}$ as determined for a hydrogel with a viscosity matching that of the cytosol (solid lines); van Stroe-Biezen et al. 1993), $\kappa = 2\ \mu m\ s^{-1}$, as determined for Jurkat T cells (Antunes and Cadenas 2000) and various values of k. The results show that although for $k = 4.5\ s^{-1}$ (the glutathione peroxidase [GPx] activity in Jurkat T cells [Antunes and Cadenas 2000] and in the order of magnitude of that in many other human cell types [Selvaggio, Coelho, and Salvador 2018]) the cytosolic H_2O_2 concentration gradient is negligible, for clearance rate constants exceeding $\approx 100\ s^{-1}$ this gradient becomes very strong in medium-sized human cells (Figure 2.2B–D). Thus, for the clearance rate constants $>10^4\ s^{-1}$ that ensue from the high reactivity ($\approx 10^7$ to $10^8\ M^{-1}s^{-1}$ [Peskin et al. 2007; Manta et al. 2009; Portillo-Ledesma et al. 2018; Carvalho et al. 2017]) and abundance [usually >100 μM, and may attain 700 μM in erythrocytes (Cho et al. 2010)] of cytosolic 2-Cys, the H_2O_2 concentration decays exponentially with distance from the supply sites, halving within just ≈100 nm (Figure 2.2D, and as per equation (2.6)). Moreover, under these conditions H_2O_2 molecules are extremely rare further from 1 μm from supply sites. *That is, the fully active cytosolic 2-Cys peroxiredoxins in human cells effectively confine H_2O_2 to microdomains around supply sites even in the absence of physical compartmentalization impeding free diffusion.*

However, two factors contribute for less extreme cytosolic H_2O_2 concentration gradients than predicted by the idealized model above. First, organelles such as mitochondria (Hoehne et al. 2021), the endoplasmic reticulum (Bestetti et al. 2020; Mishina et al. 2011; Appenzeller-Herzog et al. 2016), and endosomes (Mishina et al. 2011) may release $O_2^{\bullet-}$ and H_2O_2 at multiple points throughout the cytoplasm. Second, some H_2O_2 is generated throughout the cytosol by diffusible sources such as superoxide dismutase 1 (SOD1). Although significant superoxide concentration gradients are expected as well (Salvador, Sousa, and Pinto 2001), these should be less pronounced due to the lower concentrations of SOD1 relative to those of 2-Cys peroxiredoxins. Thus, the spatial distribution of *SOD1-generated* H_2O_2 should approximately mirror that of superoxide.

The occurrence of H_2O_2 microdomains in the cytosol despite these factors is supported by the following experimental observations using genetically encoded fluorescent H_2O_2 probes. Mishina et al. (2011) targeted HyPer probes to the cytosol-facing surfaces of the plasma membrane, endosomes, and the endoplasmic reticulum of HeLa-Kyoto cells and NIH-3T3 fibroblasts. Probes just a few micrometers apart in the same cells were not simultaneously activated when the cells were stimulated with growth factors known to cause site-specific H_2O_2 generation (Mishina et al. 2011). Mishina et al. (2019) used HyPer3 ratiometric probes targeted to the mitochondrial matrix and to the cytoskeleton of HeLa-Kyoto cells and a chemogenetic H_2O_2-generating system targeted to the nucleus. With this setup they have shown that H_2O_2 generation at the nucleus creates a H_2O_2 concentration gradient that is detectable both across the cytosol and across the mitochondrial network. The gradient was abolished by inhibition of thioredoxin reductase, but not by glutathione depletion of catalase inhibition. This is consistent with the prediction by the modeling above that typical GPx and catalase activities in human cells are insufficient to sustain a significant cytosolic H_2O_2 concentration gradient. More recent experiments in the same vein used the brighter, more sensitive, and pH-independent HyPer7 probes capable of detecting sub-nanomolar H_2O_2 concentrations. They have shown that mitochondrially generated H_2O_2 does not oxidize probes targeted to the plasma membrane or to the nucleus in HeLa Kyoto and HEK293 cells, and not even in the cytosol of the former cells, except if TrxR is inhibited (Pak et al. 2020; Hoehne et al. 2022). This occurs despite sub-micromolar H_2O_2 addition to the extracellular medium readily oxidizing probes targeted to the mitochondrial matrix, which shows that H_2O_2 easily permeates both mitochondrial membranes. Altogether, these findings indicate that the cytosolic peroxiredoxins, when not extensively oxidized, effectively confine H_2O_2 to micrometer-scale or sub-micrometer-scale domains around supply sites, again in keeping with the theoretical predictions.

Sub-cellular organelles such as mitochondria, the endoplasmic reticulum, or the Golgi complex may have very eccentric or reticular shapes, with diameters of tens of nanometers in some directions but many micrometers in others. H_2O_2 concentration gradients over the short (transversal) direction of these compartments are negligible (see Figure 2.1D, dotted blue line). However, strong gradients may still occur over the longitudinal direction if the H_2O_2 supply is not spatially homogeneous.

Extracellularly, H_2O_2 concentration gradients should be mainly determined by the permeability of the cellular membranes delimiting the extracellular space. This is a consequence of the typically large membrane area/volume ratio of the extracellular space, the lack of very active extracellular enzymatic clearance mechanisms, and the high intracellular activity of such mechanisms. Considering the 10 to 20 μm s^{-1} permeability constants determined for human erythrocytes and other cells in culture (Orrico et al. 2021; Selvaggio, Coelho, and Salvador 2018) and the morphometric characteristics of tissues such as the brain cortex (Syková and Nicholson 2008), a H_2O_2 molecule should on average diffuse no farther than 3.5 μm from the release site before being absorbed by a cell (Sousa et al. 2022). This would allow just autocrine or juxtracrine signaling. However, the following findings show that this transport distance may be much greater under some conditions and tissues. In the brain striatum of living rats, H_2O_2 has a 2.2 s half-life and a $D = 2.5 \times 10^3$ $\mu m^2 s^{-1}$ effective diffusion coefficient, allowing H_2O_2 to diffuse over 180 μm (Ledo et al. 2022). This long transport distance is presumably due to most cells in this tissue being myelinated, which makes them virtually impermeable to H_2O_2. In the zebrafish tail-wounding model, H_2O_2 concentrations decrease to 50% by ≈50 μm from the source at the wound borders, the long diffusion range being attributed to the breakdown of the intracellular peroxiredoxin pool in the surrounding cells (Jelcic et al. 2017). In the human vasculature it is estimated that H_2O_2 directly released by the endothelial cells to the lumens of the arterioles and arteries is transported on average by ≈5 μm before being absorbed, but H_2O_2 derived from released superoxide can be transported by the blood flow over 500–600 μm (Sousa et al. 2022).

2.4 H_2O_2 TRANSMEMBRANE GRADIENT

The reaction-diffusion model above also reveals several important features of the transmembrane gradient. In this case the transmembrane gradient is defined as the ratio between the extracellular H_2O_2 concentration and the cytosolic concentration immediately adjacent to the inner face of the membrane. As follows from equation (2.5), this is given by

$$\Gamma_M = c_0 / c_i(\infty,0) = 1 + \frac{\sqrt{k\,D}}{\kappa}\frac{1-e^{-L/\lambda}}{1+e^{-L/\lambda}}. \qquad (2.9)$$

For low clearance rate constants ($k << D/L^2$), this equation simplifies to

$$\Gamma_M \approx 1 + \frac{k}{\kappa/(L/2)} = 1 + k/k_{p,i} = \Gamma, \qquad (2.10)$$

as might be expected for a situation where the cytosolic gradient is negligible. However, at the opposite extreme ($k >> D/L^2$), equation (2.9) asymptotically approaches

$$\Gamma_M \approx \sqrt{k\,D}/\kappa. \qquad (2.11)$$

These results have the following implications. First, a substantial transmembrane gradient may coexist with a negligible cytosolic gradient, if $k_{p,i} << k << D/L^2$ (Figure 2.2B, and compare panels C and E). This situation is illustrated by Antunes and Cadenas (2000), where a $k_{p,i} = 0.85$ s^{-1} combines with a $k = 4.5$ s^{-1} to yield a $\Gamma = 7$ under the reported experimental conditions.

Second, although at low k the transmembrane gradient scales linearly with k, in the high-k regime it scales less steeply, as $k^{\frac{1}{2}}$ (Figure 2.2E). Because human cells are typically well within the latter regime, a substantial decrease in Γ_M and corresponding increase in the maximal cytosolic H_2O_2 concentration require a strong inhibition of the 2-Cys peroxiredoxins (e.g., 99% inhibition is required to achieve a tenfold increase in $c_i(\infty, 0)$).

This limits the efficacy of floodgate mechanisms (Karplus 2015; Wood 2003) to allow direct H_2O_2 oxidation of redox regulation targets (Travasso et al. 2017).

Third, the relatively shallow scaling of the transmembrane gradient in the high-k regime is in stark contrast with the very steep scaling of the cytosolic gradient highlighted by equation (2.8). One important operational consequence of this fact is that distinct experimental determinations of the transmembrane gradient can be obtained depending on the method used to determine the cytosolic H_2O_2 concentration. Thus, a fluorescent H_2O_2 probe targeted to the inner surface of the membrane would report on $c_i(\infty, 0)$, thereby allowing the determination of the transmembrane gradient. On the other hand, a diffusible probe would report on the average H_2O_2 over the whole cytosol (should it be sensitive enough). The apparent transmembrane gradient determined in this way is

$$\Gamma_{app} = c_o / \left(\frac{1}{L} \int_0^L c_i(\infty, x) dx \right)$$

$$= \frac{L}{2} \left(\frac{k}{\kappa} + \sqrt{\frac{k}{D}} \frac{1+e^{-L/\lambda}}{1-e^{-L/\lambda}} \right), \qquad (2.12)$$

which in the high-k regime asymptotically approaches

$$\Gamma_{app} \approx (L / 2\kappa)k = k / k_{p,i} \approx \Gamma. \qquad (2.13)$$

For $k >> k_{p,i}$, as expected to hold in most human cells, this apparent transmembrane gradient takes approximately the value obtained through the homogeneous-system treatment (equation [2.25]). However, it can overestimate the true transmembrane gradient by several orders of magnitude and it scales linearly with k (Figure 2.2E).

Fourth, although cytosolic H_2O_2 concentration gradients are much weaker in small cells (1-μm scale) such as bacterial ones (Figure 2.2C), transmembrane gradients are comparable to those in larger cells with similar high clearance rate constants (Figure 2.2E).

2.5 MEASURING H_2O_2 GRADIENTS

The preceding discussion shows that once the rate constants and abundances of the main H_2O_2 clearance systems in a cellular compartment and the rate of H_2O_2 supply are known, the concentration (distribution) of H_2O_2 can be estimated using a kinetic (respectively, reaction-diffusion) model; and if the membrane permeability is also known, then the supply rate and transmembrane gradients can also be estimated in this way. Indeed, several authors have made such estimates using kinetic and reaction-diffusion models of various levels of detail and modeling different types of cells and cell compartments (e.g., Poole 1975; Makino et al. 2004; Adimora, Jones, and Kemp 2010; Huang and Sikes 2014; Lim et al. 2015, 2016; Travasso et al. 2017; Benfeitas et al. 2014; Antunes et al. 1996; Selvaggio, Coelho, and Salvador 2018; Stein et al. 2020; Orrico et al. 2018). Recent models that already take peroxiredoxins' overwhelming contribution for H_2O_2 clearance into account point to mean cytosolic H_2O_2 concentrations in the high picomolar range, apparent transmembrane gradients in the 1000-fold range, and true transmembrane gradients in the 100-fold range (Adimora, Jones, and Kemp 2010; Lim et al. 2016; Benfeitas et al. 2014; Travasso et al. 2017; see also Figure 2.2B, E).

Two important shortcomings of the models above are that (1) they require the a priori knowledge and characterization of the processes that contribute the most for cellular H_2O_2 clearance, and (2) they cannot account for unknown or poorly characterized effectors that modulate the specific activity of the H_2O_2-clearing enzymes. The latter problem can be partly addressed by informing the models with experimental determinations of the activities of the main clearance systems. Such was the approach taken by Antunes and Cadenas (2000). These authors determined the H_2O_2 permeability of Jurkat T cells through catalase latency experiments, as well as GPx activity in homogenates and catalase activities in permeabilized cells to then estimate a sevenfold apparent transmembrane gradient for these cells. Unfortunately, this work overlooked the contribution of the peroxiredoxins, which would only be fully appreciated nearly a decade later. Huang and Sikes (2014) addressed this shortcoming through a similar approach. These authors estimated the concentration of the active cytosolic 2-Cys peroxiredoxins to H_2O_2 clearance in HeLa cells by fitting a mathematical model of the kinetics of the peroxiredoxin/thioredoxin system to NADPH consumption time courses in lysates supplemented with excess thioredoxin and thioredoxin reductase. From this determination and from an available rate constant for H_2O_2 reduction by peroxiredoxin (Prx) 2, they then inferred the contribution of these peroxiredoxins to H_2O_2 clearance as $k_{Prx} = 3.8 \times 10^9$ dm^3 s^{-1} cell^{-1}, and estimated $\Gamma_{app} = 650$. The kinetic model used for the estimate drew on the then available kinetic parameters for Prx2. But although recent studies significantly revised these kinetic parameters (Dalla Rizza et al. 2019; Portillo-Ledesma et al. 2018; Peskin et al. 2020) and revealed that the most abundant 2-Cys peroxiredoxin in HeLa cells is actually Prx1 (Geiger et al. 2012), our refit of the experimental data shown in figure 2.3B of Huang and Sikes (2014) using recently determined kinetic parameters for Prx1 yielded values in the same range. Moreover, from the data in Huang and Sikes (2014) and by treating the HeLa cells as 940 μm^3 (Fujioka et al. 2006) spheres, using a model as equation (2.4) for a spherical geometry and taking the $k_{cell} = 5.9 \times 10^{-12}$ dm^3 s^{-1} cell^{-1} (Huang and Sikes 2014) as an approximation to kA (Appendix B), we estimated that the true transmembrane gradient is $\Gamma = 110$. Although this value is sixfold lower than Γ_{app}, it implies that for a cell bathed in μM H_2O_2 the *maximal* (i.e., near the membrane) cytosolic H_2O_2 concentration is in the order of 10 nM. This concentration would not substantially oxidize a moderately reactive protein thiol ($k < 1000$ M^{-1}s^{-1}), which highlights the need for more sensitive signaling mechanisms (Travasso et al. 2017).

Estimations based on activities measured in homogenates may still miss the influence of eventual peroxiredoxin effectors that modify their activity in intact cells. The following more recent developments bypass this limitation by drawing on genetically encoded probes that can sense H_2O_2 in intact cells. Equations and protocols that include the oxidation-reduction cycle shown in reactions (2.14) and (2.15) of redox targets were developed and applied to estimate gradients and intracellular levels of H_2O_2 in the vicinity of the molecular probes (Brito and Antunes 2014; Antunes and Brito 2022).

$$\text{Target}_{rd} + H_2O_2 \rightarrow \text{Target}_{ox} + H_2O \quad (2.14)$$

$$\text{Target}_{ox} \rightarrow \text{Target}_{rd} \quad (2.15)$$

$$\frac{d[\text{Target}_{rd}]}{dt} = k_{rd}(1 - [\text{Target}_{rd}]) - k_{ox}[\text{Target}_{rd}] \quad (2.16)$$

By applying the differential equation (2.16) that describes the time course of the oxidation level of the molecular probe, Domènech et al. (2018) estimated a 300-fold apparent transmembrane gradient for *Schizosaccharomyces pombe* exposed to low-micromolar H_2O_2 concentrations. Using the same procedure, Lyublinskaya and Antunes (2019) determined a mean cytosolic H_2O_2 concentration of 2.2 nM in undisturbed K562 cells and estimated a 390-fold apparent transmembrane gradient. Based on the gradients determined in these works, subsequent experiments using more-sensitive probes also point to sub-nanomolar to low nanomolar basal cytosolic H_2O_2 concentrations in human cell lines (Pak et al. 2020) and in the yeast *S. pombe* (de Cubas et al. 2021).

2.6 REDOX SIGNALING

Oxidation of redox switches by H_2O_2 mediates transmission of information encoded in the H_2O_2 concentration. The level of oxidation of the redox switch not only depends on the rate of its oxidation but also on the rate of its reduction (reactions [2.14] and [2.15]). Two key parameters that characterize signaling transmission by redox switches and how oxidation and reduction of redox switches impact signal transmission are the input dynamic range and the response time. The former describes how the level of oxidation of the redox switch changes with the concentration of H_2O_2 in its vicinity. The input dynamic range is characteristic of each sensor as it depends on the ratio between oxidation and reduction of the sensor, as described by equation (2.17):

$$[\text{Target}_{ox}] = \frac{1}{1 + \frac{k_{switchoff}}{k_{target+H_2O_2}[H_2O_2]}}, \quad (2.17)$$

where $k_{target+H_2O_2}$ is the chemical rate constant for the oxidation of the target by H_2O_2 and $k_{switchoff}$ is the pseudo-first-order rate constant for the reduction of the target protein in the cell.

Concerning the response time, this gives the time needed for half of the variation of the oxidation of redox switch to take place (i.e., the 50% response time). The response time depends on the sum of the oxidation and redox rate constants as highlighted in equation (2.18):

$$t_{1/2} = \frac{\ln(2)}{k_{target+H_2O_2}[H_2O_2] + k_{switchoff}}. \quad (2.18)$$

By applying these two equations, one can estimate what is the level of oxidation of a redox switch and how fast that oxidation takes place, for a given local H_2O_2 concentration in the vicinity of the redox switch, provided oxidation and reduction constants are known.

In the cytosol, for which low nM basal H_2O_2 concentrations are estimated, even the most sensitive redox target (Prx1) is not oxidizable to any meaningful levels (Figure 2.3) when its fast reduction is considered. This means that it is nearly insensitive to small variations of the local H_2O_2 concentration in its vicinity, being unable to transmit information encoded in the variation of the H_2O_2 concentration. An increase of about two orders of magnitude in the low average cytosolic H_2O_2 concentration is needed to oxidize Prx1 to meaningful levels. Thus, the triggering of H_2O_2-reponsive redox switches in the cytosol in the absence of oxidative stress may be circumscribed to microdomains with H_2O_2 concentrations that are orders of magnitude higher than the basal cytosolic concentration. The formation of such microdomains may be facilitated by concentrating the H_2O_2 supply in a small number of discrete sites (Travasso et al. 2017).

In a compartment like the endoplasmic reticulum with a basal H_2O_2 concentration of near 1 μM, high reactivity targets such as Prx1 are already near fully oxidized at basal conditions, even in the presence of fast reduction (Figure 2.3). Thus, increases of H_2O_2 above basal levels have little impact in the oxidation of these high-reactivity redox targets, and consequently they would not be good signal transducers of the information encoded in H_2O_2 concentration variations. Redox signaling in ER should be sensed by redox proteins whose dynamic range is like that of Prx5 or PTP1B. Oxidation of these redox targets may be facilitated by the direct channeling from H_2O_2-producing to consuming proteins. If a H_2O_2-generating protein can interact with a redox switch in a favorable orientation, very high effective H_2O_2 concentrations can be achieved in the nanoscale cavity thus formed. Such high effective concentrations can in principle oxidize low-reactivity redox centers within the lifetime of the complex, thus accomplishing selective oxidation with little H_2O_2 leakage. Indeed, there is experimental evidence of H_2O_2 channeling in the endoplasmic reticulum from ERO1α to GPx8 (Ramming et al. 2014, 2015). Intriguingly, H_2O_2 channeling has also been observed between NOX4 to PP1 (Santos et al. 2016), in spite of the luminal topology of NOX4 active center and cytosolic localization of PP1.

H_2O_2 channeling may also operate in other compartments, and recent evidence (Montllor-Albalate et al. 2022) suggests that SOD1 can channel H_2O_2 to a variety of thiol-protein targets with which it interacts. Direct evidence for such channeling from *in vitro* studies with purified proteins is still lacking, though. A direct interaction of the cytosolic peroxiredoxins with peroxiporins could make for

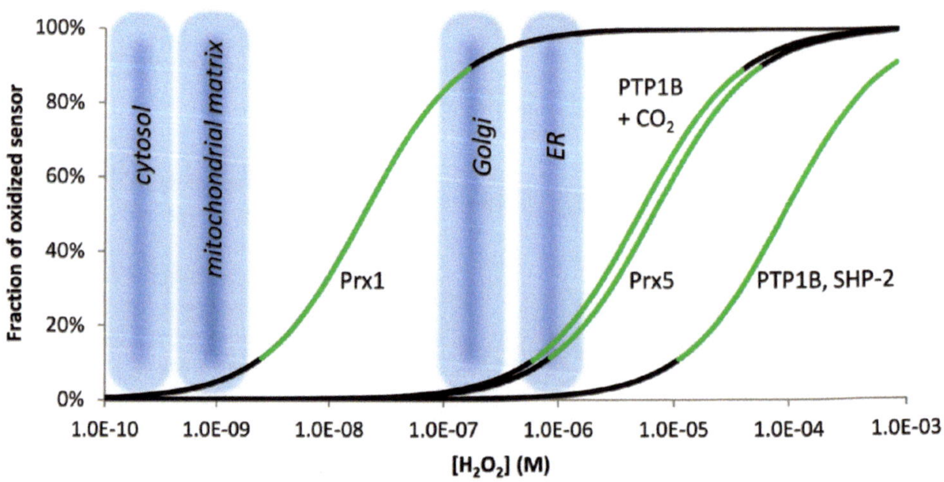

Figure 2.3 Input dynamic range for redox-sensitive proteins. Superimposed in this figure are also estimates of the H_2O_2 concentration in certain redox compartments according to Sies (2021).

extremely efficient and sensitive redox signaling. However, we are unaware of any evidence for a physical interaction between these proteins.

2.7 DO H_2O_2-DEPENDENT REDOX RELAYS FORM MICRODOMAINS?

If H_2O_2 reacts directly with the end redox target, the signal transduction process is complete. If H_2O_2 reacts with a high-reactivity center that relays the oxidation to the end target, the efficacy of the relay step must be established. In particular, whether the redox relay transmits the information encoded by its oxidation at a site far away from its oxidation is an important issue. For this analysis, the half-life and diffusion rates of the high-reactivity center must be considered.

Theoretical considerations indicate that the H_2O_2 microdomains are in turn mirrored by the localization of the oxidized forms of Prx1 and Prx2 (Travasso et al. 2017). This occurs because although these peroxiredoxin forms are much longer-lived than H_2O_2, their diffusion is also much slower. Molecules of the size of Prx1/2 decamers diffuse in the cytosol with constants in the order of 0.5 $\mu m^2 s^{-1}$ (Wojcieszyn et al. 1981). Thus, the condensation rate constants ≈11 s^{-1} and ≈0.25 s^{-1} (Dalla Rizza et al. 2019; Portillo-Ledesma et al. 2018; Peskin et al. 2020) for human Prx1 and Prx2 imply that the respective sulfenic forms will diffuse just 0.4 μm and 3 μm within their half-lives, respectively, as follows from the relationship $d_{1/2} = \sqrt{6Dt_{1/2}}$ (Einstein 1905; von Smoluchowski 1906). Likewise, considering a rate constant for reduction of the disulfide forms of Prx1 and Prx2 by human Trx1 of ≈5.5×10^5 $M^{-1}s^{-1}$ (Rodriguez et al. 2020) and typical cytosolic Trx1 concentrations in the range of 30 μM (Selvaggio, Coelho, and Salvador 2018), these forms would diffuse just 0.5 μm within their half-lives, even as dimers. As a consequence, peroxiredoxin-mediated redox relays that must operate at low H_2O_2 supply rates are spatially localized as well (Travasso et al. 2017). The effectiveness of redox relays in transferring oxidizing equivalents from H_2O_2 to end targets can be further enhanced when targets and peroxiredoxins are recruited to the membrane near sites of H_2O_2 supply (Travasso et al. 2017), such as through membrane-located protein scaffolds (Talwar, Messens, and Dick 2020).

On the other hand, microdomains likely bear no relevance for signaling under oxidative stress. The high H_2O_2 supply to the cytosol under circumstances such as wound-induced inflammation can extensively oxidize the peroxiredoxins to their disulfide or sulfinic forms (Jelcic et al. 2017; Niethammer et al. 2009). (For analysis of the conditions that prompt distinct outcomes see [Selvaggio, Coelho, and Salvador 2018].) As a consequence, the concentration gradients of H_2O_2 and of the oxidized forms of the peroxiredoxins collapse, allowing high concentrations of all these species to penetrate deeply in the cytosol. Moderately H_2O_2-reactive redox targets involved in stress adaptation may then be directly oxidized by H_2O_2 in a timely manner, such as glyceraldehyde 3-phosphate dehydrogenase (GAPDH, k = 500 $M^{-1}s^{-1}$ [Little and O'Brien 1969; Winterbourn and Hampton 2008]), Kelch-like ECH-associated protein 1 (KEAP1, k = 140 $M^{-1}s^{-1}$ [Marinho et al. 2014]), and a still unidentified target that regulates nuclear factor erythroid 2-related factor 2 (NRF2) protein synthesis ($k \geq 1300$ $M^{-1}s^{-1}$ [Marinho et al. 2014]). The recent observation that the Prx1-ASK1 (apoptosis signal-regulating kinase 1) redox relay does not require a scaffold protein (Vo et al. 2021) may also be explained by the collapse of the Prx1-SS concentration gradient at oxidative loads that trigger ASK1-dependent apoptosis.

2.8 SPECIFIC EXAMPLES

Different signaling pathways use different mechanisms of redox signaling, including direct oxidation of a target or oxidation via a one-step or a two-step redox relay (Netto and Antunes 2016). Direct and redox relay mechanisms were first observed in microorganisms with the characterization of the activation of transcription factors. OxyR is directly activated by H_2O_2 while Yap1 and Pap1 are each oxidized through thiol-disulfide reshuffling mediated by the direct H_2O_2 targets glutathione peroxidase 3 (Delaunay et al. 2002) and peroxiredoxin 1 (Bozonet et al. 2005; Vivancos et al. 2005), respectively. In mammals, redox relays include the pairs Prx1/ASK1 (Jarvis, Hughes, and Ledgerwood 2012; Vo et al. 2021), Prx2/STAT3 (Sobotta et al. 2014; Talwar, Messens, and Dick 2020), and Prx2/MEKK4 (Barata and Dick 2019). The protein tyrosine phosphatase PTP1B has a low reactivity toward H_2O_2, prompting the so far unsuccessful search for a redox relay. The reactivity of H_2O_2 toward PTP1B and other thiol proteins increases by forming peroxymonocarbonate (Zhou et al. 2011). Based on theoretical analysis the direct reaction of PTP1B with peroxymonocarbonate has been suggested as an effective means of redox signaling (Antunes and Brito 2017), and recently peroxymonocarbonate was observed to mediate PTP1B redox regulation (Dagnell et al. 2019).

2.9 CONCLUDING REMARKS

Altogether, theoretical and experimental approaches consistently support the conclusions that (1) basal mean cytosolic H_2O_2 concentrations in eukaryotic cells do not exceed the low-nanomolar range; (2) at low oxidative loads, cytosolic H_2O_2 is confined to microdomains; (3) *apparent* transmembrane gradients in cells exposed to sub-micromolar H_2O_2 are at least in the mid-hundredfold scale; (4) the low cytosolic H_2O_2 concentrations, their uneven spatial distribution, and the strong transmembrane gradients are mainly a consequence of the action of the peroxiredoxin/thioredoxin/thioredoxin reductase system; (5) the efficacy of the signal transmission by redox switches is modulated by their oxidation and reduction; (6) in the cytosol, signaling at low oxidative loads proceeds mainly by the formation of local microdomains; and (7) in cellular compartments with higher levels of H_2O_2, like the endoplasmic reticulum, channeling H_2O_2 directly to the target is expected to constitute an important mechanism underlying signal transmission.

2.10 ACKNOWLEDGMENTS

Work financed by the European Regional Development Fund, through COMPETE2020-Operational Program for Competitiveness and Internationalization, and Portuguese funds via FCT-Fundação para a Ciência e a Tecnologia, under projects UIDB/04539/2020, UIDP/04539/2020, UIDB/00313/2020, UIDP/00313/2020, UIDB/00100/2020, UIDP/00100/2020, and LA/P/0056/2020.

APPENDIX A ON THE RELATIONSHIP BETWEEN ENZYME LATENCY AND THE TRANSMEMBRANE GRADIENT

To introduce the concept of enzyme latency and examine how it relates to the transmembrane gradient, we now relax the simplifying assumption of a constant extracellular H_2O_2 concentration. Therefore, the model must now account for the dynamics of this concentration as well:

$$\frac{dc_i}{dt} = k_{p,i}(c_o - c_i) - k\,c_i$$
$$\frac{dc_o}{dt} = k_{p,o}(c_i - c_o), \tag{2.19}$$

with $k_{p,o} = \kappa n A/V_o$, the effective permeation rate constant for extracellular H_2O_2 where n is the number of cells per unit volume. Neglecting again the initial cytosolic H_2O_2 concentration, the solution describing the dynamics of the concentration in both compartments becomes

$$c_i(t) = c_o(0)F_i\left(e^{-k_{slow}t} - e^{-k_{fast}t}\right)$$
$$c_o(t) = c_o(0)\left(F_{o,1}e^{-k_{slow}t} + F_{o,2}e^{-k_{fast}t}\right), \tag{2.20}$$

with the characteristic constants

$$k_{slow} = \frac{1}{2}\left(k + k_{p,i} + k_{p,o} - \sqrt{(k + k_{p,i} + k_{p,o})^2 - 4k\,k_{p,o}}\right),$$
$$k_{fast} = \frac{1}{2}\left(k + k_{p,i} + k_{p,o} + \sqrt{(k + k_{p,i} + k_{p,o})^2 - 4k\,k_{p,o}}\right), \tag{2.21}$$

and the pre-exponential factors

$$F_i = k_{p,i}\Big/\sqrt{(k + k_{p,i} + k_{p,o})^2 - 4k\,k_{p,o}},$$
$$F_{o,1} = 1 + (k + k_{p,i} - k_{p,o})\Big/\sqrt{(k + k_{p,i} + k_{p,o})^2 - 4k\,k_{p,o}},$$
$$F_{o,2} = 1 - (k + k_{p,i} - k_{p,o})\Big/\sqrt{(k + k_{p,i} + k_{p,o})^2 - 4k\,k_{p,o}}. \tag{2.22}$$

In experiments the volume of medium typically exceeds that of cells by two to three orders of magnitude, and therefore $k_{p,o} << k_{p,i}$. Under these circumstances, the characteristic constants and pre-exponential factors are well approximated as follows:

$$k_{slow} \approx \frac{1}{\frac{1}{\gamma k} + \frac{1}{k_{p,o}}},\quad k_{fast} \approx k + k_{p,i},$$
$$F_i \approx k_{p,i}/(k + k_{p,i}),\ F_{o,1} \approx 1,\ F_{o,2} \approx 0, \tag{2.23}$$

where $\gamma = \frac{n V_i}{V_o + n V_i} \approx \frac{n V_i}{V_o}$, is the fraction of the volume occupied by the cells in the experimental setup. Moreover, the fast component typically dies out in a sub-second timescale and is thus undetectable. Therefore, over the experimentally observable timescale, equations (2.21) are well approximated by

$$c_i(t) = \frac{c_o(0)}{1+\frac{k}{k_{p,i}}} e^{-\frac{1}{\frac{1}{\gamma k}+\frac{1}{k_{p,o}}}t}$$

$$c_o(t) = c_o(0) e^{-\frac{1}{\frac{1}{\gamma k}+\frac{1}{k_{p,o}}}t}, \qquad (2.24)$$

and the transmembrane gradient is as before:

$$\Gamma(t) = \frac{c_o(t)}{c_i(t)} \approx 1 + \frac{k}{k_{p,i}}. \qquad (2.25)$$

In turn, the latency factor (Λ) is the ratio between the characteristic constant for c_o decay obtained for permeabilized cells—here corresponding to the $k_{p,o} \to \infty$ limit—and that obtained in the absence of permeabilizing agents:

$$\Lambda \approx \frac{\gamma k}{\frac{1}{\frac{1}{\gamma k}+\frac{1}{k_{p,o}}}} = 1 + \frac{\gamma k}{k_{p,o}} = 1 + \frac{k}{k_{p,i}} = \Gamma. \qquad (2.26)$$

That is, under these conditions the latency factor has approximately the same value as the transmembrane gradient. This appropriately reflects the fact that the resistance of the membrane to the passage of H_2O_2 molecules is the single factor underlying both phenomena.

APPENDIX B
ON THE RELATIONSHIP BETWEEN K_{CELL} AND H_2O_2 PERMEABILITY

Experimental determinations of H_2O_2 consumption by cell preparations are usually reported as the ratio between the observed decay constant and the number of cells per unit volume:

$$k_{cell} = \frac{k_{slow}}{\frac{n}{V_o + nV_i}} \approx \frac{1}{\frac{1}{k V_i} + \frac{1}{\kappa A}}$$
$$= \kappa A / (1 + k_{p,i} / k). \qquad (2.27)$$

As follows from this expression, k_{cell} can be viewed as a lower bound for κA. Recent estimates for a variety of cells indicate that $k_{p,i}$ is typically on the order of tens of s^{-1} (Selvaggio, Coelho, and Salvador 2018). Therefore, k_{cell} is a good approximation to κA for cells such as erythrocytes that have a high cytosolic catalase activity ($k_{cat} >$ 100 s^{-1}). However, in most cell types catalase is enclosed in peroxisomes and therefore presents a too low *effective* activity to warrant this approximation by itself. Under these circumstances, k_{cell} shows a biphasic dependence on $c_0(0)$ (Sasaki, Bannai, and Makino 1998; Makino et al. 2004, 1994; Hashida, Sakakura, and Makino 2002). This happens because k decreases with increasing $c_0(0)$ due to the progressive oxidation of the peroxiredoxin and glutathione pools, until it eventually ceases to be competitive with the efflux rate constant. Therefore, at low concentrations of added H_2O_2 (typically micromolar or lower), $k_{cell} \approx \kappa A$ holds, whereas at very high H_2O_2 concentrations (>100 μM), $k_{cell} \approx \kappa A / (1 + k_{p,i} / k^{eff}_{Cat})$, with k^{eff}_{Cat} the effective catalase activity as limited by permeation of the peroxisomal membrane. Most published k_{cell} determinations (Sobotta et al. 2013; Wagner et al. 2013; Doskey et al. 2016; e.g., Bagulho et al. 2015) were done with too strong H_2O_2 boluses for the former approximation to hold. For instance, the k_{cell} values obtained by (Doskey et al. 2016) for a variety of cell lines strongly correlate with cells' catalase activity and decrease by 1.2- to 4.6-fold upon inhibition of catalase with 3-aminotriazole (Wagner et al. 2013; Doskey et al. 2016), pointing to κA values for these cells being substantially higher than the obtained k_{cell} values.

REFERENCES

Adimora, N J, D P Jones, and M L Kemp. 2010. "A Model of Redox Kinetics Implicates the Thiol Proteome in Cellular Hydrogen Peroxide Responses." *Antioxidants and Redox Signaling* 13 (6): 731–743. https://doi.org/10.1089/ars.2009.2968.

Antunes, Fernando, and Paula Matos Brito. 2017. "Quantitative Biology of Hydrogen Peroxide Signaling." *Redox Biology* 13 (October): 1–7. https://doi.org/10.1016/j.redox.2017.04.039.

Antunes, Fernando, and Paula Matos Brito. 2022. "Data Processing to Probe the Cellular Hydrogen Peroxide Landscape." *Methods in Molecular Biology* 2385: 153–160. https://doi.org/10.1007/978-1-0716-1767-0_8.

Antunes, Fernando, and E Cadenas. 2000. "Estimation of H2O2 Gradients Across Biomembranes." *FEBS Letters* 475 (2): 121–126. https://doi.org/10.1016/S0003-9861(02)00049-8.

Antunes, Fernando, A Salvador, H S Marinho, R Alves, R E Pinto, and A S Marinho. 1996. "Lipid Peroxidation in Mitochondrial Inner Membranes. I. An Integrative Kinetic Model." *Free Radical Biology and Medicine* 21 (7): 917–943. https://doi.org/10.1016/S0891-5849(96)00185-2.

Appenzeller-Herzog, Christian, Gabor Bánhegyi, Ivan Bogeski, Kelvin J A Davies, Agnès Delaunay-Moisan, Henry Jay Forman, Agnes Görlach, et al. 2016. "Transit of H2O2 Across the Endoplasmic Reticulum Membrane Is Not Sluggish." *Free Radical Biology and Medicine* 94: 157–160. https://doi.org/10.1016/j.freeradbiomed.2016.02.030.

Bagulho, Ana, Filipe Vilas-Boas, Andreia Pena, Catarina Peneda, Filipa C Santos, Ana Jerónimo, Rodrigo F M de Almeida, and Carla Real. 2015. "The Extracellular Matrix Modulates H2O2 Degradation and Redox Signaling in Endothelial Cells." *Redox Biology* 6: 454–460. https://doi.org/10.1016/j.redox.2015.09.006.

Barata, Ana G, and Tobias P Dick. 2019. "A Role for Peroxiredoxins in H2O2- and MEKK-Dependent Activation of the P38 Signaling Pathway." *Redox Biology*, October, 101340. https://doi.org/10.1016/j.redox.2019.101340.

Benfeitas, Rui, Gianluca Selvaggio, Fernando Antunes, Pedro M B M Coelho, and Armindo Salvador. 2014. "Hydrogen Peroxide Metabolism and Sensing in Human Erythrocytes: A Validated Kinetic Model and Reappraisal of the Role of Peroxiredoxin II." *Free Radical Biology and Medicine* 74 (1): 35–49. https://doi.org/10.1016/j.freeradbiomed.2014.06.007.

Bertolotti, Milena, Stefano Bestetti, Jose M García-Manteiga, Iria Medraño-Fernandez, Andrea Dal Mas, Maria Luisa Malosio, and Roberto Sitia. 2013. "Tyrosine Kinase Signal Modulation: A Matter of H 2 O 2 Membrane Permeability?" *Antioxidants & Redox Signaling* 19 (13): 1447–1451. https://doi.org/10.1089/ars.2013.5330.

Bestetti, Stefano, Mauro Galli, Ilaria Sorrentino, Paolo Pinton, Alessandro Rimessi, Roberto Sitia, and Iria Medraño-Fernandez. 2020. "Human Aquaporin-11 Guarantees Efficient Transport of H2O2 across the Endoplasmic Reticulum Membrane." *Redox Biology* 28 (January). https://doi.org/10.1016/j.redox.2019.101326.

Bienert, Gerd P, Jan K Schjoerring, and Thomas P Jahn. 2006. "Membrane Transport of Hydrogen Peroxide." *Biochimica et Biophysica Acta* 1758 (8): 994–1003. https://doi.org/10.1016/j.bbamem.2006.02.015.

Bozonet, Stephanie M, Victoria J Findlay, Alison M Day, Jannine Cameron, Elizabeth A Veal, and Brian A Morgan. 2005. "Oxidation of a Eukaryotic 2-Cys Peroxiredoxin Is a Molecular Switch Controlling the Transcriptional Response to Increasing Levels of Hydrogen Peroxide*." *Journal of Biological Chemistry* 280 (24): 23319–23327. https://doi.org/10.1074/JBC.M502757200.

Branco, Miguel R, H Susana Marinho, Luisa Cyrne, and Fernando Antunes. 2004. "Decrease of H2O2 Plasma Membrane Permeability during Adaptation to H2O2 in Saccharomyces Cerevisiae." *Journal of Biological Chemistry* 279 (8): 6501–6506. https://doi.org/10.1074/jbc.M311818200.

Brito, Paula M, and Fernando Antunes. 2014. "Estimation of Kinetic Parameters Related to Biochemical Interactions between Hydrogen Peroxide and Signal Transduction Proteins." *Frontiers in Chemistry* 2 (October): 82. https://doi.org/10.3389/fchem.2014.00082.

Carvalho, Larissa A C, Daniela R Truzzi, Thamiris S Fallani, Simone V Alves, Jose Carlos Toledo Junior, Ohara Augusto, Luis E S Netto, and Flavia C Meotti. 2017. "Urate Hydroperoxide Oxidizes Human Peroxiredoxin 1 and Peroxiredoxin 2." *The Journal of Biological Chemistry* 292 (March): 8705–8715. https://doi.org/10.1074/jbc.M116.767657.

Chance, Britton, Helmut Sies, and Alberto Boveris. 1979. "Hydroperoxide Metabolism in Mammalian Organs." *Physiological Reviews* 59: 527–605.

Cho, Chun-Seok, S Lee, G T Lee, H A Woo, E J Choi, and S G Rhee. 2010. "Irreversible Inactivation of Glutathione Peroxidase 1 and Reversible Inactivation of Peroxiredoxin II by H2O2 in Red Blood Cells." *Antioxidants and Redox Signaling* 12 (11): 1235–1246. https://doi.org/10.1089/ars.2009.2701.

Clayton, Roderick K. 1959. "Permeability Barriers and the Assay of Catalase in Intact Cells." *Biochimica et Biophysica Acta* 36 (1): 35–39. https://doi.org/10.1016/0006-3002(59)90066-6.

Cubas, Laura de, Valeriy V. Pak, Vsevolod V. Belousov, José Ayté, and Elena Hidalgo. 2021. "The Mitochondria-to-Cytosol H2O2 Gradient Is Caused by Peroxiredoxin-Dependent Cytosolic Scavenging." *Antioxidants* 10 (5): 731. https://doi.org/10.3390/antiox10050731.

Dagnell, Markus, Qing Cheng, Syed Husain Mustafa Rizvi, Paul E Pace, Benoit Boivin, Christine C Winterbourn, and Elias S J Arnér. 2019. "Bicarbonate Is Essential for Protein Tyrosine Phosphatase 1B

(PTP1B) Oxidation and Cellular Signaling through EGF-Triggered Phosphorylation Cascades." *The Journal of Biological Chemistry* 294 (33): 12330–12338. https://doi.org/10.1074/jbc.RA119.009001.

Dalla Rizza, Joaquín, Lía M. Randall, Javier Santos, Gerardo Ferrer-Sueta, and Ana Denicola. 2019. "Differential Parameters between Cytosolic 2-Cys Peroxiredoxins, PRDX1 and PRDX2." *Protein Science* 28 (1): 191–201. https://doi.org/10.1002/pro.3520.

Delaunay, Agnès, Delphine Pflieger, Marie Bénédicte Barrault, Joelle Vinh, and Michel B Toledano. 2002. "A Thiol Peroxidase Is an H2O2 Receptor and Redox-Transducer in Gene Activation." *Cell* 111 (4): 471–481. https://doi.org/10.1016/S0092-8674(02)01048-6.

Domènech, Alba, José Ayté, Fernando Antunes, and Elena Hidalgo. 2018. "Using in Vivo Oxidation Status of One- and Two-Component Redox Relays to Determine H2O2 Levels Linked to Signaling and Toxicity." *BMC Biology* 16 (1): 61. https://doi.org/10.1186/s12915-018-0523-6.

Doskey, Claire M., Visarut Buranasudja, Brett A. Wagner, Justin G. Wilkes, Juan Du, Joseph J. Cullen, and Garry R. Buettner. 2016. "Tumor Cells Have Decreased Ability to Metabolize H2O2: Implications for Pharmacological Ascorbate in Cancer Therapy." *Redox Biology* 10: 274–284. https://doi.org/10.1016/j.redox.2016.10.010.

Duve, C De. 1965. "The Separation and Characterization of Subcellular Particles." *Harvey Lectures* 59: 49–87. www.ncbi.nlm.nih.gov/pubmed/5337823.

Einstein, Albert. 1905. "Über Die von Der Molekularkinetischen Theorie Der Wärme Geforderte Bewegung von in Ruhenden Flüssigkeiten Suspendierten Teilchen [AdP 17, 549 (1905)]." *Annalen Der Physik* 14 (S1): 182–193. https://doi.org/10.1002/andp.200590005.

Fujioka, A, K Terai, R E Itoh, K Aoki, T Nakamura, S Kuroda, E Nishida, and M Matsuda. 2006. "Dynamics of the Ras/ERK MAPK Cascade as Monitored by Fluorescent Probes." *Journal of Biological Chemistry* 281 (13): 8917–8926. https://doi.org/10.1074/jbc.M509344200.

Geiger, Tamar, Anja Wehner, Christoph Schaab, Juergen Cox, and Matthias Mann. 2012. "Comparative Proteomic Analysis of Eleven Common Cell Lines Reveals Ubiquitous But Varying Expression of Most Proteins." *Molecular & Cellular Proteomics* 11 (3): M111.014050–M111.014050. https://doi.org/10.1074/mcp.M111.014050.

Hashida, Kanae, Yuki Sakakura, and Nobuo Makino. 2002. "Kinetic Studies on the Hydrogen Peroxide Elimination by Cultured PC12 Cells: Rate Limitation by Glucose-6-Phosphate Dehydrogenase." *Biochimica et Biophysica Acta (BBA)—General Subjects* 1572 (1): 85–90. https://doi.org/10.1016/S0304-4165(02)00282-9.

Hoehne, Michaela Nicole, Lianne J H C Jacobs, Kim Jasmin Lapacz, Gaetano Calabrese, Lena Maria Murschall, Teresa Marker, Harshita Kaul, et al. 2022. "Spatial and Temporal Control of Mitochondrial H2O2 Release in Intact Human Cells." *The EMBO Journal* 41 (7): e109169. https://doi.org/10.15252/embj.2021109169.

Hoehne, Michaela Nicole, Lianne J H C Jacobs, Kim Jasmin Lapacz, Lena Maria, Teresa Marker, Bruce Morgan, Mark Fricker, Vsevolod V Belousov, and Jan Riemer. 2021. "Thioredoxin Reductase Controls the Capacity of Peroxiredoxins to Limit Mitochondrial H2O2 Release." *BioRxiv* 2021.03.04.433951. https://doi.org/10.1101/2021.03.04.433951.

Huang, Beijing K, and Hadley D Sikes. 2014. "Quantifying Intracellular Hydrogen Peroxide Perturbations in Terms of Concentration." *Redox Biology* 2: 955–962. https://doi.org/10.1016/j.redox.2014.08.001.

Jarvis, Reagan M, Stephanie M Hughes, and Elizabeth C Ledgerwood. 2012. "Peroxiredoxin 1 Functions as a Signal Peroxidase to Receive, Transduce, and Transmit Peroxide Signals in Mammalian Cells." *Free Radical Biology and Medicine* 53 (7): 1522–1530. https://doi.org/10.1016/j.freeradbiomed.2012.08.001.

Jelcic, Mark, Balázs Enyedi, João B Xavier, and Philipp Niethammer. 2017. "Image-Based Measurement of H2O2 Reaction-Diffusion in Wounded Zebrafish Larvae." *Biophysical Journal* 112 (9): 2011–2018. https://doi.org/10.1016/j.bpj.2017.03.021.

Kaplan, J G. 1955. "The Alteration of Intracellular Enzymes: I. Yeast Catalase and the Euler Effect." *Experimental Cell Research* 8 (2): 305–328. https://doi.org/10.1016/0014-4827(55)90143-9.

Karplus, P Andrew. 2015. "A Primer on Peroxiredoxin Biochemistry." *Free Radical Biology and Medicine* 80 (March): 183–190. https://doi.org/10.1016/j.freeradbiomed.2014.10.009.

Ledo, A, E Fernandes, A Salvador, J Laranjinha, and R M Barbosa. 2022. "In Vivo Hydrogen Peroxide Diffusivity in Brain Tissue Supports Volume Signaling Activity." *Redox Biology* 50 (April): 102250. https://doi.org/10.1016/j.redox.2022.102250.

Lim, Joseph B, Beijing K Huang, William M Deen, and Hadley D Sikes. 2015. "Analysis of the Lifetime and Spatial Localization of Hydrogen Peroxide Generated in the Cytosol Using a Reduced Kinetic Model." *Free Radical Biology and Medicine* 89 (July): 47–53. https://doi.org/10.1016/j.freeradbiomed.2015.07.009.

Lim, Joseph B, Troy F Langford, Beijing K Huang, William M Deen, and Hadley D Sikes. 2016.

"A Reaction-Diffusion Model of Cytosolic Hydrogen Peroxide." *Free Radical Biology and Medicine* 90 (January): 85–90. https://doi.org/10.1016/j.freeradbiomed.2015.11.005.

Little, C, and P J O'Brien. 1969. "Mechanism of Peroxide-Inactivation of the Sulphydryl Enzyme Glyceraldehyde-3-Phosphate Dehydrogenase." *European Journal of Biochemistry* 10 (3): 533–538. https://doi.org/10.1111/j.1432-1033.1969.tb00721.x.

Lyublinskaya, Olga, and Fernando Antunes. 2019. "Measuring Intracellular Concentration of Hydrogen Peroxide with the Use of Genetically Encoded H2O2 Biosensor HyPer." *Redox Biology* 24 (June): 101200. https://doi.org/10.1016/J.REDOX.2019.101200.

Ma, M, and J W Eaton. 1992. "Multicellular Oxidant Defense in Unicellular Organisms." *Proceedings of the National Academy of Sciences* 89 (17): 7924–7928. https://doi.org/10.1073/PNAS.89.17.7924.

Makino, Nobuo, Y Mochizuki, S Bannai, and Y Sugita. 1994. "Kinetic Studies on the Removal of Extracellular Hydrogen Peroxide by Cultured Fibroblasts." *Journal of Biological Chemistry* 269 (2): 1020–1025. https://doi.org/10.1016/S0021-9258(17)42214-9.

Makino, Nobuo, Kayoko Sasaki, Kanae Hashida, and Yuki Sakakura. 2004. "A Metabolic Model Describing the H2O2 Elimination by Mammalian Cells Including H2O2 Permeation through Cytoplasmic and Peroxisomal Membranes: Comparison with Experimental Data." *Biochimica et Biophysica Acta (BBA)—General Subjects* 1673 (3): 149–159. https://doi.org/10.1016/j.bbagen.2004.04.011.

Manta, Bruno, Martín Hugo, Cecilia Ortiz, Gerardo Ferrer-Sueta, Madia Trujillo, and Ana Denicola. 2009. "The Peroxidase and Peroxynitrite Reductase Activity of Human Erythrocyte Peroxiredoxin 2." *Archives of Biochemistry and Biophysics* 484 (2): 146–154. https://doi.org/10.1016/j.abb.2008.11.017.

Marinho, H Susana, Carla Real, Luísa Cyrne, Helena Soares, and Fernando Antunes. 2014. "Hydrogen Peroxide Sensing, Signaling and Regulation of Transcription Factors." *Redox Biology* 2 (January): 535–562. https://doi.org/10.1016/j.redox.2014.02.006.

Matias, A C, N Pedroso, N Teodoro, H S Marinho, F Antunes, J M F Nogueira, E Herrero, and L Cyrne. 2007. "Down-Regulation of Fatty Acid Synthase Increases the Resistance of Saccharomyces Cerevisiae Cells to H2O2." *Free Radical Biology and Medicine* 43 (10): 1458–1465. https://doi.org/10.1016/j.freeradbiomed.2007.08.003.

Miller, Evan W, Bryan C Dickinson, and Christopher J Chang. 2010. "Aquaporin-3 Mediates Hydrogen Peroxide Uptake to Regulate Downstream Intracellular Signaling." *Proceedings of the National Academy of Sciences of the U.S.A.* 107 (36): 15681–15686. https://doi.org/10.1073/pnas.1005776107.

Mishina, Natalie M, Yulia A Bogdanova, Yulia G Ermakova, Anastasiya S Panova, Daria A Kotova, Dmitry S Bilan, Benjamin Steinhorn, Elias S J Arnér, Thomas Michel, and Vsevolod V Belousov. 2019. "Which Antioxidant System Shapes Intracellular H 2 O 2 Gradients?" *Antioxidants & Redox Signaling*, 31 (9): 664–670. https://doi.org/10.1089/ars.2018.7697.

Mishina, Natalia M, Pyotr A Tyurin-Kuzmin, Kseniya N Markvicheva, Alexander V Vorotnikov, Vsevolod A Tkachuk, Vibor Laketa, Carsten Schultz, Sergey Lukyanov, and Vsevolod V Belousov. 2011. "Does Cellular Hydrogen Peroxide Diffuse or Act Locally?" *Antioxidants & Redox Signaling* 14 (1): 1–7. https://doi.org/10.1089/ars.2010.3539.

Montllor-Albalate, Claudia, Hyojung Kim, Anna E Thompson, Alex P Jonke, Matthew P Torres, and Amit R Reddi. 2022. "Sod1 Integrates Oxygen Availability to Redox Regulate NADPH Production and the Thiol Redoxome." *Proceedings of the National Academy of Sciences* 119 (1). https://doi.org/10.1073/pnas.2023328119.

Netto, Luis E, and Fernando Antunes. 2016. "The Roles of Peroxiredoxin and Thioredoxin in Hydrogen Peroxide Sensing and in Signal Transduction." *Molecules and Cells* 39 (1): 65–71. https://doi.org/10.14348/molcells.2016.2349.

Nicholls, P. 1965. "Activity of Catalase in the Red Cell." *Biochimica Et Biophysica Acta* 99: 286–297. https://doi.org/10.1016/S0926-6593(65)80125-4.

Niethammer, Philipp, Clemens Grabher, A Thomas Look, and Timothy J Mitchison. 2009. "A Tissue-Scale Gradient of Hydrogen Peroxide Mediates Rapid Wound Detection in Zebrafish." *Nature* 459 (7249): 996–999. https://doi.org/10.1038/nature08119.

Orrico, Florencia, Ana C Lopez, Daniela Saliwonczyk, Cecilia Acosta, Ismael Rodriguez-Grecco, Isabelle Mouro-Chanteloup, Mariano A Ostuni, Ana Denicola, Leonor Thomson, and Matias N Möller. 2021. "The Permeability of Human Red Blood Cell Membranes to Hydrogen Peroxide is Independent of Aquaporins." *Journal of Biological Chemistry*, December, 101503. https://doi.org/10.1016/j.jbc.2021.101503.

Orrico, Florencia, Matías N Möller, Adriana Cassina, Ana Denicola, and Leonor Thomson. 2018. "Kinetic and Stoichiometric Constraints Determine the Pathway of H2O2 Consumption by Red Blood Cells." *Free Radical Biology and Medicine*

121 (May): 231–239. https://doi.org/10.1016/j.freeradbiomed.2018.05.006.

Pak, Valeriy V, Daria Ezeriņa, Olga G Lyublinskaya, Brandán Pedre, Pyotr A Tyurin-Kuzmin, Natalie M Mishina, Marion Thauvin, et al. 2020. "Ultrasensitive Genetically Encoded Indicator for Hydrogen Peroxide Identifies Roles for the Oxidant in Cell Migration and Mitochondrial Function." *Cell Metabolism* 31 (3): 642–653.e6. https://doi.org/10.1016/j.cmet.2020.02.003.

Pedroso, Nuno, Patrícia Gomes-Alves, H Susana Marinho, Verônica B Brito, Cristina Boada, Fernando Antunes, Enrique Herrero, Deborah Penque, and Luísa Cyrne. 2012. "The Plasma Membrane-Enriched Fraction Proteome Response during Adaptation to Hydrogen Peroxide in Saccharomyces Cerevisiae." *Free Radical Research* 46 (10): 1267–1279. https://doi.org/10.3109/10715762.2012.704997.

Pedroso, Nuno, A C Matias, Luisa Cyrne, Fernando Antunes, C Borges, R Malho, R F M de Almeida, Enric Herrero, and Helena Susana Marinho. 2009. "Modulation of Plasma Membrane Lipid Profile and Microdomains by H2O2 in Saccharomyces Cerevisiae." *Free Radical Biology and Medicine* 46 (2): 289–298. https://doi.org/10.1016/j.freeradbiomed.2008.10.039.

Peskin, Alexander V, F M Low, L N Paton, G J Maghzal, M B Hampton, and C C Winterbourn. 2007. "The High Reactivity of Peroxiredoxin 2 with H(2)O(2) is Not Reflected in Its Reaction with Other Oxidants and Thiol Reagents." *The Journal of Biological Chemistry* 282 (16): 11885–11892. https://doi.org/10.1074/jbc.M700339200.

Peskin, Alexander V, Flávia C Meotti, Luiz F de Souza, Robert F Anderson, Christine C Winterbourn, and Armindo Salvador. 2020. "Intra-Dimer Cooperativity between the Active Site Cysteines during the Oxidation of Peroxiredoxin 2." *Free Radical Biology and Medicine* 158 (October): 115–125. https://doi.org/10.1016/j.freeradbiomed.2020.07.007.

Poole, Brian. 1975. "Diffusion Effects in the Metabolism of Hydrogen Peroxide by Rat Liver Peroxisomes." *Journal of Theoretical Biology* 51 (1): 149–167. https://doi.org/10.1016/0022-5193(75)90145-9.

Portillo-Ledesma, Stephanie, Lía M Randall, Derek Parsonage, Joaquín Dalla Rizza, P Andrew Karplus, Leslie B Poole, Ana Denicola, and Gerardo Ferrer-Sueta. 2018. "Differential Kinetics of Two-Cysteine Peroxiredoxin Disulfide Formation Reveal a Novel Model for Peroxide Sensing." *Biochemistry* 57 (24): 3416–3424. https://doi.org/10.1021/acs.biochem.8b00188.

Ramming, Thomas, Henning G Hansen, Kazuhiro Nagata, Lars Ellgaard, and Christian Appenzeller-Herzog. 2014. "GPx8 Peroxidase Prevents Leakage of H2O2 from the Endoplasmic Reticulum." *Free Radical Biology and Medicine* 70 (May): 106–116. https://doi.org/10.1016/J.FREERADBIOMED.2014.01.018.

Ramming, Thomas, Masaki Okumura, Shingo Kanemura, Sefer Baday, Julia Birk, Suzette Moes, Martin Spiess, et al. 2015. "A PDI-Catalyzed Thiol–Disulfide Switch Regulates the Production of Hydrogen Peroxide by Human Ero1." *Free Radical Biology and Medicine* 83 (June): 361–372. https://doi.org/10.1016/J.FREERADBIOMED.2015.02.011.

Rodriguez, Sebastian Villar, Ari Zeida, Gerardo Ferrer-Sueta, and Ana Denicola. 2020. "A Kinetic and Structural Assessment of the Reduction of Human Prx2 by Thioredoxins." *Free Radical Biology and Medicine* 159: S35. Pergamon. https://doi.org/10.1016/J.FREERADBIOMED.2020.10.102.

Salvador, Armindo, João Sousa, and Ruy E Pinto. 2001. "Hydroperoxyl, Superoxide and PH Gradients in the Mitochondrial Matrix: A Theoretical Assessment." *Free Radical Biology and Medicine* 31 (10): 1208–1215. https://doi.org/10.1016/S0891-5849(01)00707-9.

Santos, Celio X C, Anne D Hafstad, Matteo Beretta, Min Zhang, Chris Molenaar, Jola Kopec, Dina Fotinou, et al. 2016. "Targeted Redox Inhibition of Protein Phosphatase 1 by Nox4 Regulates EIF2α-Mediated Stress Signaling." *The EMBO Journal* 35 (3): 319–334. https://doi.org/10.15252/EMBJ.201592394.

Sasaki, K, S Bannai, and N Makino. 1998. "Kinetics of Hydrogen Peroxide Elimination by Human Umbilical Vein Endothelial Cells in Culture." *Biochimica et Biophysica Acta* 1380 (2): 275–288. www.ncbi.nlm.nih.gov/pubmed/9565698.

Seaver, L C, and J A Imlay. 2001. "Hydrogen Peroxide Fluxes and Compartmentalization Inside Growing Escherichia Coli." *Journal of Bacteriology* 183 (24): 7182–7189. https://doi.org/10.1128/JB.183.24.7182-7189.2001.

Selvaggio, Gianluca, Pedro M B M Coelho, and Armindo Salvador. 2018. "Mapping the Phenotypic Repertoire of the Cytoplasmic 2-Cys Peroxiredoxin—Thioredoxin System. 1. Understanding Commonalities and Differences among Cell Types." *Redox Biology* 15 (December): 297–315. https://doi.org/10.1016/j.redox.2017.12.008.

Sies, Helmut. 2021. "Oxidative Eustress: On Constant Alert for Redox Homeostasis." *Redox Biology* 41 (January): 101867. https://doi.org/10.1016/j.redox.2021.101867.

Smoluchowski, M von. 1906. "Zur Kinetischen Theorie Der Brownschen Molekularbewegung Und

Der Suspensionen." *Annalen Der Physik* 326 (14): 756–780. https://doi.org/10.1002/andp.19063261405.
Sobotta, Mirko C, Ana G Barata, Ulrich Schmidt, Sebastian Mueller, Gunda Millonig, and Tobias P Dick. 2013. "Exposing Cells to H2O2: A Quantitative Comparison between Continuous Low-Dose and One-Time High-Dose Treatments." *Free Radical Biology and Medicine* 60 (July): 325–335. https://doi.org/10.1016/j.freeradbiomed.2013.02.017.
Sobotta, Mirko C, Willy Liou, Sarah Stöcker, Deepti Talwar, Michael Oehler, Thomas Ruppert, Annette N Scharf, and Tobias P Dick. 2014. "Peroxiredoxin-2 and STAT3 Form a Redox Relay for H2O2 Signaling." *Nature Chemical Biology* 11 (1): 64–70. https://doi.org/10.1038/nchembio.1695.
Sousa, Tânia, Marcos Gouveia, Rui D M Travasso, and Armindo Salvador. 2022. "How Abundant Are Superoxide and Hydrogen Peroxide in the Vasculature Lumen, How Far Can They Reach?" *Redox Biology*, October, 102527. https://doi.org/10.1016/j.redox.2022.102527.
Stein, Kassi T, Sun Jin Moon, Athena N Nguyen, and Hadley D Sikes. 2020. "Kinetic Modeling of H2O2 Dynamics in the Mitochondria of HeLa Cells." *PLoS Computational Biology* 16 (9): e1008202. https://doi.org/10.1371/journal.pcbi.1008202.
Stroe-Biezen, S A M van, F M Everaerts, L J J Janssen, and R A Tacken. 1993. "Diffusion Coefficients of Oxygen, Hydrogen Peroxide and Glucose in a Hydrogel." *Analytica Chimica Acta* 273 (1–2): 553–560. https://doi.org/10.1016/0003-2670(93)80202-V.
Syková, Eva, and Charles Nicholson. 2008. "Diffusion in Brain Extracellular Space." *Physiological Reviews* 88 (4): 1277–1340. https://doi.org/10.1152/physrev.00027.2007.
Talwar, Deepti, Joris Messens, and Tobias P Dick. 2020. "A Role for Annexin A2 in Scaffolding the Peroxiredoxin 2–STAT3 Redox Relay Complex." *Nature Communications* 11 (1): 1–11. https://doi.org/10.1038/s41467-020-18324-9.
Travasso, Rui D M, Fernando Sampaio dos Aidos, Anahita Bayani, Pedro Abranches, and Armindo Salvador. 2017. "Localized Redox Relays as a Privileged Mode of Cytoplasmic Hydrogen Peroxide Signaling." *Redox Biology* 12: 233–245. https://doi.org/10.1016/j.redox.2017.01.003.
Vivancos, Ana P, Esther A Castillo, Benoît Biteau, Carine Nicot, J Ayté, Michel B Toledano, and Elena Hidalgo. 2005. "A Cysteine-Sulfinic Acid in Peroxiredoxin Regulates H2O2-Sensing by the Antioxidant Pap1 Pathway." *Proceedings of the National Academy of Sciences of the United States of America* 102 (25): 8875–8880. https://doi.org/10.1073/pnas.0503251102.
Vo, Trung Nghia, Julia Malo Pueyo, Khadija Wahni, Daria Ezeriņa, Jesalyn Bolduc, and Joris Messens. 2021. "Prdx1 Interacts with ASK1 upon Exposure to H 2 O 2 and Independently of a Scaffolding Protein." *Antioxidants* 10 (7): 1060. https://doi.org/10.3390/antiox10071060.
Wagner, Brett A, Jordan R Witmer, Thomas Joost van 't Erve, and Garry R Buettner. 2013. "An Assay for the Rate of Removal of Extracellular Hydrogen Peroxide by Cells." *Redox Biology* 1 (1): 210–217. https://doi.org/10.1016/j.redox.2013.01.011.
Winterbourn, C C, and M B Hampton. 2008. "Thiol Chemistry and Specificity in Redox Signaling." *Free Radical Biology and Medicine* 45 (5): 549–561. https://doi.org/10.1016/j.freeradbiomed.2008.05.004.
Wojcieszyn, John W, Robert A Schlegel, E S Wu, and Kenneth A Jacobson. 1981. "Diffusion of Injected Macromolecules within the Cytoplasm of Living Cells." *Proceedings of the National Academy of Sciences* 78 (7): 4407–4410. https://doi.org/10.1073/pnas.78.7.4407.
Wood, Zachary A. 2003. "Peroxiredoxin Evolution and the Regulation of Hydrogen Peroxide Signaling." *Science* 300 (5619): 650–653. https://doi.org/10.1126/science.1080405.
Zhou, Haiying, Harkewal Singh, Zachary D Parsons, Sarah M Lewis, Sanjib Bhattacharya, Derrick R Seiner, Jason N LaButti, Thomas J Reilly, John J Tanner, and Kent S Gates. 2011. "The Biological Buffer Bicarbonate/CO2 Potentiates H2O2-Mediated Inactivation of Protein Tyrosine Phosphatases." *Journal of the American Chemical Society* 133 (40): 15803–15805. https://doi.org/10.1021/ja2077137.

CHAPTER THREE

Diffusion and Transport of Peroxynitrite across Cell Membranes

Matías N. Möller, Karina Alleva, Ari Zeida, and Rafael Radi

ABBREVIATIONS

AQP	Aquaporin
$CO_3^{\bullet-}$	Carbonate radical
D	Diffusion coefficient
$^{\bullet}NO$	Nitric oxide
$^{\bullet}NO_2$	Nitrogen dioxide
$O_2^{\bullet-}$	Superoxide
$^{\bullet}OH$	Hydroxyl radical
$ONOO^-$	Peroxynitrite anion
ONOOH	Peroxynitrous acid
P_m	Permeability coefficient

3.1 INTRODUCTION

Peroxynitrite ($ONOO^-$) is an oxidant formed *in vivo* that can be cytotoxic at high concentration, leading to apoptotic or necrotic death, and at low concentration contributes to the damage observed in chronic degenerative diseases (Szabó et al. 2007). On the other hand, peroxynitrite is an important component of the immune response to fight invading pathogens (Radi 2013). In all cases, the action mechanism involves the oxidation of different biomolecules by peroxynitrite or its derived radicals, leading to lipid peroxidation, protein nitration, and DNA strand breaks, to name a few, which then trigger different biological responses.

In biological systems, peroxynitrite is formed mainly by the diffusion-controlled reaction between the radicals nitric oxide ($^{\bullet}NO$) and superoxide ($O_2^{\bullet-}$) (Eq. 3.1; Figure 3.1; Ferrer-Sueta et al. 2018).

$$^{\bullet}NO + O_2^{\bullet-} \rightarrow ONOO^- \quad (3.1)$$

Nitric oxide is synthesized by three isoforms of nitric oxide synthase (NOS): endothelial NOS (eNOS or NOS3), neuronal NOS (nNOS or NOS1) and inducible NOS (iNOS or NOS2; Alderton et al. 2001). These enzymes use L-arginine, oxygen, and NADPH as substrates to generate $^{\bullet}NO$ and L-citrulline as products, and are involved in the regulation of vascular tone, neurotransmission, and innate immune response, respectively (Alderton et al. 2001). $^{\bullet}NO$ can also be generated by the reduction of nitrite by deoxyhemoglobin, deoxymyoglobin, xanthine oxidoreductase, and complexes of the mitochondrial electron transport chain (Lundberg et al. 2008). Superoxide can be synthesized by NADPH oxidases, xanthine oxidase, and as a by-product of the electron transport chain in the mitochondria, among other sources (Holmström and Finkel 2014). The reaction between $^{\bullet}NO$ and $O_2^{\bullet-}$ has a rate constant in the range of $\sim 10^{10}\ M^{-1}s^{-1}$ (Huie and Padmaja 1993; Goldstein and Czapski 1995; Kissner et al. 1997), indicating that it will be formed even when the concentration of its precursors is low. There are very few reactions that can compete with peroxynitrite formation. Relevant reactions include the dismutation of $O_2^{\bullet-}$ catalyzed by the superoxide dismutases (SOD), which have a very high rate constant for $O_2^{\bullet-}$ dismutation and can effectively decrease $O_2^{\bullet-}$ concentration, and the reaction of $^{\bullet}NO$ with oxyhemoglobin (Ferrer-Sueta and Radi 2009).

DOI: 10.1201/9781003160649-3

The different physicochemical properties of $^{\bullet}NO$ and $O_2^{\bullet-}$ will influence the site of peroxynitrite formation. The small, neutral, and hydrophobic $^{\bullet}NO$ can diffuse through cell membranes virtually unhindered (Möller and Denicola 2018; Denicola et al. 1996), whereas the charged $O_2^{\bullet-}$ will face a large thermodynamic barrier that will severely limit its ability to cross membranes (Gus' kova et al. 1984; Cordeiro 2014). As a consequence, the site of peroxynitrite formation will be greatly dictated by the site of $O_2^{\bullet-}$ formation. Thus, preferred sites of peroxynitrite formation may be summarized to mitochondria, phagosomes, and the external side of the plasma membrane and caveolar regions (Figure 3.1; Ferrer-Sueta et al. 2009; Chen and Deen 2001; Daneva et al. 2021).

Macrophages provide an illustrative example of $ONOO^-$ formation by cells and how diverse factors can affect it. Macrophages can form $ONOO^-$ by the simultaneous production of $O_2^{\bullet-}$ and $^{\bullet}NO$ (Ischiropoulos et al. 1992). Large amounts of $O_2^{\bullet-}$ can be produced during phagocytosis or after an appropriate stimulus (such as PMA) by the multicomponent enzyme system NADPH oxidase (Nox2). Nox 2 is composed of several proteins including two transmembrane proteins. It consumes cytosolic NADPH to form $O_2^{\bullet-}$ in the extracellular compartment (or its topological equivalent, the intraphagosomal compartment; El-Benna et al. 2009). The inducible NOS can be induced with interferon gamma, lipopolysaccharide, and cytokines to produce large quantities of

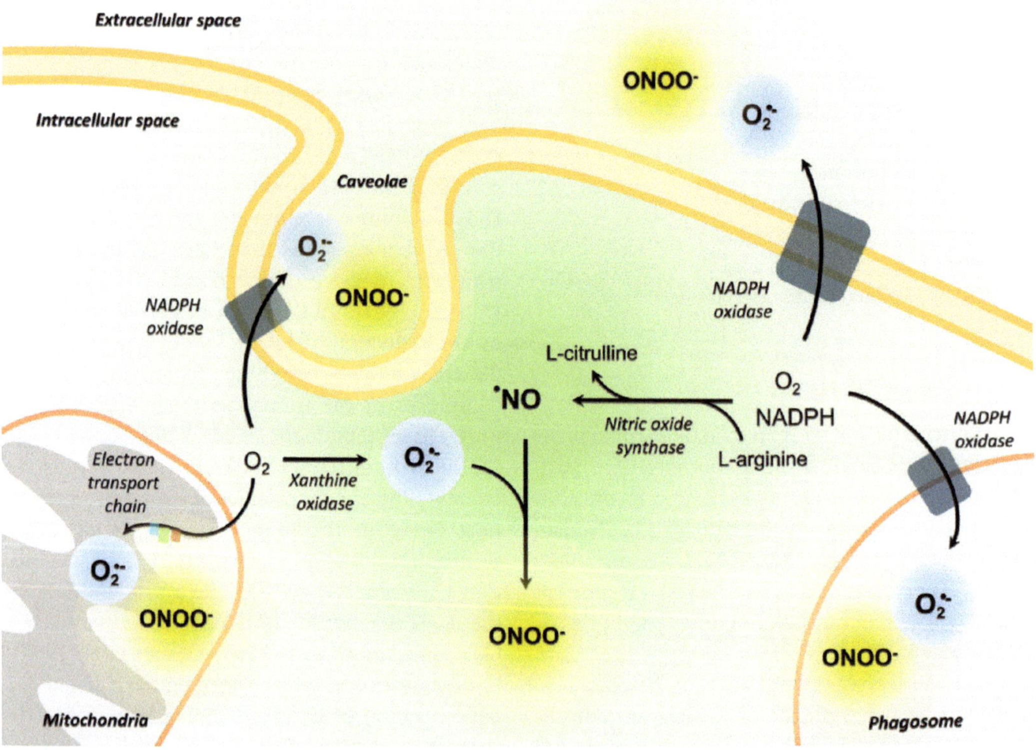

Figure 3.1 Sites of peroxynitrite formation. Peroxynitrite is formed by the diffusion-controlled reaction of $^{\bullet}NO$ with $O_2^{\bullet-}$ and is therefore formed at the sites where both radicals can meet. Whereas $^{\bullet}NO$ can diffuse relatively large distances and through cell membranes virtually unhindered, $O_2^{\bullet-}$ diffusion is more limited, both by the lower permeability of membranes to it and its smaller lifetime. Therefore, peroxynitrite formation sites will be mainly dictated by the sites of $O_2^{\bullet-}$ formation. Depending on the cell types, these sites of peroxynitrite formation may be the extracellular space, the caveolae, the phagosome, mitochondria, or cytosol.

$^{\bullet}NO$ (Pautz et al. 2010). Specific fluorescent probes for $ONOO^-$ permitted quantifying the production of $ONOO^-$ by activated RAW-264 and J774. A1 macrophages to 15–50 nM/min (Prolo et al. 2015).

The different permeability of cellular plasma membranes to $^{\bullet}NO$ and $O_2^{\bullet-}$ has been shown to regulate $ONOO^-$ formation. Red blood cells act as scavengers of $^{\bullet}NO$, because the reaction rate with oxyhemoglobin is very fast (Liu et al. 1998). Externally added red blood cells can consume a great proportion of $^{\bullet}NO$ generated in a vicinal cell. Superoxide, on the other hand, because of its anionic nature, is much less permeable and it is largely confined to the compartment where it is formed. It was shown that the extracellular formation of $ONOO^-$ by macrophages could be inhibited by adding red blood cells, which scavenge $^{\bullet}NO$ much more efficiently than peroxynitrite (Prolo et al. 2015). Intraphagosomal formation of $ONOO^-$ could not be inhibited (Prolo et al. 2015). The small volume of the phagosome ensures that larger concentrations of $O_2^{\bullet-}$ are achieved in the phagosome that will efficiently compete with $^{\bullet}NO$ diffusion to the external red blood cell (Prolo et al. 2015). The permeability of membranes to $O_2^{\bullet-}$ is greatly affected by pH, because it can protonate to the hydroperoxyl radical ($HOO^{\bullet}$) with pK_a of ~4.8. $HOO^{\bullet}$ is not charged, and can diffuse across lipid membranes much more easily (Cordeiro 2014; Gus' kova et al. 1984). Recent experiments showed that the low pH of the phagosome (pH 5.4) leads to a significant protonation of $O_2^{\bullet-}$ and an enhanced penetration through the membrane of the parasite *Trypanosoma cruzi* (Martínez et al. 2019). This low pH in the phagosome will also favor the formation of ONOOH that will diffuse across the parasite membrane more readily and it is likely an important factor in its cytotoxic properties.

3.1.1 Biological Chemistry of Peroxynitrite

Peroxynitrite is a short-lived oxidant that can react with several biomolecules by different mechanisms that will be discussed below. Protonation, reaction with CO_2 or with metals will promote the decomposition of peroxynitrite to the potent one-electron oxidants nitrogen dioxide ($^{\bullet}NO_2$), hydroxyl radical ($^{\bullet}OH$) and carbonate radical ($CO_3^{\bullet-}$; Ferrer-Sueta et al. 2018). The biological chemistry of peroxynitrite is intrinsically linked to these radicals, so their properties and the diffusion across membranes will be discussed in detail. Last, the role of transmembrane protein band 3 in facilitating peroxynitrite diffusion and the potential role of aquaporins in facilitating the diffusion of peroxynitrite are discussed.

In biochemistry, the term *peroxynitrite* includes the peroxynitrite anion ($ONOO^-$) and its conjugated acid, peroxynitrous acid (ONOOH). The IUPAC-recommended names are oxoperoxonitrate (1-) and hydrogen oxoperoxonitrate, respectively. Both $ONOO^-$ and ONOOH coexist in rapid acid–base equilibrium under biologically relevant conditions, because the pK_a is near to neutral ($pK_a = 6.8$; Figure 3.2; Ferrer-Sueta et al. 2018). For laboratory experiments, peroxynitrite can be synthesized in multiple ways and is stable at alkaline pH (Ferrer-Sueta et al. 2018). The concentration of $ONOO^-$ in alkaline solutions can be measured at 302 nm ($\varepsilon = 1700 \pm 10\ M^{-1}cm^{-1}$; Bohle et al. 1994).

Peroxynitrite is highly reactive and rapidly disappears by several possible reactions. Most of peroxynitrite biochemistry can be understood through the lens of chemical kinetics. In the absence of targets at physiological pH, ONOOH isomerizes to nitrate with pH-independent rate constants of 4.5 s^{-1} at 37°C (1.2 s^{-1} at 25°C; Koppenol et al. 1992). The isomerization occurs by homolysis to generate a geminate pair hydroxyl radical ($^{\bullet}OH$) and nitrogen dioxide ($^{\bullet}NO_2$) in a solvent cage that either recombine to give nitrate or diffuse to the bulk solution in 30% yield (Figure 3.2; Ferrer-Sueta et al. 2018). However, reaction with other targets such as CO_2, thiol, and selenol peroxidases and some transition metal centers is much faster and will effectively limit ONOOH homolysis (Ferrer-Sueta et al. 2009).

Peroxynitrite can react with other molecules by bimolecular nucleophilic substitution (S_N2; e.g., thiols) and electron transfer (ET) mechanisms (e.g., metals), or it can act as a Lewis base to form transient reactive Lewis adducts (Ferrer-Sueta et al. 2018). The main targets are dictated by kinetic properties and local concentrations.

The reaction with CO_2 accounts for a large proportion of peroxynitrite consumption in cells and tissues. The reaction occurs between $ONOO^-$ and CO_2 with pH-independent rate constants $k = 3 \times 10^4\ M^{-1}s^{-1}$ (25°C) or $5.8 \times 10^4\ M^{-1}s^{-1}$ (37°C; Lymar

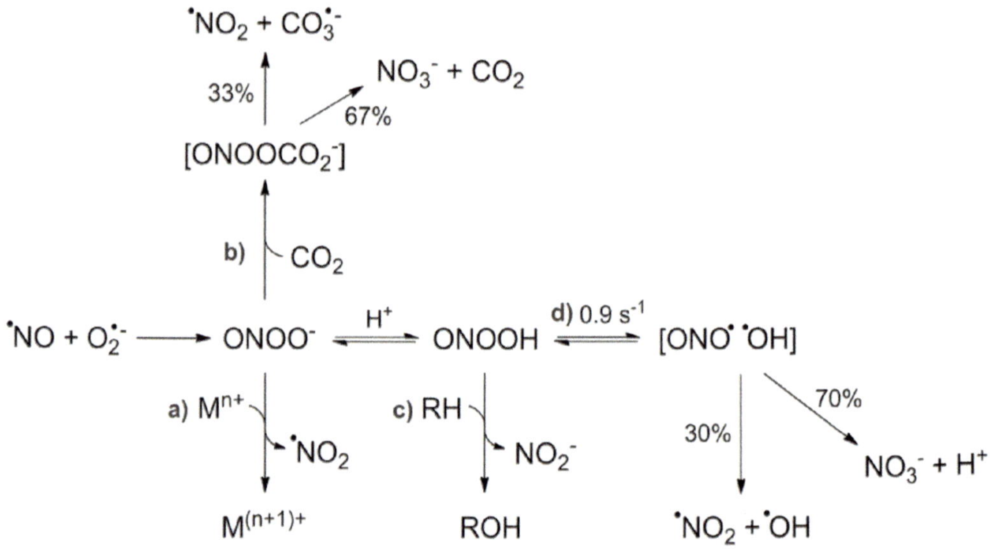

Figure 3.2 Biological chemistry of peroxynitrite. The reaction between •NO and $O_2^{\bullet-}$ generates $ONOO^-$, which is in equilibrium with ONOOH ($pK_a = 6.8$). Reactions of $ONOO^-$ include (a) the one-electron oxidation of transition metal centers (Fe, Mn, Cu); and (b) the reaction with CO_2 to generate the short-lived nitrosoperoxycarbonate ($ONOOCO_2^-$), which rapidly decomposes to secondary radicals in 33% yield and nitrate and CO_2 the remaining 67%. Reactions of ONOOH include (c) the two-electron oxidation of molecules such as thiols; and (d) homolysis at 0.9 s^{-1} to yield secondary radicals at 30% yield and the isomer nitrate at 70% yield.

Source: Modified from Radi et al. (2001).

and Hurst 1995; Denicola et al. 1996). It generates a short-lived intermediate $ONOOCO_2^-$ that homolyzes to $^{\bullet}NO_2$ and carbonate radical ($CO_3^{\bullet-}$; Figure 3.2; Bonini et al. 1999; Augusto et al. 2019). Radicals recombine in the solvent cage to yield nitrate and CO_2, but approximately 33% diffuses to the bulk solvent (Goldstein et al. 2001; Augusto et al. 2019).

The reaction with thiols is likely the first or second quantitatively most important reaction of peroxynitrite (Ferrer-Sueta et al. 2009). It is an S_N2 attack of the thiolate on ONOOH to yield the sulfenic acid and nitrite (Figure 3.2; Zeida et al. 2013). Some thiol-containing proteins react very rapidly with different hydroperoxides, including peroxynitrite. Proteins such as human peroxiredoxin 5 contain a peroxidatic thiol that usually has a low pK_a and a given protein environment that results in rate constants up to 10^8 $M^{-1}s^{-1}$ (Trujillo et al. 2007). The most reactive proteins are thiol-dependent peroxidases of the families of peroxiredoxin (Prx), glutathione peroxidase (GPx), and organic hydroperoxide resistance protein (Ohr; Ferrer-Sueta et al. 2018). The reaction is also very fast with selenocysteine-containing glutathione peroxidase 1 from ($k = [2 \pm 0.4] \times 10^6$ $M^{-1}s^{-1}$ at pH 7.4 and 25°C per monomer; Briviba et al. 1998).

In addition to cysteines, peroxynitrite reacts directly with only a few other protein residues, at much smaller rates. Direct reaction with methionine has a rate constant of 364 $M^{-1}s^{-1}$ at pH 7.4 and 37°C, in a complex mechanism that involves mainly ONOOH to generate methionine sulfoxide (Perrin and Koppenol 2000). Direct reaction with tryptophan has a rate constant of 37 $M^{-1}s^{-1}$ at pH 7.4 and 37°C (Alvarez et al. 1996), whereas the indirect reaction by radicals occurs with much higher rate constants. The main products are 6-nitrotryptophan and N-formylkynurenine (Alvarez et al. 1996).

The reaction with metalloproteins also occurs at fast rates. Peroxynitrite reacts with prostaglandin endoperoxide H synthase-1 with $k = 1.7 \times 10^7$ $M^{-1}s^{-1}$ at pH 7.0 and 8°C (Trostchansky et al. 2007), and it reacts with myeloperoxidase with $k = 6.8 \times 10^6$ $M^{-1}s^{-1}$ at pH 7.0 and 25°C (Furtmüller et al. 2005). Reaction with oxyhemoglobin is not as rapid ($k = 1.0 \times 10^4$ $M^{-1}s^{-1}$ at pH 7.4 and 25°C) but will be particularly relevant in red blood cells because of the high intracellular hemoglobin concentration (Denicola et al. 1998). The mechanisms of reaction with metalloproteins are complex and yield different products (Ferrer-Sueta et al. 2018).

Direct reactions of peroxynitrite with water-soluble antioxidants such as ascorbate or uric acid are too slow to be biologically relevant, but peroxynitrite-derived radicals react much more rapidly (Ferrer-Sueta et al. 2018), with notable effects of uric acid in the prevention of peroxynitrite-dependent biological effects *in vitro* and *in vivo*, for example.

The intrinsic instability of peroxynitrite and its high reactivity with several abundant biological targets, lead to a very short lifetime of peroxynitrite. Despite the transient existence of peroxynitrite, evidence of its formation persists in the form of tyrosine nitration (Ferrer-Sueta et al. 2018). Tyrosine nitration occurs in a two-step sequence involving tyrosine and peroxynitrite-derived radicals. The first step is the formation of tyrosyl radical from the oxidation of tyrosine by carbonate radical, followed by addition of $^{\bullet}NO_2$ to the 3-position of the tyrosyl radical (Ferrer-Sueta et al. 2018). Although tyrosine can be nitrated by other mechanisms, it is believed that peroxynitrite is one of the most important nitrating agents in biology (Ferrer-Sueta et al. 2018). Protein tyrosine nitration can result in loss- or gain-of-protein function, and is increased in different pathologies such as cardiovascular and neurodegenerative diseases, inflammation, and aging (Peluffo and Radi 2007; Souza et al. 2008).

The H^+- and CO_2-catalyzed decays of peroxynitrite are not inert events. The products of peroxynitrite decomposition are radicals even more reactive toward certain targets than its parent species, and should be considered separately.

3.1.2 Biological Chemistry of Nitrogen Dioxide

Nitrogen dioxide can be generated as a product of metal, H^+- and CO_2-catalyzed peroxynitrite decay, and also by the autoxidation of $^{\bullet}NO$ and by myeloperoxidase-catalyzed oxidation of nitrite (Ferrer-Sueta et al. 2018; Augusto et al. 2002; Möller et al. 2007; Byun et al. 1999). Furthermore, $^{\bullet}NO_2$ is an important component of ambient air pollution (Chen et al. 2007). $^{\bullet}NO_2$ reacts with itself rapidly ($k = 4.5 \times 10^8\ M^{-1}s^{-1}$) to generate N_2O_4 that is hydrolyzed in water to yield nitrite and nitrate (Huie 1994). At the low concentrations expected *in vivo* (below micromolar), $^{\bullet}NO_2$ will be present mostly in the monomeric form. $^{\bullet}NO_2$ can react with other radicals in termination reactions and with electron-rich non-radicals by electron transfer, and also hydrogen abstraction and addition. Reaction with tyrosinate anion at pH 9.5 by electron transfer has a second-order rate constant of $2.0 \times 10^7\ M^{-1}s^{-1}$, whereas the reaction with tyrosine at pH 7.4 by hydrogen abstraction has $k = 3.2 \times 10^5\ M^{-1}s^{-1}$ (Prütz et al. 1985). Both reactions generate nitrite and the tyrosyl radical. The reaction of tyrosyl radical with $^{\bullet}NO_2$ is very rapid ($k = 3.0 \times 10^9\ M^{-1}s^{-1}$), and it is the main route contributing to tyrosine nitration (Prütz et al. 1985). Reaction with unsaturated lipids arachidonic and linoleic acid ($k = 1 \times 10^6\ M^{-1}s^{-1}$ and $2 \times 10^5\ M^{-1}s^{-1}$, respectively) occurs by both hydrogen abstraction and by addition (Prütz et al. 1985). Hydrogen abstraction produces lipid radicals that can initiate lipid peroxidation chain reactions, whereas addition generates a β-nitroalkyl radical that can reverse and eliminate $^{\bullet}NO_2$ accompanied by *cis*- to *trans*-lipid isomerization (Jiang et al. 1999). Furthermore, lipids are nitrated and converted into potent signaling molecules (Freeman et al. 2017; Turell et al. 2018). $^{\bullet}NO_2$ can also react rapidly with thiols such as glutathione ($k = 2 \times 10^7\ M^{-1}s^{-1}$ at pH 7.4 and 22°C; Ford et al. 2002) and cysteine ($k = 5 \times 10^7\ M^{-1}s^{-1}$ at pH 7.4 and 22°C; Ford et al. 2002).

Other important targets of $^{\bullet}NO_2$ are ascorbate ($k = 3.5 \times 10^7\ M^{-1}s^{-1}$ at pH 6.7 and 16°C; Alfassi et al. 1990) and urate ($k = 2 \times 10^7\ M^{-1}s^{-1}$ at pH 7.4 and 22°C; Ford et al. 2002). Based on the relative abundance and rate constants, it was proposed that thiols are the dominant targets of $^{\bullet}NO_2$ in cells, whereas urate is the major scavenger in plasma (Ford et al. 2002). When respiratory tract lining fluid was exposed to $^{\bullet}NO_2$, the concentration of glutathione was slightly decreased, whereas urate and ascorbate were depleted (Kelly and Tetley 1997). This indicates not only that these three antioxidants were competing for $^{\bullet}NO_2$, but that intermediate thiol radicals could be repaired by urate and ascorbate (Ford et al. 2002).

3.1.3 Biological Chemistry of Hydroxyl Radical

Hydroxyl radical can be generated as a product of H^+-catalyzed decomposition of peroxynitrite, but also from the reaction between hydrogen peroxide (H_2O_2) and Fe^{2+} (Fenton reaction), and from the photolysis or γ-radiolysis of H_2O_2 (Koppenol and Hider 2019; Koppenol et al. 2017; Hambly and Gross 2005). Hydroxyl radical is the most oxidative species in biology and can react with virtually any molecule at very high rates (Buettner 1993). It reacts by hydrogen abstraction but also by addition. Reaction rate constants of $^{\bullet}OH$ with most organic molecules approach the diffusion limit (Buxton et al. 1988; Manion et al. 2015). For instance, the rate constant for linoleate hydrogen abstraction by $^{\bullet}OH$ is $8 \times 10^9\ M^{-1}s^{-1}$ (Patterson and Hasegawa 1978), whereas the rate constant for $^{\bullet}OH$ hydrogen abstraction and addition to tyrosine is very similar, at $0.6–7.0 \times 10^9\ M^{-1}s^{-1}$ (Solar et al. 1984). A notable exception is the reaction of $^{\bullet}OH$ with HCO_3^- to yield $CO_3^{\bullet-}$, which is in the low end of rate constants of $^{\bullet}OH$ with targets $k = 8.5 \times 10^6\ M^{-1}s^{-1}$ (Buxton and Elliot 1986).

3.1.4 Biological Chemistry of Carbonate Radical

Carbonate radical can be generated as a product of CO_2-catalyzed peroxynitrite decay and to a minor extent by carbonate or bicarbonate oxidation by $^{\bullet}OH$ (Ferrer-Sueta et al. 2018; Augusto et al. 2002; Buxton et al. 1986). It is an oxygen-centered radical, with a pK_a value <0, and therefore deprotonated and negatively charged under all physiological conditions (Czapski et al. 1999). It is a very strong one-electron oxidant, although less reactive than $^{\bullet}OH$. The most important reactions are electron transfer and hydrogen abstraction (Augusto et al. 2002). The reaction with cysteine thiolate by electron transfer has $k = 1.8 \times 10^8\ M^{-1}s^{-1}$, whereas the reaction with protonated cysteine by hydrogen abstraction has $k = 4.6 \times 10^7\ M^{-1}s^{-1}$. The reaction with tyrosine at pH 7.0 occurs with a similar rate constant, $k = 4.5 \times 10^7\ M^{-1}s^{-1}$. It reacts particularly rapidly with ascorbate ($k = 1.1 \times 10^9\ M^{-1}s^{-1}$ at pH 11) and tryptophan ($k = 7.0 \times 10^8\ M^{-1}s^{-1}$ at pH 7.0; Augusto et al. 2002).

3.2 DIFFUSION OF PEROXYNITRITE ACROSS MEMBRANES

Early experiments showed that peroxynitrite or its derived radicals could diffuse into lipid membranes. Addition of peroxynitrite to liposomes made of soybean phosphatidylcholine caused lipid peroxidation, indicating that at least the derived radicals $^{\bullet}NO_2$ and $^{\bullet}OH$ could penetrate the hydrophobic phase of membranes and oxidize lipids (Radi et al. 1991). Furthermore, addition of peroxynitrite to diverse cell cultures, including *Escherichia coli*, *Trypanosoma cruzi*, and mammalian cells resulted in cytotoxic effects (Denicola et al. 1993; Zhu et al. 1992; Estévez et al. 1995; Delaney et al. 1996) that were associated with protein tyrosine nitration and DNA strand breaks, indicating that peroxynitrite or its derived species could diffuse into the cells and cause molecular damage. More recently, it was shown that hydrophobic derivatives of tyrosine inserted in lipid membranes could be nitrated by peroxynitrite, supporting a facile diffusion of peroxynitrite and/or its derived radicals into the hydrophobic phase of lipid bilayers (Bartesaghi et al. 2006; Bartesaghi et al. 2010).

It was then demonstrated that ONOOH could diffuse across lipid membranes, whereas $ONOO^-$ can cross the membranes aided by anion channels (Denicola et al. 1998; Khairutdinov et al. 2000; Marla et al. 1997). In phosphatidylcholine liposomes, ONOOH could diffuse and oxidize intravesicular targets (Khairutdinov et al. 2000; Marla et al. 1997), and in the more complex erythrocyte membrane it was observed that ONOOH could readily diffuse through the membrane, but $ONOO^-$ diffused mostly through the bicarbonate-chloride exchange protein band 3 (Denicola et al. 1998). In erythrocytes, diffusion of peroxynitrite caused intracellular oxidation of oxyhemoglobin, which could be followed spectrophotometrically, and also hemoglobin nitration (Denicola et al. 1998). Addition of CO_2 resulted in a lower oxidation of intracellular oxyhemoglobin, which depended on the number of cells in suspension, consistent with a competition between diffusion into cells and extracellular decomposition by CO_2, and a very low permeability of $CO_3^{\bullet-}$ (Romero et al. 1999).

Determination of membrane permeability to peroxynitrite must contemplate that diffusion is competing with its decomposition to radicals, which are dependent on pH and CO_2, to name the most important. The lifetime of peroxynitrite at pH 7.4 and 37°C is 0.77 s, but in the presence of physiological concentrations of CO_2 (≥1.3 mM), the lifetime decreases to ~10 ms. The mean square distance (Δx) that peroxynitrite can diffuse in this time can be calculated using Einstein's diffusion equation, which in three dimensions is given by Eq. 3.2.

$$\Delta x = (6Dt)^{0.5} \quad (3.2)$$

Considering $D = 2 \times 10^{-5}$ cm^2s^{-1}, the same as nitrate at 37°C (calculated from Yeh and Wills, 1970), the diffusion distance at pH 7.4 in the absence of CO_2 is 96 μm (several cell diameters), whereas in the presence of 1.3 mM CO_2 it is 11 μm (about one cell radius; Figure 3.3). Therefore, the presence of CO_2 will severely limit peroxynitrite diffusion. The product $^{\bullet}NO_2$ can easily diffuse through membranes, but both $^{\bullet}OH$ and $CO_3^{\bullet -}$ will face different barriers that will limit their diffusion. Furthermore, the faster decomposition of peroxynitrite by CO_2 will also create higher transient concentration of radicals that can recombine with $^{\bullet}NO_2$ and limit its diffusion.

To deal with this complexity, we will start with a few definitions and then discuss the permeability of each component separately, including the simple diffusion of ONOOH and $^{\bullet}NO_2$, the hindered diffusion of $^{\bullet}OH$ and $CO_3^{\bullet -}$, and the facilitated diffusion of $ONOO^-$ through protein channels.

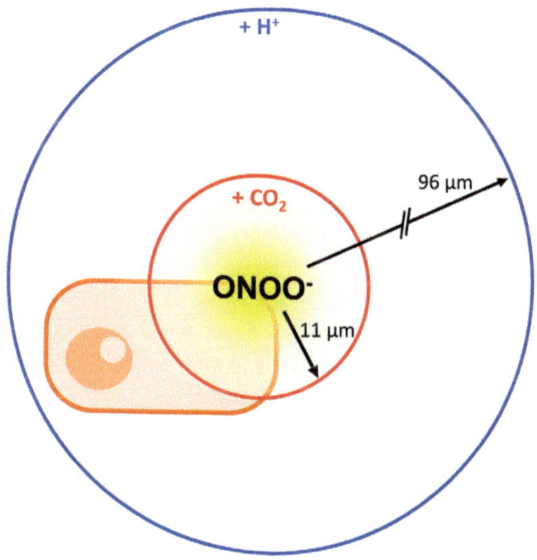

Figure 3.3 Theoretical diffusion distance of peroxynitrite. The lifetime of peroxynitrite depends on the concentration and reactivity of molecules. In the absence of targets, and only considering its decay catalyzed by protons at pH 7.4 and 37°C, peroxynitrite has a mean square distance of diffusion of 96 μm, equivalent to several cellular diameters. In the presence of CO_2 the lifetime decreases considerably, and so does the distance of diffusion to 11 μm, roughly an average cellular radius.

3.2.1 Permeability Coefficient

The permeability coefficient (P_m) is a useful term to compare the permeability of lipid membranes to different species. It is defined as a function of the number of molecules that cross through a given cross-sectional area of the membrane per unit time, the flux J (Eq. 3.3). At any given time the net flux in one direction will be given by the difference in concentration of the molecule in both compartments and P_m.

$$J = P_m \Delta C \quad (3.3)$$

To transform it into a first-order constant to be used in kinetic assays, which considers concentration of molecules, we have to consider the surface area (SA) and volume of the cell (V), as in Eq. 3.4.

$$k_{perm} = P_m SA/V \quad (3.4)$$

Permeability is closely related to diffusion across the membrane. In the solubility-diffusion model of permeability, the dependence on the solubility and diffusion of the permeant in the membrane is given by Eq. 3.5.

$$P_m = K_P D_m / d_m \quad (3.5)$$

where K_P is the partition constant between the membrane hydrophobic phase and bulk water, D_m is the diffusion constant in the membrane, and d_m is the membrane thickness (Stein 1986). It is often the case that D_m and K_P cannot be known separately but combined as an apparent diffusion coefficient in the membrane (D_{app}; Möller et al. 2016).

This model has been very useful because of its simplicity and wide range of applicability. It holds its validity for small non-electrolytes with P_m that range from 10^{-6} to 3 cm s^{-1} and K_P(hexadecane/water) from 10^{-6} to 1 (Walter and Gutknecht 1986). Further refinements take into account the microheterogeneity of lipid bilayers, such as differences in dielectric constant and free volume, that causes molecules to partition and diffuse differently at different depths within lipid bilayers. Such microheterogeneity could be exposed by using sensitive probes located at different depths of the membrane, and later by molecular dynamics simulations (Subczynski et al. 1991; Marrink and Berendsen 1996).

A note of caution is that membrane composition affects permeability, and it is usually observed that more fluid membranes are more permeable than less fluid membranes. Membrane fluidity can be increased by increasing temperature, decreasing cholesterol content, decreasing sphingomyelin content, and increasing unsaturated acyl chain content. The presence of integral protein in cellular membranes can also affect membrane fluidity but most importantly can offer alternative pathways for transport across the membrane. Such facilitated diffusion is observed for H_2O_2 by aquaporins 3 and 8, for instance, and for $ONOO^-$ and $O_2^{\bullet-}$ by erythrocyte's band 3 (Bienert and Chaumont 2014; Denicola et al. 1998; Lynch and Fridovich 1978).

3.2.2 Diffusion of Peroxynitrous Acid

Peroxynitrite has been shown to diffuse across cell membranes by both simple and facilitated diffusion. Peroxynitrous acid is a polar molecule that vaguely resembles H_2O_2, thus the solubility in the lipid phase would be very low and its permeability would be expected to be similar to that of H_2O_2 or water.

There are only a few experimental determinations of the permeability coefficient of lipid membranes to peroxynitrite. One study was done using free and liposome-encapsulated manganese porphyrins (5,10,15,20-tetrakis(N-methyl-49-pyridyl)porphinatomanganese(III) chloride, Mn(III)TMPyP), and the oxidation to oxoMn(IV) was followed by rapid mixing stopped-flow spectrophotometry within 500 ms at pH 7.4 (Marla et al. 1997). Using pseudo–first order conditions with an excess of peroxynitrite, and taking into account the spontaneous decay of peroxynitrite, the second-order rate constants for the oxidation reaction were estimated. A lipid-soluble manganese porphyrin was synthesized and found to be oxidized by peroxynitrite, albeit at lower rates, demonstrating that peroxynitrite could diffuse into the hydrophobic region of the bilayer. Liposome-entrapped porphyrin could also be oxidized, at a twofold lower rate than free porphyrin in solution. The difference in rate is given by the lower peroxynitrite concentration inside the liposome caused by a permeability barrier. Taking into account the size of the liposomes and Fick's law, a P_m of 8.0×10^{-4} cm s^{-1} was calculated for membranes composed of egg phosphatidylcholine:stearyl amine:cholesterol (7:2:1) at 22–24°C, only three times lower than the estimated P_m for water (Marla et al. 1997). Based on the low permeability of membranes to ions, it was concluded that the permeating species should be ONOOH.

A subsequent study used liposome-encapsulated ferrocyanide ($Fe(CN)_6^{4-}$) to study lipid membrane permeability to peroxynitrite (Khairutdinov et al. 2000). Membranes of different composition were assayed, including egg phosphatidylcholine, DMPC, DPPC, with or without β-arachidonoyl-γ-palmitoylphosphatidylcholine (PAPC). The reaction of peroxynitrite with free and liposome-encapsulated $Fe(CN)_6^{4-}$ was followed by rapid-mixing stopped-flow spectrophotometry and the results used to optimize a mathematical model including the diffusion steps across the membrane by ONOOH and $^{\bullet}NO_2$, as well as the inhibition by CO_2 (Khairutdinov et al. 2000). The inhibition of intraliposomal oxidation of $Fe(CN)_6^{4-}$ by CO_2 occurred because peroxynitrite decomposition to nitrate occurred at similar rates than ONOOH diffusion across the membranes and unlike $^{\bullet}NO_2$, $CO_3^{\bullet-}$ cannot diffuse across the membranes. At 21°C and pH 7.4 the authors found P_m of 1.3×10^{-3}, 0.6×10^{-3}, and 0.4×10^{-3} cm s^{-1} for membranes of egg phosphatidylcholine, DMPC, and DPPC, respectively (Khairutdinov et al. 2000). These results are consistent with an effect of membrane microviscosity on ONOOH diffusion, because in these experimental conditions DPPC is in a highly viscous gel state, whereas egg phosphatidylcholine is in a lower viscosity fluid state.

Lipid membranes are much less permeable to anions than to neutral species. For instance, P_m for chloride has been experimentally estimated to be 10^{-10} cm s^{-1}, whereas P_m for hydrochloric acid determined by different methods is 2 cm s^{-1} (Gutknecht and Walter 1981b; Khramtsov et al. 1989). The permeability to nitrate is in the same range as chloride, 10^{-9} to 10^{-10} cm s^{-1}, whereas P_m for nitric acid has been determined to be 10^{-4} to 10^{-3} cm s^{-1} (Khramtsov et al. 1989; Gutknecht and Walter 1981a). The lower P_m for nitric acid than for hydrochloric acid can be explained by the lower solubility of nitric acid in the hydrophobic region of the membrane (Walter et al.

1986). The values for nitric acid are remarkably similar to those determined for ONOOH (see above). Considering that permeability across membranes depends more on the cross-sectional area of the permeant molecule than on the volume (Xiang and Anderson 1998), it is expected that ONOOH has a slightly higher P_m than nitric acid. Note that the pK_a of nitric acid is −1.62, thus practically all nitrate will be ionized at physiological pH.

Molecular dynamics calculations have provided atomistic details on the interactions between peroxynitrite and the lipid bilayer (Cordeiro 2018). Peroxynitrous acid was found to behave as an amphiphilic molecule with the polar end given by the OH and the apolar end given by ONO, thus ONOOH locates anisotropically in the membrane with the ONO end inserted into the acyl region and the OH oriented toward the carbonyl, forming a hydrogen bond. The free energy barrier was different at different depths of the membrane. A negative free energy relative to the bulk solvent was found near the polar headgroups (−7 kJ/mol), indicating a favorable interaction, whereas a positive free energy barrier of 17 kJ/mol was found at the mid-bilayer region (relative to the minimum at the headgroup region), indicating an unfavorable solubilization in the more hydrophobic acyl region (Cordeiro 2018).

3.2.3 Diffusion of Nitrogen Dioxide

Nitrogen dioxide is more soluble in organic solvents than in water, by 5.5 times in the case of decane at 37°C (Squadrito and Postlethwait 2009). Using an approach combining quantum calculations and experimental data, it was estimated that $^{\bullet}NO_2$ would partition 1.5 times more favorably in fluid lipid membranes than in water (Signorelli et al. 2011). Based on the permeability of similar molecules and the respective partition coefficients into the acyl region, a P_m of 5 cm s^{-1} was estimated for $^{\bullet}NO_2$ (Signorelli et al. 2011). This value is in the high-end range estimated by Khairutdinov et al. (Khairutdinov et al. 2000). Molecular dynamics showed that $^{\bullet}NO_2$ faced a low energy barrier at the polar headgroup region, but that free energy at the mid-bilayer region was slightly lower than at the bulk solvent phase, consistent with a slightly favorable partition into the hydrophobic acyl region (Cordeiro 2018; Razzokov et al. 2018). Further simulations showed that $^{\bullet}NO_2$ indeed accumulated in the acyl region of the bilayer but could diffuse unimpeded into the bulk aqueous phase (Cordeiro 2018).

3.2.4 Diffusion of Carbonate Radical

When CO_2 is included in peroxynitrite permeability experiments, an apparent lower permeability is observed, consistent with both a lower lifetime of peroxynitrite and a low membrane permeability to $CO_3^{\bullet-}$ (Romero et al. 1999; Khairutdinov et al. 2000). As for other anions, the electrical charge in $CO_3^{\bullet-}$ results in a high energetic barrier opposing partition into the hydrophobic phase of the membrane and a very low permeability. Although it could potentially use anion channels, in particular the bicarbonate-chloride exchanger present in erythrocytes (band 3), evidence suggests that the high reactivity of $CO_3^{\bullet-}$ causes it to react before crossing the membrane through channels (Romero et al. 1999). When a derivative of tyrosine was incorporated in the membrane of erythrocytes, it was found that CO_2 modulated the nitration by peroxynitrite, promoting the nitration of superficial tyrosines and restricting the nitration of tyrosines located deeper in the bilayer, indicating that the diffusion of $CO_3^{\bullet-}$ across the membrane is highly limited (Romero et al. 2007).

3.2.5 Diffusion of Hydroxyl Radical

The case of $^{\bullet}OH$ is similar to that of $CO_3^{\bullet-}$. Given its similarities with water and the absence of a net electrical charge, a medium-range permeability would be expected. However, $^{\bullet}OH$ is extremely reactive and will react before crossing a membrane (see section 3.1.3). Molecular dynamics showed that $^{\bullet}OH$ interacted slightly favorably with the polar headgroup region but unfavorably with the more hydrophobic mid-bilayer region (20 kJ/mol barrier; Cordeiro 2018). Further simulations showed that if ONOOH homolysis occurred near the headgroup region, $^{\bullet}OH$ would diffuse to the aqueous phase but a significant fraction of $^{\bullet}OH$ would be retained forming hydrogen bonds with carbonyls, suggesting it could be a relevant mechanism to favor lipid acyl oxidation by a polar oxidant (Cordeiro 2018).

3.3 FACILITATED DIFFUSION MEDIATED BY TRANSMEMBRANE PROTEINS

Cell membranes contain a large number of transmembrane proteins that could facilitate the diffusion of peroxynitrite or its derived radicals. In particular, band 3 anion transport protein has been found to facilitate the diffusion of peroxynitrite anion. Besides, we also present different arguments discussing the possibility that aquaporins could also transport peroxynitrite.

3.3.1 Facilitated Diffusion Mediated by Band 3 Anion Transport Protein

As mentioned before, pioneer works established the possibility of peroxynitrite crossing mammalian cell membranes in part through anion channels. It was shown that band 3 anion transport protein (also known as anion exchange protein 1 [AE1], solute carrier family 4 member 1 [SLC4A1]) is the main route for peroxynitrite to cross the erythrocyte membrane in its anionic form because its transport was blocked by DIDS, an inhibitor of band 3 (Denicola et al. 1998; Macfadyen et al. 1999). The main function of band 3 is the exchange of bicarbonate for chloride, but other anions can also be transported, including thiocyanate, phosphoenolpyruvate, and pyridoxal-phosphate (Hamasaki et al. 1983; Jennings 1989). Superoxide has also been proposed to cross erythrocyte's membrane through band 3, as suggested by the inhibition of $O_2^{\bullet-}$ diffusion by DIDS and other band 3 inhibitors (Lynch et al. 1978).

Human erythrocyte band 3 is a 110-kDa glycoprotein with two domains, a cytosolic N-terminal domain and an integral C-terminal membrane domain (Arakawa et al. 2015). There are ~10^6 copies per cell (Bryk and Wiśniewski 2017), and each single protein can transport 5×10^4 ions per second (Brahm 1977), thus the exchange is extremely fast. Anion exchange is catalyzed by the C-terminal domain. The crystal structure of the band 3 anion exchanger domain (CTD) was solved as a dimer at 3.5 Å (Figure 3.4; Arakawa et al. 2015). Each protomer presents 14 transmembrane segments and the structure obtained was locked in an outward-facing open conformation by H_2DIDS covalently crosslinking lysine residues. There are two essential residues for anion exchange, Glu681 and Arg730. It was proposed that bicarbonate interacts electrostatically with Arg730 and hydrogen bonds with Glu681. The band 3 CTD undergoes a large conformational

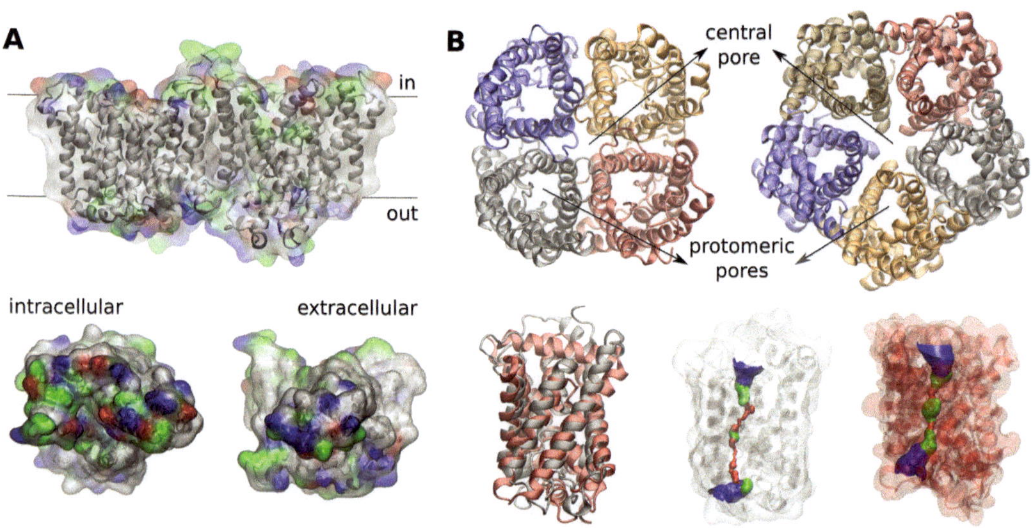

Figure 3.4 Structural characteristics of anion exchanger band 3, AQPs and FNT channels. (A) (Upper) Lateral view of human erythrocyte band 3 anion exchanger dimer (PDBid 4YZF). (Lower) Intracellular and extracellular views of band 3 monomer. Surfaces colored by residue type: white, apolar; green, polar; blue, basic; and red, acidic. (B) Structural comparison of human AQP5 (PDBid 3D9S) and nitrite channel NirC from *Salmonella typhimurium* (PDBid 4FC4). (Upper) Top view of AQP tetrameric and NirC pentameric assemblies, highlighting the presence of one central pore and protomeric individual pores. (Lower) Structural alignment of AQP (gray) and NirC (red) monomers, along with a representation of each pore topology calculated using hole (Smart et al. 1996).

change during transport, alternating the access to the binding site from the cytosol or the external medium (Arakawa et al. 2015).

Band 3 and splice variants are located in erythrocytes, cells of renal collecting duct, heart, and colon. Related AE2 and AE3 show over 50% sequence identity and are more widely distributed (Romero et al. 2013). There are other five members of the SLC4 transporter family that share approximately 30% sequence identity with band 3, which are Na^+-coupled bicarbonate transporters and are widely distributed in mammalian cells (Romero et al. 2013). The similarity in size and charge between bicarbonate and peroxynitrite suggests that all these transporters could potentially allow the transport of peroxynitrite anion (Figure 3.5).

3.3.2 Facilitated Diffusion Mediated by Aquaporins

The possibility of other proteinaceous routes in mammalian cells for peroxynitrite transport through membranes is still open. Aquaporins (AQPs) are members of the major intrinsic proteins (MIPs) family of membrane channels. In 1993, Agre, Sasaki, and Chrispeels, three of the researchers who described the first MIPs, suggested that "the functional name aquaporin will be useful for communication between scientists" as it would allow rapid recognition of proteins related to MIP whose primary function is to be "pores for water" (Agre et al. 1993). However, increasing diversity of molecules that can be transported by these channels is continuously reported (Carbrey and Agre 2009). Among the molecules transported by MIPs are relatively small solutes (e.g., glycerol, urea, lactic acid), gases (e.g., CO_2, $^{\bullet}NO$, ammonia), metalloids (e.g., arsenic, boric acid, antimony), reactive oxygen species (e.g., H_2O_2), and ions (e.g., Na^+, Cl^-, nitrate).

There are 13 AQPs (AQP0–12) in humans usually classified into groups: (1) orthodox AQPs (AQP0, 1, 2, 4, 5, 6), associated mainly with water transport; (2) the aqua-ammoniaporin (AQP8),

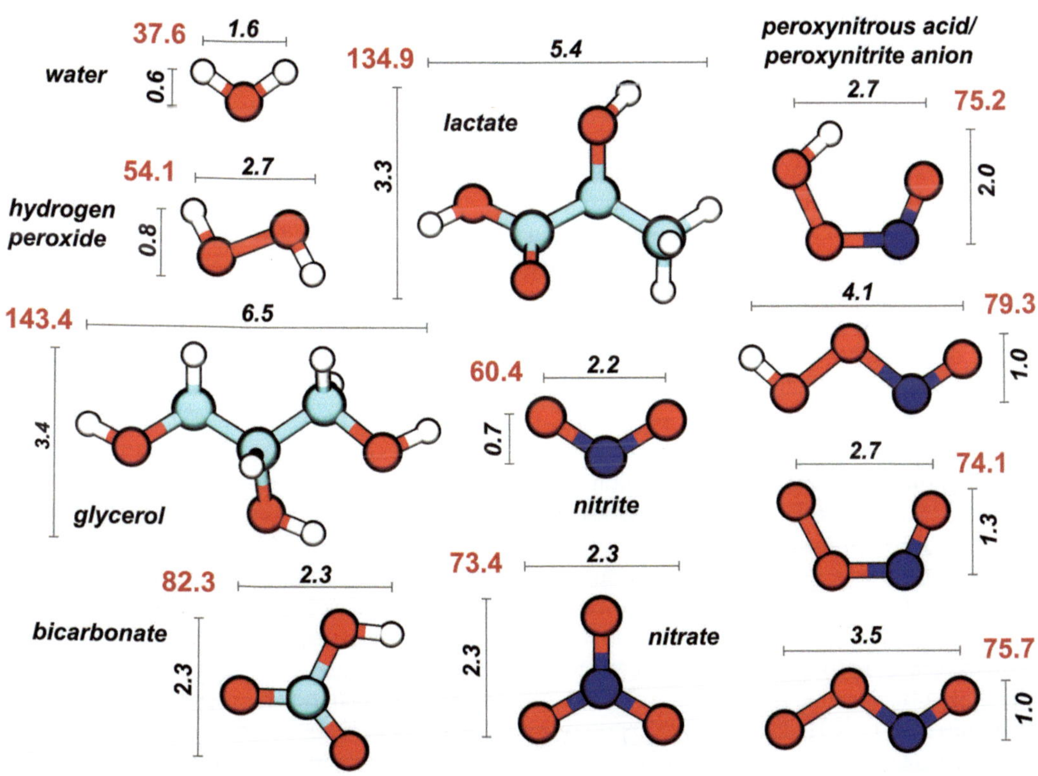

Figure 3.5 Dimensional comparison of different species known to be transported by AQPs, anion exchanger band 3 or FNT channels and peroxynitrite. Different protonation and conformational states of peroxynitrite are considered. Distances (black, Å) and volumes (red, Å^3) were calculated from electronic structure geometry optimizations at the B3LYP/6.31G++ level of theory.

also included among the orthodox AQPs in part of the literature; (3) the aquaglyceroporins (AQP3, 7, 9, 10), and (4) the unorthodox AQPs (or super-aquaporins) (AQP11–12; Finn and Cerda 2015; Abascal et al. 2014; Soto et al. 2012). Human AQPs participate in a wide variety of physiological processes but also in infectious and non-infectious diseases (e.g., renal dysfunction, cancer, neurological disorder; Azad et al. 2021). And of course, the involvement of each isoform in some of these processes depends on its permeability profile.

All AQPs preserve a typical three-dimensional structure and are organized as tetramers (Wspalz et al. 2009). Two kinds of pores are present in AQPs, the protomeric pore and the central tetrameric pore (Figure 3.4B). The main transporting path is each of the protomeric pores; however, in recent years some experimental evidence has suggested that the central pore could also be relevant. Most human AQPs have been shown to transport water at different rates by the protomeric pore. Also, other solutes have been shown to be transported through these pores. For example, AQP3-like isoforms are permeable to glycerol; AQP8 is a good ammonia and urea channel; and several isoforms such as AQP0, 3, 5, 8, 9 or 11 transport H_2O_2 (Miller et al. 2010; Watanabe et al. 2016; Almasalmeh et al. 2014; Bienert et al. 2014; Varadaraj and Kumari 2020; Rodrigues et al. 2019; Bestetti et al. 2020). Ion transport has also been reported in human AQP channels such as AQP1 and AQP6 (Yool et al. 1996; Yasui et al. 1999), but apparently the mechanisms of these two channels to transport ions are quite different. While AQP1 is supposed to transport cations through the central pore (Yu et al. 2006), AQP6 is an efficient anion channel by the protomeric pore. AQP6 is unique among other hAQPs since it is the only one hAQP permeable to a wide range of anions, presenting a high permeability to nitrate (Ikeda et al. 2002). AQP9 is an interesting case, since it is the first AQP reported as permeable to monocarboxylates, in particular lactic acid/lactate (Tsukaguchi et al. 1999). Finally, the transport of several reactive oxygen and nitrogen species has been tested by molecular dynamic simulations in hAQP1, in particular H_2O_2, $^{\bullet}OH$, $^{\bullet}NO_2$, and $^{\bullet}NO$, showing that all these species could be transported through AQP1 pore (Yusupov et al. 2019), although experimental evidence is still lacking.

AQP protomeric pores present a characteristic hourglass shape (two vestibules, located on the periplasmic and cytoplasmic ends, and a narrow pore in between) and two main constrictions (Figure 3.4B). Permeability discrimination in AQP constrictions not only depends on their size, but on a complex interplay between size, polarity and chemical properties of the transported molecule, and the filter itself (Kitchen et al. 2019; Oliva et al. 2010). At the cytoplasmic entrance of AQP channels there is the selectivity filter (SF), which executes a primary transport role controlling both size and charge of solutes transported. In orthodox AQPs SF diameters are not greater than 2.8 Å, while in AQPs that can transport, for example, glycerol or urea SF diameters are approximately 3.4 Å (Fu et al. 2000; Tajkhorshid et al. 2002; Beitz et al. 2006). This SF is also known as ar/R since most AQPs present among the four residues that constitute it an aromatic amino acid and an arginine. The ar/R region, together with the two signature NPA (Asn-Pro-Ala) motifs, which conforms to the second constriction, form an important electrostatic barrier at the central region of the pore that operates as a proton-excluding filter (Wu et al. 2009; Wree et al. 2011; Eriksson et al. 2013). It has also been reported that other structural elements that do not directly interact with the solutes, such as connecting loops, could play a role in determining AQP selectivity (Savage et al. 2010).

Given the spectra of solutes transported by human AQPs, could these channels be considered proteinaceous pores also for peroxynitrite? Peroxynitrite transport has not been tested in AQP, but in prokaryotes it is speculated that peroxynitrite can be imported by formate nitrite transport (FNT) channel NirC (Lü et al. 2013, 2012). FNT channels constitute a case of convergent evolution with AQP due to the amazing structural similarity of their pores (Figure 3.4B). The overall structure of these channels closely resembles that of AQPs, despite a lack of amino acid sequence identity (Wang et al. 2009; Waight et al. 2010; Lü et al. 2013). This high fold identity between FNT channels and aquaporins relies on the protomeric channel structure, since aquaporins are organized as tetramers and FNT members as pentamers (Wang et al. 2009). Both FNT and AQP protomeric pores are similar with an hourglass shape (two vestibules, located on the periplasmic and cytoplasmic ends, and a narrow pore in between) and two main constriction sites. Pore diameters of FNT are generally within the range calculated for

AQPs (Wang et al. 2009). The characteristic hourglass shape of the protomeric conducting pore is strikingly conserved among these two unrelated families of channels despite their divergent functional properties (Figure 3.4B). Each different FNT channel has selectivity for different permeating anions, such as nitrite, formate, acetate, pyruvate, and lactate (Lü et al. 2013). The selectivity profile is considered to reside in small structural variations at the selectivity filter level. While the selectivity filter of FNT channels is quite different from AQPs, in protozoa, FNTs evolved to transport larger substrates such as lactate/H^+ by a widened lipophilic/lysine (F/K) selectivity filter analogous to the selectivity filter (ar/R region) in the AQPs (Wiechert et al. 2017). The FNT channels have been found only in bacteria, archaea, fungi, algae, and unicellular parasites and are not present in higher eukaryotes (Rambow et al. 2014).

In summary, as some human AQPs can transport molecules such as H_2O_2, glycerol, lactate, or nitrate and are structurally similar to FNT channels that also transport molecules with similar size and/or charge, we propose that AQPs are likely candidates to facilitate the diffusion of peroxynitrite across the membranes of higher eukaryotes. Figure 3.5 shows molecules transported by AQPs, FNT channels, and band 3, emphasizing their geometric similarity. Solute permeability/exclusion of each mammalian AQP relies on a complex network of interactions (considering physical size and chemical properties) among both the pore constrictions and the molecules to be transported, but at least peroxynitrite seems to fulfill some of the requirements other solutes transported by AQP satisfied. AQP6, the nitrate channel, is a good candidate to explore $ONOO^-$ permeability, but AQP9, which transports lactate, and the peroxiporins such as AQP3 or AQP8, also can be considered as possible ONOOH channels. Future experiments will elucidate if indeed some of the 13 human AQPs are permeable to peroxynitrite.

3.4 CONCLUDING REMARKS

Cell membranes exhibit different permeabilities for different reactive species and can only compartmentalize the reactivity of some of them (Figure 3.6). Both $^{\bullet}NO$ and O_2 can diffuse across cell membranes virtually unhindered, with

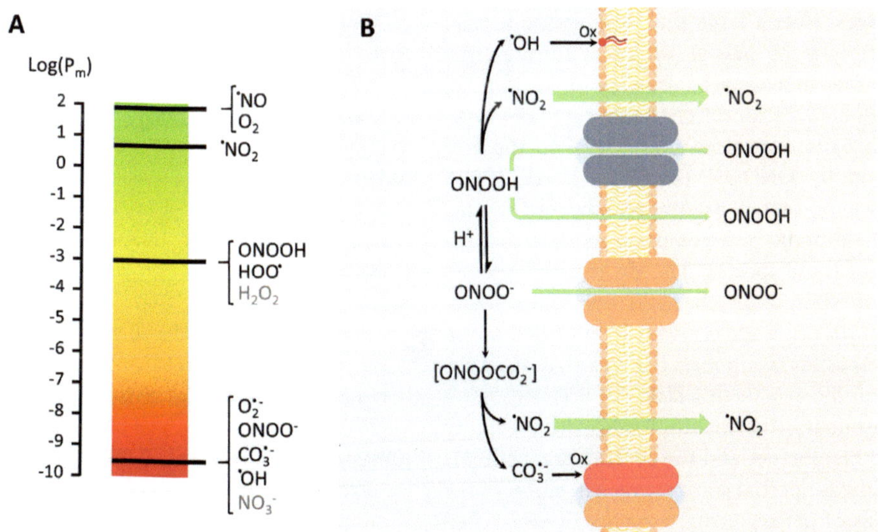

Figure 3.6 Diffusion of peroxynitrite and its derived radicals across cellular membranes. (A) Permeability coefficient range ($cm\ s^{-1}$) for peroxynitrite and related species across lipid membranes, showing that the anions have a very low P_m that prevents simple diffusion, but ONOOH has a higher P_m, and $^{\bullet}NO_2$ and $^{\bullet}NO$ a much higher P_m. (B) ONOOH may diffuse across cellular membranes at medium rates either by simple diffusion or potentially by facilitated diffusion through an AQP, whereas $ONOO^-$ can use anion channels such as band 3 (and possibly some AQP). $^{\bullet}NO_2$ can diffuse across the membrane at higher rates than ONOOH, but $^{\bullet}OH$ will be prevented from diffusing across the membrane, mostly because of its high reactivity but also because of its polarity. $CO_3^{\bullet-}$ is not able to diffuse across lipid membranes, and its high reactivity with proteins seems to prevent it from using anion channels.

diffusion rates that resemble an equally thick layer of water (Möller et al. 2018). The diffusion of $^{\bullet}NO_2$ across lipid membranes is also very rapid, at only slightly lower rates than $^{\bullet}NO$ (Möller et al. 2019; Signorelli et al. 2011; Cordeiro 2018). On the other hand, the radical products of H^+- or CO_2-catalyzed peroxynitrite decay ($^{\bullet}OH$ and $CO_3^{\bullet-}$) have very low permeability. The extremely high reactivity of $^{\bullet}OH$ makes it unlikely that it could traverse an electron-rich membrane interior unreacted. On the other hand, the electrical charge of $CO_3^{\bullet-}$ and the impossibility to form the conjugated acid prevent the diffusion of $CO_3^{\bullet-}$ through cell membranes. It could be theoretically possible for $CO_3^{\bullet-}$ to use the same band 3 channel as the other anions, but its high reactivity may prevent it from surfacing to the other side.

The diffusion properties of peroxynitrite in biological contexts are complex and depend on several factors. The permeability of membranes to ONOOH is similar to that of other polar nonelectrolytes, such as H_2O_2 and water (Figure 3.6A; Möller et al. 2019), albeit at a 50,000 times smaller rate than that estimated for $^{\bullet}NO_2$ (Khairutdinov et al. 2000; Marla et al. 1997). However, the charged $ONOO^-$, the most abundant form of peroxynitrite at physiological pH ranges, faces a large thermodynamic barrier that prevents its passive diffusion through cell membranes. But charged $ONOO^-$ diffusion can be facilitated by anion channels such as erythrocyte's band 3 at rates comparable to the free diffusion by ONOOH (Denicola et al. 1998). Also, we presented arguments suggesting that other membrane proteins may facilitate the diffusion of these species. The diversity of species transported by AQP channels and their structural similarity with previously described anion channels as FNTs supports the idea that AQPs are likely candidates for this task.

In summary, the medium range permeability of lipid membranes and the specific organelle pH will allow for partial compartmentalization of the reactions of peroxynitrite, which will be favored by the presence of CO_2 and the concomitant formation of $CO_3^{\bullet-}$. Different protein transporters have been associated with the facilitated diffusion of peroxynitrite and/or related species, but other untested transporters such as AQPs may also participate in such processes. Thus, experimental evidence regarding these aspects must be gathered to further complement our current understanding of the diffusion, transport, and consequent biological effects of peroxynitrite.

3.5 ACKNOWLEDGMENTS

Financial support was provided by Universidad de la República, Uruguay (Espacio Interdisciplinario, EI 2020, Comisión Sectorial de Investigación Científica, CSIC Grupos 2018, to RR, and CSIC I+D_2020_557 to MNM), Agencia Nacional de Investigación e Innovación, ANII, Uruguay (FCE 1_2017_1_136043 to MNM), Agencia Nacional de Promoción Científica y Tecnológica, Argentina, PICT-2019-0387 to KA. Additional support was obtained from Programa de Desarrollo de Ciencias Básicas (PEDECIBA, Uruguay) and Centro de Biología Estructural del Mercosur (CeBEM). MNM, AZ, and RR belong to the Sistema Nacional de Investigadores (SNI, Uruguay). The simulations presented in this paper were carried out using ClusterUY (https://cluster.uy).

REFERENCES

Abascal, F., I. Irisarri, and R. Zardoya. 2014. "Diversity and evolution of membrane intrinsic proteins." *Biochimica et Biophysica Acta (BBA)-General Subjects* no. 1840 (5):1468–1481.

Agre, P., S. Sasaki, and M. Chrispeels. 1993. "Aquaporins: a family of water channel proteins." *American Journal of Physiology-Renal Physiology* no. 265 (3):F461.

Alderton, W. K., C. E. Cooper, and R. G. Knowles. 2001. "Nitric oxide synthases: structure, function and inhibition." *Biochemical journal* no. 357 (3):593–615.

Alfassi, Z., R. Huie, P. Neta, and L. Shoute. 1990. "Temperature dependence of the rate constants for reaction of inorganic radicals with organic reductants." *Journal of Physical Chemistry* no. 94 (25):8800–8805.

Almasalmeh, A., D. Krenc, B. Wu, and E. Beitz. 2014. "Structural determinants of the hydrogen peroxide permeability of aquaporins." *The FEBS Journal* no. 281 (3):647–656.

Alvarez, B., H. Rubbo, M. Kirk, S. Barnes, B. A. Freeman, and R. Radi. 1996. "Peroxynitrite-dependent tryptophan nitration." *Chemical Research in Toxicology* no. 9 (2):390–396.

Arakawa, T., T. Kobayashi-Yurugi, Y. Alguel, H. Iwanari, H. Hatae, M. Iwata, Y. Abe, T. Hino, C. Ikeda-Suno, and H. Kuma. 2015. "Crystal structure of the anion exchanger domain of human erythrocyte band 3." *Science* no. 350 (6261):680–684.

Augusto, O., M. G. Bonini, A. M. Amanso, E. Linares, C. C. Santos, and S.L.L. De Menezes. 2002. "Nitrogen dioxide and carbonate radical anion: two emerging radicals in biology." *Free Radical Biology and Medicine* no. 32 (9):841–859.

Augusto, O., S. Goldstein, J. K. Hurst, J. Lind, S. V. Lymar, G. Merényi, and R. Radi. 2019. "Carbon dioxide-catalyzed peroxynitrite reactivity–the resilience of the radical mechanism after two decades of research." *Free Radical Biology and Medicine* no. 135:210–215.

Azad, A. K., T. Raihan, J. Ahmed, A. Hakim, T. H. Emon, and P. A. Chowdhury. 2021. "Human aquaporins: functional diversity and potential roles in infectious and non-infectious diseases." *Frontiers in Genetics* no. 12:344.

Bartesaghi, S., V. Valez, M. Trujillo, G. Peluffo, N. Romero, H. Zhang, B. Kalyanaraman, and R. Radi. 2006. "Mechanistic studies of peroxynitrite-mediated tyrosine nitration in membranes using the hydrophobic probe Nt-BOC-L-tyrosine tert-butyl ester." *Biochemistry* no. 45 (22):6813–6825.

Bartesaghi, S., J. Wenzel, M. Trujillo, M. López, J. Joseph, B. Kalyanaraman, and R. Radi. 2010. "Lipid peroxyl radicals mediate tyrosine dimerization and nitration in membranes." *Chemical Research in Toxicology* no. 23 (4):821–835.

Beitz, E., B. Wu, L. M. Holm, J. E. Schultz, and T. Zeuthen. 2006. "Point mutations in the aromatic/arginine region in aquaporin 1 allow passage of urea, glycerol, ammonia, and protons." *Proceedings of the National Academy of Sciences* no. 103 (2):269–274.

Bestetti, S., M. Galli, I. Sorrentino, P. Pinton, A. Rimessi, R. Sitia, and I. Medraño-Fernandez. 2020. "Human aquaporin-11 guarantees efficient transport of H2O2 across the endoplasmic reticulum membrane." *Redox Biology* no. 28:101326.

Bienert, G. P., and F. Chaumont. 2014. "Aquaporin-facilitated transmembrane diffusion of hydrogen peroxide." *Biochimica et Biophysica Acta (BBA)-General Subjects* no. 1840 (5):1596–1604.

Bohle, D. S., B. Hansert, S. C. Paulson, and B. D. Smith. 1994. "Biomimetic synthesis of the putative cytotoxin peroxynitrite, $ONOO^-$, and its characterization as a tetramethylammonium salt." *Journal of the American Chemical Society* no. 116 (16):7423–7424.

Bonini, M. G., R. Radi, G. Ferrer-Sueta, A.M.D.C. Ferreira, and O. Augusto. 1999. "Direct EPR detection of the carbonate radical anion produced from peroxynitrite and carbon dioxide." *Journal of Biological Chemistry* no. 274 (16):10802–10806.

Brahm, J. 1977. "Temperature-dependent changes of chloride transport kinetics in human red cells." *The Journal of General Physiology* no. 70 (3):283–306.

Briviba, K., R. Kissner, W. H. Koppenol, and H. Sies. 1998. "Kinetic study of the reaction of glutathione peroxidase with peroxynitrite." *Chemical Research in Toxicology* no. 11 (12):1398–1401.

Bryk, A. H., and J. R. Wiśniewski. 2017. "Quantitative analysis of human red blood cell proteome." *Journal of Proteome Research* no. 16 (8):2752–2761.

Buettner, G. R. 1993. "The pecking order of free radicals and antioxidants: lipid peroxidation, α-tocopherol, and ascorbate." *Archives of Biochemistry and Biophysics* no. 300 (2):535–543.

Buxton, G. V., and A. J. Elliot. 1986. "Rate constant for reaction of hydroxyl radicals with bicarbonate ions." *International Journal of Radiation Applications and Instrumentation. Part C. Radiation Physics and Chemistry* no. 27 (3):241–243.

Buxton, G. V., C. L. Greenstock, W. P. Helman, and A. B. Ross. 1988. "Critical review of rate constants for reactions of hydrated electrons, hydrogen atoms and hydroxyl radicals (· OH/· O− in aqueous solution." *Journal of Physical and Chemical Reference Data* no. 17 (2):513–886.

Byun, J., D. M. Mueller, J. S. Fabjan, and J. W. Heinecke. 1999. "Nitrogen dioxide radical generated by the myeloperoxidase-hydrogen peroxide-nitrite system promotes lipid peroxidation of low density lipoprotein." *FEBS Letters* no. 455 (3):243–246.

Carbrey, J. M., and P. Agre. 2009. "Discovery of the aquaporins and development of the field." In *Aquaporins*, edited by Eric Beitz, 3–28. Berlin, Heidelberg: Springer.

Chen, B., and W. M. Deen. 2001. "Analysis of the effects of cell spacing and liquid depth on nitric oxide and its oxidation products in cell cultures." *Chemical Research in Toxicology* no. 14 (1):135–147.

Chen, T.-M., W. G. Kuschner, J. Gokhale, and S. Shofer. 2007. "Outdoor air pollution: nitrogen dioxide, sulfur dioxide, and carbon monoxide health effects." *The American Journal of the Medical Sciences* no. 333 (4):249–256.

Cordeiro, R. M. 2014. "Reactive oxygen species at phospholipid bilayers: distribution, mobility and permeation." *Biochimica et Biophysica Acta (BBA)-Biomembranes* no. 1838 (1):438–444.

Cordeiro, R. M. 2018. "Reactive oxygen and nitrogen species at phospholipid bilayers: peroxynitrous acid and its homolysis products." *The Journal of Physical Chemistry B* no. 122 (34):8211–8219.

Czapski, G., S. V. Lymar, and H. A. Schwarz. 1999. "Acidity of the carbonate radical." *The Journal of Physical Chemistry A* no. 103 (18):3447–3450.

Daneva, Z., C. Marziano, M. Ottolini, Y.-L. Chen, T. M. Baker, M. Kuppusamy, A. Zhang, H. Q. Ta, C. E. Reagan, and A. D. Mihalek. 2021. "Caveolar peroxynitrite formation impairs endothelial TRPV4 channels and elevates pulmonary arterial pressure in pulmonary hypertension." *Proceedings of the National Academy of Sciences* no. 118 (17):1–12.

Delaney, C. A., B. Tyrberg, L. Bouwens, H. Vaghef, B. Hellman, and D. L. Eizirik. 1996. "Sensitivity of human pancreatic islets to peroxynitrite-induced cell dysfunction and death." *Febs Letters* no. 394 (3):300–306.

Denicola, A., B. A. Freeman, M. Trujillo, and R. Radi. 1996. "Peroxynitrite reaction with carbon dioxide/bicarbonate: kinetics and influence on peroxynitrite-mediated oxidations." *Archives of Biochemistry and Biophysics* no. 333 (1):49–58.

Denicola, A., H. Rubbo, D. Rodríguez, and R. Radi. 1993. "Peroxynitrite-mediated cytotoxicity to trypanosoma cruzi." *Archives of Biochemistry and Biophysics* no. 304 (1):279–286.

Denicola, A., J. M. Souza, and R. Radi. 1998. "Diffusion of peroxynitrite across erythrocyte membranes." *Proceedings of the National Academy of Sciences* no. 95 (7):3566–3571.

Denicola, A., J. M. Souza, R. Radi, and E. Lissi. 1996. "Nitric oxide diffusion in membranes determined by fluorescence quenching." *Archives of Biochemistry and Biophysics* no. 328 (1):208–212.

El-Benna, J., P. M.-C. Dang, M.-A. Gougerot-Pocidalo, J.-C. Marie, and F. Braut-Boucher. 2009. "p47phox, the phagocyte NADPH oxidase/NOX2 organizer: structure, phosphorylation and implication in diseases." *Experimental & Molecular Medicine* no. 41 (4):217–225.

Eriksson, U. K., G. Fischer, R. Friemann, G. Enkavi, E. Tajkhorshid, and R. Neutze. 2013. "Subangstrom resolution X-ray structure details aquaporin-water interactions." *Science* no. 340 (6138):1346–1349.

Estévez, A. G., R. Radi, L. Barbeito, J. T. Shin, J. A. Thompson, and J. S. Beckman. 1995. "Peroxynitrite-induced cytotoxicity in PC12 cells: evidence for an apoptotic mechanism differentially modulated by neurotrophic factors." *Journal of Neurochemistry* no. 65 (4):1543–1550.

Ferrer-Sueta, G., N. Campolo, M. Trujillo, S. Bartesaghi, S. Carballal, N. Romero, B. Alvarez, and R. Radi. 2018. "Biochemistry of peroxynitrite and protein tyrosine nitration." *Chemical Reviews* no. 118 (3):1338 1408.

Ferrer-Sueta, G., and R. Radi. 2009. "Chemical biology of peroxynitrite: kinetics, diffusion, and radicals." *ACS Chemical Biology* no. 4 (3):161–177.

Finn, R. N., and J. Cerda. 2015. "Evolution and functional diversity of aquaporins." *The Biological Bulletin* no. 229 (1):6–23.

Ford, E., M. N. Hughes, and P. Wardman. 2002. "Kinetics of the reactions of nitrogen dioxide with glutathione, cysteine, and uric acid at physiological pH." *Free Radical Biology and Medicine* no. 32 (12):1314–1323.

Freeman, B. A., M. Pekarova, H. Rubbo, and A. Trostchansky. 2017. "Electrophilic nitro-fatty acids: nitric oxide and nitrite-derived metabolic and inflammatory signaling mediators." In *Nitric Oxide* (Third Edition), edited by Bruce A. Freeman and Louis J. Ignarro, 213–229. Academic Press.

Fu, D., A. Libson, L. J. Miercke, C. Weitzman, P. Nollert, J. Krucinski, and R. M. Stroud. 2000. "Structure of a glycerol-conducting channel and the basis for its selectivity." *Science* no. 290 (5491):481–486.

Furtmüller, P. G., W. Jantschko, M. Zederbauer, M. Schwanninger, C. Jakopitsch, S. Herold, W. H. Koppenol, and C. Obinger. 2005. "Peroxynitrite efficiently mediates the interconversion of redox intermediates of myeloperoxidase." *Biochemical and Biophysical Research Communications* no. 337 (3):944–954.

Goldstein, S., and G. Czapski. 1995. "The reaction of NO· with O2·− and HO2·−: a pulse radiolysis study." *Free Radical Biology and Medicine* no. 19 (4):505–510.

Goldstein, S., G. Czapski, J. Lind, and G. Merényi. 2001. "Carbonate radical ion is the only observable intermediate in the reaction of peroxynitrite with CO2." *Chemical Research in Toxicology* no. 14 (9):1273–1276.

Gus' kova, R. A., I. I. Ivanov, V. K. Kol'tover, V. V. Akhobadze, and A. B. Rubin. 1984. "Permeability of bilayer lipid membranes for superoxide (O2−) radicals." *Biochimica et Biophysica Acta (BBA)-Biomembranes* no. 778 (3):579–585.

Gutknecht, J., and A. Walter. 1981a. "Hydrofluoric and nitric acid transport through lipid bilayer membranes." *Biochimica et Biophysica Acta (BBA)-Biomembranes* no. 644 (1):153–156.

Gutknecht, J., and A. Walter. 1981b. "Transport of protons and hydrochloric acid through lipid bilayer membranes." *Biochimica et Biophysica Acta (BBA)-Biomembranes* no. 641 (1):183–188.

Hamasaki, N., H. Matsuyama, and C. Hirota-Chigita. 1983. "Characterisation of phosphoenolpyruvate transport across the erythrocyte membrane: evidence for involvement of band 3 in the transport system." *European Journal of Biochemistry* no. 132 (3):531–536.

Hambly, D. M., and M. L. Gross. 2005. "Laser flash photolysis of hydrogen peroxide to oxidize protein solvent-accessible residues on the microsecond timescale." *Journal of the American Society for Mass Spectrometry* no. 16 (12):2057–2063.

Holmström, K. M., and T. Finkel. 2014. "Cellular mechanisms and physiological consequences of redox-dependent signalling." *Nature Reviews Molecular Cell Biology* no. 15 (6):411–421.

Huie, R. E. 1994. "The reaction kinetics of NO2." *Toxicology* no. 89 (3):193–216.

Huie, R. E., and S. Padmaja. 1993. "The reaction of NO with superoxide." *Free Radical Research Communications* no. 18 (4):195–199.

Ikeda, M., E. Beitz, D. Kozono, W. B. Guggino, P. Agre, and M. Yasui. 2002. "Characterization of aquaporin-6 as a nitrate channel in mammalian cells: requirement of pore-lining residue threonine 63." *Journal of Biological Chemistry* no. 277 (42):39873–39879.

Ischiropoulos, H., L. Zhu, and J. S. Beckman. 1992. "Peroxynitrite formation from macrophage-derived nitric oxide." *Archives of Biochemistry and Biophysics* no. 298 (2):446–451.

Jennings, M. 1989. "Structure and function of the red blood cell anion transport protein." *Annual Review of Biophysics and Biophysical Chemistry* no. 18 (1):397–430.

Jiang, H., N. Kruger, D. R. Lahiri, D. Wang, J.-M. Vatèle, and M. Balazy. 1999. "Nitrogen dioxide induces cis-trans-isomerization of arachidonic acid within cellular phospholipids: detection of trans-arachidonic acids in vivo." *Journal of Biological Chemistry* no. 274 (23):16235–16241.

Kelly, F. J., and T. D. Tetley. 1997. "Nitrogen dioxide depletes uric acid and ascorbic acid but not glutathione from lung lining fluid." *Biochemical Journal* no. 325 (1):95–99.

Khairutdinov, R. F., J. W. Coddington, and J. K. Hurst. 2000. "Permeation of phospholipid membranes by peroxynitrite." *Biochemistry* no. 39 (46):14238–14249.

Khramtsov, V., M. Panteleev, and L. Weiner. 1989. "ESR study of proton transport across phospholipid vesicle membranes." *Journal of Biochemical and Biophysical Methods* no. 18 (3):237–246.

Kissner, R., T. Nauser, P. Bugnon, P. G. Lye, and W. H. Koppenol. 1997. "Formation and properties of peroxynitrite as studied by laser flash photolysis, high-pressure stopped-flow technique, and pulse radiolysis." *Chemical Research in Toxicology* no. 10 (11):1285–1292.

Kitchen, P., M. M. Salman, S. U. Pickel, J. Jennings, S. Törnroth-Horsefield, M. T. Conner, R. M. Bill, and A. C. Conner. 2019. "Water channel pore size determines exclusion properties but not solute selectivity." *Scientific Reports* no. 9 (1):1–14.

Koppenol, W. H., and R. Hider. 2019. "Iron and redox cycling. Do's and don'ts." *Free Radical Biology and Medicine* no. 133:3–10.

Koppenol, W. H., J. Moreno, W. A. Pryor, H. Ischiropoulos, and J. Beckman. 1992. "Peroxynitrite, a cloaked oxidant formed by nitric oxide and superoxide." *Chemical Research in Toxicology* no. 5 (6): 834–842.

Koppenol, W. H., M. C. Vissers, M. B. Hampton, and A. J. Kettle. 2017. "Hydrogen peroxide, a molecule with a Janus face: its history, chemistry, and biology." In *Hydrogen Peroxide Metabolism in Health and Disease*, 3–16. CRC Press.

Liu, X., M. J. Miller, M. S. Joshi, H. Sadowska-Krowicka, D. A. Clark, and J. R. Lancaster. 1998. "Diffusion-limited reaction of free nitric oxide with erythrocytes." *Journal of Biological Chemistry* no. 273 (30):18709–18713.

Lü, W., J. Du, N. J. Schwarzer, T. Wacker, S. L. Andrade, and O. Einsle. 2013. "The formate/nitrite transporter family of anion channels." *Biological Chemistry* no. 394 (6):715–727.

Lü, W., N. J. Schwarzer, J. Du, E. Gerbig-Smentek, S. L. Andrade, and O. Einsle. 2012. "Structural and functional characterization of the nitrite channel NirC from Salmonella typhimurium." *Proceedings of the National Academy of Sciences* no. 109 (45):18395–18400.

Lundberg, J. O., E. Weitzberg, and M. T. Gladwin. 2008. "The nitrate–nitrite–nitric oxide pathway in physiology and therapeutics." *Nature Reviews Drug Discovery* no. 7 (2):156–167.

Lymar, S. V., and J. K. Hurst. 1995. "Rapid reaction between peroxonitrite ion and carbon dioxide: implications for biological activity." *Journal of the American Chemical Society* no. 117 (34):8867–8868.

Lynch, R. E., and I. Fridovich. 1978. "Permeation of the erythrocyte stroma by superoxide radical." *Journal of Biological Chemistry* no. 253 (13):4697–4699.

Macfadyen, A. J., C. Reiter, Y. Zhuang, and J. S. Beckman. 1999. "A novel superoxide dismutase-based trap for peroxynitrite used to detect entry of peroxynitrite into erythrocyte ghosts." *Chemical Research in Toxicology* no. 12 (3):223–229.

Manion, J., R. Huie, R. Levin, D. Burgess Jr, V. Orkin, W. Tsang, W. McGivern, J. Hudgens, V. Knyazev, and D. Atkinson. 2015. "NIST chemical kinetics database, NIST standard reference database

17, version 7.0 (Web Version), release 1.6. 8, data version 2015.12, National Institute of Standards and Technology, Gaithersburg, Maryland, 20899–8320." http://kinetics.nist.gov.

Marla, S. S., J. Lee, and J. T. Groves. 1997. "Peroxynitrite rapidly permeates phospholipid membranes." *Proceedings of the National Academy of Sciences* no. 94 (26):14243–14248.

Marrink, S. J., and H. J. Berendsen. 1996. "Permeation process of small molecules across lipid membranes studied by molecular dynamics simulations." *The Journal of Physical Chemistry* no. 100 (41):16729–16738.

Martínez, A., C. Prolo, D. Estrada, N. Rios, M. N. Alvarez, M. D. Piñeyro, C. Robello, R. Radi, and L. Piacenza. 2019. "Cytosolic Fe-superoxide dismutase safeguards Trypanosoma cruzi from macrophage-derived superoxide radical." *Proceedings of the National Academy of Sciences* no. 116 (18):8879–8888.

Miller, E. W., B. C. Dickinson, and C. J. Chang. 2010. "Aquaporin-3 mediates hydrogen peroxide uptake to regulate downstream intracellular signaling." *Proceedings of the National Academy of Sciences* no. 107 (36):15681–15686.

Möller, M. N., E. Cuevasanta, F. Orrico, A. C. Lopez, L. Thomson, and A. Denicola. 2019. "Diffusion and transport of reactive species across cell membranes." In *Bioactive Lipids in Health and Disease*, edited by Andrés Trostchansky and Homero Rubbo, 3–19. Cham: Springer.

Möller, M. N., and A. Denicola. 2018. "Diffusion of nitric oxide and oxygen in lipoproteins and membranes studied by pyrene fluorescence quenching." *Free Radical Biology and Medicine* no. 128:137–143.

Möller, M. N., Q. Li, M. Chinnaraj, H. C. Cheung, J. R. Lancaster Jr, and A. Denicola. 2016. "Solubility and diffusion of oxygen in phospholipid membranes." *Biochimica et biophysica acta (bba)-biomembranes* no. 1858 (11):2923–2930.

Möller, M. N., Q. Li, J. R. Lancaster Jr, and A. Denicola. 2007. "Acceleration of nitric oxide autoxidation and nitrosation by membranes." *IUBMB Life* no. 59 (4–5):243–248.

Oliva, R., G. Calamita, J. M. Thornton, and M. Pellegrini-Calace. 2010. "Electrostatics of aquaporin and aquaglyceroporin channels correlates with their transport selectivity." *Proceedings of the National Academy of Sciences* no. 107 (9):4135–4140.

Patterson, L., and K. Hasegawa. 1978. "Pulse radiolysis studies in model lipid systems. The influence of aggregation on kinetic behavior of OH induced radicals in aqueous sodium linoleate." *Berichte der Bunsengesellschaft für physikalische Chemie* no. 82 (9):951–956.

Pautz, A., J. Art, S. Hahn, S. Nowag, C. Voss, and H. Kleinert. 2010. "Regulation of the expression of inducible nitric oxide synthase." *Nitric Oxide* no. 23 (2):75–93.

Peluffo, G., and R. Radi. 2007. "Biochemistry of protein tyrosine nitration in cardiovascular pathology." *Cardiovascular Research* no. 75 (2):291–302.

Perrin, D., and W. H. Koppenol. 2000. "The quantitative oxidation of methionine to methionine sulfoxide by peroxynitrite." *Archives of Biochemistry and Biophysics* no. 377 (2):266–272.

Prolo, C., M. N. Álvarez, N. Ríos, G. Peluffo, R. Radi, and N. Romero. 2015. "Nitric oxide diffusion to red blood cells limits extracellular, but not intraphagosomal, peroxynitrite formation by macrophages." *Free Radical Biology and Medicine* no. 87:346–355.

Prütz, W. A., H. Mönig, J. Butler, and E. J. Land. 1985. "Reactions of nitrogen dioxide in aqueous model systems: oxidation of tyrosine units in peptides and proteins." *Archives of Biochemistry and Biophysics* no. 243 (1):125–134.

Radi, R. 2013. "Peroxynitrite, a stealthy biological oxidant." *Journal of Biological Chemistry* no. 288 (37):26464–26472.

Radi, R., J. S. Beckman, K. M. Bush, and B. A. Freeman. 1991. "Peroxynitrite-induced membrane lipid peroxidation: the cytotoxic potential of superoxide and nitric oxide." *Archives of Biochemistry and Biophysics* no. 288 (2):481–487.

Radi, R., G. Peluffo, M.A.N. Alvarez, M. Naviliat, and A. Cayota. 2001. "Unraveling peroxynitrite formation in biological systems." *Free Radical Biology and Medicine* no. 30 (5):463–488.

Rambow, J., B. Wu, D. Rönfeldt, and E. Beitz. 2014. "Aquaporins with anion/monocarboxylate permeability: mechanisms, relevance for pathogen–host interactions." *Frontiers in Pharmacology* no. 5:199.

Razzokov, J., M. Yusupov, R. M. Cordeiro, and A. Bogaerts. 2018. "Atomic scale understanding of the permeation of plasma species across native and oxidized membranes." *Journal of Physics D: Applied Physics* no. 51 (36):365203.

Rodrigues, C., C. Pimpão, A. F. Mósca, A. S. Coxixo, D. Lopes, I. V. da Silva, P. A. Pedersen, F. Antunes, and G. Soveral. 2019. "Human aquaporin-5 facilitates hydrogen peroxide permeation affecting adaptation to oxidative stress and cancer cell migration." *Cancers* no. 11 (7):932.

Romero, M. F., A.-P. Chen, M. D. Parker, and W. F. Boron. 2013. "The SLC4 family of bicarbonate

(HCO3-) transporters." *Molecular Aspects of Medicine* no. 34 (2–3):159–182.

Romero, N., A. Denicola, J. M. Souza, and R. Radi. 1999. "Diffusion of peroxynitrite in the presence of carbon dioxide." *Archives of Biochemistry and Biophysics* no. 368 (1):23–30.

Romero, N., G. Peluffo, S. Bartesaghi, H. Zhang, J. Joseph, B. Kalyanaraman, and R. Radi. 2007. "Incorporation of the hydrophobic probe N-t-BOC-L-tyrosine tert-butyl ester to red blood cell membranes to study peroxynitrite-dependent reactions." *Chemical Research in Toxicology* no. 20 (11):1638–1648.

Savage, D. F., J. D. O'Connell, L. J. Miercke, J. Finer-Moore, and R. M. Stroud. 2010. "Structural context shapes the aquaporin selectivity filter." *Proceedings of the National Academy of Sciences* no. 107 (40):17164–17169.

Signorelli, S., M. N. Möller, E. L. Coitiño, and A. Denicola. 2011. "Nitrogen dioxide solubility and permeation in lipid membranes." *Archives of Biochemistry and Biophysics* no. 512 (2):190–196.

Smart, O. S., J. G. Neduvelil, X. Wang, B. Wallace, and M. S. Sansom. 1996. "HOLE: a program for the analysis of the pore dimensions of ion channel structural models." *Journal of Molecular Graphics* no. 14 (6):354–360.

Solar, S., W. Solar, and N. Getoff. 1984. "Reactivity of hydroxyl with tyrosine in aqueous solution studied by pulse radiolysis." *The Journal of Physical Chemistry* no. 88 (10):2091–2095.

Soto, G., K. Alleva, G. Amodeo, J. Muschietti, and N. D. Ayub. 2012. "New insight into the evolution of aquaporins from flowering plants and vertebrates: orthologous identification and functional transfer is possible." *Gene* no. 503 (1):165–176.

Souza, J. M., G. Peluffo, and R. Radi. 2008. "Protein tyrosine nitration—functional alteration or just a biomarker?" *Free Radical Biology and Medicine* no. 45 (4):357–366.

Squadrito, G. L., and E. M. Postlethwait. 2009. "On the hydrophobicity of nitrogen dioxide: could there be a 'lens' effect for NO2 reaction kinetics?" *Nitric Oxide* no. 21 (2):104–109.

Stein, W. 1986. *Transport and Diffusion Across Cell Membranes.* Orlando, FL: Academic Press.

Subczynski, W. K., J. S. Hyde, and A. Kusumi. 1991. "Effect of alkyl chain unsaturation and cholesterol intercalation on oxygen transport in membranes: a pulse ESR spin labeling study." *Biochemistry* no. 30 (35):8578–8590.

Szabó, C., H. Ischiropoulos, and R. Radi. 2007. "Peroxynitrite: biochemistry, pathophysiology and development of therapeutics." *Nature Reviews Drug Discovery* no. 6 (8):662–680.

Tajkhorshid, E., P. Nollert, M. Ø. Jensen, L. J. Miercke, J. O'Connell, R. M. Stroud, and K. Schulten. 2002. "Control of the selectivity of the aquaporin water channel family by global orientational tuning." *Science* no. 296 (5567):525–530.

Trostchansky, A., V. B. O'Donnell, D. C. Goodwin, L. M. Landino, L. J. Marnett, R. Radi, and H. Rubbo. 2007. "Interactions between nitric oxide and peroxynitrite during prostaglandin endoperoxide H synthase-1 catalysis: a free radical mechanism of inactivation." *Free Radical Biology and Medicine* no. 42 (7):1029–1038.

Trujillo, M., A. Clippe, B. Manta, G. Ferrer-Sueta, A. Smeets, J.-P. Declercq, B. Knoops, and R. Radi. 2007. "Pre-steady state kinetic characterization of human peroxiredoxin 5: taking advantage of Trp84 fluorescence increase upon oxidation." *Archives of Biochemistry and Biophysics* no. 467 (1):95–106.

Tsukaguchi, H., S. Weremowicz, C. C. Morton, and M. A. Hediger. 1999. "Functional and molecular characterization of the human neutral solute channel aquaporin-9." *American Journal of Physiology-Renal Physiology* no. 277 (5):F685–F696.

Turell, L., M. Steglich, and B. Alvarez. 2018. "The chemical foundations of nitroalkene fatty acid signaling through addition reactions with thiols." *Nitric Oxide* no. 78:161–169.

Varadaraj, K., and S. S. Kumari. 2020. "Lens aquaporins function as peroxiporins to facilitate membrane transport of hydrogen peroxide." *Biochemical and Biophysical Research Communications* no. 524 (4):1025–1029.

Waight, A. B., J. Love, and D.-N. Wang. 2010. "Structure and mechanism of a pentameric formate channel." *Nature Structural & Molecular Biology* no. 17 (1):31–37.

Walter, A., and J. Gutknecht. 1986. "Permeability of small nonelectrolytes through lipid bilayer membranes." *The Journal of Membrane Biology* no. 90 (3):207–217.

Wang, Y., Y. Huang, J. Wang, C. Cheng, W. Huang, P. Lu, Y.-N. Xu, P. Wang, N. Yan, and Y. Shi. 2009. "Structure of the formate transporter FocA reveals a pentameric aquaporin-like channel." *Nature* no. 462 (7272):467–472.

Watanabe, S., C. S. Moniaga, S. Nielsen, and M. Hara-Chikuma. 2016. "Aquaporin-9 facilitates membrane transport of hydrogen peroxide in mammalian cells." *Biochemical and Biophysical Research Communications* no. 471 (1):191–197.

Wiechert, M., H. Erler, A. Golldack, and E. Beitz. 2017. "A widened substrate selectivity filter of eukaryotic formate-nitrite transporters enables high-level lactate conductance." *The FEBS Journal* no. 284 (16):2663–2673.

Wree, D., B. Wu, T. Zeuthen, and E. Beitz. 2011. "Requirement for asparagine in the aquaporin NPA sequence signature motifs for cation exclusion." *The FEBS Journal* no. 278 (5):740–748.

Wspalz, T., Y. Fujiyoshi, and A. Engel. 2009. "The AQP structure and functional implications." In *Aquaporins*, edited by Eric Beitz, 31–56. Berlin, Heidelberg: Springer.

Wu, B., C. Steinbronn, M. Alsterfjord, T. Zeuthen, and E. Beitz. 2009. "Concerted action of two cation filters in the aquaporin water channel." *The EMBO Journal* no. 28 (15):2188–2194.

Xiang, T.-X., and B. D. Anderson. 1998. "Influence of chain ordering on the selectivity of dipalmitoylphosphatidylcholine bilayer membranes for permeant size and shape." *Biophysical Journal* no. 75 (6):2658–2671.

Yasui, M., A. Hazama, T.-H. Kwon, S. Nielsen, W. B. Guggino, and P. Agre. 1999. "Rapid gating and anion permeability of an intracellular aquaporin." *Nature* no. 402 (6758):184–187.

Yeh, H., and G. Wills. 1970. "Diffusion coefficient of sodium nitrate in aqueous solution at 25. deg. as a function of concentration from 0.1 to 1.0 M." *Journal of Chemical and Engineering Data* no. 15 (1):187–189.

Yool, A. J., W. D. Stamer, and J. W. Regan. 1996. "Forskolin stimulation of water and cation permeability in aquaporin1 water channels." *Science* no. 273 (5279):1216–1218.

Yu, J., A. J. Yool, K. Schulten, and E. Tajkhorshid. 2006. "Mechanism of gating and ion conductivity of a possible tetrameric pore in aquaporin-1." *Structure* no. 14 (9):1411–1423.

Yusupov, M., J. Razzokov, R. M. Cordeiro, and A. Bogaerts. 2019. "Transport of reactive oxygen and nitrogen species across aquaporin: a molecular level picture." *Oxidative Medicine and Cellular Longevity* no. 2019.

Zeida, A., M.C.G. Lebrero, R. Radi, M. Trujillo, and D. A. Estrin. 2013. "Mechanism of cysteine oxidation by peroxynitrite: an integrated experimental and theoretical study." *Archives of Biochemistry and Biophysics* no. 539 (1):81–86.

Zhu, L., C. Gunn, and J. S. Beckman. 1992. "Bactericidal activity of peroxynitrite." *Archives of Biochemistry and Biophysics* no. 298 (2):452–457.

CHAPTER FOUR

Hydrogen Peroxide as a Substrate for Aquaporins from a Structure-Function Perspective

Eric Beitz

ABBREVIATIONS

AQP	Aquaporin
ar/R	Aromatic/arginine
CHIP28	Channel-forming integral membrane protein of 28 kDa (later AQP1)
GlpF	Glycerol facilitator
NPA	Asparagine-proline-alanine
TIP	Tonoplast intrinsic protein

4.1 INTRODUCTION

The aquaporin (AQP) protein family is named after its common classical substrate, water. The water permeability function of the aquaporins was discovered in 1992 by expressing the human red blood cell CHIP28 protein, which was later termed AQP1, in oocytes of the fresh water claw frog *Xenopus laevis* and subsequent subjection of the cells to a hypotonic buffer solution (Preston et al. 1992). The oocytes swelled and burst within 2–3 min, whereas non-expressing control cells withstood the treatment largely unaffected. This breakthrough sparked sequence searches throughout species yielding AQP homologues in all kingdoms of life. One immediate hit was the GlpF protein from *Escherichia coli* bacteria, which was named after its even earlier established transmembrane glycerol facilitating functionality (Heller et al. 1980). AQP1, however, was impermeable to glycerol (Yang and Verkman 1997). From then on it was clear that the AQP family comprises at least two functionally diverse subfamilies (Reizer et al. 1993): water-specific "orthodox" aquaporins and aquaglyceroporins (Borgnia et al. 1999).

Over the years, several more physiologically meaningful substrates were found to pass AQP channels (Beitz and Wu 2007). Urea is a small uncharged solute of the nitrogen metabolism that is conducted by many aquaglyceroporins. Other small molecules with resemblance to glycerol due to the presence of three hydroxyl moieties are the toxic antimonous and arsenous acids, $Sb(OH)_3$ and $As(OH)_3$, respectively. As very weak acid compounds, both appear mostly in their protonated, neutral form under physiological pH conditions. Aquaglyceroporins with $Sb/As(OH)_3$ permeability play roles in resistance against antimony or arsenic containing anticancer (Mukhopadhyay et al. 2011) and as antiparasitic drugs (Mukhopadhyay et al. 2011). Boric acid, $B(OH)_3$ (Diehn et al. 2019) and the larger silicic acid, $Si(OH)_4$ (Ma et al. 2010), carrying four hydroxyls have been shown to pass certain plant AQPs. Uptake of boron and silicon is required for plant growth and stress resistance. Lactic acid is a much stronger acid that deprotonates to >99% at neutral pH forming negative charged lactate anions. Nevertheless, certain aquaglyceroporins exhibit some lactic acid permeability in the physiological pH range (Bienert et al. 2013; Rothert et al. 2017; Bader and Beitz 2020). Small lipophilic gas molecules, such as carbon dioxide, CO_2, nitrogen oxide, NO, are substrates of AQPs including the orthodox human AQP1 (Musa-Aziz et al. 2009). Since lipid bilayers pose only small energetic barriers to these gases, the

DOI: 10.1201/9781003160649-4

relevance of AQP-facilitated transmembrane gas permeation is a matter of discussion (Hub et al. 2009). Gaseous ammonia, NH_3, as the neutral, deprotonated base of the ammonium ion, NH_4^+, is well conducted by certain AQPs; for example, by human AQP8 and the plant TIP family (Jahn et al. 2004). For such isoforms the term **aquaammoniaporin** has been established. Finally, the reactive oxygen species hydrogen peroxide, H_2O_2, with signaling functions, was added to the list of AQP substrates (Bienert et al. 2007; Almasalmeh et al. 2014).

With the very special exception of the chloride/nitrite/nitrate anion channel functionality of human AQP6 (Yasui et al. 1999), the remaining AQP family members facilitate the transmembrane passage of uncharged small substrate molecules. Still, individual AQP isoforms exhibit specific substrate spectra. In this chapter, we discuss mechanisms of substrate selectivity (and ion exclusion) of the AQPs in terms of structure-function relationships with a focus on hydrogen peroxide.

4.2 THE AQP FOLD AND SUBSTRATE PERMEATION PATH LAYOUT

Homotetrameric AQPs are small and compact membrane proteins that span the lipid bilayer six times (Figure 4.1A; Sui et al. 2001; Fu et al. 2000). Both termini reside in the cytosol. AQPs exhibit an internal symmetry of the first and second half that derives from a gene duplication event. The first three and the second three transmembrane helices each harbor a peculiar short helical structure, so-called half-helices, that enter the lipid bilayer, then turn and leave the membrane at the same side (Figure 4.1B). The positive ends of the two half-helices are capped by the AQP signature motifs Asn-Pro-Ala (NPA) and make contact in the center of the protein to form a pseudo-seventh transmembrane helix (Figure 4.1B). Within the NPA motifs, the asparagine position is almost perfectly conserved among the AQPs while the proline and alanine residues can be replaced by other small amino acids (e.g., serine or cysteine) without major effects on the channel structure and permeation properties (Wu et al. 2009; Wree et al. 2011). The six transmembrane spans plus the pseudo-transmembrane helix tilt in a right-handed manner (Figure 4.1B) around a central narrow tube-like substrate permeation path of about 3–4 Å in diameter and 20 Å in length (Figure 4.1C; Sui et al. 2001). The rigidity of the tube is brought about by interdigitations of the transmembrane helices at their crossing points in the center of the membrane. Highly conserved glycine residues generate dents in the helices that prevent them from sliding against each other. Four identical AQP protomer units form a tetrameric complex with four individual pores and a typically non-conducting center (Figure 4.1A). Until recently, this protein layout was considered unique. It came quite as a surprise that the sequence-wise fully unrelated protein family of weak acid formate-nitrite transporters (FNTs) assumes the same general protomer fold, yet assembles into homopentamers (Wang et al. 2009; Wiechert and Beitz 2017).

The wall of the AQP substrate permeation path is made up by mainly non-polar amino acid side chains (Figure 4.1C, right), which at first sight appears counterintuitive if one considers the highly polar nature of the water molecule as a substrate. A non-polar environment, however, is required to exclude the passage of ions through the AQP pore in order not to disturb transmembrane gradients that were actively generated under the consumption of ATP. For water molecules to enter the permeation path at a low energetic cost, a ladder of carbonyl oxygen atoms from the peptide backbone provide hydrogen bond interaction sites (Figure 4.1C, left; Sui et al. 2001). The oxygens direct the water molecules in a single file toward the center of permeation path into the NPA region. Here, the positive dipole moment of the two half-helices re-orients the water molecules and another set of carbonyl oxygen atoms lead them toward to opposite end of the permeation path (de Groot and Grubmüller 2001). The driving force of the process is the transmembrane osmotic gradient.

4.3 THE AQP SELECTIVITY FILTER

AQPs exclude inorganic ions and charged molecules (Beitz et al. 2006; Wu et al. 2009; Wree et al. 2011). Cations are repelled by the positive electrostatic field that emanates from the NPA region. Anions in turn would be attracted. Yet, the non-polar interior in combination with the narrow diameter of the permeation path and the required shedding of the ion hydration shell pose an impenetrable energy barrier. The chloride/

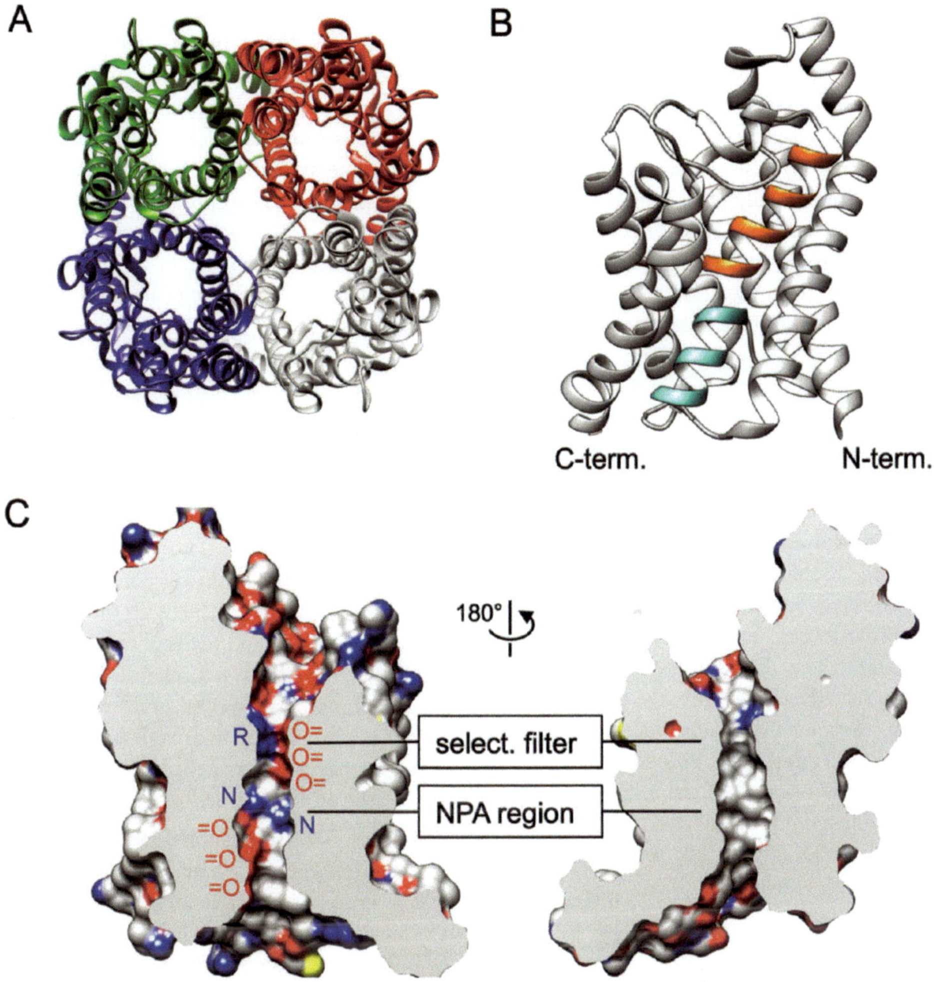

Figure 4.1 The AQP fold and permeation path (PDB# 1fx8; Fu et al. 2000). (A) Top view on the AQP homotetramer. (B) Side view on one AQP protomer with the half helices colored. (C) Surface view on two opposite sections of one AQP protomer. The left panel shows the polar side of the permeation path interior with the selectivity filter arginine (R), the NPA asparagines (N), and bands of carbonyl oxygens (=O) labeled. The right panel depicts the lipophilic side of the permeation path.

nitrite/nitrate anion permeability of AQP6 is thought to be a result of an asparagine residue instead of a glycine at the crossing point between transmembrane helix 2 and 5 (Liu et al. 2005). The larger asparagine side chain loosens the helix-helix contact allows them to slide and adapt a conformation that is compatible with the permeation of a sufficiently hydrated anion.

Protons are peculiar cations due to their small size and their inability to exist in an unbound state in aqueous solution. The exclusion of protons from passing a channel requires the breakage of water-wires, hydrogen-bonded files of water along which protons can hop, termed the Grotthuss mechanism. The major proton filter of the AQPs is constituted by the so-called aromatic/arginine region (ar/R) or selectivity filter and the positive charge of the arginine prevents hydronium ions, H_3O^+, from entering the AQP permeation path (Figure 4.1C; Beitz et al. 2006; Eriksson et al. 2013). This region is located at the extracellular entrance of the AQP pore and poses the narrowest constriction in the path. In orthodox AQPs, the diameter is about that of a water molecule (Murata et al. 2000), aquaglyceroporins are wider in the ar/R selectivity filter region to accommodate the passage of neutral solutes that are similar in size to glycerol (Fu et al. 2000).

Other than electrostatic repulsion, for neutral molecules, the ar/R selectivity filter acts by size exclusion. Additionally, the interaction intensity via hydrogen bonds seems to determine the dwell time of a substrate molecule in the selectivity filter region and, thus, the velocity at which the passage of the AQP permeation path occurs (Beitz et al. 2004; Newby et al. 2008). The timescale for water passage through an orthodox AQP is around 1 ns, whereas water molecules pass aquaglyceroporins typically one order of magnitude slower (de Groot and Grubmüller 2001; Beitz et al. 2004). The permeability rates of solutes, such as glycerol or urea, are further slowed by at least one more order of magnitude. Generally, AQP permeability rates depend directly on the steepness of the transmembrane osmotic or solute gradient. Being channel proteins with very low substrate binding affinity, AQPs facilitate the transmembrane passage of a substrate much faster than transport proteins (e.g., of the alternating access type; Bader and Beitz 2020).

Four different layouts of the selectivity filter can be discriminated that affect substrate selectivity and permeability rates (Figure 4.2; Wu and Beitz 2007):

1. "Orthodox AQP" selectivity filters (e.g., human AQP1) are characterized by an arginine flanked by an aromatic residue on one side and a histidine in juxtaposition (Murata et al. 2000; Sui et al. 2001). The fourth side typically holds an aliphatic residue (e.g., a cysteine) that acts as a binding site for mercurial AQP inhibitors. The rather polar layout offers multiple hydrogen bond interaction sites with a substrate water molecule for isolation from the aqueous bulk at low energetic cost. At the same time, cations are repelled. Due to the bulky amino acid sidechains, the diameter of the selectivity filter is small and matches that of a water molecule (≈2.4 Å).
2. "Aquaglyceroporin" selectivity filters (e.g., bacterial GlpF) lack the histidine and carry a small residue, often a glycine lacking a sidechain, as a substitute (Fu et al. 2000). This gives room for another aromatic residue (e.g., a phenylalanine) in the position that in orthodox AQPs is an aliphatic residue (Cys in AQP1). As a consequence, the selectivity filter region is wider (>3.4 Å) and amphipathic with the polar arginine at one side and hydrophobic aromatic rings forming the remainder of the constriction. This way, a glycerol molecule will orient itself with the hydroxyls pointing toward the arginine and the lipophilic back against the aromatic sides.
3. "Aquaammoniaporin" selectivity filters (e.g., human AQP8, plant TIP) exhibit a similar composition as those of orthodox AQPs. The histidine, however, is shifted to the side where orthodox AQPs often carry a phenylalanine (Kirscht et al. 2016). Both positively charged amino acids (the arginine and the histidine) are therefore positioned next

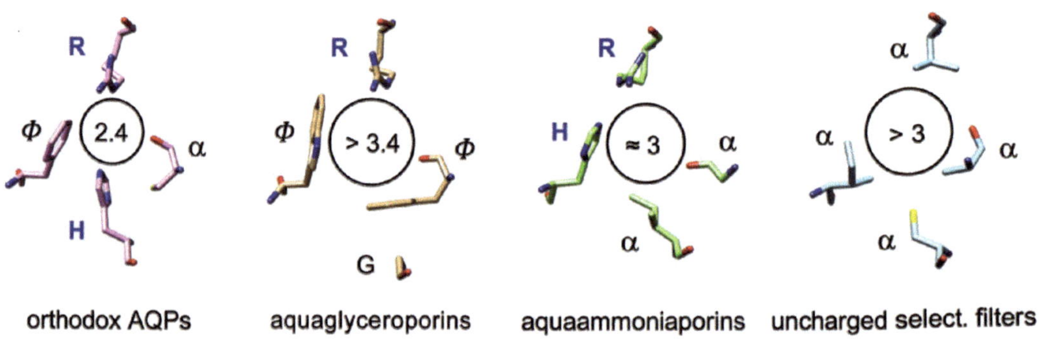

Figure 4.2 AQP selectivity filter layouts. Shown are top views on the selectivity filter regions of an orthodox AQP (AQP1, PDB# 1jn4; Sui et al. 2001); an aquaglyceroporin (GlpF, PDB# 1fx8; Fu et al. 2000); an aquaammoniaporin (TIP2;1, PDB# 5i32; Kirscht et al. 2016); and a model of an uncharged selectivity filter (*Toxoplasma gondii* AQP; Pavlovic-Djuranovic et al. 2003). The arginine (R) and histidines (H) are labeled, as well as positions that typically hold aromatic (Φ) and aliphatic residues (α). The diameters (Å) of the selectivity filters are indicated in the centers.

to each other. The position opposite the arginine often holds an aliphatic amino acid (isoleucine in human AQP8). The diameter of the aquaammoniaporin selectivity filter is intermediate between that of orthodox AQPs and aquaglyceroporins (≈3 Å).

4. "Uncharged selectivity filters" (e.g., *Toxoplasma gondii* AQP) lack an arginine altogether and instead carry an aliphatic amino acid (e.g., valine; Pavlovic-Djuranovic et al. 2003). Despite the loss of polarity and hydrogen bond interaction sites, water permeability of such AQPs can be as high as that of orthodox AQPs. This has been experimentally confirmed by introduction of point mutations into human AQP1 that exchange the selectivity filter arginine and/or the histidine to uncharged alanine residues (Beitz et al. 2006). The AQP1 selectivity filter mutants conducted water at the same rate as the wild-type protein. Apparently, a charged residue in the selectivity filter is not required to efficiently funnel water molecules into the AQP permeation path. As the selectivity filter constitutes the major barrier for protons, lacking the arginine may allow for some proton leakage. However, other than for mutants of AQP1 (Beitz et al. 2006; Eriksson et al. 2013) a proton leak has not yet been shown for natural AQPs.

Of the selectivity filter compositions described above, the aquaammoniaporin-type appears particularly suitable for passing hydrogen peroxide. Quite interestingly, in mammals, the aquaammoniaporin AQP8 was shown to be present in mitochondria (Soria et al. 2010), which are an important source of reactive oxygen species. In plants, the high hydrogen peroxide permeability of tonoplast-located TIP-type AQPs may be relevant for scavenging of hydrogen peroxide and protection of the cytosol from oxidative stress (Bienert et al. 2007).

4.4 PHYSICOCHEMICAL PROPERTIES OF HYDROGEN PEROXIDE COMPARED TO CLASSICAL AQP SUBSTRATES

The AQP selectivity filter primarily excludes charged and larger compounds. Second, optimal formation of hydrogen bonds between the substrate and the AQP permeation path facilitates efficient passage (Newby et al. 2008; Eriksson et al. 2013). In this sense it is striking how similar water and hydrogen peroxide are in terms of their physicochemical properties (Giguère and Chen 1984). Consequently, both compounds are fully miscible in each other. Water and hydrogen peroxide molecules have two hydrogen bond donor and two acceptor sites (Figure 4.3). Even the spacing of the hydrogen bonds is similar despite the extra oxygen atom in the hydrogen peroxide molecule (Figure 4.3, side view). The directionality of the hydrogen donor sites, however, is different. The 3-atom composition of the water molecule determines that the hydrogen bonds are in the same plane, whereas in the 4-atom hydrogen peroxide the angle between the two hydrogen bond vectors originating from the 2-oxygen axis is about 90° (Figure 4.3, top view). Since hydrogen bonds are formed at angles up to 180° (Torshin et al. 2002) there is still considerable overlap with the directionality in a water molecule; the bond intensity decreases with increasing angles, though. Furthermore, the strength of hydrogen bonds in hydrogen peroxide is 12%–15% weaker than in water (Giguère and Chen 1984). The affinity of a hydrogen peroxide molecule to the AQP protein should therefore be lower than that of a passing water molecule. The additional oxygen atom in hydrogen peroxide certainly affects the molecular weight (18 g mol^{-1} vs. 34 g mol^{-1}), yet the overall molecule dimensions are only moderately increased: 25% in height and 15% in diameter. The dimensions of the glycerol molecule as a substrate of the aquaglyceroporins, for comparison, are more than doubled in height and diameter (Figure 4.3).

Overall, the molecular similarity of hydrogen peroxide with water indicates that basically all water-permeable AQPs should facilitate hydrogen peroxide passage. AQPs with high water permeability concomitantly should exhibit high hydrogen peroxide permeability. In order to comply with the slightly greater spatial requirements of the hydrogen peroxide molecule, a moderately larger diameter of the selectivity filter should be beneficial. These properties are characteristic for aquaammoniaporins of the AQP8/TIP-type and can explain their particularly high hydrogen peroxide permeability. It is also thinkable that the side-by-side arrangement of the arginine and histidine in the aquaammoniaporin selectivity filter is favorable for interaction with the 90°-tilted hydrogen bond donor sites of hydrogen peroxide (Figure 4.3).

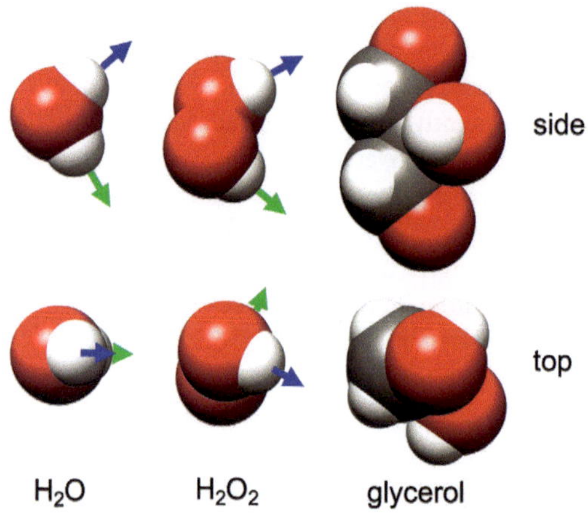

Figure 4.3 Comparison of water (H_2O), hydrogen peroxide (H_2O_2), and glycerol molecule dimensions. The molecules are shown from the side and the top along their longest axis. The directionality vectors of the two hydrogen bond donor sites of water and hydrogen peroxide are indicated by blue and green arrows.

TABLE 4.1
Effects of AQP Water Permeability and Selectivity Filter Diameter on Hydrogen Peroxide Permeability

	Rel. H_2O Permeability (%)	Select. Filter Diameter (Å)	Rel. H_2O_2 Permeability (%)
hAQP8	100	3.2	100
rAQP1	64	2.7	31
rAQP1 H180A	31	3.7	67
rAQP1 H180F	10	2.2	22

Source: Data from Almasalmeh et al. (2014).

4.5 EXPERIMENTAL EVALUATION CONFIRMS HYDROGEN PEROXIDE SELECTION PROPERTIES OF AQPS

Previously, we set out to experimentally relate the contribution to AQP hydrogen peroxide permeability of the two identified parameters, namely water permeability rate and selectivity filter diameter. We employed a set of wild-type and mutant AQPs that differ in these properties (Table 4.1; Almasalmeh et al. 2014). The AQPs were expressed in yeast for the biophysical determination of water permeability and phenotypic assessment of the hydrogen peroxide permeability. AQP8 with the highest water permeability of this set and a selectivity filter diameter between those of orthodox AQPs and aquaglyceroporins exhibited the highest hydrogen peroxide permeability. The water permeability of AQP1 was two-thirds that of AQP8, whereas the hydrogen peroxide permeability was only one-third. An increase of the selectivity filter diameter of AQP1 by introducing a H180A point mutation decreased the water permeability to about one-third that of AQP8, yet, at the same time, increased the hydrogen peroxide permeability to two-thirds. Introduction of an AQP1 H180F point mutation rendered the selectivity filter diameter smaller than that of a water molecule. Still, water permeability remained measurable at one-tenth that of AQP8 and even hydrogen peroxide appeared to pass the mutant AQP1 to some extent. This shows that dynamic sidechain movement leads to fluctuations in the selectivity filter diameter that allow larger molecules still to pass, though at a slower pace.

4.6 CONCLUSION

The theoretical considerations on the physicochemical similarities of the water/hydrogen peroxide molecules and the diameters of selectivity filters, combined with the experimental permeability data, show a clear picture: AQPs are generally fit for facilitating passage of hydrogen peroxide. Optimal conditions for hydrogen peroxide permeability are present in the aquaammoniaporin-type of the AQPs that have high water permeability and a slightly widened selectivity filter diameter.

REFERENCES

Almasalmeh, A., Krenc, D., Wu, B., and E. Beitz. 2014. Structural determinants of the hydrogen peroxide permeability of aquaporins. *FEBS J.* 281:647–656.

Bader, A., and E. Beitz. 2020. Transmembrane facilitation of lactate/H^+ instead of lactic acid is not a question of semantics but of cell viability. *Membranes* 10:E236.

Beitz, E., Pavlovic-Djuranovic, S., Yasui, M., Agre, P., and J. E. Schultz. 2004. Molecular dissection of water and glycerol permeability of the aquaglyceroporin from *Plasmodium falciparum* by mutational analysis. *Proc. Natl. Acad. Sci. USA* 101:1153–1158.

Beitz, E., Wu, B., Holm, L. M., Schultz, J. E., and T. Zeuthen. 2006. Point mutations in the aromatic/arginine region in aquaporin 1 allow passage of urea, glycerol, ammonia and protons. *Proc. Natl. Acad. Sci. USA* 103:269–274.

Bienert, G. P., Desguin, B., Chaumont, F., and P. Hols. 2013. Channel-mediated lactic acid transport: a novel function for aquaglyceroporins in bacteria. *Biochem. J.* 454:559–570.

Bienert, G. P., Møller, A. L., Kristiansen, K. A., Schulz, A., Møller, I. M., Schjoerring, J. K., and T. P. Jahn. 2007. Specific aquaporins facilitate the diffusion of hydrogen peroxide across membranes. *J. Biol. Chem.* 282:1183–1192.

Borgnia, M., Nielsen, S., Engel, A., and P. Agre. 1999. Cellular and molecular biology of the aquaporin water channels. *Annu. Rev. Biochem.* 68:425–458.

de Groot, B. L., and H. Grubmüller. 2001. Water permeation across biological membranes: mechanism and dynamics of aquaporin-1 and GlpF. *Science* 294:2353–2357.

Diehn, T. A., Bienert, M. D., Pommerrenig, B., Liu, Z., Spitzer, C., Bernhardt, N., Fuge, J., Bieber, A., Richet, N., Chaumont, F., and G. P. Bienert. 2019. Boron demanding tissues of *Brassica napus* express specific sets of functional Nodulin26-like Intrinsic Proteins and BOR1 transporters. *Plant J.* 100:68–82.

Eriksson, U. K., Fischer, G., Friemann, R., Enkavi, G., Tajkhorshid, E., and R. Neutze. 2013. Subangstrom resolution X-ray structure details aquaporin-water interactions. *Science* 340:1346–1349.

Fu, D., Libson, A., Miercke, L. J., Weitzman, C., Nollert, P., Krucinski, J., and R. M. Stroud. 2000. Structure of a glycerol-conducting channel and the basis for its selectivity. *Science* 290:481–486.

Giguère, P. A., and H. Chen. 1984. Hydrogen bonding in hydrogen peroxide and water. A Raman study of the liquid state. *J. Raman. Spectrosc.* 15:199–204.

Heller, K. B., Lin, E. C., and T. H. Wilson. 1980. Substrate specificity and transport properties of the glycerol facilitator of *Escherichia coli*. *J. Bacteriol.* 144:274–278.

Hub, J. S., Grubmüller, H., and B. L. de Groot. 2009. Dynamics and energetics of permeation through aquaporins. What do we learn from molecular dynamics simulations? *Handb. Exp. Pharmacol.* 190:57–76.

Jahn, T. P., Møller, A. L., Zeuthen, T., Holm, L. M., Klaerke, D. A., Mohsin, B., Kühlbrandt, W., and J. K. Schjoerring. 2004. Aquaporin homologues in plants and mammals transport ammonia. *FEBS Lett.* 574:31–36.

Kirscht, A., Kaptan, S. S., Bienert, G. P., Chaumont, F., Nissen, P., de Groot, B. L., Kjellbom, P., Gourdon, P., and U. Johanson. 2016. Crystal structure of an ammonia-permeable aquaporin. *PLoS Biol.* 14:e1002411.

Liu, K., Kozono, D., Kato, Y., Agre, P., Hazama, A., and M. Yasui. 2005. Conversion of aquaporin 6 from an anion channel to a water-selective channel by a single amino acid substitution. *Proc. Natl. Acad. Sci. U S A* 102:2192–2197.

Ma, J. F., Tamai, K., Yamaji, N., Mitani, N., Konishi, S., Katsuhara, M., Ishiguro, M., Mukhopadhyay, R., and E. Beitz. 2010. Metalloid transport by aquaglyceroporins: consequences in the treatment of human diseases. *Adv. Exp. Med. Biol.* 679:57–69.

Mukhopadhyay, R., Mandal, G., Atluri, V.S.R., Figarella, K., Uzcategui, N. L., Zhou, Y., Beitz, E., Ajees, A. A., and H. Bhattacharjee. 2011. The role of alanine 163 in solute permeability of *Leishmania major* aquaglyceroporin LmAQP1. *Mol. Biochem. Parasitol.* 175:83–90.

Murata, K., Mitsuoka, K., Hirai, T., Walz, T., Agre, P., Heymann, J. B., Engel, A., and Y. Fujiyoshi. 2000. Structural determinants of water permeation through aquaporin-1. *Nature* 407:599–605.

Murata, Y., and M. Yano. 2006. A silicon transporter in rice. *Nature* 440:688–691.

Musa-Aziz, R., Chen, L. M., Pelletier, M. F., and W. F. Boron. 2009. Relative CO_2/NH_3 selectivities of AQP1, AQP4, AQP5, AmtB, and RhAG. *Proc. Natl. Acad. Sci. U S A* 106:5406–5411.

Newby, Z. E., O'Connell, J., Robles-Colmenares, Y., Khademi, S., Miercke, L. J., and R. M. Stroud. 2008. Crystal structure of the aquaglyceroporin PfAQP from the malarial parasite *Plasmodium falciparum*. *Nat. Struct. Mol. Biol.* 15:619–625.

Pavlovic-Djuranovic, S., Schultz, J. E., and E. Beitz. 2003. A single aquaporin gene encodes a water/glycerol/urea facilitator in *Toxoplasma gondii* with similarity to plant tonoplast intrinsic proteins. *FEBS Lett.* 555:500–504.

Preston, G. M., Carroll, T. P., Guggino, W. B., and P. Agre. 1992. Appearance of water channels in *Xenopus* oocytes expressing red cell CHIP28 protein. *Science* 256:385–387.

Reizer, J., Reizer, A., and M. H. Saier Jr. 1993. The MIP family of integral membrane channel proteins: sequence comparisons, evolutionary relationships, reconstructed pathway of evolution, and proposed functional differentiation of the two repeated halves of the proteins. *Crit. Rev. Biochem. Mol. Biol.* 28:235–257.

Rothert, M., Rönfeldt, D., and E. Beitz. 2017. Electrostatic attraction of weak monoacid anions increases probability for protonation and passage through aquaporins. *J. Biol. Chem.* 292:9358–9364.

Soria, L. R., Fanelli, E., Altamura, N., Svelto, M., Marinelli, R. A., and G. Calamita. 2010.

Aquaporin-8-facilitated mitochondrial ammonia transport. *Biochem. Biophys. Res. Commun.* 393:217–221.

Sui, H., Han, B. G., Lee, J. K., Walian, P., and B. K. Jap. 2001. Structural basis of water-specific transport through the AQP1 water channel. *Nature* 414:872–878.

Torshin, I. Y., Weber, I. T., and R. W. Harrison. 2002. Geometric criteria of hydrogen bonds in proteins and identification of "bifurcated" hydrogen bonds. *Protein Eng.* 15:359–363.

Wang, Y., Huang, Y., Wang, J., Cheng, C., Huang, W., Lu, P., Xu, Y. N., Wang, P., Yan, N., and Y. Shi. 2009. Structure of the formate transporter FocA reveals a pentameric aquaporin-like channel. *Nature* 462:467–472.

Wiechert, M., and E. Beitz. 2017. Mechanism of formate-nitrite transporters by dielectric shift of substrate acidity. *EMBO J.* 36:949–958.

Wree, D., Wu, B., Zeuthen, T., and E. Beitz. 2011. Requirement for asparagine in the aquaporin NPA signature motifs for cation exclusion. *FEBS J.* 278:740–748.

Wu, B., and E. Beitz. 2007. Aquaporins with selectivity for unconventional permeants. *Cell. Mol. Life. Sci.* 64:2413–2421.

Wu, B., Steinbronn, C., Alsterfjord, M., Zeuthen, T., and E. Beitz. 2009. Concerted action of two cation filters in the aquaporin water channel. *EMBO J.* 28:2188–2194.

Yang, B., and A. S. Verkman. 1997. Water and glycerol permeabilities of aquaporins 1–5 and MIP determined quantitatively by expression of epitope-tagged constructs in *Xenopus* oocytes. *J. Biol. Chem.* 272:16140–16146.

Yasui, M., Hazama, A., Kwon, T. H., Nielsen, S., Guggino, W. B., and P. Agre. 1999. Rapid gating and anion permeability of an intracellular aquaporin. *Nature* 402:184–187.

CHAPTER FIVE

Use of Genetically Encoded Reporters to Quantify Intracellular H_2O_2 Levels in Eukaryotic Cells

Laura de Cubas, Fernando Antunes, Olga Lyublinskaya, and Elena Hidalgo

ABBREVIATIONS

cpYFP	Circularly permuted yellow fluorescent protein
H_2O_2	Hydrogen peroxide
Prx	Peroxiredoxin
roGFP	Reduction-oxidation-sensitive green fluorescent protein

5.1 H_2O_2 IS LINKED TO TOXICITY BUT ALSO TO PHYSIOLOGICAL SIGNALING

Hydrogen peroxide (H_2O_2), produced among others during respiration at the mitochondria, has been traditionally linked to the toxicity associated to aerobic metabolism. H_2O_2 also participates in the regulation of signaling cascades, related to physiological processes or to the activation of antioxidant cascades. This signaling role is often associated to the reversible oxidation of cysteine residues in sensor proteins, termed thiol switches. Two important families of protein sensors in signaling cascades have been described in prokaryotes and eukaryotes to respond to moderate increases of H_2O_2 to trigger antioxidant gene programs: the transcription factor OxyR in prokaryotes (Zheng et al., 1998) and the glutathione or thioredoxin peroxidases in eukaryotes (Delaunay et al., 2002; Vivancos et al., 2005).

The intracellular levels of H_2O_2 depend on synthesis but also on scavenging activities. Both generation and degradation of peroxides are also compartment-specific: biological membranes have limited permeability to H_2O_2, and compartment-specific scavenging activities further enhance the gradients across membranes. Fluctuations over certain thresholds, occurring either in space and time, can drive adaptation responses. Measuring those fluctuations has been the goal of many laboratories for decades.

5.2 HOW TO MEASURE INTRACELLULAR PEROXIDES: GENETICALLY ENCODED FLUORESCENCE REPORTERS

Permeable fluorescent dyes are still being used to measure redox fluctuations, but they are controversial due to their lack of specificity and low sensitivity toward peroxides (Winterbourn, 2014). Among them, the boronate ester/boronic acid group-based fluorescent probes have been described and improved in the last decade (Bai et al., 2019; Wang et al., 2022). To overcome these limitations, the scientific community has developed protein-based reporters to measure peroxides. The two major families of these genetically encoded redox reporters are the HyPer derivatives (Belousov et al., 2006; Pak et al., 2020) and the reduction-oxidation–sensitive green fluorescent protein (roGFP)–based proteins (Hanson et al., 2004), in which the roGFP moiety is fused to peroxidases (Carmona et al., 2019; Gutscher et al., 2009; Morgan et al., 2016). The dual excitation–single emission behavior of both families of sensors can be used for ratiometric measurements,

DOI: 10.1201/9781003160649-5

dismissing effects caused by intracellular concentrations of the biosensors.

The original HyPer is based on the *Escherichia coli* H_2O_2 sensor OxyR: the circularly permuted yellow fluorescent protein (cpYFP) is flanked by two domains of OxyR; the two cysteine residues in the transcription factor which are oxidized to a disulfide by H_2O_2 are separated by the cpYFP moiety, so that disulfide bond formation between both residues generates a conformational change in cpYFP affecting its fluorescent properties at two excitation peaks. A new family member has been recently reported, HyPer7, which is based on *Neisseria meningitides* OxyR (Pak et al., 2020). This new biosensor is also ratiometric, is not affected by pH, and its sensitivity to peroxides is significantly higher than the original HyPer (de Cubas et al., 2021).

Regarding the other family, roGFP has several substitutions on surface-exposed domains, so that disulfide bonds can be formed in response to oxidants and change the fluorescence properties of the protein in an analogous manner to HyPer, allowing ratiometric readings. The specificity and sensitivity of roGFP to sense H_2O_2 has been improved during the last years by fusing the most sensitive H_2O_2 protein sensors to this fluorescent reporter: the glutathione peroxidase Orp1 (Gutscher et al., 2009), or the peroxiredoxins (Prx) Tsa2 of *Saccharomyces cerevisiae* (Morgan et al., 2016) and Tpx1 from *Schizosaccharomyces pombe* (Carmona et al., 2019). In the case of the Prxs, they lack their resolving cysteines to enhance sensitivity: these mutations eliminate thioredoxin competition and enhance the oxidation transfer from the oxidized peroxidatic cysteine in the Prx to roGFP2. As an example, in the case of roGFP2-Tpx1.C169S, this probe can sense extracellular concentrations 5–10 times less concentrated than roGFP2-Tpx1 (Carmona et al., 2019).

The expression of some of these H_2O_2 biosensors in the same model system has allowed the determination of their relative sensitivity to oxidation by extracellular peroxides (de Cubas et al., 2021). Thus, HyPer, HyPer7, and roGFP2-Tpx1.C169S have been expressed in *S. pombe* cells, and their oxidation upon extracellular bolus of H_2O_2 has been monitored in exponentially growing cultures. The lower concentrations of extracellular peroxides induce statistically significant oxidation of HyPer, HyPer7, or roGFP2-Tpx1.C169S are 100, 5–10, and 1–2 μM, respectively (Figure 5.1).

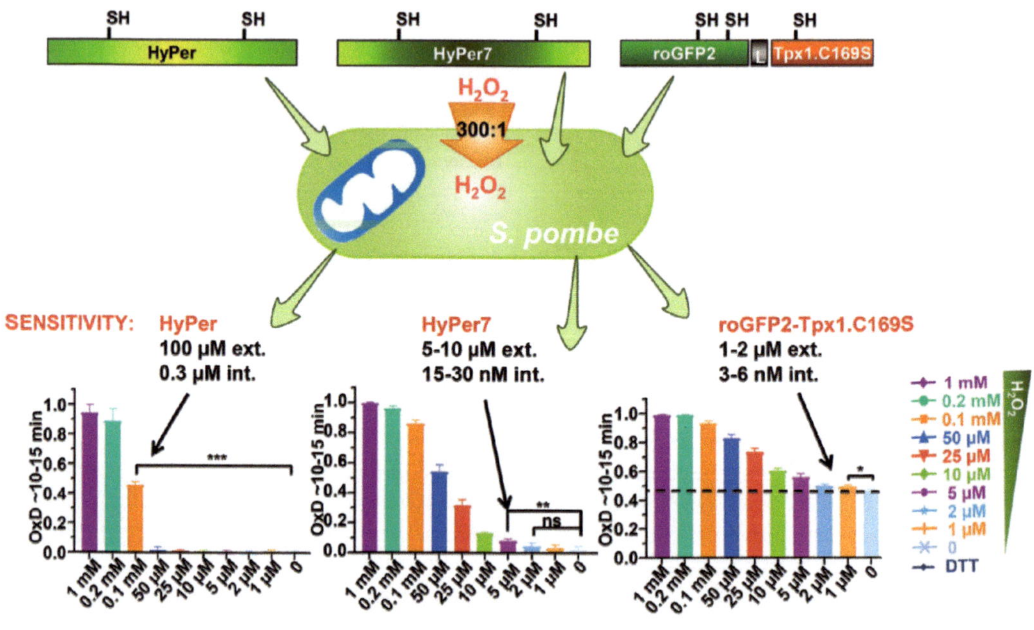

Figure 5.1 Expression of HyPer, HyPer7 and roGFP2-Tpx1.C169S in fission yeast. The minimum concentrations of extracellular (ext.) H_2O_2 capable of causing probe oxidation are indicated with arrows (de Cubas et al., 2021). For the calculation of intracellular (int.) peroxide concentrations, a gradient of 300:1 (extracellular-to-intracellular) has been applied (see text for details).

It is worth noting than the concentrations of H_2O_2 capable of triggering HyPer oxidation are already affecting the growth of fission yeast cultures, suggesting that this biosensor cannot be easily used to monitor pre-toxic signaling events in this model organism and using a fluorescence plate reader. On the contrary, HyPer7 and roGFP2-Tpx1.C169S are exquisitely designed to respond to physiological H_2O_2 fluctuations. Of note: the original HyPer has been successfully used to measure very low concentrations of H_2O_2 in a human cell line, using flow cytometry measurements and fitting the experimental curves with kinetic equations (Lyublinskaya & Antunes, 2019) (see below).

The choice of H_2O_2 biosensor should depend on the sensitivity required and on the compartment which has to be studied. While HyPer7 can be expressed in the mitochondrial matrix and is subject to regulation by oxidants and reducing agents as it is in the cytosol, roGFP2-Tpx1.C169S does not properly respond to oxidants when located at the matrix. Therefore, while roGFP2-Tpx1.C169S displays the highest sensitivity to peroxides, HyPer7 can be used to compare the relative peroxide levels between different sub-cellular compartments.

5.3 RECYCLING OF HYPER- AND ROGFP-DERIVATIVES IN HUMAN CELLS AND FISSION YEAST

Genetically encoded biosensors can be used not only to measure the intracellular peroxides but to characterize the reducing power of a cell. For example, the kinetics of HyPer biosensor signal recovery from an oxidized state offers a direct approach to determine the disulfide reduction activity in living cells (Hernandez et al., 2018). In particular, HyPer recovery kinetic from endoplasmic reticulum in HEK-293 cells was found to be slower compared to the signal recovery recorded in the cytoplasm. Besides that, using the direct pharmacological inhibition of thioredoxin-1 in TIME human endothelial cells, Hernández and colleagues showed that, under the experimental settings used, HyPer recycling depended at least partially on the antioxidant seleno-enzymes and more directly on thioredoxin-1.

Similar to that, HyPer7 was found to be also recycled by the thioredoxin system in fission yeast (de Cubas et al., 2021). On the contrary, the glutathione-glutaredoxin system is responsible for roGFP-based biosensors (Carmona et al., 2019).

5.4 USE OF HYPER DERIVATIVES AND ROGFP2-TPX1.C169S TO CALCULATE H_2O_2 LEVELS IN FISSION YEAST AND K-562 LYMPHOBLAST

Probes of the roGFP-Prx and HyPer families have been used in fission yeast and human lymphoblasts to calculate H_2O_2 steady-state cytosolic levels and peroxide gradients across the plasma membrane. We will briefly summarize the methodologies used (Figure 5.2).

In fission yeast, we first applied mathematical equations to experimental kinetic data regarding oxidation of the H_2O_2 sensor OxyR (Domenech et al., 2018). We concluded that the gradient of extracellular-to-intracellular peroxides through fission yeast membranes is around 300:1, and that 0.25–0.3 μM intracellular H_2O_2 is a toxic threshold capable of halting aerobic cell growth (this occurs in cells lacking Tpx1, the main H_2O_2 scavenger). When the sensors HyPer7 and roGFP2-Tpx1.C169S were expressed in wild-type cells, the former was fully reduced (de Cubas et al., 2021), while roGFP2-Tpx1.C169S was partially oxidized to 50% (Carmona et al., 2019). However, when expressed in Δ*tpx1* cells the probes were oxidized to 60% (de Cubas et al., 2021) and 75% (Carmona et al., 2019), respectively, setting up the values of probe oxidation related to micromolar (toxic) levels of peroxides. Calculations based on the sensitivity of roGFP2-Tpx1.C169S to extracellular peroxides and the 300:1 gradient across membranes suggests that the steady-state levels of H_2O_2 in the cytosol of wild-type cells are lower than or equal to 3–6 nM. Using roGFP2-Tpx1.C169S, we demonstrated that the extracellular-to-intracellular gradients can be decreased by expressing human aquaporin 8 in fission yeast, which facilitates H_2O_2 channeling across membranes, or by depleting the main H_2O_2 scavenger Tpx1 (Carmona et al., 2019).

The basal concentration of H_2O_2 in K-562 as well as gradients formed between external an cytosolic H_2O_2 when exposing these cells to external H_2O_2 were characterized using HyPer as a probe (Lyublinskaya & Antunes, 2019). We made these determinations by two different approaches. In the first approach, we took advantage of a linear relationship between apparent first-order rate constants for HyPer oxidation and the added external H_2O_2 concentration. In the second approach, the basal level of HyPer oxidation (i.e., in the absence of external H_2O_2) was used together with the apparent first-order rate

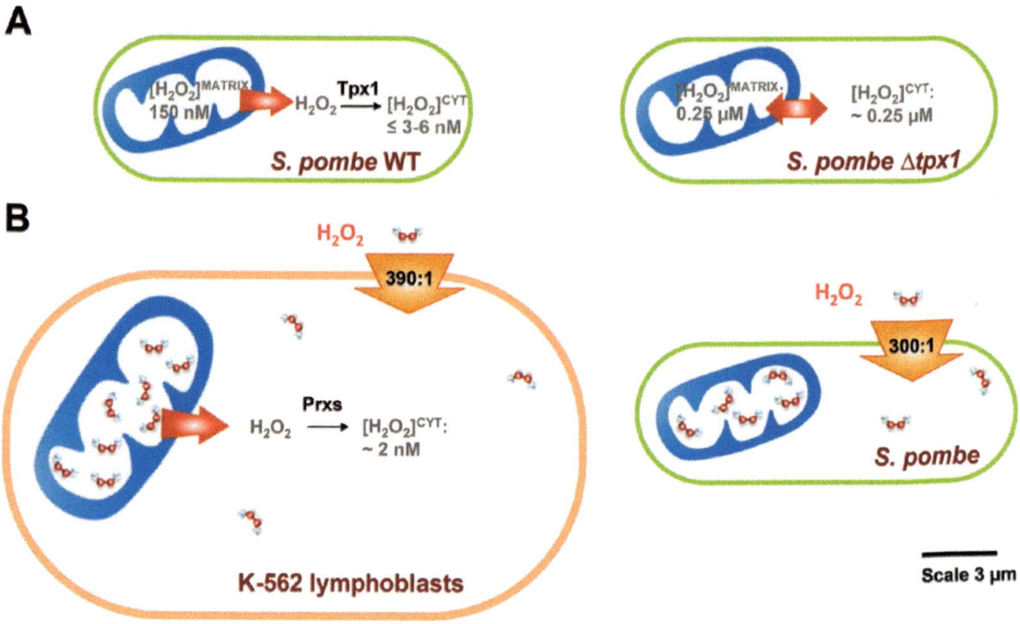

Figure 5.2 Scheme depicting the main findings of the use of the roGFP2-Tpx1.C169S and HyPer/HyPer7 reporters in the quantification of basal H_2O_2 levels and transmembrane peroxide gradients in (A) fission yeast (Carmona et al., 2019; de Cubas et al., 2021) and (B) K-562 lymphoblasts (Lyublinskaya & Antunes, 2019).

constants for HyPer recycling by the intracellular reducing systems. Similar values were obtained by the two approaches, 2 nM for the H_2O_2 basal concentration and 390 for the gradient, supporting the reliability of the methodology. If under our experimental conditions Prxs and peroxidases had been overwhelmed in the presence of added H_2O_2, the value for the basal concentration of H_2O_2 obtained by the first approach would have been much higher, at least one order of magnitude, than the second one.

To conclude, the kinetic modeling is a promising tool to quantify the redox parameters in living cells with the use of genetically encoded reporters.

5.5 CONCLUDING REMARKS

While mathematical approaches have been undertaken to infer the intracellular H_2O_2 concentrations in different organelles, we propose that the use of biosensors such as the ones described here have to be used to calculate steady-state concentrations and gradients across membranes. Thanks to the use of these reporters, we speculate that the steady-state H_2O_2 levels in the cytosol of two different eukaryotic models, fission yeast and human lymphoblast, are in the low nanomolar range, while the concentration in the mitochondrial matrix is around 50 times higher than in the cytosol, at least in fission yeast. We propose that this step gradient is caused by the scavenging activity of antioxidant cytosolic scavengers, mainly Prxs. The intracellular concentrations of peroxides linked to growth arrest, in the order of ~0.25 μM, are in the range of concentrations sensed by the original HyPer. When using this probe, mathematical approaches can help in the characterization of signaling events. The future use of these reporters in other subcellular compartments and in different model systems will be required to generalize all the findings of our work.

5.6 ACKNOWLEDGMENTS

This work is supported by grant PID2021-122837-NB-I00 to E.H., funded by MCIN/AEI/10.13039/501100011033 and by "ERDF A Way of Making Europe," by the European Union. O.L. is grateful to the Ministry of Science and Higher Education of the Russian Federation (grant agreement No. 075-15-2020-773). The Oxidative Stress and Cell Cycle group is also supported by Generalitat de Catalunya (Spain) (2017-SGR-539) and by Excellence Unit "María de Maeztu" Grant CEX2018-000792-M funded by MCIN/AEI/10.13039/501100011033. E.H. is the recipient of an ICREA Academia Award (Generalitat de Catalunya, Spain). L.C. is the recipient of a FPI fellowship from the Ministerio de Economía y Competitividad (Spain).

REFERENCES

Bai X, Ng KK, Hu JJ, Ye S, Yang D (2019) Small-molecule-based fluorescent sensors for selective detection of reactive oxygen species in biological systems. *Annu Rev Biochem* 88: 605–633

Belousov VV, Fradkov AF, Lukyanov KA, Staroverov DB, Shakhbazov KS, Terskikh AV, Lukyanov S (2006) Genetically encoded fluorescent indicator for intracellular hydrogen peroxide. *Nat Methods* 3: 281–286

Carmona M, de Cubas L, Bautista E, Moral-Blanch M, Medrano-Fernandez I, Sitia R, Boronat S, Ayte J, Hidalgo E (2019) Monitoring cytosolic H2O2 fluctuations arising from altered plasma membrane gradients or from mitochondrial activity. *Nat Commun* 10: 4526

de Cubas L, Pak VV, Belousov VV, Ayte J, Hidalgo E (2021) The mitochondria-to-cytosol H2O2 gradient is caused by peroxiredoxin-dependent cytosolic scavenging. *Antioxidants* 10

Delaunay A, Pflieger D, Barrault MB, Vinh J, Toledano MB (2002) A thiol peroxidase is an h(2)o(2) receptor and redox-transducer in gene activation. *Cell* 111: 471–481

Domenech A, Ayte J, Antunes F, Hidalgo E (2018) Using in vivo oxidation status of one- and two-component redox relays to determine H2O2 levels linked to signaling and toxicity. *BMC Biol* 16: 61

Gutscher M, Sobotta MC, Wabnitz GH, Ballikaya S, Meyer AJ, Samstag Y, Dick TP (2009) Proximity-based protein thiol oxidation by H2O2-scavenging peroxidases. *J Biol Chem* 284: 31532–31540

Hanson GT, Aggeler R, Oglesbee D, Cannon M, Capaldi RA, Tsien RY, Remington SJ (2004) Investigating mitochondrial redox potential with redox-sensitive green fluorescent protein indicators. *J Biol Chem* 279: 13044–13053

Hernandez H, Parra A, Tobar N, Molina J, Kallens V, Hidalgo M, Varela D, Martinez J, Porras O (2018) Insights into the HyPer biosensor as molecular tool for monitoring cellular antioxidant capacity. *Redox Biol* 16: 199–208

Lyublinskaya O, Antunes F (2019) Measuring intracellular concentration of hydrogen peroxide with the use of genetically encoded H2O2 biosensor HyPer. *Redox Biol* 24: 101200

Morgan B, Van Laer K, Owusu TN, Ezeriņa D, Pastor-Flores D, Amponsah PS, Tursch A, Dick TP (2016) Real-time monitoring of basal H2O2 levels with peroxiredoxin-based probes. *Nat Chem Biol* 12: 437–443

Pak VV, Ezeriņa D, Lyublinskaya OG, Pedre B, Tyurin-Kuzmin PA, Mishina NM, Thauvin M, Young D, Wahni K, Martinez Gache SA, Demidovich AD, Ermakova YG, Maslova YD, Shokhina AG, Eroglu E, Bilan DS, Bogeski I, Michel T, Vriz S, Messens J, Belousov VV (2020) Ultrasensitive genetically encoded indicator for hydrogen peroxide identifies roles for the oxidant in cell migration and mitochondrial function. *Cell Metab* 31: 642–653 e646

Vivancos AP, Castillo EA, Biteau B, Nicot C, Ayte J, Toledano MB, Hidalgo E (2005) A cysteine-sulfinic acid in peroxiredoxin regulates H2O2-sensing by the antioxidant Pap1 pathway. *Proc Natl Acad Sci U S A* 102: 8875–8880

Wang L, Hou X, Fang H, Yang X (2022) Boronate-based fluorescent probes as a prominent tool for H2O2 sensing and recognition. *Curr Med Chem* 29: 2476–2489

Winterbourn CC (2014) The challenges of using fluorescent probes to detect and quantify specific reactive oxygen species in living cells. *Biochim Biophys Acta* 1840: 730–738

Zheng M, Aslund F, Storz G (1998) Activation of the OxyR transcription factor by reversible disulfide bond formation. *Science* 279: 1718–1721

CHAPTER SIX

H_2O_2/ROS Production, Consumption, and Transport across Organelles

Tasuku Konno, Cécile Crapart, and Edward Avezov

ABBREVIATIONS

8-cpt-cAMP	8-(4-Chlorophenylthio)adenosine 3′,5′-cyclic monophosphate (8-CPT-cAMP), Cell permeable cAMP analog
8-oxo-dG	8-Oxo-2′-deoxyguanosine
AIF	Apoptosis-inducing factor
ALS	Amyotrophic lateral sclerosis
AQP	Aquaporin
ATM	Ataxia-telangiectasia mutated
ATP	Adenosine triphosphate
BCOADHC	Branched-chain 2-oxoacids dehydrogenase complex
BH4	Tetrahydrobiopterin
CaM	Calmodulin
CAT	Catalase
CHO	Chinese hamster ovary
COS	CV-1 in origin with SV40 genes
CXCL12	C-X-C motif chemokine ligand 12
CYP	Cytochrome P450
DHODH	Dihydroorotate dehydrogenase
DNA	Deoxyribonucleic acid
DUOX	Dual oxidase
EGFR	Epidermal growth factor receptor
eNOS	Endothelial nitric oxide synthase
EPHX	Epoxide hydrolase
ER	Endoplasmic reticulum
ERO1	ER oxidoreductin 1
ETC	Electron transport chain
ETF	Electron-transferring flavoprotein
ETF-QO	ETF–ubiquinone oxidoreductase system
FAD	Flavin adenine dinucleotide
FFA	Free fatty acid
FMN	Flavin mononucleotide
GFP	Green fluorescent protein
GPX	Glutathione peroxidase
GRX	Glutaredoxin
GSH	Reduced glutathione
GSSG	Oxidized glutathione
GST	Glutathione S-transferase
HEK293	Human embryonic kidney 293
HSG	Human submandibular gland
HUVEC	Human umbilical vein endothelial cell
IMCD	Inner medulla collecting duct
IP3	Inositol trisphosphate
MAM	Mitochondria-associated membrane
MAO	Monoamine oxidases
MDCK	Madin-Darby canine kidney
mGPDH	Mitochondrial glycerol-3-phosphate dehydrogenase
MMO	Mono-oxygenase
MPTP	1-methyl-4-phenyl-1,2,3,6-tetrahydropyridine
mTORC	Mammalian target of rapamycin complex
NAD^+	Nicotinamide adenine dinucleotide
NADPH	Nicotinamide adenine dinucleotide phosphate
NHSF	Normal human skin fibroblasts
NOX	NADPH oxidases
NPA	Asparagine-proline-alanine
NPC	Asparagine-proline-cysteine

DOI: 10.1201/9781003160649-6

NPR	NADPH-P450 reductase	RNA	Ribonucleic acid
NPT	Asparagine-proline-threonine	roGFP	Redox-sensitive GFP
NSC	Neural stem cell	ROS	Reactive oxygen species
OADHC	2-oxoadipate dehydrogenase complex	SDH	Succinate dehydrogenase
OGDHC	2-oxoglutarate dehydrogenase complex	SERCA	Sarco/endoplasmic reticulum calcium-ATPase
$ONOO^-$	Peroxynitrite	SMO	Spermine oxidase
PASMC	Pulmonary arterial smooth muscle cell	SOD	Superoxide dismutase
PDHC	Pyruvate dehydrogenase complex	TAT	Trans-activator of transcription
PDI	Protein disulfide isomerase	TLR	Toll-like receptor
PDK	Polycystic kidney disease	TRX	Thioredoxin
PEX	Peroxin	TRXP	Thioredoxin peroxidase
PMP20	Peroxisomal membrane protein 20	TSC	Tuberous sclerosis complex
PP1	Protein phosphatase 1	UCP	Uncoupling proteins
PRX	Peroxiredoxin	UPR	Unfolded protein response
PTP	Protein tyrosine phosphatase	UVA	Ultraviolet A
PXP	Peroxiporin	VAP27	Vesicle-associated protein 27
QSOX	Quiescin-sulfhydryl oxidase	VEGF	Vascular endothelial growth factor
		XDH	Xanthine dehydrogenase
		XO	Xanthine oxidase

6.1 SUBCELLULAR H_2O_2 SOURCES AND ITS HANDLING PATHWAYS

6.1.1 ER Sources of ROS/H_2O_2

Cell organelles must maintain an optimal redox state to ensure effective functioning while avoiding toxicity associated with oxidants. The balance between the oxidants, such as reactive oxygen species (ROS), and the reductants/antioxidants is pertinent to the characteristics of each organelle's membrane.

Unlike most organelles, the endoplasmic reticulum (ER) functions in an oxidative redox environment. The ER can probably tolerate high levels of ROS because it does not contain ROS-vulnerable genetic material (DNA/RNA). H_2O_2 is unlikely to directly interact with most protein functional groups and cause protein misfolding as this reaction is not energetically favorable (Winterbourn, 2013). The presence of abundant ROS in the ER can be explained by its role in three ROS-generating processes: oxidative protein folding, fatty acid desaturation, and xenobiotic metabolism. Estimates of the ROS levels in the ER report about 0.7 μM basal H_2O_2 concentration (Bertolotti et al., 2010; C. Gao et al., 2017; Melo et al., 2017; Shimizu & Hendershot, 2009). H_2O_2 from microsomes, a subcellular fraction consisting of sheared and reconstituted ER vesicles, accounts for 45% of its total intracellular content when measured in rat liver homogenate (Boveris et al., 1972; Chance et al., 1979), three times the amount measured in mitochondrial extracts.

6.1.1.1 *H_2O_2 Turnover during Oxidative Protein Folding*

Oxidative protein folding refers to forming intramolecular disulfide bonds to stabilize the structure of newly synthesized proteins. The proteins destined for the cell membrane or secretion (about 30% of the proteome) co-translationally enter the ER, where they undergo enzymatic oxidative folding. The oxidative ER lumen, which supports this process and correlate with high Ca^{2+} levels, provides the extracellular milieu where those proteins will ultimately function (Tu & Weissman, 2004).

The oxidative protein folding results in the large-scale generation of H_2O_2 in the ER and its usage by the protein biogenesis system. H_2O_2 is produced in the disulfide bond formation reaction, catalyzed by protein disulfide isomerase (PDI) chaperones in cooperation with enzymes of the ER oxidoreductin 1 (ERO1) family. The oxidoreductase chaperone enzymes of the PDI family are the core component of the ER oxidative folding machinery. These enzymes catalyze thiol-disulfide exchange reactions through the functional interaction of their thioredoxin domains (redox-active CXXC motifs) with their substrates (Appenzeller-Herzog & Ellgaard, 2008). The enzyme's oxidative potential is restored by ERO1, which transfers two electrons per thiol pair in PDI's active site to molecular oxygen (Gross et al., 2006). Consumption of molecular oxygen as the ultimate electron acceptor in this process is estimated at about 25% of the oxygen available in the cell (Princiotta et al., 2003; Shergalis et al., 2020; Tu & Weissman, 2004). Overall, thiol oxidation by PDI and ERO1 generates one H_2O_2 molecule for every disulfide bond formed, making up about 25% of ROS produced during protein synthesis (Tu & Weissman, 2004). Although ERO1's re-oxidizing activity can improve oxidative folding capacity, ROS hazard can emanate from excessive ERO1 activity, leading to cell death as seen in overexpression or hyperactive mutants (Haynes et al., 2004; Ramming et al., 2015). Stimuli like hypoxia and the unfolded protein response (UPR) can upregulate ERO1 gene expression (Gess et al., 2003; Han et al., 2013; Pagani et al., 2000; Takemori et al., 2006). Peroxide levels could be reduced and life span consequently extended by 32% upon knockdown of the ERO1 gene (Curran & Ruvkun, 2007). Loss of ERO1 also enhanced resistance to ER stress induced by tunicamycin in worms (Marciniak et al., 2004) and ER stress-induced apoptosis in mouse macrophages (G. Li et al., 2009).

An orthogonal pathway to recover PDI oxidative capacity is powered by the product of the ERO1 pathway, H_2O_2. Peroxiredoxin 4 (PRDX4) catalyzes this reaction and sustains the oxidative folding process (Appenzeller-Herzog et al., 2010; Zito, Chin, et al., 2010). In normal conditions, both ERO1 and PRDX4 contribute to the oxidative recycling of PDI. PRDX4 itself recycles the H_2O_2 generated from ERO1 activity to fuel its activity, with overall electrons transferring from cysteine thiols to oxygen and resulting in water production. PRDX4-mediated consumption of H_2O_2 can prevent its accumulation and consequent detrimental effects. Genetic manipulations of PRDX4 demonstrate the significance of its H_2O_2-quenching function in the ER; overexpressing PRDX4 was protective against glutamate-induced neuronal cell death and increased insulin production while inhibiting adipogenesis (Kang et al., 2020; Kim et al., 2020; Mehmeti et al., 2014). PRDX4 appears to be required at the systemic level as its knockout results in abnormal spermatogenesis and susceptibility to drug-induced colitis in mice (Iuchi et al., 2009; Takagi et al., 2019). Finally, the prognosis of patients with lung adenocarcinoma positively correlates with PRDX4 expression level (Mizutani et al., 2019). The absence of the PDI-PRDX4 pathway may explain the lethality of ERO1 deficit in yeasts, through defect of oxidative folding machinery (Frand & Kaiser, 1998; Pollard et al., 1998; Zito, Melo, et al., 2010). The evolutionary acquisition of a pathway in parallel to ERO1 may be rationalized by the need to enhance the productivity of the ER oxidative protein folding in higher organisms with elaborate secretory activities. ERO1s appear dispensable in mammals as only mild phenotypic defects arise in mice lacking all ERO1 isoforms, Ero1a and Ero1b (Tavender et al., 2010; Zito, Chin, et al., 2010). However, a triple knockout mouse model lacking Prdx4, Ero1a, and Ero1b demonstrates scurvy-like phenotypes by impaired collagen maturation (Zito et al., 2012).

The ER harbors two additional H_2O_2-utilizing enzymes, the members of glutathione peroxidase family GPX7 and GPX8, located at the ER lumen and ER membrane facing the lumen, respectively (Nguyen et al., 2011). Similarly to PRDX4, their activity is coupled to PDI oxidation and interaction with ERO1, and they contribute to the catalysis of oxidative protein folding (Mehmeti et al., 2017; Nguyen et al., 2011; Ramming et al., 2015), augmenting disulfide bond formation (Fang et al., 2018; Wei et al., 2012; Wu et al., 2019). Their activity can also protect insulin-secreting cells from saturated fatty acid (FFAs)-induced ER stress (Mehmeti et al., 2017). Sufficient levels of GPX7 seem to be protective against hyperhomocysteinemia-related vascular diseases caused by homocysteine-triggered ER stress (X. Wu et al., 2019). GPX7 upregulation is likely the mechanism underlying metformin benefits for antiaging and diminishing cognitive decline (a drug used to treat diabetes; Fang et al., 2018). On the other hand, the deficiency of GPX7 in mice resulted in systemic oxidative stress, carcinogenesis, and shortened life span (Wei et al., 2012). GPX7 provided protective roles against oxidative damage both to the ER and nuclear DNA, while GPX8 removal resulted in increased cytosolic H_2O_2 levels and potential leakage of H_2O_2 from the

ER (Ramming et al., 2014). This suggests that ROS produced in the ER can potentially affect beyond the ER membrane (Peng et al., 2012).

6.1.1.2 Disulfide Formation-Independent H_2O_2 Production

NOX4 is a membrane-localized enzyme and a part of the NOX4-p22phox complex (Zana et al., 2018). H_2O_2 derived from NOX4 acts as a signaling molecule which can, for example, oxidize the sarco/endoplasmic reticulum calcium-ATPase (SERCA) to induce ER Ca^{2+} efflux and activate Ras signaling (R. F. Wu et al., 2017; R.-F. Wu et al., 2010). This H_2O_2 production is initiated by the binding of NADPH to the NOX4's C-terminal site facing the cytosol (F. Chen et al., 2012). Intramolecular hemes transfer electrons across the membrane, through FAD (a redox cofactor), and molecular oxygen is eventually converted into H_2O_2 at the lumen-facing E-loop (Takac et al., 2011). NOX4 is expressed on several organelle membranes, but its majority is found on the ER membrane in several cell types, including human neuroblastoma SH-SY5Y cells, human hepatocyte HepG2 cells, and human umbilical vein endothelial cells (HUVEC; Auer et al., 2017; Lizák et al., 2020; Montero et al., 2013; R.-F. Wu et al., 2010). Its expression level can depend on one of its interacting factors, calnexin, an ER membrane lectin-chaperone, as reduced NOX4 expression level and ROS production were measured upon calnexin elimination (Prior et al., 2016). The activity of ER-localized NOX4 has protective effects against acute kidney injury *in vivo* through oxidative inhibition of protein phosphatase 1 (PP1), for example (Santos et al., 2016). Notably, H_2O_2 generation and/or turnover from disulfide bond formation pathways are linked to PDI, which influences the activities of NOX1 and NOX4 (Androwiki et al., 2015; Janiszewski et al., 2005).

6.1.1.3 H_2O_2 Reduction by Antioxidants

H_2O_2 levels in the ER lumen appear to be tightly controlled by several systems, as the accumulation of ER H_2O_2 could cause ER stress, activate the UPR, and lead to apoptosis. Reduced glutathione (GSH) in the ER constitutes a robust oxidant buffer, curbing H_2O_2 levels and thus preventing ROS overload in the ER (Melo et al., 2017). This was shown by measurements using a genetically encoded H_2O_2 sensor targeted to the ER. Although GSH-H_2O_2 reactivity kinetics is quite poor, it appears to be compensated by relatively high levels of ER glutathione (approximately 10–15 mM total, mostly reduced; Bass et al., 2004; Montero et al., 2013). The flux of glutathione seems to partially modulate GSH ability to regulate ER redox status, together with Ca^{2+} mobilization (Lizák et al., 2020). Importantly, GSH alone is significantly less efficient at degrading H_2O_2 than PRDX4, GPX7, and GPX8, but the chemical reduction of H_2O_2 by GSH still helps to maintain H_2O_2 levels and becomes kinetically competitive with enzymatic activity above a certain threshold concentration, within the range at which H_2O_2 acts as a second messenger (Kanemura et al., 2020; Melo et al., 2017; Zhu et al., 2014). Consequently, glutathione depletion leads to increased H_2O_2 levels in the ER. However, unlike the cytosol, the ER needs to maintain a much lower ratio of reduced (GSH) to oxidized (GSSG) glutathione to still allow for oxidative protein folding in the intraluminal ER milieu (GSH:GSSG ratio is 3:1 in the ER and 100:1 in the cytosol; Dixon et al., 2008, Zitka et al., 2012). Under reduced capacity of cytosolic antioxidants like in the pancreatic β-cell model, which natively has a catalase deficiency, cell viability is compromised. This was exacerbated by the biosynthesis of disulfide-containing substrates (e.g., proinsulin), which increased H_2O_2 production (Melo et al., 2017). Thus, GSH ability to reduce H_2O_2 is essential to control the production of pathological ROS in the ER.

6.1.1.4 H_2O_2 Production by Lipid Metabolism and Detoxification

Besides its role in protein synthesis, folding and transport, the ER is also involved in lipid metabolism and fatty acid desaturation, where double bonds are introduced between carbon atoms of a fatty acyl chain (Aguilar & de Mendoza, 2006). This process, carried by desaturases, requires one molecule of O_2 to generate unsaturated fatty acids and two molecules of H_2O_2 (Thiede et al., 1986). The electron transfer is ensured by diffusing cytochrome b5, or the cytochrome b5-like domain in some desaturase (Sayanova et al., 2006), which facilitates electron transfer between NADH cytochrome b5 reductase and the desaturase. The oxygenation and breakdown (detoxification) of endogenous and exogenous substrates also generate potentially pathogenic superoxide and H_2O_2 (Albertolle & Peter Guengerich, 2018; Bondy & Naderi, 1994; Rashba-Step & Cederbaum, 1994; Veith & Moorthy, 2018). This occurs through the microsomal mono-oxygenase (MMO) system, comprising multiple enzymes of the cytochrome P450 (CYP) family, NADPH-P450 reductase (NPR; Gillette et al., 1957), and cytochrome b5 (Davydov, 2011). The monooxygenase reaction is a multi-step process, starting with the introduction of a hydroxyl group in a lipophilic substrate by reducing O_2 with a reducing agent (Phase I). The product of this reaction (or the xenobiotic) is then conjugated with an endogenous molecule (Phase II). Crucially, cytochrome b5 can positively impact cytochrome P450 monooxygenase reaction, increasing catalysis and the rate of side product formation, such as H_2O_2 and superoxide (Bienert et al., 2006). This ROS generation process also depends on the type of substrates, oxygen concentration, and CYP isoforms involved, with CYP2E1 inducing elevated production of ROS (Bae et al., 2011; Caro & Cederbaum, 2004).

6.1.2 Non-ER Sources of ROS/H_2O_2

6.1.2.1 Mitochondrial ROS/H_2O_2

Mitochondria is a major source of intracellular ROS, emanating from the respiratory chain cascade or from the action of NAD(P)H oxidases (NOXs). These pathways are estimated to consume about 95% of cellular oxygen for energy production, with 1%–2% of the total oxygen consumption converted to ROS (Boveris et al., 1972), or 0.1%–0.2% according to a different estimation (St-Pierre et al., 2002). The mitochondrial matrix is quite reducing (redox potential of −300 mV; Jones & Go, 2010), and it contains only 15% of total H_2O_2 in cell extracts (Boveris et al., 1972; Chance et al., 1979).

Mitochondria produce energy through oxidative phosphorylation generating adenosine triphosphate (ATP). In this multi-step process, electrons are transferred from nutrition (oxidized) to electron carriers like NAD^+, FMN, and FAD (reduced), then to components of the electron transport chain (ETC, protein complexes inserted in the inner mitochondrial membrane), and finally to oxygen, reducing it to water in the process. The electron potential energy is utilized to create a proton gradient across the mitochondrial inner membrane leading to ATP synthesis when protons return into the matrix down the gradient through the mitochondrial ATP synthase (Haynes et al., 2004). Some electrons can leak from the ETC and other related enzymes, notably electron carriers (Turrens & Boveris, 1980), and partially reduce oxygen, forming superoxide anion ($O_2^{\cdot-}$) as opposed to its complete reduction forming water. Superoxide dismutase 1 (SOD1) and 2 (SOD2) convert it into the less reactive but longer lifetime oxidant H_2O_2 (which can be further turned into hydroxyl radicals by Fenton reaction) and oxygen in the mitochondrial intermembrane space and matrix, respectively (Backes & Herrmann, 2017; Finger & Riemer, 2020).

It is currently believed that 11 ETC-related enzymes generate ROS in mitochondria (Brand, 2016; Goncalves et al., 2015). Among them, complex I (NADH ubiquinone oxidoreductase) was previously considered the major source of mitochondrial ROS/H_2O_2, as well as complex III (Quinlan et al., 2013). H_2O_2 derived from complex I is then released into the mitochondrial matrix, as observed using a genetically encoded H_2O_2 sensor (Pak et al., 2020). However, ROS production during oxidative decarboxylation of 2-oxoacids catalyzed by 2-oxoacid dehydrogenase complexes may be much higher than complex I (Quinlan et al., 2014). The enzymes generating ROS in those complexes are 2-oxoglutarate dehydrogenase complex (OGDHC), pyruvate dehydrogenase complex (PDHC), branched-chain 2-oxoacids dehydrogenase complex (BCOADHC), and 2-oxoadipate dehydrogenase complex (OADHC). Finally, mitochondrial ROS can emanate from ubiquinone-linked dehydrogenases, including dihydroorotate dehydrogenase (DHODH), the electron-transferring flavoprotein (ETF)–ubiquinone oxidoreductase system (ETF-QO), mitochondrial glycerol-3-phosphate dehydrogenase (mGPDH) as observed in rat tissue (Orr et al., 2012), succinate-dependent ROS production by complex II (succinate dehydrogenase [SDH]) as observed in rat skeletal muscle (Quinlan et al., 2013), as well as from the outer ubiquinone-binding site of complex III (Turrens et al., 1985), monoamine oxidases (MAO; Kaludercic et al., 2014), and cytochrome (CYP) catalytic cycle during the metabolism of diverse organic substrates (lipids, steroid hormones, xenobiotics, monoamine neuro-transmitters, and others; Omura, 2006).

The amount of ROS generated by each of these mitochondrial sites is still unclear, although it determines the localization of $O_2^{\cdot-}$. This radical is released into the mitochondrial matrix by the generators of complex I and III, potentially causing damage to mitochondrial DNA, as well as into the intermembrane space (with facilitated access to the cytosol) by complex III and glycerol 3-phosphate dehydrogenase.

NADPH oxidase also generates H_2O_2 (Brand, 2010). Notably, NOX4 is mostly located in mitochondria in some cells, such as mouse podocytes (a renal glomerular epithelial cell), mesangial cells, and kidney cortex (Block et al., 2009; Das et al., 2014). H_2O_2-derived from mitochondrial NOX4 was implicated in aortic stiffening with age and drug resistance (Canugovi et al., 2019; Shanmugasundaram et al., 2017).

Furthermore, H_2O_2 is produced by a FAD-linked sulfhydryl oxidase ERV1 in mitochondria. ERV1 pathways resemble ERO1 as it generates H_2O_2 upon transfer of electrons to molecular oxygen during disulfide bond formation in mitochondrial protein (FAD-linked sulfhydryl oxidase). Importantly, the majority of proteins found in mitochondrial intermembrane space were post-translationally modified to introduce disulfide bonds (Backes & Herrmann, 2017). Cytochrome C scavenges H_2O_2 produced by ERV1, which may be another way to protect mitochondria from H_2O_2 toxicity (Allen et al., 2005; Bihlmaier et al., 2007; Dabir et al., 2007; Farrell & Thorpe, 2005). Uncoupling proteins (UCPs) may also regulate mitochondrial H_2O_2 production by uncoupling mitochondrial proton gradient and ATP synthesis (Dlasková et al., 2010; Jin et al., 2017; Nègre-Salvayre et al., 1997).

Mitochondria contain a wide range of antioxidant systems to prevent and defend against ROS damage: superoxide dismutases (SODs), glutathione peroxidases (GPXs), thioredoxin peroxidases (TRXPs), peroxiredoxins (PRDXs), glutathione (GSH), thioredoxin 2 (TRX2), glutaredoxin 2 (GRX2), cytochrome c oxidase (complex IV), coenzyme Q, ascorbic acid, α-tocopherol, carotene (Beyer, 1990; Ham & Liebler, 1995; Hanukoglu, 2006; Marí et al., 2009; Rabilloud et al., 2001; Rohan Fernando et al., 2006), and potentially catalase (CAT, peroxisomal H_2O_2 detoxification enzyme; Phung et al., 1994; Radi et al., 1991). Genetic manipulations of SOD2 reveal the necessity to balance ROS generation and antioxidant system activity. Indeed, ROS generated in the mitochondria were related to several diseases in human diseases like cancer, neurodegeneration, diabetes, and aging (Kirkinezos & Moraes, 2001; Kudryavtseva et al., 2016; Newsholme et al., 2007; Rimessi et al., 2016; van der Vliet et al., 2018). The damage caused by higher levels of ROS notably involves genomic and mitochondrial DNA (Księżakowska-Łakoma et al., 2014). In mice lacking SOD2 (systemic homozygous knockout), pre-weaning lethality was observed together with multiple tissue dysfunctions such as cardiomyopathy, metabolic acidosis, fatty liver, and neurodegeneration (Lebovitz et al., 1996; Y. Li et al., 1995). Protection against antiretroviral-induced oxidative stress and cardiomyopathy was provided by overexpression of both SOD2 and catalase (Kohler et al., 2009). However, SOD2 overexpression worsens cardiomyopathy caused by the ablation of desmin (intermediate filament protein). This pathology was ameliorated by catalase overexpression (Rapti et al., 2017), either because of increased mitochondrial H_2O_2 or decreased superoxide levels. Mutations in SOD1 characterize 15%–20% of familial amyotrophic lateral sclerosis (ALS; Banci et al., 2008), and potentially oxidation of cysteine residues in mitochondrial SOD1 results in aggregates forming in the motor neurons (Deng et al., 2006). However, a gain of function by misfolded SOD1 is believed to underlie its involvement in ALS.

Though initially, the notion of macromolecule-damaging oxidation by ROS led to their reputation as a hazard, ROS functions in cell signaling later emerged (Hara-Chikuma et al., 2012). H_2O_2, in particular, is an effective second messenger with features complementing other small molecules in that capacity: H_2O_2 can cause allosteric changes in proteins through cysteine residue's thiol oxidation. Disulfide bridges can be formed if one cysteine of the pair is surrounded by positive charges to impose a pKa that make it H_2O_2-reactive, a peroxidic cysteine. Moreover, the H_2O_2 effects are kinetically robust yet can be localized due to its relatively short lifetime and the abundance of peroxiredoxins in the cytosol.

Examples of functional ROS include the hypoxic signaling regulated by mitochondrial ROS through the expressions of hypoxia-inducible genes (Chandel et al., 1998), as well as hypoxic pulmonary vasoconstriction (HPV), which enhances cytosolic calcium levels in pulmonary arterial smooth muscle cells (PASMCs; Waypa et al., 2001, 2006). Mitochondria and ER may exchange ROS through mitochondria-associated membranes (MAMs; Sorrentino et al., 2022), where ERO1 and NOX4 may also localize (Gilady et al., 2010). Mitochondrial peroxide may facilitate ER Ca^{2+} efflux through the ER IP_3 receptor (Booth et al., 2016).

6.1.2.2 Peroxisomal ROS/H_2O_2

Peroxisomes are small, single membrane-enclosed organelles present in most eukaryotes. They carry out a variety of oxidation reactions such as beta-oxidation of fatty acids and alpha-oxidation, biosynthesis of cholesterol, and metabolism of amino acids and glyoxylate, purines, polyamines, and reactive oxygen and nitrogen species (including H_2O_2; de Duve & Baudhuin, 1966), transition metal ions, and others (del Río & López-Huertas, 2016; I. Singh, 1997). Many types of ROS are generated during the normal catalytic activity of several peroxisomal enzymes as by-products (Antonenkov et al., 2010), including H_2O_2, superoxides, hydroxyl radicals, as well as reactive nitrogen species (di Meo et al., 2016). Peroxisome-oxidases are considered to be a significant source of intracellular ROS and H_2O_2 (Boveris et al., 1972; Fritz et al., 2007). These are responsible for about 35% of total intracellular H_2O_2 (de Duve & Baudhuin, 1966; Schrader & Fahimi, 2006; Tripathi & Walker, 2016). Peroxisomes generate multiple types of ROS through D-amino oxidase and acyl-CoA oxidase (Angermüller et al., 2009). Both xanthine oxidase and urate oxidase can produce superoxide and H_2O_2 in the matrix (Angermüller et al., 1987). Nitric oxide (relatively unreactive) can also be produced by xanthine oxidase, which reduces nitrates and nitrites (Harrison, 2002) or L-arginine by nitric oxide synthase (del Río, 2011). The oxidation of spermidine and spermine by acetylpolyamine oxidase can generate H_2O_2 as a by-product (Hölttä, 1977) besides spermidine or putrescine. The peroxisomal membrane also contributes to superoxide production through the electron transport chain (Sandalio et al., 2013). Superoxide and nitric oxide species can react to form highly reactive species, notably the powerful oxidant peroxynitrite ($ONOO^-$), which is able to damage many biological molecules (Radi, 2013), and H_2O_2 can give rise to hydroxyl radicals via the Fenton reaction (Valko et al., 2005).

Although ROS production is abundant in peroxisomes, the steady-state peroxisomal milieu is maintained reductive, as measured by a peroxisome-targeted H_2O_2 sensor (Mehmeti et al., 2012). Some peroxisome-resident enzymes act as antioxidants, such as catalase, SOD1, SOD2, peroxiredoxins (PRDX1, PRDX5), epoxide hydrolase (EPHX2), glutathione S-transferases (GST), a peroxisomal membrane protein 20 (PMP20), and possibly a GPX as well as some non-enzymatic low-molecular-weight antioxidants (A. K. Singh et al., 1994). The antioxidant systems defend the cell from oxidative damage by controlling levels of ROS in peroxisomes. When catalase is less active, ROS production and oxidative stress (DNA damage) are increased, which favors cancer development (Bostwick et al., 2000; Cho et al., 2014; Frederiks et al., 2010; Ho et al., 2001; Lauer et al., 1999; Vuillaume et al., 1991). A subtype of GPX, Gpx1, was found in the yeast peroxisome (Ohdate & Inoue, 2012) but not yet in mammals. Catalase shifts from peroxisome to cytoplasm if both PEX5 (peroxisomal biogenesis factor) and PEX12 are deficient, which allows for the removal of cytosolic ROS preferentially and protects the cell from oxidative damage (Dubreuil et al., 2020). Thus, peroxisomes greatly limit the oxidative stress that could arise from the large production of ROS.

Peroxisomal ROS seem to cause critical oxidative stress only if multiple ROS quenching enzymes fail collectively. Despite the abundant catalase, H_2O_2 produced in the peroxisomes (from urate oxidase, for example) can readily cross the peroxisomal membrane, diffuse to the surrounding medium (20%–60%), and reach the cytosol through tubules in the crystalloid core, a denser region inside peroxisomes as apparent by electron-microscopy (Fritz et al., 2007). Peroxides can diffuse through the peroxisomal channel Pxmp2 (Rokka et al., 2009).

Peroxisomal ROS play a role in intracellular signaling, for autophagy, for example. ROS can induce autophagy by suppressing mTORC1 signaling through its negative regulators, tuberous sclerosis complex 1 and 2 (TSC1/2), on peroxisomes (J. Zhang et al., 2013). Peroxisomal ROS also induce autophagy and pexophagy (selective autophagy of peroxisomes) by activating a serine/threonine kinase responsive to DNA damage (a gene mutated ataxia-telangiectasia mutated [ATM]; J. Zhang et al., 2015; Lee et al., 2018). This may serve as a control mechanism for peroxisomal activity (Kumar et al., 2014).

6.1.2.3 *Golgi ROS/H_2O_2*

Similar to the ER, its related secretory pathway compartment, the Golgi apparatus, maintains an oxidative environment. This suggests that ROS may be generated in this compartment while it deals with secretory protein's glycosylation and sorting. However, very few studies have explored redox reactions, ROS production, and oxidative stress in this organelle (Samoylenko et al., 2013).

In the Golgi, ROS are likely produced by flavin-linked sulfhydryl oxidases, the quiescin sulfhydryl oxidases (QSOXs) playing a role in "late" disulfide bond formation for products that require tail-to-tail disulfide bond formation (Chiu & Hogg, 2019; Wagner, 1990). This process similarly generates H_2O_2 as PDI and ERO1 in the ER (Raje & Thorpe, 2003). According to immunohistochemical analysis, the Golgi contains significant levels of these oxidases (Chakravarthi et al., 2007; Ilani et al., 2013; Mairet-Coello et al., 2004; Tury et al., 2004). The transport of all QSOX isoforms from the ER to the Golgi likely occurs by default since it contains an ER signal peptide and no ER retention signals. The N-terminus of QSOX isoforms includes two thioredoxin domains with CXXC motifs and a FAD-binding domain, which enables thiol groups oxidation using oxygen as a reductant. In general, QSOXs, notably QSOX1, and Golgi ROS have been related to some (patho) physiological processes, including in the pathogenesis of preeclampsia, a pregnancy-related disorder including hypertension (J. Li et al., 2019), oxidative stress-induced cell death, cellular proliferation (Morel et al., 2007), and breast, pancreas, and prostate cancers (Katchman et al., 2013; Lake & Faigel, 2014). These links seem to depend on the level of expression, tissues, and cell types (QSOX function and the mechanisms of its actions remain unclear; Pernodet et al., 2012). Importantly, QSOX oxidative folding can compensate for incorrect folding in proteins that reach the Golgi (Chakravarthi et al., 2007), eluding the ER retention by its quality control system. The disulfide bond formation capacity was restored in Ero1p deficient yeast by exogenous expression of QSOX1 and, consequently, their viability. QSOX1 also acts extracellularly for laminin incorporation (Ilani et al., 2013). The molecular design of QSOX resembles a chimeric protein of ERO1-PDI thus may benefit disulfide-forming effectiveness outside the cells where diffusional encounters are unlikely. This further supports its intended functioning in the extracellular space.

Besides QSOX, redox is regulated by other redox enzymes in the Golgi. This includes enzymes of the glutaredoxin family which can detoxify oxidative damage, as well as GSH, GPXs, and NADPH that impair H_2O_2 production (Mailloux et al., 2016). Glutaredoxin 6 and 7 (Grx6 and Grx7) upregulation potentially defend Golgi from cellular stressors such as sodium, peroxides, or Ca^{2+}. However, they cannot prevent damage caused by heat shock and osmotic stress (Izquierdo et al., 2008). In yeasts, Grx6 and Grx7 are primarily localized in the cis-Golgi (Mesecke et al., 2008). Similar, yet to be identified, enzymes may exist in mammals to regulate the level of H_2O_2 in Golgi.

Further, one isoform of the NOS enzyme family, which catalyze the production of NO from L-arginine and consequently reactive nitrogen species, also generates H_2O_2 in the Golgi. This endothelial NOS (eNOS) preferentially locates to the Golgi membrane and plasma membrane caveolae (bulb-shaped invaginations of the plasma membrane) in endothelial cells (García-Cardeña et al., 1996; Sangwung et al., 2012; Sessa et al., 1995). NOS proteins form dimers and rely on several cofactors, including nicotinamide adenine dinucleotide phosphate (NADPH), flavin adenine dinucleotide (FAD), flavin mononucleotide (FMN), calmodulin (CaM), heme, and tetrahydrobiopterin (BH4). In conditions of L-arginine and/or BH4 deficit, eNOS uncoupling can give rise to superoxide, which after spontaneous or enzymatic (SOD) dismutation results in H_2O_2 (and molecular oxygen; Druhan et al., 2008; Sindler et al., 2009). Little is known about the role of Golgi-derived H_2O_2, but increased levels of H_2O_2 were measured in the Golgi of hypertensive mice using a novel Golgi-targeted H_2O_2 specific probe termed Np-Golgi (Wang et al., 2019).

6.1.2.4 *Nuclear ROS/H_2O_2*

The nucleus harbors the cell's "source code" stored in DNA and its integrity is vital to preserve genetic information. Its sensitivity to oxidation by ROS must be mitigated. Therefore, the nuclear milieu is the most reducing among cellular organelles (Malinouski et al., 2011). This state is maintained by reductants that can scavenge ROS in the nucleus and upstream in other compartments. These include thiol reductants such as thioredoxin-1 (TRX1), GPXs, PRDXs, GSH, and specific isoforms of glutathione S-transferases.

TRX1 preferentially localizes in the nucleus in the presence of excess ROS to alleviate oxidative stress but can also be present in the cytosol at steady state (Go & Jones, 2008). It has been proposed to facilitate the delivery of apoptosis-inducing factor (AIF) to the nucleus upon disulfide bond formation within it. AIF is released in the nucleus by reduction of the TRX1–AIF complex, where it can promote apoptosis.

GSH can directly clear nuclear ROS and monitor nucleosome stability by maintaining key cysteine residues of nuclear proteins, such as histone H3, in a reduced state (Go & Jones, 2008). By impeding the irreversible oxidation of these residues through glutathionylation (García-Giménez et al., 2013), it preserves protein's DNA binding capacities (e.g., transcription factors). GSH also protects the nucleus from oxidative stress, notably when intracellular ROS level rises during mitosis (Patterson et al., 2019). GSH is mainly concentrated in the nucleus during cell proliferation, while it is distributed over the whole cell in the resting state (Pallardó et al., 2009; Söderdahl et al., 2003). Depletion of GSH impairs cell proliferation (Markovic et al., 2009). Further, histone H3 is more glutathionylated in highly proliferating cells (e.g., cancer cells) and less in aged cells, demonstrating its role in pathophysiology.

DNA damage is partly caused by nuclear ROS originating from NOX4 located on the nuclear envelope. This is potentially relevant for neurodegeneration and other diseases like myelodysplastic syndromes and viral infection (de Mochel et al., 2010; Guida et al., 2014). Evidence of NOX4 involvement in senescence is provided by the reduction of DNA damage and expended cellular life span upon NOX4 knockdown in human endothelial cells (Lener et al., 2009). However, the regulation of nuclear ROS produced by NOX4 may be programmed in cellular signaling because NOX4 activity can be downregulated by a tyrosine kinase FYN belonging to the Src family, which is also related to cardiac remodeling (Matsushima et al., 2016). 8-Oxo-2′-deoxyguanosine (8-oxo-dG) is a product of DNA oxidation that has commonly been used as a marker of oxidative damage, but it likely has significant roles. 8-Oxo-dG is related to physical dysfunctions such as age-associated neuronal loss and susceptibility to a dopaminergic toxin 1-methyl-4-phenyl-1,2,3,6-tetrahydropyridine (MPTP; Cardozo-Pelaez et al., 2012; Minowa et al., 2000).

6.1.2.5 *Lysosomal ROS/H_2O_2*

Recycling endosomes, late endosomes, and lysosomes were shown to be oxidizing compartments by experiments using a redox-sensitive GFP (roGFP) fused to endocytic proteins (Austin et al., 2005). An acidic pH aids the activity of lysosomal hydrolases to degrade macromolecules. The proton concentration in the lysosomes is fueled by an electron transport system comprising flavins (FAD), cytochrome b, and ubiquinone, which are present at an unusually high level in this compartment (Gille & Nohl, 2000). NADH serves as a donor and O_2 as an acceptor. H_2O_2 arises from this redox chain by spontaneous dismutation of intermediate superoxide and diffusion from the cytoplasm (Nohl & Gille, 2002). However, most H_2O_2 is effectively degraded in the cytosol by catalase and glutathione peroxidase (Walton & Lewis, 1916). As lysosomes are devoid of those enzymes, H_2O_2 reacts with redox-active iron released from iron-containing macromolecules degraded locally and forms hydroxyl radical or other reactive compounds via the Fenton reaction (Eaton & Qian, 2002). Under cell oxidative stress, which antioxidant defenses cannot control, the abnormal diffusion of H_2O_2 to lysosomes and the ROS production destabilize lysosomes and can result in lysosomal rupture and release of its lytic enzymes (Kurz et al., 2008).

6.1.2.6 *Cytosolic/Plasma Membrane ROS/H_2O_2*

Any soluble cell components in the cytosol capable of undergoing redox reactions can contribute to intracellular ROS production, including thiols, hydroquinones, catecholamines, and flavins (Freeman & Crapo, 1982). Oxidation in the cytosol is often used as a signal transduction process that can initiate the translocation and action of proteins in other compartments, but also cell death in extreme cases (Cumming et al., 2004).

ROS are also generated as a by-product of the catalytic activity of some cytosolic enzymes, such as xanthine oxidase (XO). In the oxidase form, this enzyme can produce superoxide when oxidizing xanthine or hypoxanthine (transfer an electron to molecular oxygen; McCord et al., 1985), which notably happens in damaged tissues. Normally, the dehydrogenase form of the enzyme (XDH) usually catalyzes the oxidation of hypoxanthine to xanthine and then to uric acid using $NADP^+$. Other enzymes include a type of spermine oxidase (SMO), a FAD-dependent oxidase that catalyzes spermine (polyamine involved in cellular metabolism) and generates H_2O_2 as one of its by-products. SMO is located in the cytoplasm or nucleus, unlike acetylpolyamine oxidase, found in peroxisomes (Murray-Stewart et al., 2008).

Furthermore, some stimuli in the plasma membrane can lead to the oxidation of redox couples in the cytoplasm specifically. For example, Trx1 is oxidized in the cytosol upon epidermal growth factor receptor (EGFR) activation and simultaneous ROS production, while Trx2 in the nucleus and mitochondria and GSH were not affected (Halvey et al., 2005). Most NADPH oxidases are dormant in resting phagocytes, and the enzyme components are located in the cytosol and plasma membrane. Upon bacterial infection, the cytosolic components translocate to the membrane of the phagosome to associate with the catalytic subunit (cytochrome b558) and generate active NADPH oxidase (Chanock et al., 1994). ROS are also generated from the plasma membrane of nonphagocytic cells due to the presence of five NADPH oxidase (NOX) family isoforms (NOX1 NOX2, NOX3, NOX4, and NOX5) and two related dual oxidases (DUOX1 and DUOX2; Landry & Cotter, 2014). Finally, Toll-like receptors (TLRs) at the plasma membrane can interact with NOX and increase ROS generation when activating signaling pathway (transcription factor NF-κB, cytokines and chemokines) in response to recognition of conserved structures of bacteria, viruses, or fungi (Joo et al., 2012; Park et al., 2004; Yang et al., 2009).

6.2 PEROXIPORINS AND ER H_2O_2 FLUX

6.2.1 Aquaporins

Organelle membranes impose spatial and biochemical compartmentalization, effectively insulating environment-dependent processes. Therefore, the flux of substances across organelle membranes is limited. Uptake and discharge of substances are strictly controlled via various transporters and channels in the membranes. The channels obey the regulations and energy cost balance or concentration gradients (in active/passive transport, respectively). This can apply to ROS, including H_2O_2. The permanent dipole moment of H_2O_2 is similar to that of water molecules (2.26×10^{-18}, and 1.85×10^{-18} electrostatic unit, respectively). Thus, like water, H_2O_2 permeation through hydrophobic lipid bilayers is extremely low (Bienert et al., 2006). In addition, quenching by antioxidant enzymes and antioxidants, such as glutathione and ascorbic acid, promotes H_2O_2 confinement within each organelle as well as relieves oxidative stress. Several members of aquaporins (AQPs, originally discovered as a water channel) transport H_2O_2 across the plasma/organelle membranes. These findings highlight the biological significance of a regulated H_2O_2 flux.

Members of the AQP family are six-transmembrane passive channel proteins. Thirteen isozymes (AQP0–12) exist in mammals. Transport of substrates relies on the hourglass-shaped domain in the active center, which consists of asparagine-proline-alanine (NPA) in most isozymes, asparagine-proline-cysteine (NPC) in AQP11, and asparagine-proline-threonine (NPT) in AQP12. AQP genes are evolutionarily conserved between prokaryotes and eukaryotes (Lee et al., 2005; Tong et al., 2019). The development of these genes is also closely related to evolutional events such as the terrestrial adaptation of tetrapods (Finn et al., 2014). The preference of AQP for water is not stringent. They can transport glycerol, urea, and H_2O_2. AQPs are classified into several groups according to their substrates: classical AQPs (AQP0, AQP1, AQP2, AQP4, AQP5, AQP6, AQP8), aquaglyceroporins (AQP3, AQP7, AQP9, AQP10, AQP11), unorthodox/superaquaporins (AQP11, AQP12), and peroxiporins (PXPs; AQP0, AQP1, AQP3, AQP5, AQP8, AQP9, AQP11; see other chapters in the book). Interestingly, AQP isozymes differ in both tissue specificity and subcellular distribution (Table 6.1). The specificity of subcellular localization is particularly exemplified in polarized cells: based on AQP's distribution, one can distinguish the apical/basolateral side of the plasma membrane (Table 6.1). In addition, some isozymes dynamically translocate between intracellular locations and the plasma membrane in response to various stimuli, such as osmotic pressure and small molecules like cAMP and vasopressin, while accompanying post-translational modifications (phosphorylation, ubiquitylation; Conner et al., 2012; Fushimi et al., 1997; Golestaneh et al., 2008; Ishikawa et al., 2002; Jiang et al., 2021; Kamsteeg et al., 2006; Nielsen et al., 1995; Van Balkom et al., 2002; Zhang et al., 2007). The specific distribution of AQPs may play a central role in maintaining cell homeostasis as an adaptive mechanism to a rapidly changing environment inside and outside the cell.

6.2.2 Peroxiporins

H_2O_2 permeability of AQP (PXP) had been demonstrated by measuring a suppressed H_2O_2 uptake upon the blockage of plant AQP with mercury (Henzler & Steudle, 2000). Among mammal isozymes, human AQP8 was the first to be shown to carry out a similar transport together with other orthologs in *Arabidopsis* and yeast (Bienert et al., 2007). The AQP's H_2O_2 conductance has been shown for other isozymes such as AQP3, AQP9, and AQP11 (and suggested for AQP0, AQP1 and AQP5; Bertolotti et al., 2016; Bestetti et al., 2020; Miller et al., 2010; Rodrigues et al., 2019b; Varadaraj & Kumari, 2020; Watanabe et al., 2016). Due to its electric and structural similarities with the water molecule, H_2O_2 is expected to be transported by all isozymes but to a varying extent (Almasalmeh et al., 2014).

Crucially, the transport regulatory mechanism and the accessibility of H_2O_2 have been determined to depend on PXP's redox state tuned by persulfidation of the cysteine residue (Bestetti et al., 2018). Considering the role of H_2O_2 as a second messenger, its transporting channel PXPs is expected to modulate cellular signal transductions as well as the resulting cellular activities. This assumption is demonstrated by the report showing that AQP3 amplifies the AKT signal triggered by exogenous H_2O_2 in human colon cancer cells (Miller et al., 2010). AQP3 also mediates the uptake of NOX-derived H_2O_2, which is produced in response to chemokine (CXCL12) stimulus and regulates intracellular signaling to drive breast cancer migration (Satooka & Hara-Chikuma, 2016). Moreover, AQP8 is responsible for a similar process; it mediates the intracellular uptake of H_2O_2 stimulated by growth factors, such as epidermal growth factor (EGF) and vascular endothelial growth factor (VEGF) and promotes cell proliferation (Bertolotti et al., 2013; Vieceli Dalla Sega et al., 2017). The H_2O_2 production processes coupled to ligand-receptor interactions can suspend the action of negative signal regulators by oxidizing cysteine residues in the active center. For example, protein tyrosine phosphatases (PTPs), like PTEN and PTP1B, are differentially oxidized by H_2O_2 in physiological conditions (Groen et al., 2005). The reversible oxidation state of sulfenic acid (of which further oxidation is protected by sulfenyl amide) enables switching on/off its enzymatic activity in response to intracellular redox state changes (Salmeen et al., 2003; Van Montfort et al., 2003). Therefore, H_2O_2 delivered via PXPs is likely a determinant in transducing phosphorylation signaling.

On the other hand, PXPs also play a key role in suppressing the oxidative stress imposed on organelles and cells by H_2O_2. This is evident from H_2O_2 accumulation induced by simple AQP5 knockout in the lens (Varadaraj & Kumari, 2020). Further, silencing AQP8 expression results in augmented mitochondrial H_2O_2 in HepG2 and compromises cell viability following mitochondrial depolarization (of note, it is rescued by the mitochondria-targeted antioxidant MitoTempol; Marchissio et al., 2012). In a rat insulin-producing β-cell line, RINm5F, the contribution of AQP8 is also specified in mitochondria rather than other compartments such as peroxisomes and the ER, and this isozyme mediates proinflammatory cytokine-induced cytotoxicity (Krüger et al., 2021a, 2021b). The evidence that AQP8 knockdown triggers necrosis, but not apoptosis, suggests that incomplete discharge of mitochondrial H_2O_2 dysregulates cell homeostasis (Marchissio et al., 2014). In addition, AQP3 seems to protect cells from oxidative damage induced by ultraviolet A (UVA) irradiation in normal human skin fibroblasts (NHSFs; Xie et al., 2013). This is supported by the age- and sun exposure–related AQP3 downregulation in human skin, and it may partially explain the contribution of PXPs in the free-radical theory of aging (Seleit et al., 2017).

6.2.3 ER-Localizing PXPS

The physiological significance of subcellular PXPs is emphasized in regions where H_2O_2 traffic is important, such as the ER. As shown in Table 6.1, four AQP isozymes have been confirmed in the ER membrane so far (AQP4, AQP9, AQP11, and AQP12;

TABLE 6.1

Subcellular Localization of Aquaporin Isozymes in Mammal Cells and Tissues

AQP Isozyme	Localizations	Cells/Tissues	References
AQP0	plasma membrane	MDCK	(Varadaraj et al. 2008)
	plasma membrane	bovine cortical lens fiber	(Rose et al., 2006)
	microsome	rat hepatocyte	(Huebert et al. 2002)
	canalicular plasma membrane	rat hepatocyte	(Huebert et al. 2002)
	plasma membrane	human and mouse cardiac myocytes	(Montiel et al. 2020)
AQP1	plasma membrane	HT29 and SW480	(De Ieso et al. 2019)
	intracellular and plasma membrane translocation	HEK293	(Conner et al. 2012)
	cytoplasm and plasma membrane	human and mouse kidney	(Gao et al. 2012)
AQP2	plasma membrane and intracellular	rat kidney	(Christensen et al. 1998, 2000; Olesen et al. 2009)
	plasma membrane and early endosome	MDCK, CD8 cell, IMCD	(Tajika et al. 2004, 2005; Vossenkämper et al. 2007)
AQP3	basolateral membrane	rat kidney cortex	(Ishibashi et al. 1994)
	membrane	mouse kidney cortex and human erythrocyte	(Yang et al. 2001)
	basolateral membrane	mouse collecting duct epithelial	(Ma et al. 2000)
	cytoplasm to plasma membrane (by E-cadherin overexpression or Ral-A knockdown)	PC-3	(Chen et al. 2018)
	cytoplasm and plasma membrane	LNCaP	(de Almeida et al. 2021)
	cytoplasm	Du145 and PC-3	(de Almeida et al. 2021)
	basolateral plasma membrane	human ileum	(Mobasheri et al. 2004)
AQP4	ER and Golgi	NSC-derived differentiated progeny	(Cavazzin et al. 2006)
	basolateral plasma membrane	mouse kidney	(Van Hoek et al. 2000)
	plasma membrane and intracellular (hypoxia-induced calcium-, calmodulin-, and kinase PKA-dependent manner)	primary cortical rat astrocytes and HEK293	(Kitchen et al. 2015)
	plasma membrane	human primary cortical astrocytes	(Salman et al. 2017)
	plasma membrane and intracellular vesicle	rat primary cortical astrocytes	(Kitchen et al. 2020)
	ER	Hera and rat astrocyte (AQP4 delta4 splicing variant, delta4 can retain normal AQP4 on the ER when co-expressed)	(Jorgačevski et al. 2020)
	basolateral plasma membrane	mouse ileum	(Zhang et al. 2018)
AQP5	apical membrane	columnar cells of the superficial epithelium and submucosal gland acinar cells	(Kreda et al. 2001)

(Continued)

TABLE 6.1
(Continued)

AQP Isozyme	Localizations	Cells/Tissues	References
	plasma membrane	MDCK and mouse lens	(Sindhu Kumari et al. 2012)
	plasma membrane and intracellular	mouse fiber cell	(Sindhu Kumari et al. 2012)
	intracellular to plasma membrane (8-cpt-cAMP dependent manner)	MLE-12 and Beas-2b	(Sidhaye et al. 2005; Yang et al. 2003)
	apical plasma membrane	HSG	(Hasegawa et al. 2011)
AQP6	intracellular membrane	rat renal epithelia	(Yasui et al. 1999)
	cytoplasm	rat kidney collecting duct	(Ohshiro et al. 2001)
	intracellular	mouse inner ear	(Chai et al. 2010)
	intracellular vesicles	intercalated cells of rat renal collecting duct	(Yasui et al. 1999)
AQP7	plasma membrane and intracellular membrane (in response to catecholamine)	mouse adipose tissue	(Miyauchi et al. 2015)
	apical plasma membrane	rat gastrointestinal epithelium	(Laforenza et al. 2005)
	plasma membrane and slightly cytoplasm	human testis	(Yeung et al. 2010)
	releasate (dense granules)	human platelet and CHRF-288–11	(Goubau et al. 2013)
	plasma membrane	human adipocyte and endothelial cell	(Laforenza et al. 2013)
AQP8	mitochondria	rat liver, 3T3	(Calamita et al. 2005; Ikaga et al. 2015)
	canalicular plasma membrane	rat hepatocyte	(Huebert et al. 2002)
	intracellular microsomal membrane and plasma membrane (cyclic AMP dependent manner)	rat hepatocyte	(García et al., 2001)
AQP9	basolateral plasma membrane	WIF-B	(Gradilone et al., 2005)
	ER	NSC-derived differentiated progeny	(Cavazzin et al., 2006)
	plasma membrane	rat hepatocyte	(Huebert et al., 2002)
AQP10	plasma membrane	human adipocyte	(Gotfryd et al., 2018; Laforenza et al., 2013)
	apical plasma membrane	human ileum	(Mobasheri et al., 2004)
AQP11	ER	CHO-K1, mouse kidney, Hera	(Bestetti et al., 2020; Ikeda et al., 2011; Inoue et al., 2014; Morishita et al., 2005)
AQP12	intracellular (overexpressed)	HEK293 and COS7	(Itoh et al., 2005)
	rough ER	rat pancreas	(Ohta et al., 2009)

see Table 6.1), and two (AQP9 and AQP11) of them have shown H_2O_2 permeability as a PXP. It is worth noting that PXPs may function during their short residence in the early secretory pathway's compartments, where these proteins are manufactured and matured before export to their destination with the notable exception of mitochondrial PXPs (Bestetti et al., 2020). AQP11 primarily localizes on the ER, unlike the dispersed AQP9. The expression level of AQP11 is particularly abundant in testis, and it is involved in spermatogenesis together with other AQPs (Laforenza et al., 2016). AQP11 mutation lacking the ability to form stable oligomers causes susceptibility to oxidative stress (Atochina-Vasserman et al., 2013), and AQP11-deficient mice exhibit phenotypes similar to polycystic kidney disease (PDK) with lethality before weaning age (Morishita et al., 2005; Okada et al., 2008). This indicates that AQP11 is involved in kidney development from earlier stages. It has also been reported that genetic variants of AQP11 contribute to kidney disease in type 2 diabetes patients (Choma et al., 2016; Han et al., 2019). Importantly, impairments resulting from AQP11 are not compensated by other AQP isozymes expressed in the kidney (Parreira et al., 2009). This highlights the importance of ER H_2O_2 flux via AQP11. In addition, the oxidative stress associated with AQP11 deficiency also appears to be mediated by indirect pathways, such as an increase in ROS due to *Nox2* upregulation (Hoshino et al. 2019). Given that the ER is an organelle that constantly produces and consumes H_2O_2, ER-localized PXPs likely play a central role in sensing/governing ER redox gradients with an impact on the entire cell.

6.2.4 H_2O_2 Influx via ER PXPS

According to a recent study, AQP11 modulates both the influx and efflux of H_2O_2 across the ER membrane (Bestetti et al., 2020). Although AQPs are known as passive channels, the bidirectional movement of H_2O_2 across biomembranes (with an unavoidable energy expense) is predicted by the metadynamics simulation using AQP3 (Wragg et al., 2020). Also, among the multiple localizations of AQP8 in the cell, the knockdown of AQP8 diminishes the entry of exogenous H_2O_2 into the ER in plasma membrane-permeabilized cells (Bertolotti et al., 2013). As mentioned above, the physiological extent of H_2O_2 flowing into the ER remains unclear, but it most likely serves oxidative folding. Since a previous study suggested that the cells lacking ERO1, a major H_2O_2 producer in the ER, still retain the capability of H_2O_2-dependent disulfide bond formation machinery, which is driven by PRDX4, the influx of H_2O_2 via ER PXPs is presumably an additional H_2O_2 supply pathway (Konno et al., 2015). Indeed, a recent study demonstrated that the lack of H_2O_2 due to ERO1α downregulation is compensable by its supply from the mitochondrial respiratory chain, transported to the ER through AQP11 (Sorrentino et al., 2022).

The notion of ER H_2O_2 storage and regulated release draws attention to its influx in inter-organelle communications. The ER tubular network extends throughout the cell and forms contacts with virtually all organelles, including mitochondria, peroxisomes, nuclei, cytoplasm, and plasma membrane (Phillips & Voeltz, 2016; Rowland & Voeltz, 2012; Wu et al., 2018). Further, the involvement of AQPs in events at organelle contact sites is evident: in plants, AQP on the plasma membrane interacts with an ER-resident protein VAP27 (Fox et al., 2020), inhibition of AQP diminishes mitochondrial fission following ER stress (Lee et al., 2013). These reports point to the potential presence of AQP/PXP at organelle contact sites. Of particular note is the phenomenon of ER Ca^{2+}-triggered mitochondrial H_2O_2 release, which occurs at ER mitochondria contact sites (Booth et al., 2016; Yoboue et al., 2018). This mechanism suggests that the ER may collect H_2O_2 emitted from other organelles. In the ER, H_2O_2 can be utilized as a fuel for its organelle function rather than a risk factor of oxidative stress. Therefore, it may not be irrational to envisage that ER-resident PXPs mediate the recycling of unwanted H_2O_2 from other organelles.

6.2.5 H_2O_2 Efflux from the ER

The redox status of non-secretory organelles is kept at a relatively reductive poise compared to the ER, suggesting that the efflux of ER H_2O_2 seems to be strictly controlled and counterbalanced by cytoplasmic redox machinery (Belousov et al., 2006; Markvicheva et al., 2011). Indeed, H_2O_2 leakage to the cytoplasm does not occur even when ER H_2O_2 is increased using ER stressors such as tunicamycin or HIV-1 trans-activator of transcription (TAT; Santos et al., 2016; Wu et al., 2010). However, knocking down GPX8, a gatekeeper of ROS localized on the luminal side of the ER membrane, increases cytosol H_2O_2 (Ramming et al., 2014), meaning that perturbation of ER's antioxidative capacity (rather than an increase in ER H_2O_2) causes ER H_2O_2 leakage.

On the other hand, despite such robust confinement of ER H_2O_2 leakage, some physiological effluxes of H_2O_2 through the ER membrane have been reported. Thus, the decrease of cytosolic H_2O_2 using RNA interference targeting AQP11 demonstrates H_2O_2 efflux from the ER at the steady state (Bestetti et al., 2020). Notably, NOX4 oxidizes the active metal group at the core of serine/threonine phosphatase PP1 on the cytosol side of the ER membrane (without changing the oxidation ratio of cytosolic redox). It suppresses PP1 activity (Santos et al., 2016). This allows sustaining phosphorylation of eIF2α to attenuate protein synthesis while coping with ER stress by UPR.

Moreover, inhibition of physiological ER oxidative folding machinery with an ERO1 inhibitor EN460 reduced cytosolic H_2O_2 and resulted in perturbed intracellular calcium homeostasis via the KEAP1-NRF2 ROS-responsive signaling pathway (Granatiero et al., 2019). These point to the homeostatic significance of ER H_2O_2 efflux. Thus, EN460 and PDI inhibitors (CCF642, LOC14) are explored as therapeutic agents for cancer progression as well as neurodegenerative diseases, with benefits overweighing the effect on ER's oxidative folding capacity (Hayes et al., 2019; Kamarehei et al., 2020; Zhou et al., 2018).

6.3 CONCLUSIONS AND PERSPECTIVES

ROS production inevitably occurs due to aerobic metabolism throughout the cell and its organelles (Figure 6.1). In most cases, they are shortly scavenged by antioxidants prior to unwanted oxidation of sensitive cell material. Thus, the physiological roles of ROS as a second messenger are strictly controlled with a limited spatiotemporal range.

Numerous studies to date have demonstrated that such oxidative signaling is essential to drive cell activities such as proliferation, migration, and differentiation, whereas beyond a certain threshold, increased ROS levels activate different signaling pathways and result in stress responses or cell death. ROS levels exceeding local antioxidative capacity can only be equilibrated by discharging from their sources. PXPs may be responsible for sensing the H_2O_2 gradient across the plasma/organelle membranes and determining the cell states by tuning intracellular redox equilibrium. Indeed, simultaneous dysfunction/dysregulation of AQPs/PXPs and oxidative stress are often reported in aging and diseases

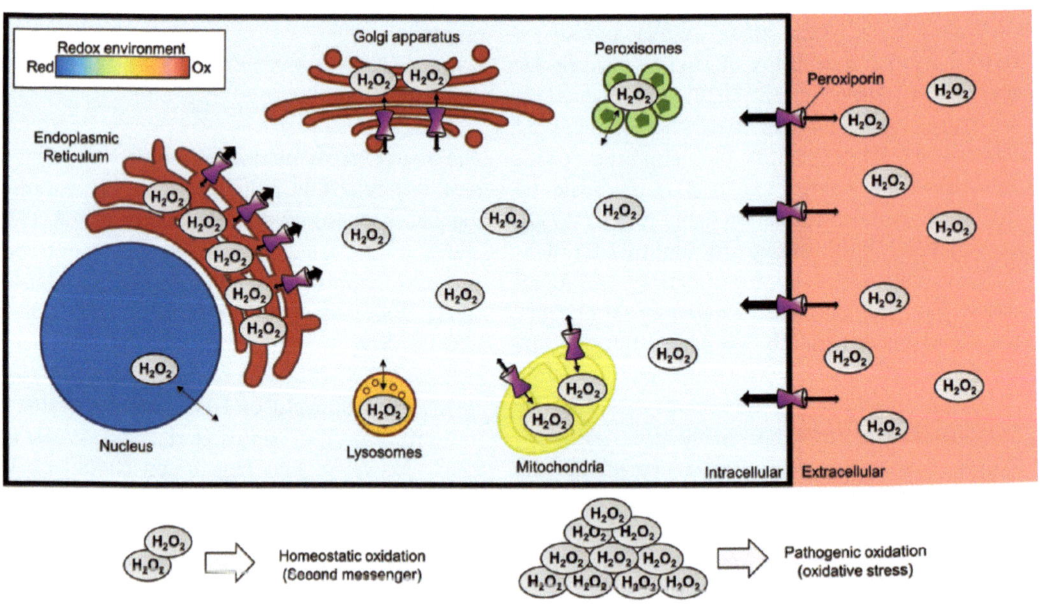

Figure 6.1 Organelle redox and H_2O_2 flux. Redox environment in respective organelles is color coded based on the reported measurements (Austin et al., 2005; Hassinen et al., 2019; Malinouski et al., 2011). The organelles produce H_2O_2 purposely or by metabolic reactions (as a by-product). Spatiotemporal gradients of H_2O_2 determine its flux direction, and local amount of H_2O_2 varies based on its consumption through homeostatic/pathogenic oxidation. The presence of H_2O_2 channel peroxiporins can tune cellular redox.

(Atochina-Vasserman et al., 2013; Liu et al., 2012; Rodrigues et al., 2019a; Sada et al., 2016; Wen et al., 2018; Xie et al., 2013).

In terms of therapeutic strategy, targeting AQPs is being explored against the pathologies of cancers, renal dysfunction, and neurodegenerative diseases (Azad et al., 2021; Verkman et al., 2014). For instance, a monoclonal anti-AQP4 antibody (aquaporumab) has been engineered for a therapy of neuromyelitis optica spectrum disorders (an autoimmune disease carrying autoantibody productions against AQP4 as well as AQP1; Duan et al., 2020; Tzartos et al., 2013). Also, neutralization of AQP3 in macrophages using a monoclonal antibody relieves liver injury by preventing the transport of extracellular H_2O_2 (Hara-Chikuma et al., 2020), and a blockage of H_2O_2 uptake via AQP1 suppresses the pathological myocardial remodeling, which causes cardiac hypertrophy, providing a new sight to target PXPs for therapeutics (Montiel et al., 2020). Of note, inhibition of PXPs seems to represent two sides of the same coin because H_2O_2 permeation through PXPs appears essential for cancer cell susceptibility to specific therapies accompanying ROS production to exclude cancer cells (Erudaitius et al., 2017). Therefore, the development of flexible AQPs/PXPs manipulations can be expected for on-demand therapeutics. Moreover, most drugs targeting AQPs/PXPs assume to have an inhibitory action on the cell surface, such as antibody therapeutics. Despite the development of numerous chemical-based AQP inhibitors (Abir-Awan et al., 2019), their effects on intracellular isozymes remain elusive. Recently, the prevention of AQP4's translocation from the endomembranes to the plasma membrane has been achieved by a chemical that inhibits the interaction between calmodulin and AQP4 intracellularly, and this chemical has shown a promise to cure edema (swelling caused by excess fluid) in the central nervous system (Kitchen et al. 2020). This encourages that intracellular AQPs/PXPs can also be a potential therapeutic target.

Among intracellular AQPs/PXPs, the biological relevance of ER PXPs remains obscure despite its apparent contributions to cell homeostasis and various disorders, as described above. The ER is a reservoir and a consumer of H_2O_2, so unnecessary H_2O_2 efflux should not occur at the steady state. Therefore, this implies that the physiological export of ER H_2O_2 through ER PXPs is likely conducted only for crucial cell activities. The augmented H_2O_2 puffs from the ER can be transduced immediately as an emergency alert of ER dysfunctions. The stationing of some phosphatases on the ER membrane likely supports this scenario and can also postulate that the amplitude and sustainability of specific signal transduction events rely on ER PXPs availability (Frangioni et al., 1992; Saito et al., 2008; Santos et al., 2016). Studies on AQP11, an ER specific PXP, will expectedly tow this field. In addition, given that AQP12 null mice exhibit pancreatic dysfunctions (Ohta et al., 2009), new findings about this ER isozyme may provide further insights into ROS handling. However, the H_2O_2 permeability of AQP12 as a PXP is yet to be established. Further research will shed light on the spatiotemporal regulation of ER PXPs for a better understanding of ROS biology and its harnessing for novel therapeutics.

REFERENCES

Abir-Awan, M., Kitchen, P., Salman, M. M., Conner, M. T., Conner, A. C., & Bill, R. M. (2019). Inhibitors of mammalian aquaporin water channels. *International Journal of Molecular Sciences*, 20(7). https://doi.org/10.3390/IJMS20071589

Aguilar, P. S., & de Mendoza, D. (2006). Control of fatty acid desaturation: A mechanism conserved from bacteria to humans. *Molecular Microbiology*, 62(6). https://doi.org/10.1111/j.1365-2958.2006.05484.x

Albertolle, M. E., & Peter Guengerich, F. (2018). The relationships between cytochromes P450 and H2O2: Production, reaction, and inhibition. *Journal of Inorganic Biochemistry*, 186, 228. https://doi.org/10.1016/J.JINORGBIO.2018.05.014

Allen, S., Balabanidou, V., Sideris, D. P., Lisowsky, T., & Tokatlidis, K. (2005). Erv1 mediates the Mia40-dependent protein import pathway and provides a functional link to the respiratory chain by shuttling electrons to cytochrome c. *Journal of Molecular Biology*, 353(5), 937–944. https://doi.org/10.1016/J.JMB.2005.08.049

Almasalmeh, A., Krenc, D., Wu, B., & Beitz, E. (2014). Structural determinants of the hydrogen peroxide permeability of aquaporins. *The FEBS Journal*, 281(3), 647–656. https://doi.org/10.1111/FEBS.12653

Androwiki, A.C.D., Camargo, L. de L., Sartoretto, S., Couto, G. K., Ribeiro, I.M.R., Veríssimo-Filho, S., Rossoni, L. V., & Lopes, L. R. (2015). Protein disulfide isomerase expression increases in resistance arteries during hypertension development.

Effects on Nox1 NADPH oxidase signaling. *Frontiers in Chemistry*, 3(Mar). https://doi.org/10.3389/fchem.2015.00024

Angermüller, S., Bruder, G., Völkl, A., Wesch, H., & Fahimi, H. D. (1987). Localization of xanthine oxidase in crystalline cores of peroxisomes. A cytochemical and biochemical study. *European Journal of Cell Biology*, 45(1).

Angermüller, S., Islinger, M., & Völkl, A. (2009). Peroxisomes and reactive oxygen species, a lasting challenge. *Histochemistry and Cell Biology*, 131(4). https://doi.org/10.1007/s00418-009-0563-7

Antonenkov, V. D., Grunau, S., Ohlmeier, S., & Hiltunen, J. K. (2010). Peroxisomes are oxidative organelles. *Antioxidants and Redox Signaling*, 13(4). https://doi.org/10.1089/ars.2009.2996

Appenzeller-Herzog, C., & Ellgaard, L. (2008). The human PDI family: Versatility packed into a single fold. *Biochimica et Biophysica Acta—Molecular Cell Research*, 1783(4). https://doi.org/10.1016/j.bbamcr.2007.11.010

Appenzeller-Herzog, C., Riemer, J., Zito, E., Chin, K. T., Ron, D., Spiess, M., & Ellgaard, L. (2010). Disulphide production by Ero1α-PDI relay is rapid and effectively regulated. *The EMBO Journal*, 29(19), 3318–3329. https://doi.org/10.1038/EMBOJ.2010.203

Atochina-Vasserman, E. N., Biktasova, A., Abramova, E., Cheng, D. S., Polosukhin, V. V., Tanjore, H., Takahashi, S., Sonoda, H., Foye, L., Venkov, C., Ryzhov, S. V., Novitskiy, S., Shlonimskaya, N., Ikeda, M., Blackwell, T. S., Lawson, W. E., Gow, A. J., Harris, R. C., Dikov, M. M., & Tchekneva, E. E. (2013). Aquaporin 11 insufficiency modulates kidney susceptibility to oxidative stress. *American Journal of Physiology. Renal Physiology*, 304(10), 1295–1307. https://doi.org/10.1152/AJPRENAL.00344.2012

Auer, S., Rinnerthaler, M., Bischof, J., Streubel, M. K., Breitenbach-Koller, H., Geisberger, R., Aigner, E., Cadamuro, J., Richter, K., Sopjani, M., Haschke-Becher, E., Felder, T. K., & Breitenbach, M. (2017). The human NADPH oxidase, Nox4, regulates cytoskeletal organization in two cancer cell lines, HepG2 and SH-SY5Y. *Frontiers in Oncology*, 7(May). https://doi.org/10.3389/fonc.2017.00111

Austin, C. D., Wen, X., Gazzard, L., Nelson, C., Scheller, R. H., & Scales, S. J. (2005). Oxidizing potential of endosomes and lysosomes limits intracellular cleavage of disulfide-based antibody–drug conjugates. *Proceedings of the National Academy of Sciences of the United States of America*, 102(50), 17987. https://doi.org/10.1073/PNAS.0509035102

Azad, A. K., Raihan, T., Ahmed, J., Hakim, A., Emon, T. H., & Chowdhury, P. A. (2021). Human aquaporins: Functional diversity and potential roles in infectious and non-infectious diseases. *Frontiers in Genetics*, 12. https://doi.org/10.3389/FGENE.2021.654865

Backes, S., & Herrmann, J. M. (2017). Protein translocation into the intermembrane space and matrix of mitochondria: Mechanisms and driving forces. *Frontiers in Molecular Biosciences*, 4(Dec), 83. https://doi.org/10.3389/FMOLB.2017.00083/BIBTEX

Bae, Y. S., Oh, H., Rhee, S. G., & Yoo, Y. do. (2011). Regulation of reactive oxygen species generation in cell signaling. *Molecules and Cells*, 32(6). https://doi.org/10.1007/s10059-011-0276-3

Banci, L., Bertini, I., Boca, M., Girotto, S., Martinelli, M., Valentine, J. S., & Vieru, M. (2008). SOD1 and amyotrophic lateral sclerosis: Mutations and oligomerization. *PLoS ONE*, 3(2). https://doi.org/10.1371/journal.pone.0001677

Bass, R., Ruddock, L. W., Klappa, P., & Freedman, R. B. (2004). A major fraction of endoplasmic reticulum-located glutathione is present as mixed disulfides with protein. *Journal of Biological Chemistry*, 279(7). https://doi.org/10.1074/jbc.M304951200

Belousov, V. V., Fradkov, A. F., Lukyanov, K. A., Staroverov, D. B., Shakhbazov, K. S., Terskikh, A. V., & Lukyanov, S. (2006). Genetically encoded fluorescent indicator for intracellular hydrogen peroxide. *Nature Methods*, 3(4), 281–286. https://doi.org/10.1038/NMETH866

Bertolotti, M., Bestetti, S., García-Manteiga, J. M., Medraño-Fernandez, I., Dal Mas, A., Malosio, M. L., & Sitia, R. (2013). Tyrosine kinase signal modulation: a matter of H2O2 membrane permeability? *Antioxidants & Redox Signaling*, 19(13), 1447–1451. https://doi.org/10.1089/ARS.2013.5330

Bertolotti, M., Farinelli, G., Galli, M., Aiuti, A., & Sitia, R. (2016). AQP8 transports NOX2-generated H2O2 across the plasma membrane to promote signaling in B cells. *Journal of Leukocyte Biology*, 100(5), 1071–1079. https://doi.org/10.1189/JLB.2AB0116-045R

Bertolotti, M., Yim, S. H., Garcia-Manteiga, J. M., Masciarelli, S., Kim, Y. J., Kang, M. H., Iuchi, Y., Fujii, J., Vené, R., Rubartelli, A., Rhee, S. G., & Sitia, R. (2010). B-to plasma-cell terminal differentiation entails oxidative stress and profound reshaping of the antioxidant responses. *Antioxidants and Redox Signaling*, 13(8). https://doi.org/10.1089/ars.2009.3079

Bestetti, S., Galli, M., Sorrentino, I., Pinton, P., Rimessi, A., Sitia, R., & Medraño-Fernandez, I.

(2020). Human aquaporin-11 guarantees efficient transport of H 2 O 2 across the endoplasmic reticulum membrane. *Redox Biology*, 28. https://doi.org/10.1016/J.REDOX.2019.101326

Bestetti, S., Medraño-Fernandez, I., Galli, M., Ghitti, M., Bienert, G. P., Musco, G., Orsi, A., Rubartelli, A., & Sitia, R. (2018). A persulfidation-based mechanism controls aquaporin-8 conductance. *Science Advances*, 4(5). https://doi.org/10.1126/SCIADV.AAR5770

Beyer, R. E. (1990). The participation of coenzyme Q in free radical production and antioxidation. *Free Radical Biology and Medicine*, 8(6). https://doi.org/10.1016/0891-5849(90)90154-B

Bienert, G. P., Møller, A.L.B., Kristiansen, K. A., Schulz, A., Møller, I. M., Schjoerring, J. K., & Jahn, T. P. (2007). Specific aquaporins facilitate the diffusion of hydrogen peroxide across membranes. *The Journal of Biological Chemistry*, 282(2), 1183–1192. https://doi.org/10.1074/JBC.M603761200

Bienert, G. P., Schjoerring, J. K., & Jahn, T. P. (2006). Membrane transport of hydrogen peroxide. *Biochimica et Biophysica Acta—Biomembranes*, 1758(8). https://doi.org/10.1016/j.bbamem.2006.02.015

Bihlmaier, K., Mesecke, N., Terziyska, N., Bien, M., Hell, K., & Herrmann, J. M. (2007). The disulfide relay system of mitochondria is connected to the respiratory chain. *The Journal of Cell Biology*, 179(3), 389. https://doi.org/10.1083/JCB.200707123

Block, K., Gorin, Y., & Abboud, H. E. (2009). Subcellular localization of Nox4 and regulation in diabetes. *Proceedings of the National Academy of Sciences of the United States of America*, 106(34). https://doi.org/10.1073/pnas.0906805106

Bondy, S. C., & Naderi, S. (1994). Contribution of hepatic cytochrome P450 systems to the generation of reactive oxygen species. *Biochemical Pharmacology*, 48(1). https://doi.org/10.1016/0006-2952(94)90235-6

Booth, D. M., Enyedi, B., Geiszt, M., Várnai, P., & Hajnóczky, G. (2016). Redox nanodomains are induced by and control calcium signaling at the ER-mitochondrial interface. *Molecular Cell*, 63(2). https://doi.org/10.1016/j.molcel.2016.05.040

Bostwick, D. G., Alexander, E. E., Singh, R., Shan, A., Qian, J., Santella, R. M., Oberley, L. W., Yan, T., Zhong, W., Jiang, X., & Oberley, T. D. (2000). Antioxidant enzyme expression and reactive oxygen species damage in prostatic intraepithelial neoplasia and cancer. *Cancer*, 89(1). https://doi.org/10.1002/1097-0142(20000701)89:1<123::AID-CNCR17>3.0.CO;2-9

Boveris, A., Oshino, N., & Chance, B. (1972). The cellular production of hydrogen peroxide. *The Biochemical Journal*, 128(3), 617–630. https://doi.org/10.1042/BJ1280617

Brand, M. D. (2010). The sites and topology of mitochondrial superoxide production. *Experimental Gerontology*, 45(7–8). https://doi.org/10.1016/j.exger.2010.01.003

Brand, M. D. (2016). Mitochondrial generation of superoxide and hydrogen peroxide as the source of mitochondrial redox signaling. *Free Radical Biology and Medicine*, 100. https://doi.org/10.1016/j.freeradbiomed.2016.04.001

Calamita, G., Ferri, D., Gena, P., Liquori, G. E., Cavalier, A., Thomas, D., & Svelto, M. (2005). The inner mitochondrial membrane has aquaporin-8 water channels and is highly permeable to water. *The Journal of Biological Chemistry*, 280(17), 17149–17153. https://doi.org/10.1074/JBC.C400595200

Canugovi, C., Stevenson, M. D., Vendrov, A. E., Hayami, T., Robidoux, J., Xiao, H., Zhang, Y. Y., Eitzman, D. T., Runge, M. S., & Madamanchi, N. R. (2019). Increased mitochondrial NADPH oxidase 4 (NOX4) expression in aging is a causative factor in aortic stiffening. *Redox Biology*, 26. https://doi.org/10.1016/j.redox.2019.101288

Cardozo-Pelaez, F., Sanchez-Contreras, M., & Nevin, A.B.C. (2012). Ogg1 null mice exhibit age-associated loss of the nigrostriatal pathway and increased sensitivity to MPTP. *Neurochemistry International*, 61(5). https://doi.org/10.1016/j.neuint.2012.06.013

Caro, A. A., & Cederbaum, A. I. (2004). Oxidative stress, toxicology, and pharmacology of CYP2E1. *Annual Review of Pharmacology and Toxicology*, 44. https://doi.org/10.1146/annurev.pharmtox.44.101802.121704

Cavazzin, C., Ferrari, D., Facchetti, F., Russignan, A., Vescovi, A. L., La Porta, C.A.M., & Gritti, A. (2006). Unique expression and localization of aquaporin-4 and aquaporin-9 in murine and human neural stem cells and in their glial progeny. *Glia*, 53(2), 167–181. https://doi.org/10.1002/GLIA.20256

Chai, K., Kitamura, K., McCann, A., & Wu, X. R. (2010). The epithelium—molecular landscaping for an interactive barrier. *Journal of Biomedicine & Biotechnology*, 2010. https://doi.org/10.1155/2010/870506

Chakravarthi, S., Jessop, C. E., Willer, M., Stirling, C. J., & Bulleid, N. J. (2007). Intracellular catalysis of disulfide bond formation by the human sulfhydryl oxidase, QSOX1. *Biochemical Journal*, 404(3). https://doi.org/10.1042/BJ20061510

Chance, B., Sies, H., & Boveris, A. (1979). Hydroperoxide metabolism in mammalian organs. *Physiological Reviews*, 59(3). https://doi.org/10.1152/physrev.1979.59.3.527

Chandel, N. S., Maltepe, E., Goldwasser, E., Mathieu, C. E., Simon, M. C., & Schumacker, P. T. (1998). Mitochondrial reactive oxygen species trigger hypoxia-induced transcription. *Proceedings of the National Academy of Sciences of the United States of America*, 95(20). https://doi.org/10.1073/pnas.95.20.11715

Chanock, S. J., el Benna, J., Smith, R. M., & Babior, B. M. (1994). The respiratory burst oxidase. *Journal of Biological Chemistry*, 269(40).

Chen, F., Haigh, S., Barman, S., & Fulton, D.J.R. (2012). From form to function: the role of Nox4 in the cardiovascular system. *Frontiers in Physiology*, 3. https://doi.org/10.3389/FPHYS.2012.00412

Chen, Q., Zhu, L., Zong, H., Song, X., Wang, L., Wang, X., Yang, D., & Wang, J. (2018). Subcellular localization of aquaporin 3 in prostate cancer is regulated by RalA. *Oncology Reports*, 39(5), 2171–2177. https://doi.org/10.3892/OR.2018.6308

Chiu, J., & Hogg, P. J. (2019). Allosteric disulfides: Sophisticated molecular structures enabling flexible protein regulation. *Journal of Biological Chemistry*, 294(8). https://doi.org/10.1074/jbc.REV118.005604

Cho, M. Y., Cheong, J. Y., Lim, W., Jo, S., Lee, Y., Wang, H. J., Han, K. H., & Cho, H. (2014). Prognostic significance of catalase expression and its regulatory effects on hepatitis B virus X protein (HBx) in HBV-related advanced hepatocellular carcinomas. *Oncotarget*, 5(23). https://doi.org/10.18632/oncotarget.2625

Choma, D. P., Vanacore, R., Naylor, H., Zimmerman, I. A., Pavlichenko, A., Pavlichenko, A., Foye, L., Carbone, D. P., Harris, R. C., Dikov, M. M., & Tchekneva, E. E. (2016). Aquaporin 11 variant associates with kidney disease in type 2 diabetic patients. *American Journal of Physiology. Renal Physiology*, 310(5), F416–F425. https://doi.org/10.1152/AJPRENAL.00295.2015

Christensen, B. M., Marples, D., Jensen, U. B., Frøkiaer, J., Sheikh-Hamad, D., Knepper, M., & Nielsen, S. (1998). Acute effects of vasopressin V2-receptor antagonist on kidney AQP2 expression and subcellular distribution. *The American Journal of Physiology*, 275(2). https://doi.org/10.1152/AJPRENAL.1998.275.2.F285

Christensen, B. M., Zelenina, M., Aperia, A., & Nielsen, S. (2000). Localization and regulation of PKA-phosphorylated AQP2 in response to V(2)-receptor agonist/antagonist treatment. *American Journal of Physiology. Renal Physiology*, 278(1). https://doi.org/10.1152/AJPRENAL.2000.278.1.F29

Conner, M. T., Conner, A. C., Bland, C. E., Taylor, L.H.J., Brown, J.E.P., Parri, H. R., & Bill, R. M. (2012). Rapid aquaporin translocation regulates cellular water flow: mechanism of hypotonicity-induced subcellular localization of aquaporin 1 water channel. *The Journal of Biological Chemistry*, 287(14), 11516–11525. https://doi.org/10.1074/JBC.M111.329219

Cumming, R. C., Andon, N. L., Haynes, P. A., Park, M., Fischer, W. H., & Schubert, D. (2004). Protein disulfide bond formation in the cytoplasm during oxidative stress. *Journal of Biological Chemistry*, 279(21), 21749–21758. https://doi.org/10.1074/JBC.M312267200

Curran, S. P., & Ruvkun, G. (2007). Lifespan regulation by evolutionarily conserved genes essential for viability. *PLoS Genetics*, 3(4). https://doi.org/10.1371/journal.pgen.0030056

Dabir, D. V., Leverich, E. P., Kim, S. K., Tsai, F. D., Hirasawa, M., Knaff, D. B., & Koehler, C. M. (2007). A role for cytochrome c and cytochrome c peroxidase in electron shuttling from Erv1. *The EMBO Journal*, 26(23), 4801–4811. https://doi.org/10.1038/SJ.EMBOJ.7601909

Das, R., Xu, S., Quan, X., Nguyen, T. T., Kong, I. D., Chung, C. H., Lee, E. Y., Cha, S. K., & Park, K. S. (2014). Upregulation of mitochondrial Nox4 mediates TGF-β-induced apoptosis in cultured mouse podocytes. *American Journal of Physiology—Renal Physiology*, 306(2). https://doi.org/10.1152/ajprenal.00438.2013

Davydov, D. R. (2011). Microsomal monooxygenase as a multienzyme system: The role of P450-P450 interactions. *Expert Opinion on Drug Metabolism and Toxicology*, 7(5). https://doi.org/10.1517/17425255.2011.562194

de Almeida, A., Parthimos, D., Dew, H., Smart, O., Wiltshire, M., & Errington, R. J. (2021). Aquaglyceroporin-3's expression and cellular localization is differentially modulated by hypoxia in prostate cancer cell lines. *Cells*, 10(4). https://doi.org/10.3390/CELLS10040838

de Duve, C., & Baudhuin, P. (1966). Peroxisomes (microbodies and related particles). *Physiological Reviews*, 46(2), 323–357. https://doi.org/10.1152/PHYSREV.1966.46.2.323

De Ieso, M. L., Pei, J. V., Nourmohammadi, S., Smith, E., Chow, P. H., Kourghi, M., Hardingham, J. E., & Yool, A. J. (2019). Combined pharmacological administration of AQP1 ion channel blocker AqB011 and water channel blocker Bacopaside II amplifies inhibition of colon cancer cell migration. *Scientific Reports*, 9(1). https://doi.org/10.1038/S41598-019-49045-9

de Mochel, N.S.R., Seronello, S., Wang, S. H., Ito, C., Zheng, J. X., Liang, T. J., Lambeth, J. D., & Choi, J. (2010). Hepatocyte NAD(P)H oxidases as an endogenous source of reactive oxygen species during hepatitis C virus infection. *Hepatology*, 52(1). https://doi.org/10.1002/hep.23671

del Río, L. A. (2011). Peroxisomes as a cellular source of reactive nitrogen species signal molecules. *Archives of Biochemistry and Biophysics*, 506(1). https://doi.org/10.1016/j.abb.2010.10.022

del Río, L. A., & López-Huertas, E. (2016). ROS generation in peroxisomes and its role in cell signaling. *Plant and Cell Physiology*, 57(7). https://doi.org/10.1093/pcp/pcw076

Deng, H. X., Shi, Y., Furukawa, Y., Zhai, H., Fu, R., Liu, E., Gorrie, G. H., Khan, M. S., Hung, W. Y., Bigio, E. H., Lukas, T., Dal Canto, M. C., O'Halloran, T. v., & Siddique, T. (2006). Conversion to the amyotrophic lateral sclerosis phenotype is associated with intermolecular linked insoluble aggregates of SOD1 in mitochondria. *Proceedings of the National Academy of Sciences of the United States of America*, 103(18). https://doi.org/10.1073/pnas.0602046103

di Meo, S., Reed, T. T., Venditti, P., & Victor, V. M. (2016). Role of ROS and RNS sources in physiological and pathological conditions. *Oxidative Medicine and Cellular Longevity*, 2016. https://doi.org/10.1155/2016/1245049

Dixon, B. M., Heath, S.H.D., Kim, R., Suh, J. H., & Hagen, T. M. (2008). Assessment of endoplasmic reticulum glutathione redox status is confounded by extensive ex vivo oxidation. *Antioxidants and Redox Signaling*, 10(5). https://doi.org/10.1089/ars.2007.1869

Dlasková, A., Clarke, K. J., & Porter, R. K. (2010). The role of UCP 1 in production of reactive oxygen species by mitochondria isolated from brown adipose tissue. *Biochimica et Biophysica Acta—Bioenergetics*, 1797(8). https://doi.org/10.1016/j.bbabio.2010.04.008

Druhan, L. J., Forbes, S. P., Pope, A. J., Chen, C. A., Zweier, J. L., & Cardounel, A. J. (2008). Regulation of eNOS-derived superoxide by endogenous methylarginines. *Biochemistry*, 47(27). https://doi.org/10.1021/bi702377a

Duan, T., Tradtrantip, L., Phuan, P. W., Bennett, J. L., & Verkman, A. S. (2020). Affinity-matured "aquaporumab" anti-aquaporin-4 antibody for therapy of seropositive neuromyelitis optica spectrum disorders. *Neuropharmacology*, 162. https://doi.org/10.1016/J.NEUROPHARM.2019.107827

Dubreuil, M. M., Morgens, D. W., Okumoto, K., Honsho, M., Contrepois, K., Lee-McMullen, B., Traber, G.M.A., Sood, R. S., Dixon, S. J., Snyder, M. P., Fujiki, Y., & Bassik, M. C. (2020). Systematic identification of regulators of oxidative stress reveals non-canonical roles for peroxisomal import and the pentose phosphate pathway. *Cell Reports*, 30(5), 1417–1433.e7. https://doi.org/10.1016/J.CELREP.2020.01.013

Eaton, J. W., & Qian, M. (2002). Molecular bases of cellular iron toxicity. *Free Radical Biology and Medicine*, 32(9). https://doi.org/10.1016/S0891-5849(02)00772-4

Erudaitius, D., Huang, A., Kazmi, S., Buettner, G. R., & Rodgers, V.G.J. (2017). Peroxiporin expression is an important factor for cancer cell susceptibility to therapeutic H2O2: implications for pharmacological ascorbate therapy. *PLoS ONE*, 12(1). https://doi.org/10.1371/JOURNAL.PONE.0170442

Fang, J., Yang, J., Wu, X., Zhang, G., Li, T., Wang, X., Zhang, H., Wang, C.-C., Liu, G. H., & Wang, L. (2018). Metformin alleviates human cellular aging by upregulating the endoplasmic reticulum glutathione peroxidase 7. *Aging Cell*, 17(4). https://doi.org/10.1111/acel.12765

Farrell, S. R., & Thorpe, C. (2005). Augmenter of liver regeneration: a flavin-dependent sulfhydryl oxidase with cytochrome c reductase activity. *Biochemistry*, 44(5), 1532–1541. https://doi.org/10.1021/BI0479555

Finger, Y., & Riemer, J. (2020). Protein import by the mitochondrial disulfide relay in higher eukaryotes. *Biological Chemistry*, 401(6–7). https://doi.org/10.1515/hsz-2020-0108

Finn, R. N., Chauvigné, F., Hlidberg, J. B., Cutler, C. P., & Cerdà, J. (2014). The lineage-specific evolution of aquaporin gene clusters facilitated tetrapod terrestrial adaptation. *PLoS ONE*, 9(11). https://doi.org/10.1371/JOURNAL.PONE.0113686

Fox, A. R., Scochera, F., Laloux, T., Filik, K., Degand, H., Morsomme, P., Alleva, K., & Chaumont, F. (2020). Plasma membrane aquaporins interact with the endoplasmic reticulum resident VAP27 proteins at ER-PM contact sites and endocytic structures. *The New Phytologist*, 228(3), 973–988. https://doi.org/10.1111/NPH.16743

Frand, A. R., & Kaiser, C. A. (1998). The ERO1 gene of yeast is required for oxidation of protein dithiols in the endoplasmic reticulum. *Molecular Cell*, 1(2). https://doi.org/10.1016/S1097-2765(00)80017-9

Frangioni, J. V., Beahm, P. H., Shifrin, V., Jost, C. A., & Neel, B. G. (1992). The nontransmembrane tyrosine phosphatase PTP-1B localizes to the endoplasmic reticulum via its 35 amino acid C-terminal sequence. *Cell*, 68(3), 545–560. https://doi.org/10.1016/0092-8674(92)90190-N

Frederiks, W. M., Bosch, K. S., Hoeben, K. A., van Marle, J., & Langbein, S. (2010). Renal cell carcinoma and oxidative stress: The lack of peroxisomes. *Acta Histochemica*, 112(4). https://doi.org/10.1016/j.acthis.2009.03.003

Freeman, B. A., & Crapo, J. D. (1982). Biology of disease. Free radicals and tissue injury. *Laboratory Investigation*, 47(5).

Fritz, R., Bol, J., Hebling, U., Angermüller, S., Völkl, A., Fahimi, H. D., & Mueller, S. (2007). Compartment-dependent management of H2O2 by peroxisomes. *Free Radical Biology and Medicine*, 42(7). https://doi.org/10.1016/j.freeradbiomed.2007.01.014

Fushimi, K., Sasaki, S., & Marumo, F. (1997). Phosphorylation of serine 256 is required for cAMP-dependent regulatory exocytosis of the aquaporin-2 water channel. *The Journal of Biological Chemistry*, 272(23), 14800–14804. https://doi.org/10.1074/JBC.272.23.14800

Gao, C., Tian, Y., Zhang, R., Jing, J., & Zhang, X. (2017). Endoplasmic reticulum-directed ratiometric fluorescent probe for quantitive detection of basal H2O2. *Analytical Chemistry*, 89(23). https://doi.org/10.1021/acs.analchem.7b03809

Gao, J., Tan, M., Gu, M., Marshall, C., Ding, J., Hu, G., & Xiao, M. (2012). Cellular localization of aquaporin-1 in the human and mouse trigeminal systems. *PLoS ONE*, 7(9). https://doi.org/10.1371/JOURNAL.PONE.0046379

García, F., Kierbel, A., Larocca, M. C., Gradilone, S. A., Splinter, P., LaRusso, N. F., & Marinelli, R. A. (2001). The water channel aquaporin-8 is mainly intracellular in rat hepatocytes, and its plasma membrane insertion is stimulated by cyclic AMP. *The Journal of Biological Chemistry*, 276(15), 12147–12152. https://doi.org/10.1074/JBC.M009403200

García-Cardeña, G., Oh, P., Liu, J., Schnitzer, J. E., & Sessa, W. C. (1996). Targeting of nitric oxide synthase to endothelial cell caveolae via palmitoylation: Implications for nitric oxide signaling. *Proceedings of the National Academy of Sciences of the United States of America*, 93(13). https://doi.org/10.1073/pnas.93.13.6448

García-Giménez, J. L., Olaso, G., Hake, S. B., Bönisch, C., Wiedemann, S. M., Markovic, J., Dasí, F., Gimeno, A., Pérez-Quilis, C., Palacios, Ò., Capdevila, M., Viña, J., & Pallardó, F. V. (2013). Histone H3 glutathionylation in proliferating mammalian cells destabilizes nucleosomal structure. *Antioxidants and Redox Signaling*, 19(12). https://doi.org/10.1089/ars.2012.5021

Gess, B., Hofbauer, K. H., Wenger, R. H., Lohaus, C., Meyer, H. E., & Kurtz, A. (2003). The cellular oxygen tension regulates expression of the endoplasmic oxidoreductase ERO1-Lα. *European Journal of Biochemistry*, 270(10). https://doi.org/10.1046/j.1432-1033.2003.03590.x

Gilady, S. Y., Bui, M., Lynes, E. M., Benson, M. D., Watts, R., Vance, J. E., & Simmen, T. (2010). Ero1α requires oxidizing and normoxic conditions to localize to the mitochondria-associated membrane (MAM). *Cell Stress and Chaperones*, 15(5). https://doi.org/10.1007/s12192-010-0174-1

Gille, L., & Nohl, H. (2000). The existence of a lysosomal redox chain and the role of ubiquinone. *Archives of Biochemistry and Biophysics*, 375(2). https://doi.org/10.1006/abbi.1999.1649

Gillette, J. R., Brodie, B. B., & la Du, B. N. (1957). The oxidation of drugs by liver microsomes: on the role of TPNH and oxygen. *The Journal of Pharmacology and Experimental Therapeutics*, 119(4), 532–540.

Go, Y. M., & Jones, D. P. (2008). Redox compartmentalization in eukaryotic cells. *Biochimica et Biophysica Acta*, 1780(11), 1273–1290. https://doi.org/10.1016/J.BBAGEN.2008.01.011

Golestaneh, N., Fan, J., Zelenka, P., & Chepelinsky, A. B. (2008). PKC putative phosphorylation site Ser235 is required for MIP/AQP0 translocation to the plasma membrane. *Molecular Vision*, 14, 1006.

Goncalves, R.L.S., Quinlan, C. L., Perevoshchikova, I. V., Hey-Mogensen, M., & Brand, M. D. (2015). Sites of superoxide and hydrogen peroxide production by muscle mitochondria assessed ex vivo under conditions mimicking rest and exercise. *Journal of Biological Chemistry*, 290(1). https://doi.org/10.1074/jbc.M114.619072

Gotfryd, K., Mósca, A. F., Missel, J. W., Truelsen, S. F., Wang, K., Spulber, M., Krabbe, S., Hélix-Nielsen, C., Laforenza, U., Soveral, G., Pedersen, P. A., & Gourdon, P. (2018). Human adipose glycerol flux is regulated by a pH gate in AQP10. *Nature Communications*, 9(1). https://doi.org/10.1038/S41467-018-07176-Z

Goubau, C., Jaeken, J., Levtchenko, E. N., Thys, C., Di Michele, M., Martens, G. A., Gerlo, E., De Vos, R., Buyse, G. M., Goemans, N., Van Geet, C., & Freson, K. (2013). Homozygosity for aquaporin 7 G264V in three unrelated children with hyperglyceroluria and a mild platelet secretion defect. *Genetics in Medicine: Official Journal of the American College of Medical Genetics*, 15(1), 55–63. https://doi.org/10.1038/GIM.2012.90

Gradilone, S. A., Tietz, P. S., Splinter, P. L., Marinelli, R. A., & LaRusso, N. F. (2005). Expression and subcellular localization of aquaporin water channels in the polarized hepatocyte cell line, WIF-B. *BMC Physiology*, 5. https://doi.org/10.1186/1472-6793-5-13

Granatiero, V., Konrad, C., Bredvik, K., Manfredi, G., & Kawamata, H. (2019). Nrf2 signaling links ER oxidative protein folding and calcium homeostasis in health and disease. *Life Science Alliance*, 2(5). https://doi.org/10.26508/LSA.201900563

Groen, A., Lemeer, S., Van Der Wijk, T., Overvoorde, J., Heck, A.J.R., Ostman, A., Barford, D., Slijper, M., & Den Hertog, J. (2005). Differential oxidation of protein-tyrosine phosphatases. *The Journal of Biological Chemistry*, 280(11), 10298–10304. https://doi.org/10.1074/JBC.M412424200

Gross, E., Sevier, C. S., Heldman, N., Vitu, E., Bentzur, M., Kaiser, C. A., Thorpe, C., & Fass, D. (2006). Generating disulfides enzymatically: Reaction products and electron acceptors of the endoplasmic reticulum thiol oxidase Ero1p. *Proceedings of the National Academy of Sciences of the United States of America*, 103(2). https://doi.org/10.1073/pnas.0506448103

Guida, M., Maraldi, T., Beretti, F., Follo, M. Y., Manzoli, L., & de Pol, A. (2014). Nuclear Nox4-derived reactive oxygen species in myelodysplastic syndromes. *BioMed Research International*, 2014. https://doi.org/10.1155/2014/456937

Halvey, P. J., Watson, W. H., Hansen, J. M., Go, Y. M., Samali, A., & Jones, D. P. (2005). Compartmental oxidation of thiol–disulphide redox couples during epidermal growth factor signalling. *Biochemical Journal*, 386(Pt 2), 215. https://doi.org/10.1042/BJ20041829

Ham, A.J.L., & Liebler, D. C. (1995). Vitamin E oxidation in rat liver mitochondria. *Biochemistry*, 34(17). https://doi.org/10.1021/bi00017a007

Han, B., Wu, X., Huang, P. P., Zhu, F. X., & Liu, S. (2019). Aquaporin 11 rs2276415 variant and progression of chronic kidney disease. *Nephrology, Dialysis, Transplantation : Official Publication of the European Dialysis and Transplant Association—European Renal Association*, 34(6), 970–973. https://doi.org/10.1093/NDT/GFY219

Han, J., Back, S. H., Hur, J., Lin, Y. H., Gildersleeve, R., Shan, J., Yuan, C. L., Krokowski, D., Wang, S., Hatzoglou, M., Kilberg, M. S., Sartor, M. A., & Kaufman, R. J. (2013). ER-stress-induced transcriptional regulation increases protein synthesis leading to cell death. *Nature Cell Biology*, 15(5). https://doi.org/10.1038/ncb2738

Hanukoglu, I. (2006). Antioxidant protective mechanisms against reactive oxygen species (ROS) generated by mitochondrial P450 systems in steroidogenic cells. *Drug Metabolism Reviews*, 38(1–2). https://doi.org/10.1080/03602530600570040

Hara-Chikuma, M., Chikuma, S., Sugiyama, Y., Kabashima, K., Verkman, A. S., Inoue, S., & Miyachi, Y. (2012). Chemokine-dependent T cell migration requires aquaporin-3-mediated hydrogen peroxide uptake. *Journal of Experimental Medicine*, 209(10). https://doi.org/10.1084/jem.20112398

Hara-Chikuma, M., Tanaka, M., Verkman, A. S., & Yasui, M. (2020). Inhibition of aquaporin-3 in macrophages by a monoclonal antibody as potential therapy for liver injury. *Nature Communications*, 11(1). https://doi.org/10.1038/S41467-020-19491-5

Harrison, R. (2002). Structure and function of xanthine oxidoreductase: Where are we now? *Free Radical Biology and Medicine*, 33(6). https://doi.org/10.1016/S0891-5849(02)00956-5

Hasegawa, T., Azlina, A., Javkhlan, P., Yao, C., Akamatsu, T., & Hosoi, K. (2011). Novel phosphorylation of aquaporin-5 at its threonine 259 through cAMP signaling in salivary gland cells. *American Journal of Physiology. Cell Physiology*, 301(3). https://doi.org/10.1152/AJPCELL.00058.2011

Hassinen, A., Khoder-Agha, F., Khosrowabadi, E., Mennerich, D., Harrus, D., Noel, M., Dimova, E. Y., Glumoff, T., Harduin-Lepers, A., Kietzmann, T., & Kellokumpu, S. (2019). A Golgi-associated redox switch regulates catalytic activation and cooperative functioning of ST6Gal-I with B4GalT-I. *Redox Biology*, 24. https://doi.org/10.1016/J.REDOX.2019.101182

Hayes, K. E., Batsomboon, P., Chen, W. C., Johnson, B. D., Becker, A., Eschrich, S., Yang, Y., Robart, A. R., Dudley, G. B., Geldenhuys, W. J., & Hazlehurst, L. A. (2019). Inhibition of the FAD containing ER oxidoreductin 1 (Ero1) protein by EN-460 as a strategy for treatment of multiple myeloma. *Bioorganic & Medicinal Chemistry*, 27(8), 1479–1488. https://doi.org/10.1016/J.BMC.2019.02.016

Haynes, C. M., Titus, E. A., & Cooper, A. A. (2004). Degradation of misfolded proteins prevents ER-derived oxidative stress and cell death. *Molecular Cell*, 15(5). https://doi.org/10.1016/j.molcel.2004.08.025

Henzler, T., & Steudle, E. (2000). Transport and metabolic degradation of hydrogen peroxide in Chara corallina: model calculations and measurements with the pressure probe suggest transport of H(2)O(2) across water channels. *Journal of*

Experimental Botany, 51(353), 2053–2066. https://doi.org/10.1093/JEXBOT/51.353.2053

Ho, J. C. M., Zheng, S., Comhair, S.A.A., Erzurum, S. C., & Farver, C. (2001). Differential expression of manganese superoxide dismutase and catalase in lung cancer. *Cancer Research*, *61*(23).

Hölttä, E. (1977). Oxidation of spermidine and spermine in rat liver: Purification and properties of polyamine oxidase. *Biochemistry*, 16(1). https://doi.org/10.1021/bi00620a015

Hoshino, Y., Sonoda, H., Nishimura, R., Mori, K., Ishibashi, K., & Ikeda, M. (2019). Involvement of the NADPH oxidase 2 pathway in renal oxidative stress in Aqp11 -/- mice. *Biochemistry and Biophysics Reports*, *17*, 169–176. https://doi.org/10.1016/J.BBREP.2019.01.003

Huebert, R. C., Splinter, P. L., Garcia, F., Marinelli, R. A., & Larusso, N. F. (2002). Expression and localization of aquaporin water channels in rat hepatocytes. Evidence for a role in canalicular bile secretion. *The Journal of Biological Chemistry*, 277(25), 22710–22717. https://doi.org/10.1074/JBC.M202394200

Ikaga, R., Namekata, I., Kotiadis, V. N., Ogawa, H., Duchen, M. R., Tanaka, H., & Iida-Tanaka, N. (2015). Knockdown of aquaporin-8 induces mitochondrial dysfunction in 3T3-L1 cells. *Biochemistry and Biophysics Reports*, 4, 187–195. https://doi.org/10.1016/J.BBREP.2015.09.009

Ikeda, M., Andoo, A., Shimono, M., Takamatsu, N., Taki, A., Muta, K., Matsushita, W., Uechi, T., Matsuzaki, T., Kenmochi, N., Takata, K., Sasaki, S., Ito, K., & Ishibashi, K. (2011). The NPC motif of aquaporin-11, unlike the NPA motif of known aquaporins, is essential for full expression of molecular function. *The Journal of Biological Chemistry*, 286(5), 3342–3350. https://doi.org/10.1074/JBC.M110.180968

Ilani, T., Alon, A., Grossman, I., Horowitz, B., Kartvelishvily, E., Cohen, S. R., & Fass, D. (2013). A secreted disulfide catalyst controls extracellular matrix composition and function. *Science*, 341(6141). https://doi.org/10.1126/science.1238279

Inoue, Y., Sohara, E., Kobayashi, K., Chiga, M., Rai, T., Ishibashi, K., Horie, S., Su, X., Zhou, J., Sasaki, S., & Uchida, S. (2014). Aberrant glycosylation and localization of polycystin-1 cause polycystic kidney in an AQP11 knockout model. *Journal of the American Society of Nephrology : JASN*, 25(12), 2789–2799. https://doi.org/10.1681/ASN.2013060614

Ishibashi, K., Sasaki, S., Fushimi, K., Uchida, S., Kuwahara, M., Saito, H., Furukawa, T., Nakajima, K., Yamaguchi, Y., Gojobori, T., & Marumo, F. (1994). Molecular cloning and expression of a member of the aquaporin family with permeability to glycerol and urea in addition to water expressed at the basolateral membrane of kidney collecting duct cells. *Proceedings of the National Academy of Sciences of the United States of America*, 91(14), 6269–6273. https://doi.org/10.1073/PNAS.91.14.6269

Ishikawa, Y., Iida, H., & Ishida, H. (2002). The muscarinic acetylcholine receptor-stimulated increase in aquaporin-5 levels in the apical plasma membrane in rat parotid acinar cells is coupled with activation of nitric oxide/cGMP signal transduction. *Molecular Pharmacology*, 61(6), 1423–1434. https://doi.org/10.1124/MOL.61.6.1423

Itoh, T., Rai, T., Kuwahara, M., Ko, S.B.H., Uchida, S., Sasaki, S., & Ishibashi, K. (2005). Identification of a novel aquaporin, AQP12, expressed in pancreatic acinar cells. *Biochemical and Biophysical Research Communications*, 330(3), 832–838. https://doi.org/10.1016/J.BBRC.2005.03.046

Iuchi, Y., Okada, F., Tsunoda, S., Kibe, N., Shirasawa, N., Ikawa, M., Okabe, M., Ikeda, Y., & Fujii, J. (2009). Peroxiredoxin 4 knockout results in elevated spermatogenic cell death via oxidative stress. *Biochemical Journal*, 419(1). https://doi.org/10.1042/BJ20081526

Izquierdo, A., Casas, C., Mühlenhoff, U., Lillig, C. H., & Herrero, E. (2008). Saccharomyces cerevisiae Grx6 and Grx7 are monothiol glutaredoxins associated with the early secretory pathway. *Eukaryotic Cell*, 7(8). https://doi.org/10.1128/EC.00133-08

Janiszewski, M., Lopes, L. R., Carmo, A. O., Pedro, M. A., Brandes, R. P., Santos, C.X.C., & Laurindo, F.R.M. (2005). Regulation of NAD(P)H oxidase by associated protein disulfide isomerase in vascular smooth muscle cells. *The Journal of Biological Chemistry*, 280(49), 40813–40819. https://doi.org/10.1074/JBC.M509255200

Jiang, Y., Wang, C., Ma, R., Zhao, Y., Ma, X., Wan, J., Li, C., Chen, F., Fang, F., & Li, M. (2021). Aquaporin 1 mediates early responses to osmotic stimuli in endothelial cells via the calmodulin pathway. *FEBS Open Bio*, 11(1), 75–84. https://doi.org/10.1002/2211-5463.13020

Jin, X., Yu, M. S., Huang, Y., Xiang, Z., & Chen, Y. P. (2017). MiR-30e-UCP2 pathway regulates alcoholic hepatitis progress by influencing ATP and hydrogen peroxide expression. *Oncotarget*, 8(38). https://doi.org/10.18632/oncotarget.19729

Jones, D. P., & Go, Y. M. (2010). Redox compartmentalization and cellular stress. *Diabetes, Obesity and Metabolism*, 12(Suppl. 2). https://doi.org/10.1111/j.1463-1326.2010.01266.x

Joo, J. H., Ryu, J. H., Kim, C. H., Kim, H. J., Suh, M. S., Kim, J. O., Chung, S. Y., Lee, S. N., Kim, H. M., Bae, Y. S., & Yoon, J. H. (2012). Dual oxidase 2 is essential for the toll-like receptor 5-mediated inflammatory response in airway mucosa. *Antioxidants and Redox Signaling*, 16(1). https://doi.org/10.1089/ars.2011.3898

Jorgačevski, J., Zorec, R., & Potokar, M. (2020). Insights into cell surface expression, supramolecular organization, and functions of aquaporin 4 isoforms in astrocytes. *Cells*, 9(12). https://doi.org/10.3390/CELLS9122622

Kaludercic, N., Mialet-Perez, J., Paolocci, N., Parini, A., & di Lisa, F. (2014). Monoamine oxidases as sources of oxidants in the heart. *Journal of Molecular and Cellular Cardiology*, 73, 34–42. https://doi.org/10.1016/J.YJMCC.2013.12.032

Kamarehei, M., Pejman, S., Kaboudanian Ardestani, S., Zahednasab, H., Firouzi, M., & Harirchian, M. H. (2020). Inhibition of protein disulfide isomerase has neuroprotective effects in a mouse model of experimental autoimmune encephalomyelitis. *International Immunopharmacology*, 82. https://doi.org/10.1016/J.INTIMP.2020.106286

Kamsteeg, E. J., Hendriks, G., Boone, M., Konings, I.B.M., Oorschot, V., Van Der Sluijs, P., Klumperman, J., & Deen, P.M.T. (2006). Short-chain ubiquitination mediates the regulated endocytosis of the aquaporin-2 water channel. *Proceedings of the National Academy of Sciences of the United States of America*, 103(48), 18344–18349. https://doi.org/10.1073/PNAS.0604073103

Kanemura, S., Sofia, E. F., Hirai, N., Okumura, M., Kadokura, H., & Inaba, K. (2020). Characterization of the endoplasmic reticulum-resident peroxidases GPx7 and GPx8 shows the higher oxidative activity of GPx7 and its linkage to oxidative protein folding. *The Journal of Biological Chemistry*, 295(36), 12772–12785. https://doi.org/10.1074/JBC.RA120.013607

Kang, J. H., Kim, M. H., Lee, H. J., Huh, J. W., Lee, H. S., & Lee, D. S. (2020). Peroxiredoxin 4 attenuates glutamate-induced neuronal cell death through inhibition of endoplasmic reticulum stress. *Free Radical Research*, 54(4). https://doi.org/10.1080/10715762.2020.1745201

Katchman, B. A., Ocal, I. T., Cunliffe, H. E., Chang, Y. H., Hostetter, G., Watanabe, A., LoBello, J., & Lake, D. F. (2013). Expression of quiescin sulfhydryl oxidase 1 is associated with a highly invasive phenotype and correlates with a poor prognosis in Luminal B breast cancer. *Breast Cancer Research*, 15(2). https://doi.org/10.1186/bcr3407

Kim, J. Y., Kim, M. H., Lee, H. J., Huh, J. W., Lee, S. R., Lee, H. S., & Lee, D. S. (2020). Peroxiredoxin 4 inhibits insulin-induced adipogenesis through regulation of ER stress in 3T3-L1 cells. *Molecular and Cellular Biochemistry*, 468(1–2). https://doi.org/10.1007/s11010-020-03714-w

Kirkinezos, I. G., & Moraes, C. T. (2001). Reactive oxygen species and mitochondrial diseases. *Seminars in Cell and Developmental Biology*, 12(6). https://doi.org/10.1006/scdb.2001.0282

Kitchen, P., Day, R. E., Taylor, L.H.J., Salman, M. M., Bill, R. M., Conner, M. T., & Conner, A. C. (2015). Identification and molecular mechanisms of the rapid tonicity-induced relocalization of the aquaporin 4 channel. *The Journal of Biological Chemistry*, 290(27), 16873–16881. https://doi.org/10.1074/JBC.M115.646034

Kitchen, P., Salman, M. M., Halsey, A. M., Clarke-Bland, C., MacDonald, J. A., Ishida, H., Vogel, H. J., Almutiri, S., Logan, A., Kreida, S., Al-Jubair, T., Winkel Missel, J., Gourdon, P., Törnroth-Horsefield, S., Conner, M. T., Ahmed, Z., Conner, A. C., & Bill, R. M. (2020). Targeting aquaporin-4 subcellular localization to treat central nervous system edema. *Cell*, 181(4), 784–799.e19. https://doi.org/10.1016/J.CELL.2020.03.037

Kohler, J. J., Cucoranu, I., Fields, E., Green, E., He, S., Hoying, A., Russ, R., Abuin, A., Johnson, D., Hosseini, S. H., Raper, C. M., & Lewis, W. (2009). Transgenic mitochondrial superoxide dismutase and mitochondrially targeted catalase prevent antiretroviral-induced oxidative stress and cardiomyopathy. *Laboratory Investigation*, 89(7). https://doi.org/10.1038/labinvest.2009.39

Konno, T., Melo, E. P., Lopes, C., Mehmeti, I., Lenzen, S., Ron, D., & Avezov, E. (2015). ERO1-independent production of H2O2 within the endoplasmic reticulum fuels Prdx4-mediated oxidative protein folding. *The Journal of Cell Biology*, 211(2), 253–259. https://doi.org/10.1083/JCB.201506123

Kreda, S. M., Gynn, M. C., Fenstermacher, D. A., Boucher, R. C., & Gabriel, S. E. (2001). Expression and localization of epithelial aquaporins in the adult human lung. *American Journal of Respiratory Cell and Molecular Biology*, 24(3), 224–234. https://doi.org/10.1165/AJRCMB.24.3.4367

Krüger, C., Jörns, A., Kaynert, J., Waldeck-Weiermair, M., Michel, T., Elsner, M., & Lenzen, S. (2021a). The importance of aquaporin-8 for cytokine-mediated toxicity in rat insulin-producing cells. *Free Radical Biology & Medicine*, 174, 135–143. https://doi.org/10.1016/J.FREERADBIOMED.2021.08.003

Krüger, C., Waldeck-Weiermair, M., Kaynert, J., Pokrant, T., Komaragiri, Y., Otto, O., Michel, T., & Elsner, M. (2021b). AQP8 is a crucial H 2 O 2 transporter in insulin-producing RINm5F cells. *Redox Biology*, 43. https://doi.org/10.1016/J.REDOX.2021.101962

Księżakowska-Łakoma, K., Zyła, M., & Wilczyński, J. R. (2014). Mitochondrial dysfunction in cancer. *Przeglad Menopauzalny*, *18*(2), 136–144. Termedia Publishing House Ltd. https://doi.org/10.5114/pm.2014.42717

Kudryavtseva, A. V., Krasnov, G. S., Dmitriev, A. A., Alekseev, B. Y., Kardymon, O. L., Sadritdinova, A. F., Fedorova, M. S., Pokrovsky, A. V., Melnikova, N. V., Kaprin, A. D., Moskalev, A. A., & Snezhkina, A. V. (2016). Mitochondrial dysfunction and oxidative stress in aging and cancer. *Oncotarget*, *7*(29). https://doi.org/10.18632/oncotarget.9821

Kumar, S., Kawałek, A., & van der Klei, I. J. (2014). Peroxisomal quality control mechanisms. *Current Opinion in Microbiology*, 22. https://doi.org/10.1016/j.mib.2014.09.009

Kurz, T., Terman, A., Gustafsson, B., & Brunk, U. T. (2008). Lysosomes in iron metabolism, ageing and apoptosis. *Histochemistry and Cell Biology*, *129*(4). https://doi.org/10.1007/s00418-008-0394-y

Laforenza, U., Gastaldi, G., Grazioli, M., Cova, E., Tritto, S., Faelli, A., Calamita, G., & Ventura, U. (2005). Expression and immunolocalization of aquaporin-7 in rat gastrointestinal tract. *Biology of the Cell*, *97*(8), 605–613. https://doi.org/10.1042/BC20040090

Laforenza, U., Pellavio, G., Marchetti, A. L., Omes, C., Todaro, F., & Gastaldi, G. (2016). Aquaporin-mediated water and hydrogen peroxide transport is involved in normal human spermatozoa functioning. *International Journal of Molecular Sciences*, *18*(1). https://doi.org/10.3390/IJMS18010066

Laforenza, U., Scaffino, M. F., & Gastaldi, G. (2013). Aquaporin-10 represents an alternative pathway for glycerol efflux from human adipocytes. *PLoS ONE*, *8*(1). https://doi.org/10.1371/JOURNAL.PONE.0054474

Lake, D. F., & Faigel, D. O. (2014). The emerging role of qsox1 in cancer. *Antioxidants and Redox Signaling*, *21*(3). https://doi.org/10.1089/ars.2013.5572

Landry, W. D., & Cotter, T. G. (2014). ROS signalling, NADPH oxidases and cancer. *Biochemical Society Transactions*, *42*(4). https://doi.org/10.1042/BST20140060

Lauer, C., Völkl, A., Riedl, S., Fahimi, H. D., & Beier, K. (1999). Impairment of peroxisomal biogenesis in human colon carcinoma. *Carcinogenesis*, *20*(6). https://doi.org/10.1093/carcin/20.6.985

Lebovitz, R. M., Zhang, H., Vogel, H., Cartwright, J., Dionne, L., Lu, N., Huang, S., & Matzuk, M. M. (1996). Neurodegeneration, myocardial injury, and perinatal death in mitochondrial superoxide dismutase-deficient mice. *Proceedings of the National Academy of Sciences of the United States of America*, *93*(18). https://doi.org/10.1073/pnas.93.18.9782

Lee, J. K., Kozono, D., Remis, J., Kitagawa, Y., Agre, P., & Stroud, R. M. (2005). Structural basis for conductance by the archaeal aquaporin AqpM at 1.68 A. *Proceedings of the National Academy of Sciences of the United States of America*, *102*(52), 18932–18937. https://doi.org/10.1073/PNAS.0509469102

Lee, J. N., Dutta, R. K., Maharjan, Y., Liu, Z. Qiang, Lim, J. Y., Kim, S. J., Cho, D. H., So, H. S., Choe, S. K., & Park, R. (2018). Catalase inhibition induces pexophagy through ROS accumulation. *Biochemical and Biophysical Research Communications*, *501*(3). https://doi.org/10.1016/j.bbrc.2018.05.050

Lee, J. S., Hou, X., Bishop, N., Wang, S., Flack, A., Cho, W. J., Chen, X., Mao, G., Taatjes, D. J., Sun, F., Zhang, K., & Jena, B. P. (2013). Aquaporin-assisted and ER-mediated mitochondrial fission: a hypothesis. *Micron (Oxford, England : 1993)*, *47*, 50–58. https://doi.org/10.1016/J.MICRON.2013.01.005

Lener, B., Kozieł, R., Pircher, H., Hütter, E., Greussing, R., Herndler-Brandstetter, D., Hermann, M., Unterluggauer, H., & Jansen-Dürr, P. (2009). The NADPH oxidase Nox4 restricts the replicative lifespan of human endothelial cells. *Biochemical Journal*, *423*(3). https://doi.org/10.1042/BJ20090666

Li, G., Mongillo, M., Chin, K. T., Harding, H., Ron, D., Marks, A. R., & Tabas, I. (2009). Role of ERO1-α-mediated stimulation of inositol 1,4,5-triphosphate receptor activity in endoplasmic reticulum stress-induced apoptosis. *Journal of Cell Biology*, *186*(6). https://doi.org/10.1083/jcb.200904060

Li, J., Tong, C., Xu, P., Wang, L., Han, T. li, Wen, L., Luo, X., Tan, B., Zhu, F., Gui, S., Gao, R., Qi, H., & Baker, P. N. (2019). QSOX1 regulates trophoblastic apoptosis in preeclampsia through hydrogen peroxide production. *Journal of Maternal-Fetal and Neonatal Medicine*, *32*(22). https://doi.org/10.1080/14767058.2018.1471459

Li, Y., Huang, T. T., Carlson, E. J., Melov, S., Ursell, P. C., Olson, J. L., Noble, L. J., Yoshimura, M. P., Berger, C., Chan, P. H., Wallace, D. C., & Epstein, C. J. (1995). Dilated cardiomyopathy and neonatal lethality in mutant mice lacking manganese superoxide dismutase. *Nature Genetics*, *11*(4). https://doi.org/10.1038/ng1295-376

Liu, L., Lu, Y., Kong, H., Li, L., Marshall, C., Xiao, M., Ding, J., Gao, J., & Hu, G. (2012). Aquaporin-4 deficiency exacerbates brain oxidative damage and memory deficits induced by long-term ovarian hormone deprivation and D-galactose injection. *The International Journal of Neuropsychopharmacology*, 15(1), 55–68. https://doi.org/10.1017/S1461145711000022

Lizák, B., Birk, J., Zana, M., Kosztyi, G., Kratschmar, D. V., Odermatt, A., Zimmermann, R., Geiszt, M., Appenzeller-Herzog, C., & Bánhegyi, G. (2020). Ca2+ mobilization-dependent reduction of the endoplasmic reticulum lumen is due to influx of cytosolic glutathione. *BMC Biology*, 18(1). https://doi.org/10.1186/s12915-020-0749-y

Ma, T., Song, Y., Yang, B., Gillespie, A., Carlson, E. J., Epstein, C. J., & Verkman, A. S. (2000). Nephrogenic diabetes insipidus in mice lacking aquaporin-3 water channels. *Proceedings of the National Academy of Sciences of the United States of America*, 97(8), 4386–4391. https://doi.org/10.1073/PNAS.080499597

Mailloux, R. J., Craig Ayre, D., & Christian, S. L. (2016). Induction of mitochondrial reactive oxygen species production by GSH mediated S-glutathionylation of 2-oxoglutarate dehydrogenase. *Redox Biology*, 8. https://doi.org/10.1016/j.redox.2016.02.002

Mairet-Coello, G., Tury, A., Esnard-Feve, A., Fellmann, D., Risold, P. Y., & Griffond, B. (2004). FAD-linked sulfhydryl oxidase QSOX: Topographic, cellular, and subcellular immunolocalization in adult rat central nervous system. *Journal of Comparative Neurology*, 473(3). https://doi.org/10.1002/cne.20126

Malinouski, M., Zhou, Y., Belousov, V. V., Hatfield, D. L., & Gladyshev, V. N. (2011). Hydrogen peroxide probes directed to different cellular compartments. *PLoS ONE*, 6(1). https://doi.org/10.1371/journal.pone.0014564

Marchissio, M. J., Francés, D.E.A., Carnovale, C. E., & Marinelli, R. A. (2012). Mitochondrial aquaporin-8 knockdown in human hepatoma HepG2 cells causes ROS-induced mitochondrial depolarization and loss of viability. *Toxicology and Applied Pharmacology*, 264(2), 246–254. https://doi.org/10.1016/J.TAAP.2012.08.005

Marchissio, M. J., Francés, D.E.A., Carnovale, C. E., & Marinelli, R. A. (2014). Evidence for necrosis, but not apoptosis, in human hepatoma cells with knockdown of mitochondrial aquaporin-8. *Apoptosis: An International Journal on Programmed Cell Death*, 19(5), 851–859. https://doi.org/10.1007/S10495-014-0966-3

Marciniak, S. J., Yun, C. Y., Oyadomari, S., Novoa, I., Zhang, Y., Jungreis, R., Nagata, K., Harding, H. P., & Ron, D. (2004). CHOP induces death by promoting protein synthesis and oxidation in the stressed endoplasmic reticulum. *Genes and Development*, 18(24). https://doi.org/10.1101/gad.1250704

Marí, M., Morales, A., Colell, A., García-Ruiz, C., & Fernández-Checa, J. C. (2009). Mitochondrial glutathione, a key survival antioxidant. *Antioxidants and Redox Signaling*, 11(11). https://doi.org/10.1089/ars.2009.2695

Markovic, J., Mora, N. J., Broseta, A. M., Gimeno, A., de-la-Concepción, N., Viña, J., & Pallardó, F. V. (2009). The depletion of nuclear glutathione impairs cell proliferation in 3t3 fibroblasts. *PLoS ONE*, 4(7). https://doi.org/10.1371/journal.pone.0006413

Markvicheva, K. N., Bilan, D. S., Mishina, N. M., Gorokhovatsky, A. Y., Vinokurov, L. M., Lukyanov, S., & Belousov, V. V. (2011). A genetically encoded sensor for H2O2 with expanded dynamic range. *Bioorganic & Medicinal Chemistry*, 19(3), 1079–1084. https://doi.org/10.1016/J.BMC.2010.07.014

Matsushima, S., Kuroda, J., Zhai, P., Liu, T., Ikeda, S., Nagarajan, N., Oka, S. I., Yokota, T., Kinugawa, S., Hsu, C. P., Li, H., Tsutsui, H., & Sadoshima, J. (2016). Tyrosine kinase FYN negatively regulates NOX4 in cardiac remodeling. *Journal of Clinical Investigation*, 126(9). https://doi.org/10.1172/JCI85624

McCord, J. M., Roy, R. S., & Schaffer, S. W. (1985). Free radicals and myocardial ischemia. The role of xanthine oxidase. *Advances in Myocardiology*, 5.

Mehmeti, I., Lortz, S., Avezov, E., Jörns, A., & Lenzen, S. (2017). ER-resident antioxidative GPx7 and GPx8 enzyme isoforms protect insulin-secreting INS-1E β-cells against lipotoxicity by improving the ER antioxidative capacity. *Free Radical Biology and Medicine*, 112. https://doi.org/10.1016/j.freeradbiomed.2017.07.021

Mehmeti, I., Lortz, S., Elsner, M., & Lenzen, S. (2014). Peroxiredoxin 4 improves insulin biosynthesis and glucose-induced insulin secretion in insulin-secreting INS-1E cells. *Journal of Biological Chemistry*, 289(39). https://doi.org/10.1074/jbc.M114.568329

Mehmeti, I., Lortz, S., & Lenzen, S. (2012). The H2O2-sensitive HyPer protein targeted to the endoplasmic reticulum as a mirror of the oxidizing thiol-disulfide milieu. *Free Radical Biology & Medicine*, 53(7), 1451–1458. https://doi.org/10.1016/J.FREERADBIOMED.2012.08.010

Melo, E. P., Lopes, C., Gollwitzer, P., Lortz, S., Lenzen, S., Mehmeti, I., Kaminski, C. F., Ron, D., & Avezov, E. (2017). TriPer, an optical probe tuned to the endoplasmic reticulum tracks changes in luminal H2O2. *BMC Biology*, 15(1). https://doi.org/10.1186/s12915-017-0367-5

Mesecke, N., Spang, A., Deponte, M., & Herrmann, J. M. (2008). A novel group of glutaredoxins in the cis-Golgi critical for oxidative stress resistance. *Molecular Biology of the Cell, 19*(6). https://doi.org/10.1091/mbc.E07-09-0896

Miller, E. W., Dickinson, B. C., & Chang, C. J. (2010). Aquaporin-3 mediates hydrogen peroxide uptake to regulate downstream intracellular signaling. *Proceedings of the National Academy of Sciences of the United States of America, 107*(36), 15681–15686. https://doi.org/10.1073/PNAS.1005776107

Minowa, O., Arai, T., Hirano, M., Monden, Y., Nakai, S., Fukuda, M., Itoh, M., Takano, H., Hippou, Y., Aburatani, H., Masumura, K. I., Nohmi, T., Nishimura, S., & Noda, T. (2000). Mmh/Ogg1 gene inactivation results in accumulation of 8-hydroxyguanine in mice. *Proceedings of the National Academy of Sciences of the United States of America, 97*(8). https://doi.org/10.1073/pnas.050404497

Miyauchi, T., Yamamoto, H., Abe, Y., Yoshida, G. J., Rojek, A., Sohara, E., Uchida, S., Nielsen, S., & Yasui, M. (2015). Dynamic subcellular localization of aquaporin-7 in white adipocytes. *FEBS Letters, 589*(5), 608–614. https://doi.org/10.1016/J.FEBSLET.2015.01.025

Mizutani, K., Guo, X., Shioya, A., Zhang, J., Zheng, J., Kurose, N., Ishibashi, H., Motono, N., Uramoto, H., & Yamada, S. (2019). The impact of PRDX4 and the EGFR mutation status on cellular proliferation in lung adenocarcinoma. *International Journal of Medical Sciences, 16*(9). https://doi.org/10.7150/ijms.36071

Mobasheri, A., Shakibaei, M., & Marples, D. (2004). Immunohistochemical localization of aquaporin 10 in the apical membranes of the human ileum: a potential pathway for luminal water and small solute absorption. *Histochemistry and Cell Biology, 121*(6), 463–471. https://doi.org/10.1007/S00418-004-0657-1

Montero, D., Tachibana, C., Rahr Winther, J., & Appenzeller-Herzog, C. (2013). Intracellular glutathione pools are heterogeneously concentrated. *Redox Biology, 1*(1). https://doi.org/10.1016/j.redox.2013.10.005

Montiel, V., Bella, R., Michel, L.Y.M., Esfahani, H., De Mulder, D., Robinson, E. L., Deglasse, J.-P., Tiburcy, M., Chow, P. H., Jonas, J.-C., Gilon, P., Steinhorn, B., Michel, T., Beauloye, C., Bertrand, L., Farah, C., Dei Zotti, F., Debaix, H., Bouzin, C., … Balligand, J.-L. (2020). Inhibition of aquaporin-1 prevents myocardial remodeling by blocking the transmembrane transport of hydrogen peroxide. *Science Translational Medicine, 12*(564). https://doi.org/10.1126/SCITRANSLMED.AAY2176

Morel, C., Adami, P., Musard, J. F., Duval, D., Radom, J., & Jouvenot, M. (2007). Involvement of sulfhydryl oxidase QSOX1 in the protection of cells against oxidative stress-induced apoptosis. *Experimental Cell Research, 313*(19). https://doi.org/10.1016/j.yexcr.2007.09.003

Morishita, Y., Matsuzaki, T., Hara-chikuma, M., Andoo, A., Shimono, M., Matsuki, A., Kobayashi, K., Ikeda, M., Yamamoto, T., Verkman, A., Kusano, E., Ookawara, S., Takata, K., Sasaki, S., & Ishibashi, K. (2005). Disruption of aquaporin-11 produces polycystic kidneys following vacuolization of the proximal tubule. *Molecular and Cellular Biology, 25*(17), 7770–7779. https://doi.org/10.1128/MCB.25.17.7770-7779.2005

Murray-Stewart, T., Wang, Y., Goodwin, A., Hacker, A., Meeker, A., & Casero, R. A. (2008). Nuclear localization of human spermine oxidase isoforms—Possible implications in drug response and disease etiology. *FEBS Journal, 275*(11). https://doi.org/10.1111/j.1742-4658.2008.06419.x

Nègre-Salvayre, A., Hirtz, C., Carrera, G., Cazenave, R., Troly, M., Salvayre, R., Pénicaud, L., & Casteilla, L. (1997). A role for uncoupling protein-2 as a regulator of mitochondrial hydrogen peroxide generation. *The FASEB Journal, 11*(10). https://doi.org/10.1096/fasebj.11.10.9271366

Newsholme, P., Haber, E. P., Hirabara, S. M., Rebelato, E.L.O., Procopio, J., Morgan, D., Oliveira-Emilio, H. C., Carpinelli, A. R., & Curi, R. (2007). Diabetes associated cell stress and dysfunction: Role of mitochondrial and non-mitochondrial ROS production and activity. *Journal of Physiology, 583*(1). https://doi.org/10.1113/jphysiol.2007.135871

Nguyen, V. D., Saaranen, M. J., Karala, A. R., Lappi, A. K., Wang, L., Raykhel, I. B., Alanen, H. I., Salo, K.E.H., Wang, C. C., & Ruddock, L. W. (2011). Two endoplasmic reticulum PDI peroxidases increase the efficiency of the use of peroxide during disulfide bond formation. *Journal of Molecular Biology, 406*(3), 503–515. https://doi.org/10.1016/J.JMB.2010.12.039

Nielsen, S., Chou, C. L., Marples, D., Christensen, E. I., Kishore, B. K., & Knepper, M. A. (1995). Vasopressin increases water permeability of kidney collecting duct by inducing translocation of aquaporin-CD water channels to plasma membrane. *Proceedings of the National Academy of Sciences of the United States of America, 92*(4), 1013–1017. https://doi.org/10.1073/PNAS.92.4.1013

Nohl, H., & Gille, L. (2002). The bifunctional activity of ubiquinone in lysosomal membranes. *Biogerontology*, 3(1–2). https://doi.org/10.1023/A:1015288220217

Ohdate, T., & Inoue, Y. (2012). Involvement of glutathione peroxidase 1 in growth and peroxisome formation in Saccharomyces cerevisiae in oleic acid medium. *Biochimica et Biophysica Acta—Molecular and Cell Biology of Lipids*, 1821(9). https://doi.org/10.1016/j.bbalip.2012.05.004

Ohshiro, K., Yaoita, E., Yoshida, Y., Fujinaka, H., Matsuki, A., Kamiie, J., Kovalenko, P., & Yamamoto, T. (2001). Expression and immunolocalization of AQP6 in intercalated cells of the rat kidney collecting duct. *Archives of Histology and Cytology*, 64(3), 329–338. https://doi.org/10.1679/AOHC.64.329

Ohta, E., Itoh, T., Nemoto, T., Kumagai, J., Ko, S.B.H., Ishibashi, K., Ohno, M., Uchida, K., Ohta, A., Sohara, E., Uchida, S., Sasaki, S., & Rai, T. (2009). Pancreas-specific aquaporin 12 null mice showed increased susceptibility to caerulein-induced acute pancreatitis. *American Journal of Physiology. Cell Physiology*, 297(6). https://doi.org/10.1152/AJPCELL.00117.2009

Okada, S., Misaka, T., Tanaka, Y., Matsumoto, I., Ishibashi, K., Sasaki, S., & Abe, K. (2008). Aquaporin-11 knockout mice and polycystic kidney disease animals share a common mechanism of cyst formation. *FASEB Journal: Official Publication of the Federation of American Societies for Experimental Biology*, 22(10), 3672–3684. https://doi.org/10.1096/FJ.08-111872

Olesen, E.T.B., De Seigneux, S., Wang, G., Lütken, S. C., Frøkiaer, J., Kwon, T. H., & Nielsen, S. (2009). Rapid and segmental specific dysregulation of AQP2, S256-pAQP2 and renal sodium transporters in rats with LPS-induced endotoxaemia. *Nephrology, Dialysis, Transplantation: Official Publication of the European Dialysis and Transplant Association—European Renal Association*, 24(8), 2338–2349. https://doi.org/10.1093/NDT/GFP011

Omura, T. (2006). Mitochondrial P450s. *Chemico-Biological Interactions*, 163(1–2). https://doi.org/10.1016/j.cbi.2006.06.008

Orr, A. L., Quinlan, C. L., Perevoshchikova, I. V., & Brand, M. D. (2012). A refined analysis of superoxide production by mitochondrial sn-glycerol 3-phosphate dehydrogenase. *Journal of Biological Chemistry*, 287(51). https://doi.org/10.1074/jbc.M112.397828

Pagani, M., Fabbri, M., Benedetti, C., Fassio, A., Pilati, S., Bulleid, N. J., Cabibbo, A., & Sitia, R. (2000). Endoplasmic reticulum oxidoreductin 1-Lβ (ERO1-Lβ), a human gene induced in the course of the unfolded protein response. *Journal of Biological Chemistry*, 275(31). https://doi.org/10.1074/jbc.M003061200

Pak, V. V, Ezeriņa, D., Lyublinskaya, O. G., Pedre, B., Tyurin-Kuzmin, P. A., Mishina, N. M., Thauvin, M., Young, D., Wahni, K., Martínez Gache, S. A., Demidovich, A. D., Ermakova, Y. G., Maslova, Y. D., Shokhina, A. G., Eroglu, E., Bilan, D. S., Bogeski, I., Michel, T., Vriz, S., … Belousov, V. V. (2020). Ultrasensitive genetically encoded indicator for hydrogen peroxide identifies roles for the oxidant in cell migration and mitochondrial function. *Cell Metabolism*, 31(3). https://doi.org/10.1016/j.cmet.2020.02.003

Pallardó, F. V., Markovic, J., García, J. L., & Viña, J. (2009). Role of nuclear glutathione as a key regulator of cell proliferation. *Molecular Aspects of Medicine*, 30(1–2). https://doi.org/10.1016/j.mam.2009.01.001

Park, H. S., Jung, H. Y., Park, E. Y., Kim, J., Lee, W. J., & Bae, Y. S. (2004). Cutting edge: direct interaction of TLR4 with NAD(P)H Oxidase 4 Isozyme is essential for lipopolysaccharide-induced production of reactive oxygen species and activation of NF-κB. *The Journal of Immunology*, 173(6). https://doi.org/10.4049/jimmunol.173.6.3589

Parreira, K. S., Debaix, H., Cnops, Y., Geffers, L., & Devuyst, O. (2009). Expression patterns of the aquaporin gene family during renal development: Influence of genetic variability. *Pflugers Archiv: European Journal of Physiology*, 458(4), 745–759. https://doi.org/10.1007/S00424-009-0667-X

Patterson, J. C., Joughin, B. A., van de Kooij, B., Lim, D. C., Lauffenburger, D. A., & Yaffe, M. B. (2019). ROS and oxidative stress are elevated in mitosis during asynchronous cell cycle progression and are exacerbated by mitotic arrest. *Cell Systems*, 8(2). https://doi.org/10.1016/j.cels.2019.01.005

Peng, D. F., Belkhiri, A., Hu, T. L., Chaturvedi, R., Asim, M., Wilson, K. T., Zaika, A., & El-Rifai, W. (2012). Glutathione peroxidase 7 protects against oxidative DNA damage in oesophageal cells. *Gut*, 61(9). https://doi.org/10.1136/gutjnl-2011-301078

Pernodet, N., Hermetet, F., Adami, P., Vejux, A., Descotes, F., Borg, C., Adams, M., Pallandre, J. R., Viennet, G., Esnard, F., Jouvenot, M., & Despouy, G. (2012). High expression of QSOX1 reduces tumorogenesis, and is associated with a better outcome for breast cancer patients. *Breast Cancer Research*, 14(5). https://doi.org/10.1186/bcr3341

Phillips, M. J., & Voeltz, G. K. (2016). Structure and function of ER membrane contact sites with other organelles. *Nature Reviews. Molecular Cell Biology*, 17(2), 69–82. https://doi.org/10.1038/NRM.2015.8

Phung, C. D., Ezieme, J. A., & Turrens, J. F. (1994). Hydrogen peroxide metabolism in skeletal muscle mitochondria. *Archives of Biochemistry and Biophysics*, 315(2). https://doi.org/10.1006/abbi.1994.1528

Pollard, M. G., Travers, K. J., & Weissman, J. S. (1998). Ero1p: A novel and ubiquitous protein with an essential role in oxidative protein folding in the endoplasmic reticulum. *Molecular Cell*, 1(2). https://doi.org/10.1016/S1097-2765(00)80018-0

Princiotta, M. F., Finzi, D., Qian, S. B., Gibbs, J., Schuchmann, S., Buttgereit, F., Bennink, J. R., & Yewdell, J. W. (2003). Quantitating protein synthesis, degradation, and endogenous antigen processing. *Immunity*, 18(3). https://doi.org/10.1016/S1074-7613(03)00051-7

Prior, K. K., Wittig, I., Leisegang, M. S., Groenendyk, J., Weissmann, N., Michalak, M., Jansen-Dürr, P., Shah, A. M., & Brandes, R. P. (2016). The endoplasmic reticulum chaperone calnexin is a NADPH oxidase NOX4 interacting protein. *Journal of Biological Chemistry*, 291(13). https://doi.org/10.1074/jbc.M115.710772

Quinlan, C. L., Goncalves, R.L.S., Hey-Mogensen, M., Yadava, N., Bunik, V. I., & Brand, M. D. (2014). The 2-oxoacid dehydrogenase complexes in mitochondria can produce superoxide/hydrogen peroxide at much higher rates than complex I. *Journal of Biological Chemistry*, 289(12). https://doi.org/10.1074/jbc.M113.545301

Quinlan, C. L., Perevoshchikova, I. V., Hey-Mogensen, M., Orr, A. L., & Brand, M. D. (2013). Sites of reactive oxygen species generation by mitochondria oxidizing different substrates. *Redox Biology*, 1(1). https://doi.org/10.1016/j.redox.2013.04.005

Rabilloud, T., Heller, M., Rigobello, M. P., Bindoli, A., Aebersold, R., & Lunardi, J. (2001). The mitochondrial antioxidant defence system and its response to oxidative stress. *Proteomics*, 1(9). https://doi.org/10.1002/1615-9861(200109)1:9<1105::AID-PROT1105>3.0.CO;2-M

Radi, R. (2013). Peroxynitrite, a stealthy biological oxidant. *Journal of Biological Chemistry*, 288(37). https://doi.org/10.1074/jbc.R113.472936

Radi, R., Turrens, J. F., Chang, L. Y., Bush, K. M., Crapo, J. D., & Freeman, B. A. (1991). Detection of catalase in rat heart mitochondria. *Journal of Biological Chemistry*, 266(32). https://doi.org/10.1016/s0021-9258(18)54740-2

Raje, S., & Thorpe, C. (2003). Inter-domain redox communication in flavoenzymes of the quiescin/sulfhydryl oxidase family: Role of a thioredoxin domain in disulfide bond formation. *Biochemistry*, 42(15). https://doi.org/10.1021/bi030003z

Ramming, T., Hansen, H. G., Nagata, K., Ellgaard, L., & Appenzeller-Herzog, C. (2014). GPx8 peroxidase prevents leakage of H2O2 from the endoplasmic reticulum. *Free Radical Biology and Medicine*, 70. https://doi.org/10.1016/j.freeradbiomed.2014.01.018

Ramming, T., Okumura, M., Kanemura, S., Baday, S., Birk, J., Moes, S., Spiess, M., Jenö, P., Bernèche, S., Inaba, K., & Appenzeller-Herzog, C. (2015). A PDI-catalyzed thiol-disulfide switch regulates the production of hydrogen peroxide by human Ero1. *Free Radical Biology and Medicine*, 83. https://doi.org/10.1016/j.freeradbiomed.2015.02.011

Rapti, K., Diokmetzidou, A., Kloukina, I., Milner, D. J., Varela, A., Davos, C. H., & Capetanaki, Y. (2017). Opposite effects of catalase and MnSOD ectopic expression on stress induced defects and mortality in the desmin deficient cardiomyopathy model. *Free Radical Biology and Medicine*, 110. https://doi.org/10.1016/j.freeradbiomed.2017.06.010

Rashba-Step, J., & Cederbaum, A. I. (1994). Generation of reactive oxygen intermediates by human liver microsomes in the presence of NADPH or NADH. *Molecular Pharmacology*, 45(1).

Rimessi, A., Previati, M., Nigro, F., Wieckowski, M. R., & Pinton, P. (2016). Mitochondrial reactive oxygen species and inflammation: Molecular mechanisms, diseases and promising therapies. *International Journal of Biochemistry and Cell Biology*, 81. https://doi.org/10.1016/j.biocel.2016.06.015

Rodrigues, C., Milković, L., Bujak, I. T., Tomljanovic, M., Soveral, G., & Cipak Gasparovic, A. (2019a). Lipid profile and aquaporin expression under oxidative stress in breast cancer cells of different malignancies. *Oxidative Medicine and Cellular Longevity*, 2019. https://doi.org/10.1155/2019/2061830

Rodrigues, C., Pimpão, C., Mósca, A. F., Coxixo, A. S., Lopes, D., Da Silva, I. V., Pedersen, P. A., Antunes, F., & Soveral, G. (2019b). Human aquaporin-5 facilitates hydrogen peroxide permeation affecting adaption to oxidative stress and cancer cell migration. *Cancers*, 11(7). https://doi.org/10.3390/CANCERS11070932

Rohan Fernando, M., Lechner, J. M., Löfgren, S., Gladyshev, V. N., Lou, M. F., Rohan Fernando, M., Lechner, J. M., Löfgren, S., Gladyshev, V. N., & Lou, M. F. (2006). Mitochondrial thioltransferase (glutaredoxin 2) has GSH-dependent and thioredoxin reductase-dependent peroxidase activities in vitro and in lens epithelial cells. *The FASEB Journal*, 20(14). https://doi.org/10.1096/fj.06-5919fje

Rokka, A., Antonenkov, V. D., Soininen, R., Immonen, H. L., Pirilä, P. L., Bergmann, U., Sormunen, R. T.,

Weckström, M., Benz, R., & Hiltunen, J. K. (2009). Pxmp2 is a channel-forming protein in mammalian peroxisomal membrane. *PLoS ONE*, 4(4). https://doi.org/10.1371/journal.pone.0005090

Rose, K.M.L., Gourdie, R. G., Prescott, A. R., Quinlan, R. A., Crouch, R. K., & Schey, K. L. (2006). The C terminus of lens aquaporin 0 interacts with the cytoskeletal proteins filensin and CP49. *Investigative Ophthalmology & Visual Science*, 47(4), 1562–1570. https://doi.org/10.1167/IOVS.05-1313

Rowland, A. A., & Voeltz, G. K. (2012). Endoplasmic reticulum-mitochondria contacts: Function of the junction. *Nature Reviews. Molecular Cell Biology*, 13(10), 607–615. https://doi.org/10.1038/NRM3440

Sada, K., Nishikawa, T., Kukidome, D., Yoshinaga, T., Kajihara, N., Sonoda, K., Senokuchi, T., Motoshima, H., Matsumura, T., & Araki, E. (2016). Hyperglycemia induces cellular hypoxia through production of mitochondrial ROS followed by suppression of aquaporin-1. *PLoS ONE*, 11(7). https://doi.org/10.1371/JOURNAL.PONE.0158619

Saito, S., Matsui, H., Kawano, M., Kumagai, K., Tomishige, N., Hanada, K., Echigo, S., Tamura, S., & Kobayashi, T. (2008). Protein phosphatase 2Cepsilon is an endoplasmic reticulum integral membrane protein that dephosphorylates the ceramide transport protein CERT to enhance its association with organelle membranes. *The Journal of Biological Chemistry*, 283(10), 6584–6593. https://doi.org/10.1074/JBC.M707691200

Salman, M. M., Kitchen, P., Woodroofe, M. N., Brown, J. E., Bill, R. M., Conner, A. C., & Conner, M. T. (2017). Hypothermia increases aquaporin 4 (AQP4) plasma membrane abundance in human primary cortical astrocytes via a calcium/transient receptor potential vanilloid 4 (TRPV4)- and calmodulin-mediated mechanism. *The European Journal of Neuroscience*, 46(9), 2542–2547. https://doi.org/10.1111/EJN.13723

Salmeen, A., Andersen, J. N., Myers, M. P., Meng, T. C., Hinks, J. A., Tonks, N. K., & Barford, D. (2003). Redox regulation of protein tyrosine phosphatase 1B involves a sulphenyl-amide intermediate. *Nature*, 423(6941), 769–773. https://doi.org/10.1038/NATURE01680

Samoylenko, A., Hossain, J. al, Mennerich, D., Kellokumpu, S., Hiltunen, J. K., & Kietzmann, T. (2013). Nutritional countermeasures targeting reactive oxygen species in cancer: From mechanisms to biomarkers and clinical evidence. *Antioxidants and Redox Signaling*, 19(17). https://doi.org/10.1089/ars.2012.4662

Sandalio, L. M., Rodríguez-Serrano, M., Romero-Puertas, M. C., & del Río Luis, A. (2013). Role of peroxisomes as a source of reactive oxygen species (ROS) signaling molecules. *Subcellular Biochemistry*, 69. https://doi.org/10.1007/978-94-007-6889-5_13

Sangwung, P., Greco, T. M., Wang, Y., Ischiropoulos, H., Sessa, W. C., & Iwakiri, Y. (2012). Proteomic identification of S-nitrosylated Golgi proteins: New insights into endothelial cell regulation by eNOS-derived NO. *PLoS ONE*, 7(2). https://doi.org/10.1371/journal.pone.0031564

Santos, C. X., Hafstad, A. D., Beretta, M., Zhang, M., Molenaar, C., Kopec, J., Fotinou, D., Murray, T. V, Cobb, A. M., Martin, D., Zeh Silva, M., Anilkumar, N., Schröder, K., Shanahan, C. M., Brewer, A. C., Brandes, R. P., Blanc, E., Parsons, M., Belousov, V., … Shah, A. M. (2016). Targeted redox inhibition of protein phosphatase 1 by Nox4 regulates eIF2α-mediated stress signaling. *The EMBO Journal*, 35(3), 319–334. https://doi.org/10.15252/EMBJ.201592394

Satooka, H., & Hara-Chikuma, M. (2016). Aquaporin-3 controls breast cancer cell migration by regulating hydrogen peroxide transport and its downstream cell signaling. *Molecular and Cellular Biology*, 36(7), 1206–1218. https://doi.org/10.1128/MCB.00971-15

Sayanova, O., Haslam, R., Qi, B., Lazarus, C. M., & Napier, J. A. (2006). The alternative pathway C20 Δ8-desaturase from the non-photosynthetic organism Acanthamoeba castellanii is an atypical cytochrome b5-fusion desaturase. *FEBS Letters*, 580(8). https://doi.org/10.1016/j.febslet.2006.02.062

Schrader, M., & Fahimi, H. D. (2006). Peroxisomes and oxidative stress. *Biochimica et Biophysica Acta*, 1763(12), 1755–1766. https://doi.org/10.1016/J.BBAMCR.2006.09.006

Seleit, I., Bakry, O. A., El Rebey, H. S., El-Akabawy, G., & Hamza, G. (2017). Is aquaporin-3 a determinant factor of intrinsic and extrinsic aging? An immunohistochemical and morphometric study. *Applied Immunohistochemistry & Molecular Morphology : AIMM*, 25(1), 49–57. https://doi.org/10.1097/PAI.0000000000000265

Sessa, W. C., García-Cardeña, G., Liu, J., Keh, A., Pollock, J. S., Bradley, J., Thiru, S., Braverman, I. M., & Desai, K. M. (1995). The Golgi association of endothelial nitric oxide synthase is necessary for the efficient synthesis of nitric oxide. *Journal of Biological Chemistry*, 270(30). https://doi.org/10.1074/jbc.270.30.17641

Shanmugasundaram, K., Nayak, B. K., Friedrichs, W. E., Kaushik, D., Rodriguez, R., & Block, K.

(2017). NOX4 functions as a mitochondrial energetic sensor coupling cancer metabolic reprogramming to drug resistance. *Nature Communications*, 8(1). https://doi.org/10.1038/s41467-017-01106-1

Shergalis, A. G., Hu, S., Bankhead, A., & Neamati, N. (2020). Role of the ERO1-PDI interaction in oxidative protein folding and disease. *Pharmacology and Therapeutics*, 210. https://doi.org/10.1016/j.pharmthera.2020.107525

Shimizu, Y., & Hendershot, L. M. (2009). Oxidative folding: cellular strategies for dealing with the resultant equimolar production of reactive oxygen species. *Antioxidants and Redox Signaling*, 11(9), 2317–2331. https://doi.org/10.1089/ars.2009.2501

Sidhaye, V., Hoffert, J. D., & King, L. S. (2005). cAMP has distinct acute and chronic effects on aquaporin-5 in lung epithelial cells. *The Journal of Biological Chemistry*, 280(5), 3590–3596. https://doi.org/10.1074/JBC.M411038200

Sindhu Kumari, S., Varadaraj, M., Yerramilli, V. S., Menon, A. G., & Varadaraj, K. (2012). Spatial expression of aquaporin 5 in mammalian cornea and lens, and regulation of Its localization by phosphokinase A. *Molecular Vision*, 18, 957.

Sindler, A. L., Delp, M. D., Reyes, R., Wu, G., & Muller-Delp, J. M. (2009). Effects of ageing and exercise training on eNOS uncoupling in skeletal muscle resistance arterioles. *Journal of Physiology*, 587(15). https://doi.org/10.1113/jphysiol.2009.172221

Singh, A. K., Dhaunsi, G. S., Gupta, M. P., Orak, J. K., Asayama, K., & Singh, I. (1994). Demonstration of glutathione peroxidase in rat liver peroxisomes and its intraorganellar distribution. *Archives of Biochemistry and Biophysics*, 315(2). https://doi.org/10.1006/abbi.1994.1508

Singh, I. (1997). Biochemistry of peroxisomes in health and disease. *Molecular and Cellular Biochemistry*, 167(1–2). https://doi.org/10.1023/A:1006883229684

Söderdahl, T., Enoksson, M., Lundberg, M., Holmgren, A., Ottersen, O. P., Orrenius, S., Bolcsfoldi, G., & Cotgreave, I. A. (2003). Visualization of the compartmentalization of glutathione and protein-glutathione mixed disulfides in cultured cells. *The FASEB Journal: Official Publication of the Federation of American Societies for Experimental Biology*, 17(1). https://doi.org/10.1096/fj.02-0259fje

Sorrentino, I., Galli, M., Medraño-Fernandez, I., & Sitia, R. (2022). Transfer of H 2 O 2 from Mitochondria to the endoplasmic reticulum via Aquaporin-11. *Redox Biology*, 55. https://doi.org/10.1016/J.REDOX.2022.102410

St-Pierre, J., Buckingham, J. A., Roebuck, S. J., & Brand, M. D. (2002). Topology of superoxide production from different sites in the mitochondrial electron transport chain. *Journal of Biological Chemistry*, 277(47). https://doi.org/10.1074/jbc.M207217200

Tajika, Y., Matsuzaki, T., Suzuki, T., Ablimit, A., Aoki, T., Hagiwara, H., Kuwahara, M., Sasaki, S., & Takata, K. (2005). Differential regulation of AQP2 trafficking in endosomes by microtubules and actin filaments. *Histochemistry and Cell Biology*, 124(1), 1–12. https://doi.org/10.1007/S00418-005-0010-3

Tajika, Y., Matsuzaki, T., Suzuki, T., Aoki, T., Hagiwaka, H., Kuwahara, M., Sasaki, S., & Takata, K. (2004). Aquaporin-2 is retrieved to the apical storage compartment via early endosomes and phosphatidylinositol 3-kinase-dependent pathway. *Endocrinology*, 145(9), 4375–4383. https://doi.org/10.1210/EN.2004-0073

Takac, I., Schröder, K., Zhang, L., Lardy, B., Anilkumar, N., Lambeth, J. D., Shah, A. M., Morel, F., & Brandes, R. P. (2011). The E-loop is involved in hydrogen peroxide formation by the NADPH oxidase Nox4. *The Journal of Biological Chemistry*, 286(15), 13304–13313. https://doi.org/10.1074/JBC.M110.192138

Takagi, T., Homma, T., Fujii, J., Shirasawa, N., Yoriki, H., Hotta, Y., Higashimura, Y., Mizushima, K., Hirai, Y., Katada, K., Uchiyama, K., Naito, Y., & Itoh, Y. (2019). Elevated ER stress exacerbates dextran sulfate sodium-induced colitis in PRDX4-knockout mice. *Free Radical Biology and Medicine*, 134. https://doi.org/10.1016/j.freeradbiomed.2018.12.024

Takemori, Y., Sakaguchi, A., Matsuda, S., Mizukami, Y., & Sakurai, H. (2006). Stress-induced transcription of the endoplasmic reticulum oxidoreductin gene ERO1 in the yeast Saccharomyces cerevisiae. *Molecular Genetics and Genomics*, 275(1). https://doi.org/10.1007/s00438-005-0065-9

Tavender, T. J., Springate, J. J., & Bulleid, N. J. (2010). Recycling of peroxiredoxin IV provides a novel pathway for disulphide formation in the endoplasmic reticulum. *The EMBO Journal*, 29(24), 4185. https://doi.org/10.1038/EMBOJ.2010.273

Thiede, M. A., Ozols, J., & Strittmatter, P. (1986). Construction and sequence of cDNA for rat liver stearyl coenzyme A desaturase. *Journal of Biological Chemistry*, 261(28). https://doi.org/10.1016/s0021-9258(18)69295-6

Tong, H., Hu, Q., Zhu, L., & Dong, X. (2019). Prokaryotic aquaporins. *Cells*, 8(11). https://doi.org/10.3390/CELLS8111316

Tripathi, D. N., & Walker, C. L. (2016). The peroxisome as a cell signaling organelle. *Current Opinion in Cell Biology*, 39. https://doi.org/10.1016/j.ceb.2016.02.017

Tu, B. P., & Weissman, J. S. (2004). Oxidative protein folding in eukaryotes: Mechanisms and consequences. *The Journal of Cell Biology*, 164(3), 341–346. https://doi.org/10.1083/JCB.200311055

Turrens, J. F., Alexandre, A., & Lehninger, A. L. (1985). Ubisemiquinone is the electron donor for superoxide formation by complex III of heart mitochondria. *Archives of Biochemistry and Biophysics*, 237(2). https://doi.org/10.1016/0003-9861(85)90293-0

Turrens, J. F., & Boveris, A. (1980). Generation of superoxide anion by the NADH dehydrogenase of bovine heart mitochondria. *The Biochemical Journal*, 191(2). https://doi.org/10.1042/bj1910421

Tury, A., Mairet-Coello, G., Poncet, F., Jacquemard, C., Risold, P. Y., Fellmann, D., & Griffond, B. (2004). QSOX sulfhydryl oxidase in rat adenohypophysis: localization and regulation by estrogens. *Journal of Endocrinology*, 183(2). https://doi.org/10.1677/joe.1.05842

Tzartos, J. S., Stergiou, C., Kilidireas, K., Zisimopoulou, P., Thomaidis, T., & Tzartos, S. J. (2013). Anti-aquaporin-1 autoantibodies in patients with neuromyelitis optica spectrum disorders. *PLoS ONE*, 8(9). https://doi.org/10.1371/JOURNAL.PONE.0074773

Valko, M., Morris, H., & Cronin, M. (2005). Metals, toxicity and oxidative stress. *Current Medicinal Chemistry*, 12(10). https://doi.org/10.2174/0929867053764635

Van Balkom, B.W.M., Savelkoul, P.J.M., Markovich, D., Hofman, E., Nielsen, S., Van Der Sluijs, P., & Deen, P.M.T. (2002). The role of putative phosphorylation sites in the targeting and shuttling of the aquaporin-2 water channel. *The Journal of Biological Chemistry*, 277(44), 41473–41479. https://doi.org/10.1074/JBC.M207525200

van der Vliet, A., Janssen-Heininger, Y.M.W., & Anathy, V. (2018). Oxidative stress in chronic lung disease: From mitochondrial dysfunction to dysregulated redox signaling. *Molecular Aspects of Medicine*, 63. https://doi.org/10.1016/j.mam.2018.08.001

Van Hoek, A. N., Ma, T., Yang, B., Verkman, A. S., & Brown, D. (2000). Aquaporin-4 is expressed in basolateral membranes of proximal tubule S3 segments in mouse kidney. *American Journal of Physiology. Renal Physiology*, 278(2). https://doi.org/10.1152/AJPRENAL.2000.278.2.F310

Van Montfort, R.L.M., Congreve, M., Tisi, D., Carr, R., & Jhoti, H. (2003). Oxidation state of the active-site cysteine in protein tyrosine phosphatase 1B. *Nature*, 423(6941), 773–777. https://doi.org/10.1038/NATURE01681

Varadaraj, K., & Kumari, S. S. (2020). Lens aquaporins function as peroxiporins to facilitate membrane transport of hydrogen peroxide. *Biochemical and Biophysical Research Communications*, 524(4), 1025–1029. https://doi.org/10.1016/J.BBRC.2020.02.031

Varadaraj, K., Kumari, S. S., Patil, R., Wax, M. B., & Mathias, R. T. (2008). Functional characterization of a human aquaporin 0 mutation that leads to a congenital dominant lens cataract. *Experimental Eye Research*, 87(1), 9–21. https://doi.org/10.1016/J.EXER.2008.04.001

Veith, A., & Moorthy, B. (2018). Role of cytochrome P450s in the generation and metabolism of reactive oxygen species. *Current Opinion in Toxicology*, 7. https://doi.org/10.1016/j.cotox.2017.10.003

Verkman, A. S., Anderson, M. O., & Papadopoulos, M. C. (2014). Aquaporins: Important but elusive drug targets. *Nature Reviews. Drug Discovery*, 13(4), 259–277. https://doi.org/10.1038/NRD4226

Vieceli Dalla Sega, F., Prata, C., Zambonin, L., Angeloni, C., Rizzo, B., Hrelia, S., & Fiorentini, D. (2017). Intracellular cysteine oxidation is modulated by aquaporin-8-mediated hydrogen peroxide channeling in leukaemia cells. *BioFactors (Oxford, England)*, 43(2), 232–242. https://doi.org/10.1002/BIOF.1340

Vossenkämper, A., Nedvetsky, P. I., Wiesner, B., Furkert, J., Rosenthal, W., & Klussmann, E. (2007). Microtubules are needed for the perinuclear positioning of aquaporin-2 after its endocytic retrieval in renal principal cells. *American Journal of Physiology. Cell Physiology*, 293(3). https://doi.org/10.1152/AJPCELL.00628.2006

Vuillaume, I., Decroix, Y., Calvayrac, R., True, L., Daniel, B., Vincens, P., Tarroux, P., & Poitout, P. (1991). Catalase-associated abnormalities and H2O2 increase in pre-neoplastic and neoplastic lesions of the human lower female genital tract and their near adjacent epithelia. *Biomedicine and Pharmacotherapy*, 45(10). https://doi.org/10.1016/0753-3322(91)90044-T

Wagner, D. D. (1990). Cell biology of von Willebrand factor. *Annual Review of Cell Biology*, 6. https://doi.org/10.1146/annurev.cellbio.6.1.217

Walton, J. H., & Lewis, H. A. (1916). The partition coefficients of hydrogen peroxide between water and certain organic solvents. *Journal of the American Chemical Society*, 38(3). https://doi.org/10.1021/ja02260a013

Wang, H., He, Z., Yang, Y., Zhang, J., Zhang, W., Zhang, W., Li, P., & Tang, B. (2019). Ratiometric fluorescence imaging of Golgi H2O2 reveals a correlation between Golgi oxidative stress and hypertension. *Chemical Science*, 10(47). https://doi.org/10.1039/c9sc04384e

Watanabe, S., Moniaga, C. S., Nielsen, S., & Hara-Chikuma, M. (2016). Aquaporin-9 facilitates membrane transport of hydrogen peroxide in mammalian cells. *Biochemical and Biophysical Research*

Communications, 471(1), 191–197. https://doi.org/10.1016/J.BBRC.2016.01.153

Waypa, G. B., Chandel, N. S., & Schumacker, P. T. (2001). Model for hypoxic pulmonary vasoconstriction involving mitochondrial oxygen sensing. *Circulation Research*, 88(12). https://doi.org/10.1161/hh1201.091960

Waypa, G. B., Guzy, R., Mungai, P. T., Mack, M. M., Marks, J. D., Roe, M. W., & Schumacker, P. T. (2006). Increases in mitochondrial reactive oxygen species trigger hypoxia-induced calcium responses in pulmonary artery smooth muscle cells. *Circulation Research*, 99(9). https://doi.org/10.1161/01.RES.0000247068.75808.3f

Wei, P. C., Hsieh, Y. H., Su, M. I., Jiang, X., Hsu, P. H., Lo, W. T., Weng, J. Y., Jeng, Y. M., Wang, J. M., Chen, P. lang, Chang, Y. C., Lee, K. F., Tsai, M. D., Shew, J. Y., & Lee, W. H. (2012). Loss of the oxidative stress sensor NPGPx compromises GRP78 chaperone activity and induces systemic disease. *Molecular Cell*, 48(5). https://doi.org/10.1016/j.molcel.2012.10.007

Wen, J., Wang, Y., Gao, C., Zhang, G., You, Q., Zhang, W., Zhang, Z., Wang, S., Peng, G., & Shen, L. (2018). Helicobacter pylori infection promotes Aquaporin 3 expression via the ROS-HIF-1α-AQP3-ROS loop in stomach mucosa: A potential novel mechanism for cancer pathogenesis. *Oncogene*, 37(26), 3549–3561. https://doi.org/10.1038/S41388-018-0208-1

Winterbourn, C. C. (2013). The biological chemistry of hydrogen peroxide. *Methods in Enzymology*, 528. https://doi.org/10.1016/B978-0-12-405881-1.00001-X

Wragg, D., Leoni, S., & Casini, A. (2020). Aquaporin-driven hydrogen peroxide transport: A case of molecular mimicry? *RSC Chemical Biology*, 1(5), 390–394. https://doi.org/10.1039/D0CB00160K

Wu, H., Carvalho, P., & Voeltz, G. K. (2018). Here, there, and everywhere: The importance of ER membrane contact sites. *Science (New York, N.Y.)*, 361(6401). https://doi.org/10.1126/SCIENCE.AAN5835

Wu, R. F., Liao, C., Hatoum, H., Fu, G., Ochoa, C. D., & Terada, L. S. (2017). RasGRF couples Nox4-dependent endoplasmic reticulum signaling to Ras. *Arteriosclerosis, Thrombosis, and Vascular Biology*, 37(1). https://doi.org/10.1161/ATVBAHA.116.307922

Wu, R.-F., Ma, Z., Liu, Z., & Terada, L. S. (2010). Nox4-derived H2O2 mediates endoplasmic reticulum signaling through local Ras activation. *Molecular and Cellular Biology*, 30(14), 3553–3568. https://doi.org/10.1128/MCB.01445-09

Wu, X., Zhang, L., Miao, Y., Yang, J., Wang, X., Wang, C. Chen, Feng, J., & Wang, L. (2019). Homocysteine causes vascular endothelial dysfunction by disrupting endoplasmic reticulum redox homeostasis. *Redox Biology*, 20. https://doi.org/10.1016/j.redox.2018.09.021

Xie, H., Liu, F., Liu, L., Dan, J., Luo, Y., Yi, Y., Chen, X., & Li, J. (2013). Protective role of AQP3 in UVA-induced NHSFs apoptosis via Bcl2 up-regulation. *Archives of Dermatological Research*, 305(5), 397–406. https://doi.org/10.1007/S00403-013-1324-Y

Yang, B., Ma, T., & Verkman, A. S. (2001). Erythrocyte water permeability and renal function in double knockout mice lacking aquaporin-1 and aquaporin-3. *The Journal of Biological Chemistry*, 276(1), 624–628. https://doi.org/10.1074/JBC.M008664200

Yang, C.-S., Shin, D.-M., Kim, K.-H., Lee, Z.-W., Lee, C.-H., Park, S. G., Bae, Y. S., & Jo, E.-K. (2009). NADPH oxidase 2 interaction with TLR2 is required for efficient innate immune responses to mycobacteria via cathelicidin expression. *The Journal of Immunology*, 182(6). https://doi.org/10.4049/jimmunol.0802217

Yang, F., Kawedia, J. D., & Menon, A. G. (2003). Cyclic AMP regulates aquaporin 5 expression at both transcriptional and post-transcriptional levels through a protein kinase a pathway. *The Journal of Biological Chemistry*, 278(34), 32173–32180. https://doi.org/10.1074/JBC.M305149200

Yasui, M., Kwon, T. H., Knepper, M. A., Nielsen, S., & Agre, P. (1999). Aquaporin-6: An intracellular vesicle water channel protein in renal epithelia. *Proceedings of the National Academy of Sciences of the United States of America*, 96(10), 5808–5813. https://doi.org/10.1073/PNAS.96.10.5808

Yasui, M., Hazama, A., Kwon, T. H., Nielsen, S., Guggino, W. B., & Agre, P. (1999). Rapid gating and anion permeability of an intracellular aquaporin. *Nature*, 402(6758), 184–187. https://doi.org/10.1038/46045

Yeung, C. H., Callies, C., Tüttelmann, F., Kliesch, S., & Cooper, T. G. (2010). Aquaporins in the human testis and spermatozoa—identification, involvement in sperm volume regulation and clinical relevance. *International Journal of Andrology*, 33(4), 629–641. https://doi.org/10.1111/J.1365-2605.2009.00998.X

Yoboue, E. D., Sitia, R., & Simmen, T. (2018). Redox crosstalk at endoplasmic reticulum (ER) membrane contact sites (MCS) uses toxic waste to deliver messages. *Cell Death & Disease*, 9(3). https://doi.org/10.1038/S41419-017-0033-4

Zana, M., Péterfi, Z., Kovács, H. A., Tóth, Z. E., Enyedi, B., Morel, F., Paclet, M. H., Donkó, Á., Morand, S., Leto, T. L., & Geiszt, M. (2018). Interaction between p22 phox and Nox4 in the endoplasmic reticulum suggests a unique mechanism of NADPH oxidase complex formation. *Free Radical Biology and Medicine*, *116*. https://doi.org/10.1016/j.freeradbiomed.2017.12.031

Zhang, D., Yang, L., Su, W., Zhao, Y., Ma, X., Zhou, H., Xu, B., Zhang, K., & Ma, H. (2018). Aquaporin-4 is downregulated in the basolateral membrane of ileum epithelial cells during enterotoxigenic Escherichia coli-induced diarrhea in mice. *Frontiers in Microbiology*, 8(Jan). https://doi.org/10.3389/FMICB.2017.02655

Zhang, J., Kim, J., Alexander, A., Cai, S., Tripathi, D. N., Dere, R., Tee, A. R., Tait-Mulder, J., di Nardo, A., Han, J. M., Kwiatkowski, E., Dunlop, E. A., Dodd, K. M., Folkerth, R. D., Faust, P. L., Kastan, M. B., Sahin, M., & Walker, C. L. (2013). A tuberous sclerosis complex signalling node at the peroxisome regulates mTORC1 and autophagy in response to ROS. *Nature Cell Biology*, *15*(10). https://doi.org/10.1038/ncb2822

Zhang, J., Tripathi, D. N., Jing, J., Alexander, A., Kim, J., Powell, R. T., Dere, R., Tait-Mulder, J., Lee, J. H., Paull, T. T., Pandita, R. K., Charaka, V. K., Pandita, T. K., Kastan, M. B., & Walker, C. L. (2015). ATM functions at the peroxisome to induce pexophagy in response to ROS. *Nature Cell Biology*, *17*(10). https://doi.org/10.1038/ncb3230

Zhang, W., Zitron, E., Hömme, M., Kihm, L., Morath, C., Scherer, D., Hegge, S., Thomas, D., Schmitt, C. P., Zeier, M., Katus, H., Karle, C., & Schwenger, V. (2007). Aquaporin-1 channel function is positively regulated by protein kinase C. *The Journal of Biological Chemistry*, *282*(29), 20933–20940. https://doi.org/10.1074/JBC.M703858200

Zhou, X., Li, G., Kaplan, A., Gaschler, M. M., Zhang, X., Hou, Z., Jiang, M., Zott, R., Cremers, S., Stockwell, B. R., & Duan, W. (2018). Small molecule modulator of protein disulfide isomerase attenuates mutant huntingtin toxicity and inhibits endoplasmic reticulum stress in a mouse model of Huntington's disease. *Human Molecular Genetics*, *27*(9), 1545–1555. https://doi.org/10.1093/HMG/DDY061

Zhu, L., Yang, K., Wang, X., Wang, X., & Wang, C. C. (2014). A novel reaction of peroxiredoxin 4 towards substrates in oxidative protein folding. *PLoS ONE*, *9*(8). https://doi.org/10.1371/JOURNAL.PONE.0105529

Zitka, O., Skalickova, S., Gumulec, J., Masarik, M., Adam, V., Hubalek, J., Trnkova, L., Kruseova, J., Eckschlager, T., & Kizek, R. (2012). Redox status expressed as GSH: GSSG ratio as a marker for oxidative stress in paediatric tumour patients. *Oncology Letters*, *4*(6). https://doi.org/10.3892/ol.2012.931

Zito, E., Chin, K. T., Blais, J., Harding, H. P., & Ron, D. (2010). ERO1-β, a pancreas-specific disulfide oxidase, promotes insulin biogenesis and glucose homeostasis. *Journal of Cell Biology*, *188*(6). https://doi.org/10.1083/jcb.200911086

Zito, E., Hansen, H. G., Yeo, G.S.H., Fujii, J., & Ron, D. (2012). Endoplasmic reticulum thiol oxidase deficiency leads to ascorbic acid depletion and noncanonical scurvy in mice. *Molecular Cell*, *48*(1). https://doi.org/10.1016/j.molcel.2012.08.010

Zito, E., Melo, E. P., Yang, Y., Wahlander, Å., Neubert, T. A., & Ron, D. (2010). Oxidative protein folding by an endoplasmic reticulum-localized peroxiredoxin. *Molecular Cell*, *40*(5), 787–797. https://doi.org/10.1016/J.MOLCEL.2010.11.010

CHAPTER SEVEN

The Mystery behind Hydrogen Peroxide Permeation across the Peroxisomal Membrane

Cláudio F. Costa, Celien Lismont, Marco Anteghini, Iulia Revenco, Hongli Li, and Marc Fransen

ABBREVIATIONS

AQP	Aquaporin
ER	Endoplasmic reticulum
MCS	Membrane contact site
PEX	Peroxin
PRDX	Peroxiredoxin
PXMP	Peroxisomal membrane protein
Redox	Reduction-oxidation
ROS	Reactive oxygen species

7.1 REDOX REGULATION: A REQUIREMENT FOR LIFE

To sense, process, and respond to intracellular and environmental cues, cells have developed sophisticated safeguard mechanisms. This phenomenon, known as cell signaling, represents a series of biological events that initiate and drive sets of biochemical reactions, which in turn transduce the initial signals into appropriate cellular responses. Effective signaling networks are vital for proper cell functioning and disruptions in such networks can lead to disease (Nair et al., 2019).

Hydrogen peroxide (H_2O_2) is a non-radical reactive oxygen species (ROS) produced in living cells as a product of aerobic metabolism. Nowadays, this oxidant is widely recognized as one of the most important redox signaling molecules in human (patho)physiology (Sies et al., 2022). For example, the significance of H_2O_2 as a pleiotropic physiological signaling agent in the human body is reflected in several aspects of cell behavior, such as proliferation, migration, differentiation, and survival or death (Sies and Jones, 2020). The major mechanism by which H_2O_2 functions as a signaling messenger involves the oxidative modification of particular cysteine thiol residues in target proteins (Poole, 2015). H_2O_2 can directly oxidize redox-sensitive protein thiols or oxidatively modify redox-regulated proteins indirectly through thiol peroxidase-based redox relays (Netto and Antunes, 2016; Stöcker et al., 2018). This in turn can result in structural changes (e.g., through intra- or intermolecular disulfide bond formation) that may modulate macromolecular interactions, protein localization, function, activity, and/or stability (Fransen and Lismont, 2019). Importantly, a key feature of thiol-redox switches is their reversibility, a necessity to ensure adaptive homeostasis. Indeed, in highly oxidative environments, protein thiols can be irreversibly modified to sulfonic acids, a process often associated with injurious consequences (Sies and Jones, 2020).

Under steady-state conditions, H_2O_2 levels are tightly regulated, being maintained from a low picomolar to a high nanomolar range (Lim et al., 2015; Gao et al., 2017). The precise levels may vary considerably depending on the physiological state, function, and environment of the cell. In addition, the occurrence of H_2O_2 concentration gradients between different subcellular locations highlights the importance of sustainable local redox transients to optimally

DOI: 10.1201/9781003160649-7

perform a diverse array of functions (Lim et al., 2015). For example, the lumen of the endoplasmic reticulum (ER) provides a redox environment favoring oxidative protein folding, a process that involves the formation and isomerization of disulfide bonds and is intimately intertwined with H_2O_2 metabolism and signaling (Delaunay-Moisan and Appenzeller-Herzog, 2015). In contrast, in the nucleus, such high levels of H_2O_2 may interfere with protein function (e.g., the DNA binding activity of cysteine-containing transcription factors; Go and Jones, 2008) and cause DNA damage (Konat, 2003). In addition, circumstances wherein the (local) H_2O_2 levels are below the physiological range, a condition sometimes referred to as "reductive stress," can also be detrimental to cells (Rajasekaran, 2021). Thus, sequestration of different H_2O_2-producing and H_2O_2-scavenging enzymes in specific subcellular compartments (Figure 7.1), along with a tight regulation of H_2O_2 permeation across organellar membranes, is essential to ensure redox homeostasis and preserve cellular fitness (Sies et al., 2017).

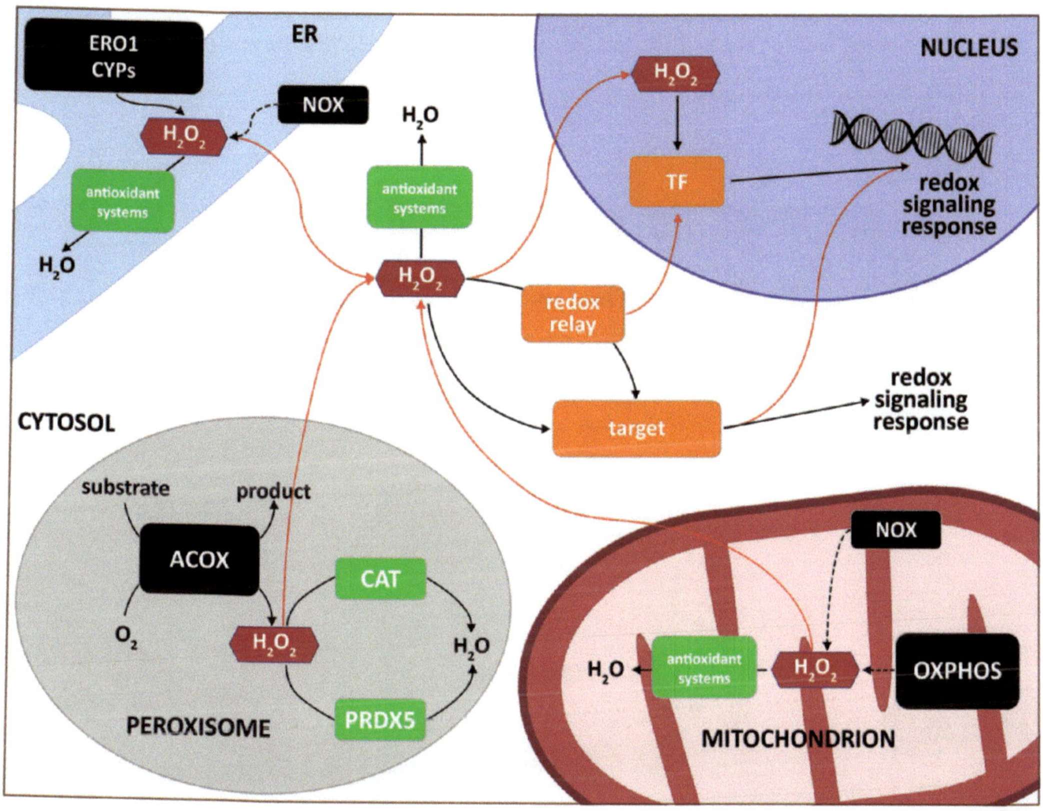

Figure 7.1 Major intracellular sources and sinks of hydrogen peroxide in mammalian cells. Major intracellular sources of H_2O_2 in living cells include a myriad of flavin-dependent enzymes (e.g., the cytochrome P-450 (CYPs) and oxidoreductase 1 (ERO1) enzymes in the endoplasmic reticulum (ER); the acyl-coenzyme A oxidases (ACOX) in peroxisomes) and superoxide dismutases. The latter class of enzymes (not depicted in the graph) catalyzes the dismutation of superoxide radicals, which are predominantly produced by the mitochondrial electron transport chain (OXPHOS) and membrane-associated NADPH oxidases (NOX), to O_2 and H_2O_2 (depicted by dashed black lines). Major H_2O_2-scavenging enzymes include catalase (CAT), a predominantly peroxisomal enzyme, and the thiol peroxidases of the glutathione peroxidase and peroxiredoxin (PRDX) families (outside peroxisomes depicted as "antioxidant systems"). H_2O_2 can function as a pleiotropic signaling agent, either directly or through peroxidase-based redox relays, and this may alter the localization and activity of potential target proteins (target). For example, in case the target is a transcription factor (TF), this may elicit a retrograde signaling response. For more details, we refer to the text (and references therein). The red and black arrows indicate inter- and intra-compartmental events, respectively.

As for many other biological processes, intracellular membranes are important elements for compartmentalization of the cellular redox landscape. Indeed, although H_2O_2 is a neutral molecule, its polarity (dipole moment: $\pm 7.337 \times 10^{-30}$ C·m) is higher than that of water (dipole moment: $\pm 6.185 \times 10^{-30}$ C·m; Yu and Yang, 2011). This implies that efficient transport of H_2O_2 across biological membranes requires the presence of specific channels (Bienert and Chaumont, 2014), referred to as "peroxiporins" (Sies, 2014), or other non-selective pore-forming proteins (Figure 7.2). Currently, several mammalian aquaporins (AQP) have been shown to facilitate H_2O_2 diffusion across membranes. These include AQP0, AQP1, AQP3, AQP5, AQP6, AQP8, and AQP9 in the plasma membrane (Miller et al., 2010; Bertolotti et al., 2016; Watanabe et al., 2016; Rodrigues et al., 2019; Varadaraj and Kumari, 2020; Krüger et al., 2021; Pellavio et al., 2022); AQP8 in the inner mitochondrial membrane (Marchissio et al., 2012; Krüger et al., 2021); and AQP11 in the ER (Bestetti et al., 2020). Given that at least some of these aquaporins (e.g., AQP0, AQP1, and AQP5) can funnel H_2O_2 across the membrane in a bidirectional fashion (Varadaraj and Kumari, 2020), it is plausible to assume that the directionality of H_2O_2 transport is driven by its concentration gradient. Also, a study on AQP8 has demonstrated that the transport of H_2O_2 through this channel is reversibly gated by a two-step sulfenylation-persulfidation process of Cys-53, thereby providing cells with a powerful means to tune local H_2O_2 concentration gradients (Bestetti et al., 2018).

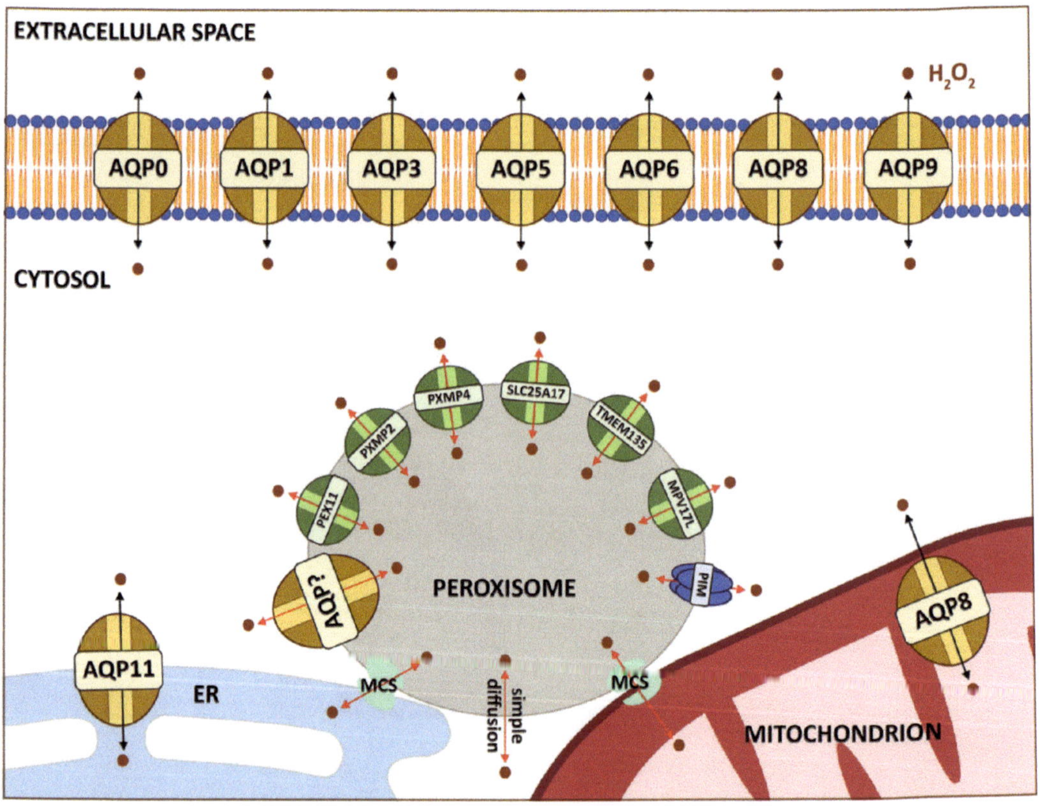

Figure 7.2 Schematic of hydrogen peroxide transport across cellular membranes, with a focus on mammalian peroxisomes. Efficient passage of H_2O_2 across biological membranes requires the presence of proteinaceous channels, pores, peroxisomal matrix protein import machinery (PIM) components, or membrane contact sites (MCS). The best characterized example includes a select set of aquaporins (AQP), referred to as peroxiporins, which are located in multiple subcellular locations (e.g., the plasma membrane, the mitochondrial membrane, and the endoplasmic reticulum [ER] membrane). However, the molecular mechanisms and players governing H_2O_2 diffusion across the peroxisomal membrane remain unclear. Possibilities discussed in the chapter are depicted. Brown dots represent H_2O_2, and red and black arrows indicate hypothetical and experimentally verified H_2O_2 transport activity, respectively. The directionality of H_2O_2 transport is most likely driven by its concentration gradient.

7.2 PEROXISOMES: KEY PLAYERS FOR CELLULAR H_2O_2 HOMEOSTASIS

Peroxisomes are dynamic and functionally versatile organelles traditionally known for their role in cellular lipid metabolism (Van Veldhoven, 2010). However, as their name suggests, they are also important sites of H_2O_2 metabolism (de Duve and Baudhuin, 1966). This is perhaps best illustrated by the fact that human peroxisomes are equipped with at least ten different flavin-dependent oxidases that produce H_2O_2 as a by-product, the H_2O_2-producing enzyme superoxide dismutase, and the H_2O_2-scavenging enzymes catalase and peroxiredoxin 5 (PRDX5) (Fransen and Lismont, 2018; Yifrach et al., 2018) (Figure 7.1). Catalase, the most abundant peroxisomal matrix enzyme, can break down H_2O_2 in a catalatic ($2\ H_2O_2 \rightarrow 2\ H_2O + O_2$) or peroxidatic ($H_2O_2 + AH_2 \rightarrow A + 2\ H_2O$; AH_2 and A represent the reduced and oxidized forms of suitable electron donors, respectively) manner, and the route chosen depends on substrate availability (Kirkman and Gaetani, 2007); and PRDX5 is a cysteine-based peroxidase that can reduce H_2O_2 into H_2O and O_2 by a process that is coupled to intramolecular disulfide bond formation (Smeets et al., 2008). Given that (1) catalase and PRDX5 exhibit distinct H_2O_2 scavenging kinetics (Lismont et al., 2019a, and references therein) and that (2) PRDX5 also displays a high affinity for peroxynitrite and organic peroxides (Knoops et al., 2011), the peroxisomal pools of these enzymes most likely perform non-overlapping functions. Both catalase and PRDX5 contain a relatively weak peroxisomal targeting signal and, depending on the cell type and environmental context, differing amounts of these enzymes can also be found in other subcellular compartments, including the cytosol, the nucleus, and the mitochondria (Yamamoto et al., 1992; Knoops et al., 2011; Rindler et al., 2013). For example, it has been demonstrated that catalase can be retained in the cytosol in times of oxidative stress (Walton et al., 2017; Fujiki and Bassik, 2021).

7.3 PEROXISOMAL H_2O_2 METABOLISM AND DISEASE

Depending on their local environment, peroxisomes can act as net sources or sinks of H_2O_2. As such, it comes as no surprise that these organelles are increasingly recognized as critical modulators of both physiological and pathophysiological processes (Fransen and Lismont, 2019, and references therein). A prime example is cellular and organismal aging (Terlecky et al., 2006). Indeed, since the middle of the previous century, biological aging is considered to be a multifactorial process characterized by a progressive and irreversible accumulation of oxidative damage (Harman, 1956), and today it is widely accepted that ROS, including H_2O_2, function as critical regulators of aging hallmarks such as genomic instability, telomere attrition, epigenetic alterations, loss of proteostasis, deregulated nutrient sensing, mitochondrial dysfunction, cellular senescence, stem cell exhaustion, and altered intercellular communication (van der Rijt et al., 2020). In the context of this chapter, it is of note to highlight that alterations in peroxisomal H_2O_2 metabolism have been associated with a broad range of cellular processes (e.g., gene expression, energy metabolism, autophagy, inflammation, proliferation, differentiation, senescence, and cell death) (Lismont et al., 2019a, and references therein) as well as a myriad of age-related pathologies, including cancer (Dahabieh et al., 2018), diet-induced obesity (Diano et al., 2011), diabetes (Elsner et al., 2011), heart (Wölkart et al., 2006; Chen et al., 2012) and kidney (Vasko, 2016) disease, noise-induced hearing loss (Defourny et al., 2019), and neurodegeneration (Fransen et al., 2021). However, given that the aim of this chapter is not to extensively describe all the (potential) molecular mechanisms underlying many of these associations, we refer interested readers to the papers and reviews cited.

7.4 THE PEROXISOMAL MEMBRANE: THE PORTAL OF H_2O_2 ENTRY AND EXIT

Like other biological membranes, the peroxisomal membrane acts as a physical barrier for H_2O_2. This concept is strongly supported by the observation that, when cells are exposed to an external source of this oxidant, H_2O_2 gradients are not only formed across the plasma membrane but also across the peroxisomal membrane (Makino et al., 1994; Antunes and Cadenas, 2000; Fritz et al., 2007; Laporte et al., 2020). Importantly, these and other studies (e.g., Ho et al., 2004) also provided compelling evidence that externally added H_2O_2 can permeate across the peroxisomal membrane. Likewise, it has been demonstrated that H_2O_2 production inside peroxisomes can result in an efficient release of this molecule into the extraperoxisomal environment, both *in vitro* (Boveris et al., 1972; Mueller et al., 2002; Fritz et al., 2007) and in cellulo (Lismont et al., 2019b). However, the underlying molecular mechanisms remain obscure.

7.5 CAN THE PEROXISOMAL MEMBRANE PROTEOME DISCLOSE THE SECRET KEY?

In the search for peroxisomal membrane proteins (PXMPs) with peroxiporin-like properties, the million-dollar question is whether this is a known PXMP with an already attributed biological function or a protein that still needs to be discovered. Unfortunately, there is no easy answer. Given that at least 8 AQPs endowed with peroxiporin activity have already been identified in the membranes of other cellular compartments (see earlier), one can argue that it would be surprising that no such proteins would be present in the peroxisomal membrane. However, as key cell functions are often mediated through a repertoire of structurally distinct but functionally redundant proteins, it cannot be ruled out that PXMPs with or without any assigned pore-forming properties, or even membrane contact sites, may function as sites for H_2O_2 exchange. In the following subsections, we further elaborate on these concepts.

7.6 PEROXISOMAL PEROXIPORINS: MISSING OR THE MISSING LINK?

Contrary to the situation in other subcellular locations, no bona fide peroxiporins have yet been identified in the peroxisomal membrane proteome (Yifrach et al., 2018; Krüger et al., 2021). In addition, despite the fact that the peroxisomal membrane contains multiple proteins of which the function remains to be elucidated, none of these PXMPs contains the two highly conserved pore-forming asparagine-proline-alanine (NPA) signature repeats characteristic of AQPs (Ishibashi, 2006; Benga, 2012) or display any significant sequence similarity or structural homology with AQPs, as predicted by BLAST+ (Camacho et al., 2009) and i-Tasser (Yang and Zhang, 2015) analyses using default parameters (our unpublished results). However, a large interactome screening recently revealed that the human orthologue of AQP6, a protein mainly expressed in kidneys, interacts with the peroxins (PEXs) PEX11G and PEX16 (Luck et al., 2020), two peroxisomal membrane proteins involved in peroxisome biogenesis. Despite the fact that this observation certainly offers perspectives for future research, the possible role of AQP6 as H_2O_2 transporter in human kidney peroxisomes remains to be fully established.

7.6.1 PXMPs with Assigned (Non-Specific) Pore-Forming Properties

PXMP2: Until recently, the non-selective channel-forming PXMP2 was considered the best candidate to mediate H_2O_2 transport across the peroxisomal membrane (Lismont et al., 2019c). Indeed, measurements of its pore-forming activity upon reconstitution of the purified protein into liposomes, using organic anions as electrolytes, allowed to predict that this PXMP forms a relatively wide (estimated diameter: 1.4 nm), water-filled channel with long-lasting open states that allows solutes with a mass smaller than 200–300 Da to diffuse nearly freely across the peroxisomal membrane (Rokka et al., 2009). In addition, disruption of PXMP2 in mice leads to a partial restriction of peroxisomal membrane permeability to solutes *in vitro* and *in vivo* (Rokka et al., 2009). However, by employing a ΔPXMP2 Flp-In T-REx 293 cell line in which the production of peroxisomal H_2O_2 can be regulated in a dose- and time-dependent manner, it was shown that PXMP2 is not essential for H_2O_2 permeation across the peroxisomal membrane (Lismont et al., 2019c). Importantly, these findings do not rule out that PXMP2 is able to do so. In this context, it is of interest to note that, while male mice deficient in PXMP2 present with only slightly elevated uric acid levels in the serum and urine (1.4- and 3.1-fold, respectively), female mice also exhibited disturbances of lipid metabolism in the mammary fat pad (Rokka et al., 2009; Vapola et al., 2014), a phenomenon that was explained by cell-type specific functional redundancy of PXMP2 and another transporter.

Members of the PEX11 protein family: Members of the PEX11 family are generally thought to play a central role in peroxisome proliferation through membrane tubulation (Opaliński et al., 2011). However, it has been shown that this group of PXMPs, which displays sequence homology with transient receptor potential cation-selective channels, can also form non-selective pores (diameter: 1.2–1.6 nm) permeable to molecules up to 300–400 Da (Gualdron-López et al., 2012; Mindthoff et al., 2016). In mammals, three *PEX11*-related proteins have been identified: PEX11A is a relatively low abundant PXMP, predominantly expressed in liver, whose expression levels can dramatically increase in response to stimuli that trigger peroxisome abundance; PEX11B is a PXMP constitutively expressed in most tissues; and PEX11G is a constitutively expressed, predominantly liver-specific PXMP that may share some functional redundancy with PEX11B (Schrader and Fahimi, 2006). Recently, it has been shown that H_2O_2 can efficiently permeate across the peroxisomal membrane in PEX11B-deficient human embryonic kidney 293 (HEK-293) cells, both in a control and ΔPXMP2 background (Lismont et al., 2019c). This finding clearly demonstrates that also PEX11B is not essential for this permeation process. Of course, PEX11A and PEX11G may be alternative candidates. However, given that both PXMPs are apparently hardly expressed in HEK-293 cells (Geiger et al., 2012; Lismont et al., 2019c), this possibility is rather unlikely.

The peroxisomal matrix protein import machinery: Unlike the protein import machinery of the endoplasmic reticulum and mitochondria, the peroxisomal matrix protein import allows membrane passage of folded proteins (Francisco et al., 2017). A central player in this process is PEX5, a predominantly cytosolic protein that functions as the cycling import receptor for peroxisomal matrix proteins. Interestingly, upon docking at the peroxisomal membrane, PEX5 may change its topology to form a transient pore. Indeed, it has been described that the yeast orthologue of membrane-associated PEX5, together with its docking partner, PEX14, can form a gated ion-conducting channel that can be opened to a diameter of about 9 nm by the cytosolic receptor-cargo complex (Meinecke et al., 2010). This finding, combined with the observations that PEX5 can be oxidatively modified by peroxisome-derived H_2O_2 (Lismont et al., 2019b) and accumulates in the peroxisomal membrane in times of oxidative stress (Legakis et al., 2002), makes it tempting to speculate that PEX5 can also operate as a non-selective pore for H_2O_2. However, in case this hypothesis is correct, this phenomenon is likely to represent an adaptive peroxisomal stress or retrograde signaling response rather than a physiological equilibrium. Another plausible candidate is the PEX2/PEX10/PEX12 ubiquitin ligase complex, which is part of the PEX5 docking/translocation module at the peroxisomal membrane (Francisco et al., 2017). Very recently, structure determination studies in combination with biochemical and *in vivo* experiments have revealed that the transmembrane segments of the PEX2/PEX10/PEX12 ligase complex from *Thermothelomyces thermophilus* can form a hydrophilic pore with a diameter of approximately 1 nm (Feng et al., 2022), which is sufficiently wide to allow the passage of H_2O_2. Once again, this hypothesis needs to be experimentally verified.

7.6.2 PXMPs without Assigned Pore-Forming Properties

PXMP4: Peroxisomal membrane protein 24, officially designated as "PXMP4," is a widely expressed peroxisomal membrane protein that belongs to the translocase of the inner membrane 17 (TIM17) protein family, which contains proteins that mediate different functions in multiple subcellular compartments, including mitochondria, plastids, and peroxisomes (Reguenga et al., 1999; Žárský and Doležal, 2016). The function of PXMP4 is not yet clear, but its expression has been linked to hepatic lipid metabolism (Blankestijn et al., 2022), anti-tumor activity (Wu and Ho, 2004; Zhang et al., 2010; Zhang et al., 2020), natural killer T cell function (Fletcher et al., 2008), and obesity (Padilla et al., 2014). In addition, it has been suggested that PXMP4 may play a role in the transport of glycerol 3-phosphate and/or dihydroxyacetone-3-phosphate across the peroxisomal membrane, although supporting evidence is currently lacking (Chornyi et al., 2021). Albeit speculative, given that PXMP4 displays 29% identity with a putative bacterial permease belonging to the drug/metabolite transporter superfamily (Visser et al., 2007), it may also have the potential to facilitate H_2O_2 permeation across the peroxisomal membrane.

TMEM135: Transmembrane protein 135 (TMEM135) is another member of the TIM17 protein family (Žárský and Doležal, 2016) that has been found to localize both in peroxisomes (Islinger et al., 2007; Wiese et al., 2007; Renquist et al., 2019) and mitochondria (Lee et al., 2016). Until now, this protein has been implicated in various biological processes, including peroxisomal metabolism (Renquist et al., 2019), mitochondrial dynamics (Lee et al., 2016), intracellular cholesterol trafficking (Maharjan et al., 2020), and fat storage and longevity regulation (Exil et al., 2010). However, the precise molecular function of TMEM135 is currently unknown and, albeit unlikely, it cannot be ruled out that the protein may play a role in H_2O_2 transport across the peroxisomal membrane.

SLC25A17: Solute carrier family 25 member 17 (SLC25A17) is a peroxisomal member of the "mitochondrial" carrier family (Wylin et al., 1998). Over the last two decades, it was reported that this protein has potential transporter activity for nucleotide-based molecules such as AMP, ADP, ATP, coenzyme A, FAD, FMN, and NAD^+ (Visser et al., 2002; Agrimi et al., 2012; Kim et al., 2020). In addition, given that dietary phytol administration to SLC25A17 knockout mice causes hepatic lipid accumulation, this PXMP might be involved either in the transport of cofactors required for phytanic acid α-oxidation and/or pristanic acid β-oxidation, or in the export of their metabolites (Van Veldhoven et al., 2020). The definitive function and ligand specificity of SLC25A17 remain uncertain, but indications so far do not support a role in H_2O_2 transport.

MPV17L: Mitochondrial inner membrane protein 17-like (MPV17L), a protein detected with high frequency in the peroxisomal proteome of mice and men (Iida et al., 2003; Yifrach et al., 2018, and references therein), was found to regulate intracellular ROS levels through the transcriptional regulation of the genes coding for catalase, glutathione peroxidase 3, and superoxide dismutase 2 (Iida et al., 2006). According to other studies, MPV17L is located in the mitochondrial inner membrane, where it exerts a protective effect against mitochondrial oxidative stress and apoptosis (Krick et al., 2008). The reason for the apparently conflicting localization data is not clear. In addition, given that, according to i-Tasser prediction models (Yang and Zhang, 2015), MPV17L displays structural homology to microbial rhodopsins (our unpublished results), this membrane protein is unlikely to function in H_2O_2 transport.

7.6.3 Dually Localized Membrane Proteins

Multiple proteins identified in the peroxisomal membrane (e.g., MFF, MAVS, BCL2, B2CL1) are also located in other organellar membranes (e.g., mitochondria, the ER), and this list of proteins is continuously growing (Costello et al., 2018). As such, it cannot be ruled out with 100% certainty that H_2O_2 permeation across the peroxisomal membrane occurs through transporters (e.g., AQPs) that have currently not yet been identified in the peroxisomal membrane. In this context, it is interesting to note that (1) as mentioned before, AQP6 was found to interact with PEX16 (Luck et al., 2020), a peroxin required for insertion of proteins into the peroxisomal membrane, and (2) under specific conditions, the proapoptotic BCL2 effector protein BAK can be translocated to peroxisomes, thereby regulating peroxisomal membrane permeability (Hosoi et al., 2017).

7.6.4 Membrane Contact Sites

In order to efficiently exert their function, organelles must exchange information with each other. This communication often occurs through physical connections, called membrane contact sites (MCSs). Peroxisomes are no exception to this rule, and it has been well documented that these organelles can contact mitochondria, the ER, lipid droplets, lysosomes, and the Golgi apparatus (Valm et al., 2017; Schrader et al., 2020). Importantly, increasing evidence suggests that MCSs between different compartment-enclosing membranes may act as regulated gatekeepers of local H_2O_2 exchange (Yoboue et al., 2018). However, despite the fact that peroxisome-derived H_2O_2 can function as an interorganellar redox messenger (Fransen and Lismont, 2019), it remains to be established if and to what extent MCSs facilitate H_2O_2 influx and efflux across the peroxisomal membrane.

7.7 DO PEROXISOMES REALLY NEED A H_2O_2 TRANSPORTER?

Currently, it cannot be excluded that the exchange of H_2O_2 between peroxisomes and the cytosol is determined by the physicochemical properties of the peroxisomal membrane. Indeed, given that (1) the peroxisomal membrane is characterized by high phosphatidylcholine and low cholesterol contents (Van Veldhoven and Mannaerts, 1994); (2) the presence of oxidized phosphatidylcholines can lead to the formation of nanoscopic cavities that permit small molecules such as H_2O_2 to pass a phospholipid bilayer, thereby significantly deteriorating the membrane permeability barrier (Ouchi et al., 2019); and (3) cholesterol significantly increases the order of lipid packing, thereby lowering membrane permeability and protecting phospholipids from oxidation (Zhang et al., 2018), it is plausible to hypothesize that the relatively high levels of H_2O_2 that can be found inside peroxisomes (Lismont et al., 2019b) can destabilize the peroxisomal lipid bilayer, thereby triggering H_2O_2 leakage (Figure 7.2). Here, it is interesting to note that a recent report claimed that the main mechanism of H_2O_2 permeation across human red blood cell membranes involves simple diffusion across the lipid fraction, and not movement through aquaporins (Orrico et al., 2022). However, the authors do not rule out that other proteins may be involved.

7.8 CONCLUDING REMARKS AND PERSPECTIVES

Over the last decades, the scientific community has witnessed tremendous progress in understanding the cell biology of peroxisomes. Yet, our knowledge of how redox-active metabolites, including H_2O_2, are transported across the peroxisomal membrane is still rudimentary. One of the main conclusions of this chapter is that the peroxisomal membrane is unlikely to contain bona fide peroxiporins of the AQP family. In addition, in silico sequence and structural similarity searches are not helpful either to identify potential "golden nuggets." However, it is clear that mammalian peroxisomes contain multiple PXMPs (e.g., PXMP2 and PEX11B) that exhibit non-selective pore-forming activity, thereby allowing the passage of molecules up to 400 Da (Rokka et al., 2009; Gualdron-López et al., 2012). Importantly, albeit these PXMPs are currently the best candidates for H_2O_2 transport across the peroxisomal membrane, none of the proteins investigated until now appears to be crucial, at least not in HEK-293 cells (Lismont et al., 2019c). In addition, nothing is known about how their gating activity is regulated. As such, it remains an open question which other PXMPs or mechanisms may be involved, and successful identification of PXMPs essential for H_2O_2 permeation across the peroxisomal membrane may require double or even triple gene knockout models. In addition, it is worth considering that H_2O_2 permeation across the peroxisomal membrane may also be (partially) controlled by the physicochemical properties of this membrane (Figure 7.2).

Unambiguous identification and characterization of lipid- and/or protein-based mechanisms underlying H_2O_2 permeation across the peroxisomal membrane is key to fully understanding how peroxisomes are embedded in the cellular H_2O_2 signaling network. Providing a satisfactory answer to this long-lasting dilemma will offer the biomedical and medical science field new opportunities to exploit peroxisomal H_2O_2 metabolism as a promising target for new therapeutic strategies.

7.9 ACKNOWLEDGMENTS

This work has received funding from the European Union's Horizon 2020 Research and Innovation Programme under the Marie Skłodowska-Curie grant agreement No 812968, the KU Leuven (C14/18/088), and the Research Foundation–Flanders (G091819N). CL was supported by the Research Foundation–Flanders (1213620N), and HL was a recipient of a doctoral fellowship from the China Scholarship Council (201906790005).

REFERENCES

Agrimi, G., Russo, A., Scarcia, P., & Palmieri, F. (2012). The human gene SLC25A17 encodes a peroxisomal transporter of coenzyme A, FAD and NAD^+. *Biochemical Journal*, 443(1), 241–247. https://doi.org/10.1042/BJ20111420

Antunes, F., & Cadenas, E. (2000). Estimation of H_2O_2 gradients across biomembranes. *FEBS Letters*, 475(2), 121–126. https://doi.org/10.1016/S0014-5793(00)01638-0

Benga, G. (2012). On the definition, nomenclature and classification of water channel proteins (aquaporins and relatives). *Molecular Aspects of Medicine*, 33(5–6), 514–517. https://doi.org/10.1016/j.mam.2012.04.003

Bertolotti, M., Farinelli, G., Galli, M., Aiuti, A., & Sitia, R. (2016). AQP8 transports NOX2-generated H_2O_2 across the plasma membrane to promote signaling in B cells. *Journal of Leukocyte Biology*, 100(5), 1071–1079. https://doi.org/10.1189/jlb.2AB0116-045R

Bestetti, S., Galli, M., Sorrentino, I., Pinton, P., Rimessi, A., Sitia, R., & Medraño-Fernandez, I. (2020). Human aquaporin-11 guarantees efficient transport of H_2O_2 across the endoplasmic reticulum membrane. *Redox Biology*, 28, 101326. https://doi.org/10.1016/j.redox.2019.101326

Bestetti, S., Medraño-Fernandez, I., Galli, M., Ghitti, M., Bienert, G. P., Musco, G., Orsi, A., Rubartelli,

A., & Sitia, R. (2018). A persulfidation-based mechanism controls aquaporin-8 conductance. *Science Advances*, 4(5), eaar5770. https://doi.org/10.1126/sciadv.aar5770

Bienert, G. P., & Chaumont, F. (2014). Aquaporin-facilitated transmembrane diffusion of hydrogen peroxide. *Biochimica et Biophysica Acta*, 1840(5), 1596–1604. https://doi.org/10.1016/j.bbagen.2013.09.017

Blankestijn, M., Bloks, V. W., Struik, D., Huijkman, N., Kloosterhuis, N., Wolters, J. C., Wanders, R.J.A., Vaz, F. M., Islinger, M., Kuipers, F., van de Sluis, B., Groen, A. K., Verkade, H. J., & Jonker, J.W. (2022) Mice with a deficiency in Peroxisomal Membrane Protein 4 (PXMP4) display mild changes in hepatic lipid metabolism. *Scientific Reports*, 12(1), 2512. https://doi.org/10.1038/s41598-022-06479-y

Boveris, A., Oshino, N., & Chance, B. (1972). The cellular production of hydrogen peroxide. *Biochemical Journal*, 128(3), 617–630. http://doi.org/10.1042/bj1280617

Camacho, C., Coulouris, G., Avagyan, V., Ma, N., Papadopoulos, J., Bealer, K., & Madden, T. L. (2009). BLAST+: architecture and applications. *BMC Bioinformatics*, 10, 421. https://doi.org/10.1186/1471-2105-10-421

Chen, T., Jin, X., Crawford, B. H., Cheng, H., Saafir, T. B., Wagner, M. B., Yuan, Z., & Ding, G. (2012). Cardioprotection from oxidative stress in the newborn heart by activation of PPARγ is mediated by catalase. *Free Radical Biology & Medicine*, 53(2), 208–215. https://doi.org/10.1016/j.freeradbiomed.2012.05.014

Chornyi, S., IJlst, L., van Roermund, C.W.T., Wanders, R.J.A., & Waterham, H. R. (2021) Peroxisomal metabolite and cofactor transport in humans. *Frontiers in Cell and Developmental Biology*, 8, 613892. https://doi.org/10.3389/fcell.2020.613892

Costello, J. L., Passmore, J. B., Islinger, M., & Schrader, M (2018) Multi-localized proteins: the peroxisome-mitochondria connection. *Subcellular Biochemistry*, 89, 383–415. http://doi.org/10.1007/978-981-13-2233-4_17

Dahabieh, M. S., Di Pietro, E., Jangal, M., Goncalves, C., Witcher, M., Braverman, N. E., & Del Rincón, S. V. (2018). Peroxisomes and cancer: the role of a metabolic specialist in a disease of aberrant metabolism. *Biochimica et Biophysica Acta—Reviews on Cancer*, 1870(1), 103–121. https://doi.org/10.1016/j.bbcan.2018.07.004

de Duve, C., & Baudhuin, P. (1966). Peroxisomes (microbodies and related particles). *Physiological Reviews*, 46(2), 323–357. https://doi.org/10.1152/physrev.1966.46.2.323

Defourny, J., Aghaie, A., Perfettini, I., Avan, P., Delmaghani, S., & Petit, C. (2019). Pejvakin-mediated pexophagy protects auditory hair cells against noise-induced damage. *Proceedings of the National Academy of Sciences of the United States of America*, 116(16), 8010–8017. https://doi.org/10.1073/pnas.1821844116

Delaunay-Moisan, A., & Appenzeller-Herzog, C. (2015). The antioxidant machinery of the endoplasmic reticulum: protection and signaling. *Free Radical Biology and Medicine*, 83, 341–351. https://doi.org/10.1016/j.freeradbiomed.2015.02.019

Diano, S., Liu, Z. W., Jeong, J. K., Dietrich, M. O., Ruah, H. B, Kim, E., Suyama, S., Kelly, K., Gyengesi, E., Arbiser, J. L., Belsham, D. D., Sarruf, D. A., Schwartz, M. W., Bennett, A. M., Shanabrough, M., Mobbs, C. V., Yang, X., Gao, X. B., & Horvath, T. L. (2011). Peroxisome proliferation-associated control of reactive oxygen species sets melanocortin tone and feeding in diet-induced obesity. *Nature Medicine*, 17(9), 1121–1127. https://doi.org/10.1038/nm.2421

Elsner, M., Gehrmann, W., & Lenzen, S. (2011). Peroxisome-generated hydrogen peroxide as important mediator of lipotoxicity in insulin-producing cells. *Diabetes*, 60(1), 200–208. https://doi.org/10.2337/db09-1401

Exil, V. J., Avila, D. A, Benedetto, A., Exil, E. A., Adams, M. R., Au, C., & Aschner, M. (2010). Stressed-induced TMEM135 protein is part of a conserved genetic network involved in fat storage and longevity regulation in *Caenorhabditis elegans*. *PLoS ONE*, 5(12), e14228. https://doi.org/10.1371/journal.pone.0014228

Feng, P., Wu, X., Erramilli, S. K., Paulo, J. A., Knejski, P., Gygi, S. P., Kossiakoff, A. A., & Rapoport, T. A. (2022). A peroxisomal ubiquitin ligase complex forms a retrotranslocation channel. *Nature*, 607(7918), 374–380. https://doi.org/10.1038/s41586-022-04903-x

Fletcher, J. M., Jordan, M. A., Snelgrove, S. L., Slattery, R. M., Dufour, F. D., Kyparissoudis, K., Besra, G. S., Godfrey, D. I., & Baxter, A. G. (2008). Congenic analysis of the NKT cell control gene Nkt2 implicates the peroxisomal protein Pxmp4. *Journal of Immunology*, 181(5), 3400–3412. https://doi.org/10.4049/jimmunol.181.5.3400

Francisco, T., Rodrigues, T. A., Dias, A. F., Barros-Barbosa, A., Bicho, D., & Azevedo, J. E. (2017). Protein transport into peroxisomes: Knowns and unknowns. *BioEssays*, 39(10), 1700047. https://doi.org/10.1002/bies.201700047

Fransen, M., & Lismont, C. (2018). Peroxisomes and cellular oxidant/antioxidant balance: Protein redox modifications and impact on inter-organelle communication. *Subcellular Biochemistry*, 89, 435–461. https://doi.org/10.1007/978-981-13-2233-4_19

Fransen, M., & Lismont, C. (2019). Redox signaling from and to peroxisomes: Progress, challenges, and prospects. *Antioxidants & Redox Signaling*, 30(1), 95–112. https://doi.org/10.1089/ars.2018.7515

Fransen, M., Revenco, I., Li, H., Costa, C. F., Lismont, C., & Van Veldhoven, P. P. (2021). Peroxisomal dysfunction and oxidative stress in neurodegenerative disease: a bidirectional crosstalk. *Advances in Experimental Medicine and Biology*, 1299, 19–30. https://doi.org/10.1007/978-3-030-60204-8_2

Fritz, R., Bol, J. Hebling, U., Angermüller, S., Völkl, A., Fahimi, H. D., & Mueller, S. (2007) Compartment-dependent management of H_2O_2 by peroxisomes. *Free Radial Biology and Medicine*, 42(7), 1119–1129. https://doi.org/10.1016/j.freeradbiomed.2007.01.014

Fujiki, Y., & Bassik, M. C. (2021). A new paradigm in catalase research. *Trends in Cell Biology*, 31(3) 148–151. https://doi.org/10.1016/j.tcb.2020.12.006

Gao, C., Tian, Y., Zhang, R., Jing, J., & Zhang, X. (2017). Endoplasmic reticulum-directed ratiometric fluorescent probe for quantitive detection of basal H_2O_2. *Analytical Chemistry*, 89(23), 12945–12950. https://doi.org/10.1021/acs.analchem.7b03809

Geiger, T., Wehner, A., Schaab, C., Cox, J., & Mann, M. (2012). Comparative proteomic analysis of eleven common cell lines reveals ubiquitous but varying expression of most proteins. *Molecular and Cellular Proteomics*, 11(3), M111.014050. https://doi.org/10.1074/mcp.M111.014050

Go, Y. M., & Jones, D. P. (2008). Redox compartmentalization in eukaryotic cells. *Biochimica et Biophysica Acta*, 1780(11), 1273–1290. https://doi.org/10.1016/j.bbagen.2008.01.011

Gualdron-López, M., Vapola, M. H., Miinalainen, I. J., Hiltunen, J. K., Michels, P.A.M., & Antonenkov, V. D. (2012). Channel-forming activities in the glycosomal fraction from the bloodstream form of *Trypanosoma brucei*. *PLoS ONE*, 7(4), e34530. https://doi.org/10.1371/journal.pone.0034530

Harman, D. (1956). Aging: a theory based on free radical and radiation chemistry. *Journal of Gerontology*, 11(3), 298–300. https://doi.org/10.1093/geronj/11.3.298

Ho, Y. S., Xiong, Y., Ma, W., Spector, A., & Ho, D. S. (2004). Mice lacking catalase develop normally but show differential sensitivity to oxidant tissue injury. *Journal of Biological Chemistry*, 279(31), 32804–32812. https://doi.org/10.1074/jbc.M404800200

Hosoi, K. I., Miyata, N., Mukai, S., Furuki, S., Okumoto, K., Cheng, E. H., & Fujiki, Y. (2017) The VDAC2-BAK axis regulates peroxisomal membrane permeability. *Journal of Cell Biology*, 216(3), 709–722. https://doi.org/10.1083/jcb.201605002

Iida, R., Yasuda, T., Tsubota, E., Takatsuka, H., Masuyama, M., Matsuki, T., & Kishi, K. (2003). M-LP, Mpv17-like protein, has a peroxisomal membrane targeting signal comprising a transmembrane domain and a positively charged loop and up-regulates expression of the manganese superoxide dismutase gene. *Journal of Biological Chemistry*, 278(8), 6301–6306. https://doi.org/10.1074/jbc.M210886200

Iida, R., Yasuda, T., Tsubota, E., Takatsuka, H., Matsuki, T., & Kishi, K. (2006). Human Mpv17-like protein is localized in peroxisomes and regulates expression of antioxidant enzymes. *Biochemical and Biophysical Research Communications*, 344(3), 948–954. https://doi.org/10.1016/j.bbrc.2006.04.008

Ishibashi, K. (2006). Aquaporin subfamily with unusual NPA boxes. *Biochimica et Biophysica Acta*, 1758(8), 989–993. https://doi.org/10.1016/j.bbamem.2006.02.024

Islinger, M., Lüers, G. H., Li, K. W., Loos, M., & Völkl, A. (2007). Rat liver peroxisomes after fibrate treatment. A survey using quantitative mass spectrometry. *Journal of Biological Chemistry*, 282(32), 23055–23069. https://doi.org/10.1074/jbc.M610910200

Kim, Y. I., Nam, I. K., Lee, D. K., Bhandari, S., Charton, L., Kwak, S. A., Lim, J. Y., Hong, K. H., Kim, S. J., Lee, J. N., Kwon, S. W., So, H. S., Linka, N., Park, R., & Choe, S. K. (2020). Slc25a17 acts as a peroxisomal coenzyme a transporter and regulates multiorgan development in zebrafish. *Journal of Cellular Physiology*, 235(1), 151–165. https://doi.org/10.1002/jcp.28954

Kirkman, H. N., & Gaetani, G. F. (2007). Mammalian catalase: a venerable enzyme with new mysteries. *Trends in Biochemical Sciences*, 32(1), 44–50. https://doi.org/10.1016/j.tibs.2006.11.003

Knoops, B., Goemaere, J., Van der Eecken, V., & Declercq, J. P. (2011). Peroxiredoxin 5: structure, mechanism, and function of the mammalian atypical 2-Cys peroxiredoxin. *Antioxidants and Redox Signaling*, 15(3), 817–829. https://doi.org/10.1089/ars.2010.3584

Konat, G. W. (2003). H_2O_2-induced higher order chromatin degradation: a novel mechanism of oxidative genotoxicity. *Journal of Biosciences*, 28(1), 57–60. https://doi.org/10.1007/BF02970132

Krick, S., Shi, S., Ju, W., Faul, C., Tsai, S. Y., Mundel, P., & Böttinger, E. P. (2008). Mpv17l protects against mitochondrial oxidative stress and apoptosis by activation of Omi/HtrA2 protease. *Proceedings of the National Academy of Sciences of the United States of America*,

105(37), 14106–14111. https://doi.org/10.1073/pnas.0801146105

Krüger, C., Waldeck-Weiermair, M., Kaynert, J., Pokrant, T., Komaragiri, Y., Otto, O., Michel, T., & Elsner, M. (2021). AQP8 is a crucial H_2O_2 transporter in insulin-producing RINm5F cells. *Redox Biology*, 43, 101962. https://doi.org/10.1016/j.redox.2021.101962

Laporte, A., Lortz, S., Schaal, C., Lenzen, S., & Elsner, M. (2020). Hydrogen peroxide permeability of cellular membranes in insulin-producing cells. *Biochimica et Biophysica Acta*, 1862(2), 183093. https://doi.org/10.1016/j.bbamem.2019.183096

Lee, W. H., Higuchi, H., Ikeda, S., Macke, E. L., Takimoto, T. Pattnaik, B. R., Liu, C., Chu, L. F., Siepka, S. M., Krentz, K. J., Rubinstein, C. D., Kalejta, R. F., Thomson, J. A., Mullins, R. F., Takahashi, J. S., Pinto, L. H., & Ikeda, A. (2016). Mouse Tmem135 mutation reveals a mechanism involving mitochondrial dynamics that leads to age-dependent retinal pathologies. *eLife*, 5, e19264. https://doi.org/10.7554/eLife.19264.001

Legakis, J. E., Koepke, J. E., Jedeszko, C., Barlaskar, F., Terlecky, L. J., Edwards, H. J., Walton, P. A., & Terlecky, S. R. (2002). Peroxisome senescence in human fibroblasts. *Molecular Biology of the Cell*, 13(12), 4243–4255. https://doi.org/10.1091/mbc.e02-06-0322

Lim, J. B., Huang, B. K., Deen, W. M., & Sikes, H. D. (2015). Analysis of the lifetime and spatial localization of hydrogen peroxide generated in the cytosol using a reduced kinetic model. *Free Radical Biology and Medicine*, 89, 47–53. https://doi.org/10.1016/j.freeradbiomed.2015.07.009

Lismont, C., Revenco, I., & Fransen, M. (2019a). Peroxisomal hydrogen peroxide metabolism and signaling in health and disease. *International Journal of Molecular Sciences*, 20(15), 3673. https://doi.org/10.3390/ijms20153673

Lismont, C., Nordgren, M., Brees, C., Knoops, B., Van Veldhoven, P. P., & Fransen, M. (2019b). Peroxisomes as modulators of cellular protein thiol oxidation: a new model system. *Antioxidants & Redox Signaling*, 30(1), 22–39. https://doi.org/10.1089/ars.2017.6997

Lismont, C., Koster, J., Provost, S., Baes, M., Van Veldhoven, P. P., Waterham, H. R., & Fransen, M. (2019c). Deciphering the potential involvement of PXMP2 and PEX11B in hydrogen peroxide permeation across the peroxisomal membrane reveals a role for PEX11B in protein sorting. *Biochimica et Biophysica Acta*, 1861(10), 182991. https://doi.org/10.1016/j.bbamem.2019.05.013

Luck, K., Kim, D. K., Lambourne, L., et al. (2020). A reference map of the human binary protein interactome. *Nature*, 580(7803), 402–408. https://doi.org/10.1038/s41586-020-2188-x

Maharjan, Y., Lee, J. N., Kwak, S. A., Dutta, R. K. Park, C., Choe, S. K., & Park, R. (2020). TMEM135 regulates primary ciliogenesis through modulation of intracellular cholesterol distribution. *EMBO Reports*, 21(5), e48901. https://doi.org/10.15252/embr.201948901

Makino, N., Mochizuki, Y., Bannai, S., & Sugita, Y. (1994). Kinetic studies on the removal of extracellular hydrogen peroxide by cultured fibroblasts. *Journal of Biological Chemistry*, 269(2), 1020–1025. https://doi.org/10.1016/S0021-9258(17)42214-9

Marchissio, M. J., Francés, D.E.A., Carnovale, C. E., & Marinelli, R. A. (2012). Mitochondrial aquaporin-8 knockdown in human hepatoma HepG2 cells causes ROS-induced mitochondrial depolarization and loss of viability. *Toxicology and Applied Pharmacology*, 264(2), 246–254. https://doi.org/10.1016/j.taap.2012.08.005

Meinecke, M., Cizmowski, C., Schliebs, W., Krüger, V., Beck, S., Wagner, R., & Erdmann R. (2010). The peroxisomal importomer constitutes a large and highly dynamic pore. *Nature Cell Biology*, 12(3), 273–277. https://doi.org/10.1038/ncb2027

Miller, E. W., Dickinson, B. W., & Chang, C. J. (2010). Aquaporin-3 mediates hydrogen peroxide uptake to regulate downstream intracellular signaling. *Proceedings of the National Academy of Sciences of the United States of America*, 107(36), 15681–15686. https://doi.org/10.1073/pnas.1005776107

Mindthoff, S., Grunau, S., Steinfort, L. L., Girzalsky, W., Hiltunen, J. K., Erdmann, R., & Antonenkov, V. D. (2016). Peroxisomal Pex11 is a pore-forming protein homologous to TRPM channels. *Biochimica et Biophysica Acta*, 1863(3), 271–283. https://doi.org/10.1016/j.bbamcr.2015.11.013

Mueller, S., Weber, A., Fritz, R., Mütze, S., Rost, D., Walczak, H., Völkl, A., & Stremmel, W. (2002). Sensitive and real-time determination of H_2O_2 release from intact peroxisomes. *Biochemical Journal*, 363(3), 483–491. https://doi.org/10.1042/bj3630483

Nair, A., Chauhan, P., Saha, B., & Kubatzky, K. F. (2019). Conceptual evolution of cell signaling. *International Journal of Molecular Sciences*, 20(13), 3292. https://doi.org/10.3390/ijms20133292

Netto, L. E. S, & Antunes, F. (2016). The roles of peroxiredoxin and thioredoxin in hydrogen peroxide sensing and in signal transduction. *Molecules and Cells*, 39(1), 65–71. https://doi.org/10.14348/molcells.2016.2349

Opaliński, L., Kiel, J.A.K.W., Williams, C., Veenhuis, M., & van der Klei, I. J. (2011). Membrane curvature during peroxisome fission requires Pex11. *EMBO Journal*, 30(1), 4–16. https://doi.org/10.1038/emboj.2010.299

Orrico, F., Lopez, A. C., Saliwonczyk, D., Acosta, C., Rodriguez-Grecco, I., Mouro-Chanteloup, I., Ostuni, M. A., Denicola, A., Thomson, L., & Möller, M. N. (2022). The permeability of human red blood cell membranes to hydrogen peroxide is independent of aquaporins. *Journal of Biological Chemistry*, 298(1), 101503. https://doi.org/10.1016/j.jbc.2021.101503

Ouchi, Y., Unoura, K., & Nabika, H. (2019). Role of oxidized lipids in permeation of H_2O_2 through a lipid membrane: molecular mechanism of an inhibitor to promoter switch. *Scientific Reports*, 9(1), 12497. https://doi.org/10.1038/s41598-019-48954-z

Padilla, J., Jenkins, N. T., Thorne, P. K., Martin, J. S., Rector, R. S., Davis, J. W., & Laughlin, M. H. (2014). Identification of genes whose expression is altered by obesity throughout the arterial tree. *Physiological Genomics*, 45(22), 821–832. https://doi.org/10.1152/physiolgenomics.00091.2014

Pellavio, G., Martinotti, S., Patrone, M., Ranzato, E., & Laforenza, U. (2022). Aquaporin-6 may increase the resistance to oxidative stress of malignant pleural mesothelioma cells. *Cells*, 11(12), 1892. https://doi.org/10.3390/cells11121892

Poole, L. B. (2015). The basics of thiols and cysteines in redox biology and chemistry. *Free Radical Biology and Medicine*, 80, 148–157. https://doi.org/10.1016/j.freeradbiomed.2014.11.013

Rajasekaran, N. S. (2021). Reductive stress: neglected science. *Antioxidants & Redox Signaling* (in press). https://doi.org/10.1089/ars.2020.8114

Reguenga, C., Oliveira, M. E., Gouveia, A.M., Eckerskorn, C., Sá-Miranda, C., & Azevedo, J. E. (1999). Identification of a 24 kDa intrinsic membrane protein from mammalian peroxisomes. *Biochimica et Biophysica Acta*, 1445(3), 337–341. https://doi.org/10.1016/S0167-4781(99)00061-5

Renquist, B. J., Madanayake, T. W., Ghimire, S., Geisler, C. E., & Xu, Y. (2019). TMEM135 is an LXR-inducible regulator of peroxisomal metabolism. *BioRxiv*, 334979. https://doi.org/10.1101/334979

Rindler, P. M., Plafker, S. M., Szweda, L.I., & Kinter, M. (2013). High dietary fat selectively increases catalase expression within cardiac mitochondria. *Journal of Biological Chemistry*, 288(3), 1979–1990. https://doi.org/10.1074/jbc.M112.412890

Rodrigues, C., Pimpão, C., Mósca, A. F., Coxixo, A. S., Lopes, D., da Silva, I. V., Pedersen, P. A., Antunes, F., & Soveral, G. (2019). Human aquaporin-5 facilitates hydrogen peroxide permeation affecting adaption to oxidative stress and cancer cell migration. *Cancers*, 11(7), 932. https://doi.org/10.3390/cancers11070932

Rokka, A., Antonenkov, V. D., Soininen, R., Immonen, H. L., Pirilä, P. L., Bergmann, U., Sormunen, R. T., Weckström, M., Benz, R., & Hiltunen, J. K. (2009). Pxmp2 is a channel-forming protein in mammalian peroxisomal membrane. *PLoS ONE*, 4(4), e5090. https://doi.org/10.1371/journal.pone.0005090

Schrader, M., & Fahimi, D. (2006). Growth and division of peroxisomes. *International Review of Cytology*, 255, 237–290. https://doi.org/10.1016/S0074-7696(06)55005-3

Schrader, M., Kamoshita, M., & Islinger, M. (2020). Organelle interplay-peroxisome interactions in health and disease. *Journal of Inherited Metabolic Disease*, 43(1), 71–89. https://doi.org/10.1002/jimd.12083

Sies, H. (2014). Role of metabolic H_2O_2 generation: redox signaling and oxidative stress. *Journal of Biological Chemistry*, 289(13), 8735–8741. https://doi.org/10.1074/jbc.R113.544635

Sies, H., Belousov, V. V., Chandel, N. S., Davies, M. J., Jones, D. P., Mann, G. E., Murphy, M. P., Yamamoto, M., & Winterbourn, C. (2022) Defining roles of specific reactive oxygen species (ROS) in cell biology and physiology. *Nature Reviews Molecular Cell Biology*, 23(7), 499–515. https://doi.org/10.1038/s41580-022-00456-z

Sies, H., Berndt, C., & Jones, D. P. (2017). Oxidative stress. *Annual Review of Biochemistry*, 86, 715–748. https://doi.org/10.1146/annurev-biochem-061516-045037

Sies, H., & Jones, D. P. (2020). Reactive oxygen species (ROS) as pleiotropic physiological signaling agents. *Nature Reviews Molecular Cell Biology*, 21(7), 363–383. https://doi.org/10.1038/s41580-020-0230-3

Smeets, A., Marchand, C., Linard, D., Knoops, B., & Declercq, J. P. (2008). The crystal structures of oxidized forms of human peroxiredoxin 5 with an intramolecular disulfide bond confirm the proposed enzymatic mechanism for atypical 2-Cys peroxiredoxins. *Archives of Biochemistry and Biophysics*, 477(1), 98–104. https://doi.org/10.1016/j.abb.2008.04.036

Stöcker, S., Van Laer, K., Mijuskovic, A., & Dick, T. P. (2018). The conundrum of hydrogen peroxide signaling and the emerging role of peroxiredoxins as redox relay hubs. *Antioxidants & Redox Signaling*, 28(7), 558–573. https://doi.org/10.1089/ars.2017.7162

Terlecky, S. R., Koepke, J. I., & Walton, P. A. (2006). Peroxisomes and aging. *Biochimica et Biophysica Acta*, 1763(12), 1749–1754. https://doi.org/10.1016/j.bbamcr.2006.08.017

Valm, A. M., Cohen, S. Legant, W. R., Melunis, J., Hershberg, U., Wait, U., Cohen, A. R. Davidson, M. W., Betzig, E., & Lippincott-Schwartz, J. (2017) Applying systems-level spectral imaging and analysis to reveal the organelle interactome. *Nature*, 546(7656), 162–167. https://doi.org/10.1038/nature22369

van der Rijt, S., Molenaars, M., McIntyre, R. L., Janssens, G. E., & Houtkooper, R. H. (2020). Integrating the hallmarks of aging throughout the tree of life: a focus on mitochondrial dysfunction. *Frontiers in Cell and Developmental Biology*, 8, 594416. https://doi.org/10.3389/fcell.2020.594416

Van Veldhoven, P. P. (2010). Biochemistry and genetics of inherited disorders of peroxisomal fatty acid metabolism. *Journal of Lipid Research*, 51(10), 2863–2895. https://doi.org/10.1194/jlr.R005959

Van Veldhoven, P. P., de Schryver, E., Young, S. G., Zwijsen, A., Fransen, M., Espeel, M., Baes, M., & Van Ael, E. (2020). Slc25a17 gene trapped mice: PMP34 plays a role in the peroxisomal degradation of phytanic and pristanic acid. *Frontiers in Cell and Developmental Biology*, 8, 144. https://doi.org/10.3389/fcell.2020.00144

Van Veldhoven, P. P., & Mannaerts, G. P. (1994). Assembly of the peroxisomal membrane. *Subcellular Biochemistry*, 22, 231–261. https://doi.org/10.1007/978-1-4615-2401-4_8

Vapola, M. H., Rokka, A., Sormunen, R. T., Alhonen, L., Schmitz, W., Conzelmann, E., Wärri, A., Grunau, S., Antonenkov, V. D., & Hiltunen, J. K. (2014). Peroxisomal membrane channel Pxmp2 in the mammary fat pad is essential for stromal lipid homeostasis and for development of mammary gland epithelium in mice. *Developmental Biology*, 391(1), 66–80. https://doi.org/10.1016/j.ydbio.2014.03.022

Varadaraj, K., & Kumari, S. S. (2020). Lens aquaporins function as peroxiporins to facilitate membrane transport of hydrogen peroxide. *Biochemical and Biophysical Research Communications*, 524(4), 1025–1029. https://doi.org/10.1016/j.bbrc.2020.02.031

Vasko, R. (2016). Peroxisomes and kidney injury. *Antioxidants & Redox Signaling*, 25(4), 217–231. https://doi.org/10.1089/ars.2016.6666

Visser, W. F., van Roermund, C.W.T., IJlst, L., Waterham, H. R., & Wanders, R.J.A. (2007). Metabolite transport across the peroxisomal membrane. *Biochemical Journal*, 401(2), 365–375. https://doi.org/10.1042/BJ20061352

Visser, W. F., van Roermund, C.W.T., Waterham, H. R., & Wanders, R.J.A. (2002). Identification of human PMP34 as a peroxisomal ATP transporter. *Biochemical and Biophysical Research Communications*, 299(3), 494–497. https://doi.org/10.1016/S0006-291X(02)02663-3

Walton, P. A., Brees, C., Lismont, C., Apanasets, O., & Fransen, M. (2017). The peroxisomal import receptor PEX5 functions as a stress sensor, retaining catalase in the cytosol in times of oxidative stress. *Biochimica et Biophysica Acta—Molecular Cell Research*, 1864(10), 1833–1843. https://doi.org/10.1016/j.bbamcr.2017.07.013

Watanabe, S., Moniaga, C. S., Nielsen, S., & Hara-Chikuma, M. (2016). Aquaporin-9 facilitates membrane transport of hydrogen peroxide in mammalian cells. *Biochemical and Biophysical Research Communications*, 471(1), 191–197. https://doi.org/10.1016/j.bbrc.2016.01.153

Wiese, S., Gronemeyer, T., Ofman, R., Kunze, M., Grou, C. P., Almeida, J. A., Eisenacher, M., Hayen, C.S.H., Schollenberger, L., Korosec, T., Waterham, H. R., Schliebs, W., Erdmann, R., Berger, J., Meyer, H. E., Just, W., Azevedo, J. E., Wanders, R.J.A., & Warscheid, B. (2007). Proteomics characterization of mouse kidney peroxisomes by tandem mass spectrometry and protein correlation profiling. *Molecular and Cellular Proteomics*, 6(12), 2045–2057. https://doi.org/10.1074/mcp.M700169-MCP200

Wölkart, G., Kaber, G., Kojda, G., & Brunner, F. (2006). Role of endogenous hydrogen peroxide in cardiovascular ischaemia/reperfusion function: studies in mouse hearts with catalase-overexpression in the vascular endothelium. *Pharmacological Research*, 54(1), 50–56. https://doi.org/10.1016/j.phrs.2006.02.005

Wu, M., & Ho, S. M. (2004). PMP24, a gene identified by MSRF, undergoes DNA hypermethylation-associated gene silencing during cancer progression in an LNCaP model. *Oncogene*, 23(1), 250–259. https://doi.org/10.1038/sj.onc.1207076

Wylin, T., Baes, M., Brees, C., Mannaerts, G. P., Fransen, M., & Van Veldhoven, P. P. (1998). Identification and characterization of human PMP34, a protein closely related to the peroxisomal integral membrane protein PMP47 of *Candida boidinii*. *European Journal of Biochemistry*, 258(2), 332–338. https://doi.org/10.1046/j.1432-1327.1998.2580332.x

Yamamoto, K., Völkl A., & Fahimi, H. D. (1992). Investigation of peroxisomal lipid beta-oxidation enzymes in guinea pig liver peroxisomes by immunoblotting and immunocytochemistry. *Journal of*

Histochemistry and Cytochemistry, 40(12), 1909–1918. https://doi.org/10.1177/40.12.1360481

Yang, J., & Zhang, Y. (2015). I-TASSER server: new development for protein structure and function predictions. *Nucleic Acids Research*, 43(W1), W174–W181. https://doi.org/10.1093/nar/gkv342

Yifrach, E., Fischer, S., Oeljeklaus, D., Schuldiner, M., Zalckvar, E., & Warscheid, B. (2018). Defining the mammalian peroxisomal proteome. *Subcellular Biochemistry*, 89, 47–66. https://doi.org/10.1007/978-981-13-2233-4_2

Yoboue, E. D., Sitia, R., & Simmen, T. (2018). Redox crosstalk at endoplasmic reticulum (ER) membrane contact sites (MCS) uses toxic waste to deliver messages. *Cell Death and Disease*, 9(3), 331. https://doi.org/10.1038/s41419-017-0033-4

Yu, C. Y., & Yang, Z. Z. (2011). A systemic investigation of hydrogen peroxide clusters $(H_2O_2)_n$ (n = 1–6) and liquid-state hydrogen peroxide: based on atom-bond electronegativity equalization method fused into molecular mechanics and molecular dynamics. *Journal of Physical Chemistry*, 115(12), 2615–2626. https://doi.org/10.1021/jp111284t

Žárský, V., & Doležal, P. (2016). Evolution of the Tim17 protein family. *Biology Direct*, 11(1), 54. https://doi.org/10.1186/s13062-016-0157-y

Zhang, X., Barraza, K. M., & Beauchamp, J. L. (2018). Cholesterol provides nonsacrificial protection of membrane lipids from chemical damage at air-water interface. *Proceedings of the National Academy of Sciences of the United States of America*, 115(13), 3255–3260. https://doi.org/10.1073/pnas.1722323115

Zhang, X., Wu, M., Xiao, H., Lee, M. T. Levin, L., Leung, Y. K., & Ho, S. M. (2010). Methylation of a single intronic CpG mediates expression silencing of the PMP24 gene in prostate cancer. *Prostate*, 70(7), 765–776. https://doi.org/10.1002/pros.21109

Zhang, X., Yang, H., Zhang, J., Gao, F., & Dai, L. (2020). HSD17B4, ACAA1, and PXMP4 in peroxisome pathway are down-regulated and have clinical significance in non-small cell lung cancer. *Frontiers in Genetics*, 11, 273. https://doi.org/10.3389/fgene.2020.00273

CHAPTER EIGHT

Role of Mammalian Aquaporin inside the Cell

AN UPDATE

Kenichi Ishibashi, Yoshiyuki Morishita, and Yasuko Tanaka-Mochizuki

ABBREVIATIONS

AQP	Aquaporin
CKD	Chronic kidney disease
ER	Endoplasmic reticulum
FNT	Formate/nitrite channels
GlpF	Glycerol facilitator
KO	Knockout
MAMs	Mitochondria-ER associated membranes
MIP	Major intrinsic protein
NAC	N-acetylcysteine
NOX	NADPH oxidase
NPA	Asparagine–proline–arginine
PDI	Protein disulfide isomerase
PXP	Peroxiporin
ROS	Reactive oxygen species
SGLT2	Sodium-dependent glucose transport 2
SNP	Single-nucleotide polymorphism
SOD	Superoxide dismutase
TGF	Transforming growth factor
TIP	Tonoplast intrinsic protein
TM	Transmembrane
TNF	Tumor necrosis factor
UPR	Unfolded protein response

8.1 INTRODUCTION

After the discovery of AQP1 as a water channel (Preston et al. 1992), the research on aquaporins (AQP) has been directed mainly to their role in water homeostasis by facilitating the passage of water through the plasma membrane driven by osmotic gradients. As water is vital for all living organisms, the importance of AQP seemed to be obvious and every organism was expected to have some of this family. Indeed, AQP are expressed in a wide range of tissues and organs participating in fluid secretion and absorption including the kidney, lung, eye, exocrine gland, and central nervous system. However, this relatively biased research interest is now faced with a couple of "inconvenient truths." First, the phenotypes of mammalian AQP knockout (KO) mice and humans are generally mild (Verkman 2005) or even normal (Roudier et al. 2002). Second, AQP are absent in many microorganisms as revealed by extensive genome projects (Tong et al. 2019a). Third, other than plasma membrane, AQP have also been identified in the membrane of intracellular organelles including tonoplasts, mitochondria, and the endoplasmic reticulum (ER), where small osmotic gradients are present and the intracellular membrane itself has sufficient water permeability even in the absence of water channel due to the small size of intracellular organelles with a high surface-to-volume ratio, while the regulation of osmotic substances will be more important to control their volumes as water moves secondarily toward to the produced or accumulated osmotically active substances inside the intracellular organelles. Fourth,

DOI: 10.1201/9781003160649-8

another channel family similar to the AQP structure has been identified in bacteria as formate/nitrite channels (FNT), although they are absent in eukaryotes (Wang et al. 2009). This discovery indicates that AQP are just another member of such solute channel families. This shift will be particularly relevant to AQP of small-sized intracellular organelles and microorganisms including bacteria and possibly parasites with recently identified AQPX (Tesan et al. 2021).

Before the identification of CHIP28 (Preston et al. 1991), several similar channels were cloned with unknown function: the major intrinsic protein (MIP) from the junction membrane of bovine lens fibers (Gorin et al. 1984), the plant nodulin-26 (NOD) from the peribacteroid membrane of root nodules (Sandal and Marcker 1988), the glycerol facilitator (GlpF) from *Escherichia coli* (Muramatsu et al. 1989), the tonoplast intrinsic protein (TIP) of higher plant seeds (Johnson et al. 1990), and a protein named Big Brain (BIB) from *Drosophila* (Rao et al. 1990). All are membrane proteins with six transmembrane domains in two tandem repeats suggesting their production by internal duplication during evolution, each of which is made of a hemi-channel structure containing a highly conserved asparagine–proline–arginine (NPA) motif, the signature sequence for MIP family (Baker and Saier 1990; Pao et al. 1991). The NPA motif is also functionally important as a selectivity filter and a barrier for the passage of inorganic cations (such as Na^+ and K^+) and protons (H^+; Wu et al. 2009; Wree et al. 2011). CHIP28 is also a member of the MIP family and unexpectedly displayed a dramatic water permeability in a heterologous expression system; exposing cRNA-injected *Xenopus laevis* oocytes to a hypotonic osmotic shock resulted in rapid swell and rupture within 5 min (Preston et al. 1992). Hence the name, aquaporin, was coined from this unique function of CHIP28, which was renamed as AQP1 to distinguish it from other functionally undefined MIP family (Agre et al. 1993; Agre 1997).

Before the discovery of water channel function by CHIP28, bacterial glycerol facilitator GlpF was postulated to transport glycerol *in vivo* (Heller et al. 1980), which was later more clearly shown in a *Xenopus* oocyte expression system without water transport (Maurel et al. 1994). Hence, the name GlpF is still used because of the absence of water channel activity while MIP itself was cleverly renamed as AQP0 because MIP was cloned before AQP1. As MIP family has a set of highly conserved NPA motif, it was cloned relatively easily by PCR homology cloning using degenerative primers for each NPA motif. Soon, AQP2 (Fushimi et al. 1993) and AQP3 (Echevarria et al. 1994; Ishibashi et al. 1994; Ma et al. 1994) were cloned and functionally characterized in a *Xenopus* oocyte expression system. While AQP2 is AQP1-like and expressed at the apical membrane of the principal cell in the collecting duct, AQP3 is expressed at the basolateral membrane (Ishibashi et al. 1994) and a mammalian homolog of GlpF which also permeates glycerol, urea, and surprisingly water as well (Knepper 1994). Subsequent homology cloning efforts have identified only AQP1-likes (AQP4–AQP6) from water transporting organs such as the kidney and submandibular gland until the discovery of the second AQP3-like, AQP7, cloned from tissues seemingly unrelated to water metabolism, the testis and adipose tissue (Ishibashi et al. 1997a; Kishida et al. 2000), although it may be important for glycerol metabolism. After the identification of AQP7, the presence of two major subgroups of AQP in mammals became evident on the basis of phylogeny: AQP1-like and AQP3-like (Ishibashi et al. 1998). The latter was later named aquaglyceroporin based on its function (Borgnia et al. 1999). However, the phylogenetic naming will be appropriate because the function surpasses this dichotomy to permeate several solutes as well as water. Indeed, the classification based on the transporting properties is blurred and confusing because some AQP1-likes permeate glycerol such as entomoglyceroporin (Eglp) in insects (Finn et al. 2015; Promeneur et al. 2018; Linhares Lino de Souza et al. 2020). On the other hand, phylogenetic naming should avoid such confusions. Soon, another AQP family, AQP11-like, was identified with very low homology of less than 20% to AQP1-like and AQP3-like at the level of supergene family, hence they were named super-aquaporin or AQPX (Ishibashi et al. 2000). Notably, they have highly deviated primary sequences around two NPA motifs which had made PCR homology cloning difficult. As NPA motif is critical for pore formation, these unique NPA motifs may suggest the presence of an unknown solute selectivity profile, hence named originally AQPX1 and AQPX2, but later renamed as AQP11 and AQP12 to follow the nomenclature of AQP family. However, their intracellular localization challenged functional studies to identify water permeability (Morishita et al. 2005; Itoh et al. 2005; Gorelick et al. 2006), although some reports suggest that AQP11 has similar solute transport characteristics with other AQP (reviewed in Ishibashi et al. 2021).

This review will focus on the update of our previous review on this topic (Nozaki et al. 2008), but it also emphasizes a new role as a H_2O_2 channel, peroxiporin (PXP) (Bienert and Chaumont 2014), with associated diseases (Prata et al. 2019). Hopefully, this review would shift the research focus of AQP away from water business, especially for intracellular AQP whose water channel activity is not so much expected with possible exception of plant TIP in tonoplasts.

8.2 AQP/MIP AS A SOLUTE CHANNEL

From an evolutional perspective, AQP were already present in some bacteria as AQP1-like and AQP3-like but not as AQP11-like (Ishibashi et al. 2020). However, their function as a water channel is controversial. In fact, AQP3-like, GlpF has no water permeability but dominant glycerol permeability (Maurel et al. 1994) and is located at a functional operon with *glpK* encoding glycerol kinase, while archaeal AQP3-like, AqpM weakly permeates water but no glycerol with a channel size of ~2.8 Å, resembling the size of a water molecule (Kozono et al. 2003; Lee et al. 2005). The results suggest that AqpM could represent a primitive non-specialized AQP, a possible ancestor of specialized AQP, AQP1-like, and AQP3-like. Alternatively, it may indicate the changing phase of AQP3-like to AQP1-like because archaea are more advanced than eubacteria. On the other hand, mammalian AQP3-like has evolved to permeate both glycerol and water efficiently (Moss et al. 2020; de Maré et al. 2020).

As a facilitated water transport may be harmful for bacteria to aggravate the swelling in hypotonic environment, the emergence of water channels will be detrimental. Moreover, small-sized bacteria with a high surface-to-volume ratio may not require a water channel as is the case with intracellular organelles. In fact, many bacteria do not have AQP, especially archaea which live in harsh environments such as hot springs and deep oceans near volcanos with much higher temperatures and water pressures. A recent extensive genome survey revealed that there are many bacteria without AQP (Tong et al. 2019a). From the sequenced genomes of 5294 bacteria and 299 archaea, orthologs of AQP1-like and AQP3-like were found in 3315 (60%) prokaryotic species. Of these, 977 bacterial species encode both AQP1-like and AQP3-like, whereas 698 bacteria encode only AQP1-like and 1552 encode only AQP3-like, indicating AQP3-like is the predominant subfamily. On the other hand, 2067 (40%) bacterial species do not have any AQP homologues in their genomes, especially in animal pathogenic microorganisms and extreme-environment inhabitants, suggesting that they do not need water channels but another apparatus for solute transport. Of note, archaeal phyla rarely have AQP. Accordingly, AQP in bacteria may well be used for solute transport but not for water channel. For example, in the lactic acid bacterium *Lactobacillus plantarum*, the AQP family is composed of six AQP3-likes, GlpF1–GlpF6, which export the metabolic end products such as lactic acid, especially by GlpF1 and GlpF4 (Bienert et al. 2013).

Interestingly, another channel protein, FocA, a member of the FNT family has been identified in bacteria which transport various anions during anaerobic bacterial growth and located at a functional operon. FocA does not conduct cations or divalent anions but exports major end products of anerobic mixed-acid fermentation such as formate, lactate, and acetate to prevent their intracellular accumulation (Rü et al. 2012). Among anions, HS^- binds to FNT with a higher affinity than formate, Cl^-, or NO_2^- (Czyzewski and Wang 2012). Both FNT and AQP have two structural repeats in six transmembrane (TM) segments: TM1–TM3 and TM4–TM6. Similar to AQP, the second transmembrane segment of FNT in each repeat (TM2 or TM5) is disrupted by a highly conserved loop named L2 and L5, respectively, but with different sequences from NPA motif of AQP. Moreover, FocA is organized in a homopentamer in contrast to a homotetramer of AQP (Wang et al. 2009). While FocA-loaded proteoliposomes displayed no water channel activity, AqpZ, AQP1-like in *Escherichia coli*, permeated formate albeit less efficiently (Wang et al. 2009), suggesting that these structures are well suited for solute transport rather than water transport.

Previously, it was hypothesized that AQP may facilitate H_2S transport across cell membranes because of the structural similarity of H_2S and H_2O. Although H_2S permeability of lipid bilayers is quite high (Mathai et al. 2009), its ionized form as HS^- occupies more than 80% in a biological system. Thus, which channel functions as an HS^- facilitator is an intriguing problem because FNT are absent in eukaryotes. It turned out that HS^- is a substrate for anion exchanger 1 (AE1) across the human erythrocyte membrane; the Cl/HS^- exchange rate is about one-third as rapid as $Cl^-/HCO3^-$ exchange (Jennings 2013). Accordingly, intracellular AQP may not be a channel for biologically abundant HS^-.

Speculatively, ancestor AQP might share some permeation characteristics with FNT in bacteria. AQP may have acquired a water channel function when evolving to eukaryotes whereas FNT family has been lost in the eukaryotic lineage. If FNT were evolutionally transferred to eukaryotes, they might have acquired ability to conduct a variety of solutes, even water, as AQP. Conversely, AQP could revive or change their ancestral function even in mammals possibly with some modifications. For example, a certain function of bacterial AQP has been altered in eukaryotes as shown in plant NIP: bacterial arsenic detoxification function has been converted to nutrient uptake in plants (Pommerrenig et al. 2020).

Indeed, AQP are not water specific because most members of this huge family are able to permeate other molecules: non-charged or partially polar solutes such as glycerol (Rodriguez et al. 2020) as well as urea, NH_3, polyols, purines, pyrimidines, silicon, and trivalent inorganic forms of arsenic and antimony. Water permeation of AQP seems to have been acquired secondarily for the purpose of the rapid balancing for osmotic gradients built up by the solute passage through the same AQP. As the size and the electro-chemical properties of solutes are the main factors determining the diffusion through AQP, permeating substances through AQP have been predicted from the primary sequence as detailed herein.

In the 3D structure of AQP, two pore constrictions as selectivity filters have been identified: the NPA region and the so-called ar/R (aromatic residue [ar] and arginine [R]) region in TM2 and TM5 (Tajkhorshid et al. 2002). While the NPA region also serves as a large electrostatic barrier, the ar/R region is located in the outer channel entrance to form a strong selectivity filter to determine the size of permeating solutes. The pore size of ~2.8 Å in AQP1-like just fits the diameter of a water molecule while the pore size in AQP3-like is wider, ~3.4 Å or more (Kreida et al. 2015). The pore size in AQP11-like is still unknown.

Besides these two filters, the glycerol permeation is regulated by other molecules. Molecular dynamics simulations demonstrate that an aromatic residue (F) in AQP7 at the entrance of channel undergoes large and rapid conformational changes to allow glycerol molecules to permeate without orientational restriction with rotating movement through the channel (de Maré et al. 2020). On the other hand, GlpF has a different corresponding residue (W) which induces rare but long-lasting conformational changes to block the pore for water and even glycerol. Moreover, GlpF has additional more rigid orientational constraints on glycerol molecules passing through the selective filter to limit the glycerol passage (Moss et al. 2020; Pluhackova et al. 2022).

Interestingly, small gas molecules seem to permeate AQP. The involvement of plant AQP in CO_2 delivery for photosynthesis has been shown by monitoring cellular carbonic acid accumulation or measuring stable-isotope labeled CO_2 (Uehlein et al. 2003; Michenkova et al. 2021). Motivated by the hypoxia-induced expression of AQP1 (Zhang et al. 2013), O_2 permeation was tested in heterologous AQP1 expression system of yeast plasma membranes by myoglobin oxygenation as an indicator for O_2 permeability (Zwiazek et al. 2017). The results have indicated that AQP1 transports O_2 although its physiological and pathological roles are still unknown (Hu et al. 2021).

Although homologues of AQP1-like and AQP3-like in lower animals are surprisingly well conserved, such is not the case with AQP11-like. Salmon lice have unusual members of AQP11-like with very low homology (11%–18%) which are localized intracellularly and difficult to be expressed in *Xenopus oocytes* (Stavang et al. 2015). Three more deviated AQP11-likes are also found in the *Caenorhabditis elegans* genome, which have not yet been tested in in the *Xenopus* oocyte expression system. Of note, the first half of AQP11-like in lower animals is more deviated from AQP1-like and AQP3-like (Ishibashi et al. 2020). As the second half mainly defines the entrance selective filter as the ar/R region, it will be conserved with evolutional constraints. On the other hand, the first half may change more freely without such a restriction and could obtain new function. The comprehensive analysis of phenotypes in genetically modified animals with these deviated AQP11-likes will be a clue to identify new function as well as novel permeating molecules.

The revival of the initial name, MIP family, may well lead to unbiased research in the field of functional studies of AQP to explore their pleiotropic nature including a controversial ion channel function (Ikeda et al. 2002; Yool et al. 2012; De Ieso et al. 2019). A more open-minded approach should be encouraged based on the following observations: AQP1 is mechanosensitive (Ozu et al. 2013), and AQP2 is a cell-attaching molecule (Chen et al. 2012).

8.3 INTRACELLULAR AQP/MIP: CLUES FOR INTRACELLULAR WATER AND SOLUTE HOMEOSTASIS?

As plants have no circulation systems, such as the blood and lymph vessels as observed in mammals, they have developed a new transcellular water and solute transport system: the symplastic transport (through cytoplasm) and the apoplastic transport (through cell wall). To facilitate the symplastic transport, multiple AQP are present at the plasma membrane and at the membrane of intracellular organelles, especially plant-specific tonoplasts with abundant TIP. The reason for the need of many AQP in plants will be that water and solutes must cross cells in series, especially through central vacuoles (tonoplasts) which occupy as much as 90% of the cell volume.

On the other hand, animals may also need intracellular AQP as well as plasma membrane AQP to facilitate transcellular water transport even though they have well-developed circulation systems. However, as shown by AQP KO studies, only the collecting duct–specific AQP2 is critical for transcellular water transport for the maintenance of body water homeostasis because of the limited paracellular water transport in this nephron segment (Yang et al. 2006a). Moreover, as AQP2-expressing principal cells do not have specialized intracellular AQP other than plasma membrane AQP2, 3, 4 (Matsuzaki et al. 2017), intracellular AQP may not be needed to conduct a rapid transcellular water transport at least in the distal nephron. On the other hand, other AQP may not be so important in global water metabolism because their KO mice display few problems in water homeostasis caused by a presumed defective transcellular water transport.

Of note is a dramatic phenotype of AQP11 KO mice which develop giant intracellular vacuoles in the proximal tubules with a progression to cyst-like tubular dilatation, and terminated by fatal renal failure around 1 month old (Morishita et al. 2005; Rützler et al. 2017). This is the first demonstration of a pathophysiological role of intracellular AQP *in vivo*. However, a conditional AQP11 KO after postnatal day 8 (P8) produced no kidney phenotype (Rützler et al. 2017), suggesting that its critical role is limited to the developmental stage of the kidney and not in the matured kidney. Currently, whether the mechanism for the kidney phenotype of AQP11 KO mice is related to the defective water or solute permeation is unclear. In the first place, the AQP11 permeability of water and solutes has not been well characterized due to its intracellular localization. Overexpression of AQP11 in CHO-K1 cells or 3T3-L1 adipocytes, however, has made AQP11 detectable at the plasma membrane to display water or glycerol transport by cell swelling assay (Ikeda et al. 2011) or by a volume-sensitive fluorophore assay (Madeira et al. 2014). The results indicated the increase of osmotic water permeability by 1.6-fold and glycerol permeability by 2-fold induced by AQP11 expression. Another method also revealed the osmotic water permeability of AQP11 is 8-fold lower than that of AQP1: measuring the osmotic water permeability of membrane vesicles with AQP11 produced by insect Sf9 cells to preclude the effect of detergents on the water conductance (Yakata et al. 2011). Thus, AQP11 seems to be a less efficient water channel, if any. However, the results of heterologous expression systems may be influenced by unknown regulatory factors stabilizing its open or closed states. Thus, definitive conclusions will be obtained only by measuring water and solute permeation inside the cell, preferably in *in vivo* conditions. As AQP11 is widely detected almost ubiquitously by RT-PCR, its basic function may be for cellular homeostasis but not so critical for cell survival.

Another outstanding phenotype of AQP KO mice and even humans is cataracts in AQP0 KO, where the exact mechanism is still unclear as AQP0 has low water permeability (Yang and Verkman 1997). Interestingly, AQP0 opens with acidic pH to permeate water (Nemeth-Cahalan and Hall 2000), and its C-terminal end has been shown to be coupled with gap junction channels (Varadaraj et al. 2019). More recently, AQP0 has been shown to be PXP, suggesting the accumulation of damaging H_2O_2 in the absence of AQP0 as a cause of cataracts (Varadaraj et al. 2020).

The compensation by other proteins and developmental adaptation are always caveats for the interpretation of whole animal KO. However, few papers on conditional AQP KO mice have been reported possibly due to unremarkable phenotypes of AQP KO mice and humans. Obviously, systemic roles of AQP cannot be evaluated by AQP inhibition through iRNA in cell culture studies. Currently, no specific inhibitor for each AQP is available. Moreover, poor qualities of commercially available antibodies make the localization of AQP, especially intracellular AQP, difficult by immunohistochemical studies and by immunoelectron microscopy. While heterologous expression of AQP fusion proteins may provide an insight into the intracellular localization (Beitz et al. 2006), the results from protein overexpression and bulky proteins may not reflect *in vivo* conditions. Due to these difficulties, the role of AQP is still elusive after all KO mice studies of mammal AQP with the exception of AQP2 playing a critical role in urine concentration.

AQP12 is another AQP11-like, but its expression is limited to the pancreas and retina. AQP12 is also predominantly localized at the ER and partially in the membrane of zymogen granules near the ER in CCK-8 treated rats, as revealed by immunogold-labeled immunoelectron microscopy (Ohta et al. 2009). Due to the intracellular localization, functional studies in the *Xenopus* oocyte system were difficult (Itoh et al. 2005). To overcome this problem, the osmotic water permeability of isolated ER vesicles from the pancreases of wild and AQP12 KO mice was compared. The results showed no difference, indicating that AQP12 has no water channel function (Ohta et al. 2009). Although no abnormal phenotypes were observed in AQP12 KO mice, large intracellular vacuoles were detected in the acinar cell of the pancreas by a strong CCK-8 stimulation (Ohta et al. 2009). These vacuoles seemed to be fused exocytotic vesicles, probably a well-known hallmark for early pancreatitis caused by aberrant endocytosis of large post-exocytic structures. Alternatively, these vacuoles may represent cell damage caused by ROS because a massive synthesis of digestive enzymes in the ER of stimulated acinar cells may produce huge amounts of H_2O_2 which needs to be exported out of the ER through AQP12 functioning as PXP. As this is the second demonstration of the pathophysiological role of intracellular AQP *in vivo*, the mechanism for proper secretion of pancreatic fluid regulated by AQP12 in an extreme condition needs to be clarified. Of note, even under a strong secretagogue stimulation, AQP12 did not display trafficking to the plasma membrane from the ER but rather shifting to early zymogen granules (Ohta et al. 2009).

AQP6 is present in vesicles within intercalated cells of the collecting duct colocalized with H-pumps, suggesting its presence at the endosome and a role in the acidification of these vesicles (Yasui et al. 1999a). Nonetheless, AQP6 is expressed in the plasma membrane when overexpressed in *Xenopus* oocytes, where permeation of ions occurs in response to acidic pH or Hg^{2+} (Yasui et al. 1999b). When tagged with GFP at its N terminus, AQP6 was directed to the plasma membrane of cultured mammalian cells where nitrate channel activity was demonstrated although AQP rarely permeate ions (Ikeda et al. 2002). Interestingly, AQP6 seemed to be present in synaptic vesicles to facilitate vesicle swelling which will be needed for secretary vesicles to conduct the regulated expulsion of intra-vesicular neuro-transmitters by nerve stimulation (Jeremic et al. 2005). Currently, however, the role of AQP6 *in vivo* is unclear and even the phenotype of AQP6 KO mice has not yet been published.

AQP10 has been shown to be present at the apical membrane of intestinal epithelia (Mobasheri et al. 2004), the granular vesicles of enterochromaffin cells (Li et al. 2005), adipocytes, and pancreatic β cells (Laforenza et al. 2013). These differences may be due to the different specificities of the antibodies, which is difficult to test in the absence of a negative control such as KO mice because AQP10 is not expressed as a pseudogene in rodents (Morinaga et al. 2002). Interestingly, AQP10 has also turned to be a pseudogene in some Artiodactyla such as cattle, sheep, and whale excluding swine (Tanaka et al. 2015), indicating that AQP10 is dispensable for their development and survival. The role of AQP10 as intracellular AQP is not well characterized in enterochromaffin cells as human materials are not readily available. The fact that authentic AQP10 has been mutated to be a pseudogene in some animals may suggest the presence of too many AQP in GI tracts especially in ruminants with longer intestine and possibly in adipocytes with multiple AQP3-likes for glycerol transport.

Although some AQP seem to be present inside the cell as well at the plasma membrane, the function of intracellular AQP may be different from that of AQP at the plasma membrane. Therefore, the functional studies should be conducted in *in situ* conditions to clarify this difference especially in the presence of unknown regulatory factors. Moreover, some exhibit a translocation from the intracellular space to the plasma membrane such as AQP2 (Olesen and Fenton 2017), AQP4 (Kitchen et al. 2020), and AQP8 (Garcia et al. 2001) stimulated by cyclic AMP. The translocation of AQP1 under a hypotonic stimulus and AQP7/AQP10 stimulated by isoproterenol have also been reported (Conner et al. 2012; Laforenza et al. 2013). In fact, almost all AQP except AQP11-like seem to be conducted by regulated trafficking targeted to the plasma membrane with some controversies with AQP5 (Hosoi et al. 2020). The purpose of such translocations will be to control the amount of AQP in the plasma membrane and a specialized intracellular role may not be present even if they function similarly inside the cell. The biological factors to stimulate AQP expression in the plasma membrane will be a clue to speculate the permeating molecules and roles of AQP as suggested by the possible O_2 permeation by AQP1 whose expression is stimulated by hypoxia (Zhang et al. 2013). Conversely, in plants, oxidative stress has been shown to induce redistribution of a plasma membrane AQP1-like to intracellular membranes such as endosomes without further degradation (Wudick et al. 2015). This H_2O_2-induced endocytosis suggests that plasma membrane-residing AQP may be sequestered inside the cell or, alternatively, it may turn to intracellular AQP functioning as a PXP inside the cell as a part of oxidative stress responses. The stimulator for this reverse mode of trafficking will also provide a clue for AQP function. Moreover, targeting such trafficking signals will be useful therapeutic interventions to modulate the expression levels of AQP at the plasma membrane as shown with AQP4 in brain edema (Kitchen et al. 2020).

AQP in bacteria may have the characteristics of intracellular AQP for the much smaller size similar to intracellular organelles in eukaryotes. As bacterial channels and transporters are sometimes found in operons as a part of a functional unit, as is the case with GlpF and FocA, the documentation of regulatory factors for AQP gene expression and modulating enzymes may provide a clue to identify new permeating molecules.

More analogous to intracellular AQP will be the AQP of parasitic protozoans. For example, the human malaria parasite *Plasmodium falciparum* has only a single AQP, PfAQP, as AQP3-like to permeate a wide range of small, uncharged molecules to supply biochemical precursors or to release metabolic wastes and side products (Hansen et al. 2002). Recently, a new AQP cluster from trypanosomatid has been identified, which is deviated far away from the known MIP with a wider selectivity filter (Tesan et al. 2021). These are proposed as members of a new AQP family, AQPX, whose homologues are already present in prokaryotes and previously designated as TIP-like similar to plant intracellular AQP. Since the function of AQPX is unknown with no permeability to water or glycerol, AQPX may permeate novel molecules because of the different solute selectivity profile with rare selectivity filters (Tesan et al. 2021). Hopefully, the knowledge on AQP from these lower animals may provide clues to identify novel permeating molecules in mammalian intracellular AQP.

8.4 INTRACELLULAR PEROXIPORINS: SCAVENGER OR SIGNALING MODULATORS

Major intracellular sources of reactive oxygen species (ROS) are mitochondria, the ER, lysosomes, peroxisomes, and enzymes of the cytoplasm or plasma membrane. In addition, ROS is also produced by extracellular metals. Among ROS, hydrogen peroxide (H_2O_2) will be the main effector in redox signaling based on its stability and diffusibility from the production sources. Interestingly, H_2O_2 plays paradoxical dual roles: H_2O_2 can function as a signaling molecule at low concentrations both as an autocrine (including subcellular compartments) and a paracrine (intercellular) signal (Lismont et al. 2019), while it damages cells and tissues at higher concentrations depending on the physiological state and the duration of exposure (Miller et al. 2010; Pizzino et al. 2017; Nordzieke and Medraño-Fermandez 2018). To protect a cell, peroxidases and catalase convert H_2O_2 into oxygen and water to detoxicate, and catalase is present in all aerobic cells to reduce ROS and prevent lipid peroxidation.

Most ROS are generated during mitochondrial oxidative phosphorylation linked to aerobic energy metabolism while smaller amounts are generated by the NADPH-oxidase system. In the presence of mitochondrial superoxide dismutase (SOD), superoxide is converted to H_2O_2, then exiting mitochondria into the cytoplasm (Figure 8.1).

NADPH oxidases also produce superoxide by transferring electrons from NADPH which react with oxygen to form superoxide, then rapidly converted to H_2O_2. NADPH oxidase resides in the plasma membrane to generate superoxide for redox signaling leading to cell proliferation and immune reactions such as phagocytotic bacterial killing. H_2O_2 also inhibits phosphatases to enhance kinase-induced signaling. Moreover, NADPH oxidase isoforms colocalize with signaling molecules to create localized high concentrations of H_2O_2 to restrict its action (Augsburger et al. 2019).

The ER synthesizes proteins and subsequently forms the disulfide bonds promoted by its reducing/oxidizing folding environment, which generates ROS as a by-product of protein folding and oxidation (Malhotra and Kaufman 2007). Therefore, protein folding and generation of ROS in the ER are closely linked, as shown in highly active cells in which rapid increases in H_2O_2 within the ER are observed as a result of oxidative protein folding catalyzed by a family of ER oxidoreductases, including PDI (protein disulfide isomerase). If uncontrolled, the increased ROS may cause ER stress and activate the unfolded protein response (UPR), leading to apoptosis/autophagy (Malhotra and Kaufman 2007; Chen et al. 2007). The protein-folding machinery of the ER is composed of protein chaperones, proteins for catalyzing protein folding, and sensors for the detection of misfolded or unfolded proteins. Misfolded proteins are either retained within the ER lumen in complex with chaperones or are transferred to proteasome for degradation using ATP (ER-associated degradation [ERAD]) or to be

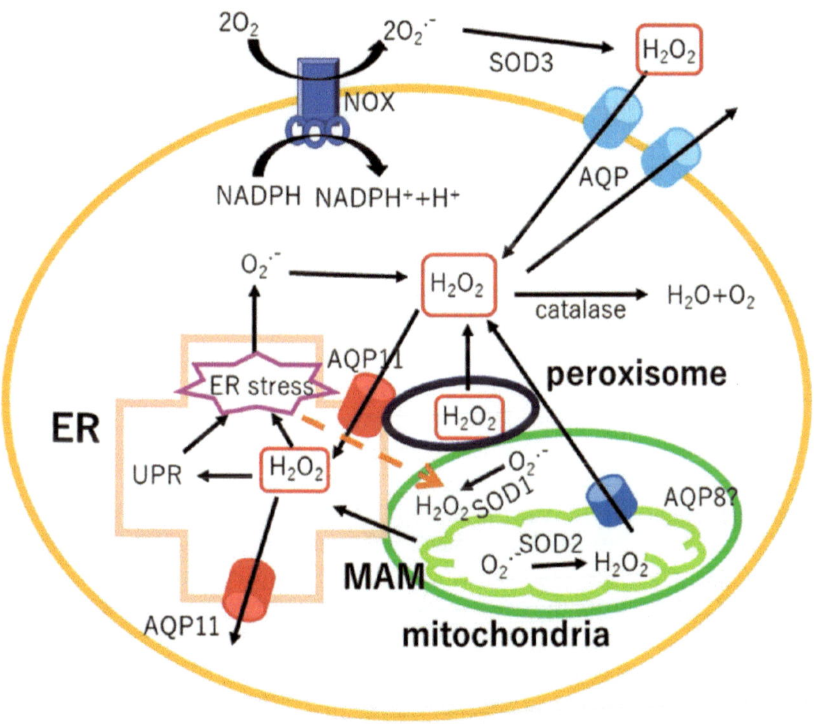

Figure 8.1 Pathways for H_2O_2 production and flux in the cell. ER-residing AQP11 takes up H_2O_2 to control cytosolic and mitochondrial H_2O_2 levels. As indicated by a dotted arrow, ER stress induces mitochondrial superoxide dismutase (SOD) to produce H_2O_2 and may be exported to the cytoplasm through AQP8. NADPH oxidase (NOX) at the plasma membrane also produces H_2O_2 which is transported into the cell by AQP. Moreover, H_2O_2 is accumulated in the cytoplasm by the transfer from peroxisomes. The ER may regulate the cytoplasmic H_2O_2 through AQP11. H_2O_2 also enters the ER directly from mitochondria through mitochondria-ER associated membrane (MAM) through AQP11 (Sorrentino et al. 2022). Catalase converts H_2O_2 into oxygen and water.

processed by autophagy. The ER also stores Ca^{2+}, which is required for both protein-folding reaction and protein chaperone function. Persistent oxidative stress and protein misfolding initiate apoptotic cascades to cell death, which underlies the pathophysiology of various diseases such as diabetes, atherosclerosis, and neurodegenerative diseases.

H_2O_2 has long been believed to permeate biological membranes freely by simple diffusion. However, as a signal molecule in various cellular processes, it is hypothesized that some membrane regulators such as transporters or channels may be necessary to control the transport of H_2O_2 across the membrane. As H_2O_2 and H_2O share similar sizes and electrochemical properties, AQP will be good candidates for being a H_2O_2-permeating membrane protein (Almasalmeh et al. 2014). In fact, a *Streptococcus* AQP1-like whose expression is induced by H_2O_2 has recently been shown to facilitate the efflux of endogenous H_2O_2, which not only contributes to detoxification protecting the bacterium from oxidant attacks but also enhances species competitiveness (Tong et al. 2019b). More recently, highly water-permeable intracellular plant AQP, TIP, have been shown to permeate H_2O_2 (Sudhakaran et al. 2021), suggesting that TIP could also have a role in scavenging H_2O_2 into the vacuole, which could be detoxified by peroxidases using flavonoids as an electron donor.

AQP0, 1, 3, 5, 8, 9, and 11 have been demonstrated to permeate H_2O_2 in mammalian cells (Miller et al. 2010; Watanabe et al. 2016; Rodrigues et al. 2019; Montiel et al. 2020; Varadaraj and Kumari 2020; Bestetti et al. 2020). Although all water-permeable AQP seem to become PXP, H_2O_2 permeability varies with the isoform of AQP partially depending on the presence of a highly conserved cysteine residue in TM1 such as AQP8 (C53), AQP11-like, and all AQP3-like except AQP7 (Almasalmeh et al. 2014; Medraño-Fernandez et al. 2016; Bestetti et al. 2020) suggesting the presence of a common gating mechanism for H_2O_2 permeation and AQP7 as a least efficient PXP. Among them, AQP8, 9, 10, 11, and 12 have been localized inside the cell and their roles in intracellular H_2O_2 homeostasis are intriguing.

AQP8 has been reported to be present in the inner mitochondrial membrane of the liver, kidney, and neural stem cells (Calamita et al. 2005). However, the measurement of water and glycerol permeability of the inner membrane of mitochondria showed the absence of functional AQP in rat brain, liver, and kidney (Yang et al. 2006b), and our initial survey failed to detect AQP8 in the kidney (Ishibashi et al. 1997b), although AQP8 seems to be induced by metabolic acidosis in the kidney (Molinas et al. 2012). These confronting data need to be solved by further studies. By Western blot analysis, AQP8 was detected in hepatocytes with two bands; the smaller band of 28 kDa was observed in the inner mitochondrial membrane-enriched fraction, and the larger glycosylated band of 34 kDa in the plasma membrane fractions (Garcia et al. 2001). Although the antibody against AQP8 seems to be specific, more stringent testing for the specificity will be required by using AQP8 KO mice as a negative control with the application for the histochemical identification by electron microscopy. If AQP8 should be expressed in mitochondria, it will be plausible that AQP8 permeates H_2O_2 generated in the mitochondria, where AQP8 is involved in redox-related signaling to stop the flow of H_2O_2 under a redox stress and protects mitochondria by facilitating H_2O_2 efflux from mitochondria especially following cytokine attacks (Krüger et al. 2021). The effluxed H_2O_2 may need to be sequestered in the ER and/or expelled from the cell, in which the balanced coordination of PXP by modulation and cross-talk might be important to avoid cell damage (Figure 8.1). The fact that AQP8, AQP9, and AQP12 KO mice appear normal suggests that these AQP may play roles in pathological conditions such as H_2O_2 overproduction and its excessive accumulation in the cell. Such studies are scarce especially in *in vivo* conditions with the exception of AQP12, whose disruption in mice predisposes to severe acute pancreatitis after strong secretagogue stimulations (Ohta et al. 2009).

Fluorescent dyes and the genetically encoded fluorescent protein HyPer have been developed to specifically detect intracellular H_2O_2. Compared with fluorescent dyes (Rezende et al. 2018), HyPer estimates the cellular H_2O_2 levels in a reversible and real-time manner (Bilan and Belousov 2018; Bestetti et al. 2020). These organelle-targeted sensors to document H_2O_2 transport have identified AQP11 as PXP in the ER (Bestetti et al. 2020). If AQP11 is important for H_2O_2 homeostasis in the ER, its disruption may lead to ER stress especially when the cell is actively producing massive proteins such as enzymes and transporters.

In AQP11 KO mice, intracellular large vacuoles derived from the ER are observed in the kidney and liver (Morishita et al. 2005). Similar phenotypes have been observed with a point mutation of cysteine residue downstream of second NPA motif (Tchekneva et al. 2008) which is critical for AQP11 function and the signature sequence for AQP11-like (Ishibashi et al. 2017). This cysteine residue may act to form a disulfide bond with another unknown cysteine residue of AQP11 or as a redox sensor of PXP facing the lumen of the ER. Interestingly, the vacuoles in the kidney proximal tubules are developmentally regulated (Rützler et al. 2017), while those in the liver are protein-intake dependent (Rojek et al. 2013).

The proximal tubular cell vacuolization developed only in mice with the AQP11 gene disrupted before P8 when nephron maturation still continues at the outer layer of kidney cortex (Hartman et al. 2007). On the other hand, AQP11 gene deletion from P12 on did not result in abnormal renal development with vacuoles or cyst formation (Rützler et al. 2017). A similar phenomenon of developmental stage dependency has been reported in polycystic kidney disease model mice with pkd1 conditional KO (Piontek et al. 2007; Lantinga-van Leeuwen et al. 2007). With normal appearance of mitochondria, the vacuoles may be caused by ER stress (Okada et al. 2008; Tanaka et al. 2016) due to accumulated H_2O_2 produced by active protein synthesis during tubulogenesis with the failed efflux in the absence of AQP11. Subsequently, autophagy may be induced to stimulate cell proliferation (Tanaka et al. 2016; Chen et al. 2007). Alternatively, reduced arginosuccinate synthase 1 (ASS1) expression in the early AQP11 KO kidney (Saito et al. 2017) may decrease the synthesis of arginine, a conditionally essential amino acid, which may lead to the accumulation of aspartate for pyrimidine synthesis to facilitate cyst growth as shown in pkd1 KO kidneys (Trott et al. 2018). Hence, enhanced epithelial proliferation was observed in kidney cortex of AQP11 KO mice during early postnatal development without obvious apoptosis. The initial damage during tubulogenesis should suppress proper tubule maturation followed by continued and misguided cell division, resulting in tubular dilation with increased diameter simulating cysts (Karner et al. 2009; Rützler et al. 2017). Most likely, the tubular dilatation may be produced by defective glycosylation and trafficking of polycystin-1 (PC-1), the product of pkd1, from AQP11-deficient ER to the plasma membrane simulating the pathogenesis of ADPKD with unregulated epithelial proliferation (Inoue et al. 2014). After dilated tubule formation, AQP11 KO mice suffer from decreased levels of mitochondrial proteins (Saito et al. 2017) and a dramatic increase of ROS, mainly due to NOX2 proteins produced by infiltrating macrophages (Hoshino et al. 2019). Both may contribute to the rapid progression of renal failure leading to the death within a month. Thus, the primary role of AQP11 in the kidney appears to facilitate early postnatal development and not to sustain the cellular homeostasis in the matured kidney.

In contrast, liver-specific AQP11 KO mice have normal liver function and structure under unstressed conditions. However, refeeding of amino acids following 24 hours of fasting produced intracellular giant vacuoles limited to periportal hepatocytes similar to kidney proximal tubules although the liver damage was mild and seems to be reversible (Rojek et al. 2012). Indeed, amino acids as building blocks for proteins play an important role in protein synthesis and the following protein folding will produce H_2O_2 to disturb redox homeostasis (Yang et al. 2021), which will be the reason for the production of vacuoles by refeeding amino acids but not by fasting. Accordingly, ER stress may play a role in the development of vacuoles as documented by enhanced expression of ER stress markers and decreased glutathione levels caused by refeeding challenges with apparently normal mitochondria without any signs of swelling or damage. Alternatively, although blood levels of ammonia were not significantly different (Rojek et al. 2013), acute rise in intracellular ammonia by rapid deamination of supplied amino acids may cause ammonia accumulation and/or pH change leading to disfunction of cellular and/or ER homeostasis, where the role of AQP11 as PXP may be important.

Of note, both proximal tubules and hepatocytes are highly sensitive to ER stress as revealed by tunicamycin, an N-glycosylation inhibitor, which produces huge vacuoles in the proximal tubule and periportal hepatocytes similar to the phenotype of AQP11 KO mice (Finnie et al. 2004). As active protein synthesis of transporters and secreted proteins are conducted in the kidney and liver, H_2O_2 will be inevitably produced during the process of 3D structure formation by disulfide bonds, which should be removed from the ER possibly through AQP11 to suppress ER stress reactions.

8.5 DISEASES ASSOCIATED WITH INTRACELLULAR PEROXIPORINS

Within the nephron, the proximal tubule is a major site for ROS generation due to the large amounts of ATP production by consuming oxygen to support massive active solute transports. Glucose transport is one of them which is potentiated in diabetes leading to more ROS production. The excessive uptake of glucose may also enhance glycolysis to produce a high mitochondrial proton gradient accompanied by elevated production and release of mitochondrial superoxide and ROS. Moreover, increased glucose uptake also stimulates PKC-dependent activation of NADPH oxidase, resulting in excessive production of superoxide by NADPH oxidase (Sachse and Wolf 2007). Thus, hyperglycemia-induced ROS play a major role in tubular injury which can be prevented by antioxidant therapy as well as by inhibiting glucose uptake. In fact, the inhibition of sodium-dependent glucose transport 2 (SGLT2) in the proximal tubule is emerging as a new therapeutic strategy for the treatment of diabetic nephropathy (Giorgino et al. 2020) and even chronic kidney disease (CKD; Mima 2021). On the other hand, disappointing clinical results have been reported on the treatment in patients with CKD by N-acetylcysteine (NAC) which replenishes intracellular glutathione and also directly scavenges free radicals (Renke et al. 2008), even though NAC has been shown to reduce lipid peroxidation, to maintain mitochondrial membrane potential, and to prevent apoptosis in various animal models (Abdelrahman et al. 2010). In the acute setting, however, NAC seems to be beneficial as shown by the improvement of early outcomes in kidney transplant recipients (Danilovic et al. 2011).

As hyperglycemia induces increased production of H_2O_2 in mitochondria, the enhanced susceptibility of the proximal tubule to oxidative stress derived from glucose uptake may be responsible for the kidney injury in AQP11-insufficient mice (Atochina-Vasserman et al. 2013). A glucose transport caused more damage in the kidney of AQP11 heterozygote mice expressing a half amount of AQP11 than wild-type mice. When the renal levels of superoxide are increased, scavenging H_2O_2 into the ER through AQP11 may alleviate the proximal tubular damage caused by increased ROS production. This safeguard mechanism may be insufficient in the cell with decreased AQP11, leading to the induction of ER stress, UPR, and apoptosis. In agreement with this hypothesis, the glucose-induced renal dysfunction with decreased AQP11 was ameliorated by antioxidant therapy as well as by inhibiting glucose transport through a SGLT inhibitor (Atochina-Vasserman et al. 2013). Of note is the absence of intracellular vacuoles in these cells, suggesting a different mechanism of cellular damage in decreased AQP11 from that in total absence of AQP11. Indeed, the expression of AQP11 was increased by glucose in the kidney and cultured cells, which was prevented by phlorizin, a SGLT inhibitor. The results suggest the associated and probably adaptive increase of AQP11 by a glucose load (Atochina-Vasserman et al. 2013). Cisplatin also directly damages the kidney mostly due to the induction of ROS in the proximal tubule (Chirino et al. 2009). We have preliminary results indicating that AQP11 heterozygous KO mice suffered more damage in the proximal tubule than wild mice by cisplatin. This damage was attenuated by D-cysteine pre-administration which may also increase intracellular H_2S production to serve as an antioxidant and anti-inflammation agent (Shibuya et al. 2013).

As a human disease related to aberrant expression of AQP11 has not yet been identified, a single-nucleotide polymorphism (SNP) of AQP11 may provide an insight into the role of AQP11 in human. AQP11 rs2276415 variant is identified as a genetic factor predisposing to a greater risk for the development of kidney disease in diabetes (Choma et al. 2016), CKD progression (Han et al. 2018), and graft failure in kidney transplantation (Park et al. 2015). Interestingly, this SNP is a missense mutation causing a G102S amino acid substitution located just following the pore-forming first NPA motif, resulting in NPCS instead of NPCG. This mutation may affect the expression and function of AQP11 since a previous site-directed mutagenesis at the first NPA motif from NPC to NPA reduced oligomerization and osmotic water permeability (Ikeda et al. 2011). Alternatively, it is possible that the mutated protein is unstable or its translation is inefficient since donor kidneys with the minor allele (~1%) had a half expression levels of AQP11 protein compared with the major allele similar to the case of AQP11 heterozygous KO mice (Park et al. 2015). Obviously, additional studies are required to characterize this single nucleotide polymorphism in the progression of kidney disease and possibly in diseases of other organs as AQP11 is widely expressed (Koike et al. 2016).

AQP11 has also been shown to be preferentially located intracellularly in the vicinity of lipid droplets possibly at the ER in cultured human adipocytes (Madeira et al. 2014) and at the ER of human visceral adipocytes (Frühbeck et al. 2020). Moreover, the expression of AQP11 in adipocytes is regulated by hormones and cytokines: lipolysis-inducing isoproterenol stimulates AQP11 expression possibly to support glycerol export, whereas proinflammatory factors transforming growth factor β1 (TGF-β1) and tumor necrosis factor-alpha (TNF-α) inhibit AQP11 expression and enhance the subcellular distribution surrounding lipid droplets (Frühbeck et al. 2020). As both cytokines are involved in obesity-associated inflammation and ER stress, this downregulation of AQP11 indeed aggravated TGF-β1–induced ER stress in visceral adipocytes (Frühbeck et al. 2020). Since obesity induces mitochondrial dysfunction in visceral adipocytes with the subsequent oxidative stress resulting in H_2O_2 accumulation in the cell, the role of AQP11 may well be to sequester H_2O_2 into the ER in adipocytes, which will be enhanced by the overexpression of AQP11 in obese visceral adipose tissue (Frühbeck et al. 2020). AQP11 may also attenuate H_2O_2 accumulation in mitochondria injured by lipotoxicity through mitochondria-ER associated membranes (MAMs) by facilitating the efflux of H_2O_2 into the ER (Sorrentino et al. 2022). As the overaccumulation of H_2O_2 in the ER may facilitate ER stress, the expression of AQP11 should be regulated to balance the redox homeostasis. Further *in vivo* studies of AQP11 on adipose tissues will be needed. Similar regulation may also be present in diabetic nephropathy and fatty liver.

AQP12 is expressed at the ER of acinar cells in rat exocrine pancreas but contributes little to pancreatic fluid secretion in a normal state as is the case with AQP1, AQP5, and AQP8 shown by KO mice. However, AQP12 KO mice develop severe acute pancreatitis after a strong secretagogue stimulation (CCK-8; Ohta et al. 2009), suggesting that the stimulated synthesis of digestive enzymes leads to the accumulation of H_2O_2 in the ER without AQP12 functioning as PXP leading to cellular damage. AQP12 is also shown to be expressed in β-cells of rat islets and rat RIN-m5Fβ cell line (da Silva et al. 2020) although we did not observe AQP12 in β-cells of rat islets (Ohta et al. 2009). The different results may come from the variable specificity of AQP12 antibodies, which were very difficult to produce in our hands (Ohta et al. 2009). Our *in situ* hybridization in mouse pancreas also showed limited expression of AQP12 in acinar cells (Itoh et al. 2005). Nonetheless, they have reported interesting observations: TNF-α and lipopolysaccharides (LPS) upregulated AQP12 in β-cells which was also enhanced in obese rats (da Silva et al. 2020). They also showed that AQP12-overexpressing cells were more resistant to inflammation, possibly due to the efflux of H_2O_2 from the ER by AQP12 as PXP, although its contribution to insulin release is still unclear.

The research on intracellular PXP in disease states is difficult to conduct without reliable measurement and monitoring of intracellular and organellar H_2O_2 levels. With the advance of such techniques, more clear data will be obtained for intracellular PXP committed to initiation and progression of diseases already modified by plasma membrane PXP with possible cross-talks among these PXP.

8.6 CONCLUDING REMARKS

Although the role of intracellular AQP as PXP is still obscure, some intracellular AQP seem to prevent ER stress by the efflux of H_2O_2 produced in the ER. Conversely, H_2O_2 may also be accumulated in the ER from the cytosol by the reverse transport mode of ER-located AQP. This bidirectional transport of H_2O_2 by intracellular AQP needs to be controlled to prevent the accumulation of H_2O_2 in the ER, possibly by sensing the concentration of H_2O_2 and a regulated gating mechanism. Moreover, other than intracellular organelles, H_2O_2 also comes from outside the cell imported by PXP at the plasma membrane. The possible cross-talk between intracellular AQP and plasma membrane AQP for intracellular H_2O_2 homeostasis is also intriguing.

Accumulating evidence indicates the involvement of intracellular AQP in the cellular H_2O_2 homeostasis but the knowledge of organelle interactions is needed to clarify the role especially in disease states. Unfortunately, poor specificities of commercially available antibodies against

AQP6–AQP12 have confused the results of cellular and subcellular localization of AQP. Studies on the function and physiological role of AQP in the cell and tissue is difficult without full knowledge of their localization. Organelle-targeted sensors to document H_2O_2 transport will be a promising technique to examine the function of AQP at defined locations, although *in vivo* measurement will be challenging (Bilan et al. 2018; Bestetti et al. 2020). The exploration from the results of cultured cells to a whole body is another challenge for the clarification of the role of intracellular AQP.

From a clinical point of view, intracellular AQP potentially represent a new therapeutic target for the treatment of diseases by modulating the function and expression levels. The fact that KO mice of intracellular AQP (AQP8, 9, 10, and 12 in whole animal KO, and AQP11 in conditional KO) display little phenotype may indicate less adverse effects by their inhibition for therapeutic purposes. In the absence of specific stimulators and inhibitors, manipulating microRNAs will be a promising therapeutic strategy: the expression of AQP11 is enhanced by specific microRNAs (Xi et al. 2018; Sakurai et al. 2019). Clarifying the full spectrum of functional roles and regulatory pathways of intracellular AQP will lead to exploring novel strategies for the treatment of various diseases.

REFERENCES

Abdelrahman, A.M. et al., 2010, N-acetylcysteine improves renal hemodynamics in rats with cisplatin-induced nephrotoxicity. *J Appl Toxicol*, 30, 15–21.

Agre, P., 1997, Molecular physiology of water transport: aquaporin nomenclature workshop. Mammalian aquaporins. *Biol Cell*, 89, 255–257.

Agre, P. et al., 1993, Aquaporin CHIP: the archetypal molecular water channel. *Am J Physiol*, 265, F463–F476.

Almasalmeh, A. et al., 2014, Structural determinants of the hydrogen peroxide permeability of aquaporins. *FEBS J*, 281, 647–656.

Atochina-Vasserman, E.N. et al., 2013, Aquaporin 11 insufficiency modulates kidney susceptibility to oxidative stress. *Am J Physiol Renal Physiol*, 304 (10), F1295–F1307.

Augsburger, F., Filippova, A., and Jaquet, V., 2019, Methods for detection of NOX-derived superoxide radical anion and hydrogen peroxide in cells. *Methods Mol Biol*, 1982, 233–241.

Baker, M.E. and Saier Jr., M.H., 1990, A common ancestor for bovine lens fiber major intrinsic protein, soybean nodulin-26 protein, and E. coli glycerol facilitator. *Cell*, 26, 185–186.

Beitz, E. et al., 2006, Determinants of AQP6 trafficking to intracellular sites versus the plasma membrane in transfected mammalian cells. *Biol Cell*, 98(2), 101–109.

Bestetti, S. et al., 2020, Human aquaporin-11 guarantees efficient transport of H_2O_2 across the endoplasmic reticulum membrane. *Redox Biol*, 28, 101326.

Bienert, G.P. et al., 2013, Channel-mediated lactic acid transport: a novel function for aquaglyceroporins in bacteria. *Biochem J*, 454, 559–570.

Bienert, G.P. and Chaumont, F., 2014, Aquaporin-facilitated transmembrane diffusion of hydrogen peroxide. *Biochim Biophys Acta*, 1840, 1596–1604.

Bilan, D.S. and Belousov, V.V., 2018, In vivo imaging of hydrogen peroxide with HyPer probes. *Antioxid Redox Signal*, 29, 569–584.

Borgnia, M. et al., 1999, Cellular and molecular biology of the aquaporin water channels. *Annu Rev Biochem*, 68, 425–458.

Calamita, G. et al., 2005, The inner mitochondrial membrane has aquaporin-8 water channels and is highly permeable to water. *J Biol Chem*, 280, 17149–17153.

Chen, Y. et al., 2007, Mitochondrial electron-transport-chain inhibitors of complexes I and II induce autophagic cell death mediated by reactive oxygen species. *J Cell Sci*, 120, 4155–4166.

Chen, Y. et al., 2012, Aquaporin 2 promotes cell migration and epithelial morphogenesis. *J Am Soc Nephrol*, 23(9), 1506–1517.

Chirino, Y.I. and Pedraza-Chaverri, J., 2009, Role of oxidative and nitrosative stress in cisplatin-induced nephrotoxicity. *Exp Toxicol Pathol*, 61, 223–242.

Choma, D.P. et al., 2016, Aquaporin 11 variant associates with kidney disease in type 2 diabetic patients. *Am J Physiol Renal Physiol*, 310(5), F416–F425.

Conner, M.T. et al., 2012, Rapid aquaporin translocation regulates cellular water flow: the mechanism of hypotonicity-induced sub-cellular localization of the aquaporin 1 water channel. *J Biol Chem*, 287, 11516–11525.

Czyzewski, B.K. and Wang, D.N., 2012, Identification and characterization of a bacterial hydrosulphide ion channel. *Nature*, 483(7390), 494–497.

Danielli, M. et al., 2019, Data of H_2O_2 release from AQP8-knockdown rat hepatocyte mitochondria. *Data Brief*, 23, 103722.Danilovic, A. et al., 2011, Protective effect of N-acetylcysteine on early outcomes of deceased renal transplantation. *Transplant Proc*, 43, 1443–1449.

Da Silva, I.V. et al., 2020, Aquaporin-7 and aquaporin-12 modulate the inflammatory phenotype of endocrine pancreatic beta-cells. *Arch Biochem Biophys*, 691, 108481.

De Ieso, M.L. et al., 2019, Combined pharmacological administration of AQP1 ion channel blocker AqB011 and water channel blocker Bacopaside II amplifies inhibition of colon cancer cell migration. *Sci Rep*, 9(1), 12635.

De Maré, S.W. et al., 2020, Structural basis for glycerol efflux and selectivity of human aquaporin 7. *Structure*, 28(2), 215–222.e3.

Echevarria, M. et al., 1994, Cloning and expression of AQP3, a water channel from the medullary collecting duct of rat kidney. *Proc Natl Acad Sci U S A*, 91(23), 10997–11001.

Finn, R.N. et al., 2015, Insect glycerol transporters evolved by functional co-option and gene replacement. *Nat Commun*, 6, 7814.

Finnie, J.W., Read S.H., and Swift J.G., 2004, Apoptosis in liver damage produced by tunicamycin. *Aust Vet J*, 82(1–2), 87–90.

Frühbeck, G. et al., 2020, Aquaporin-11 contributes to TGF-β1-induced endoplasmic reticulum stress in human visceral adipocytes: role in obesity-associated inflammation. *Cells*, 9(6), 1403.

Fushimi, K. et al., 1993, Cloning an d expression of apical membrane water channel of rat kidney collecting tubule. *Nature*, 361(6412), 549–552.

Garcia, F. et al., 2001, The water channel aquaporin-8 is mainly intracellular in rat hepatocytes, and its plasma membrane insertion is stimulated by cyclic AMP. *J Biol Chem*, 276, 12147–12152.

Giorgino, F. et al., 2020, Renoprotection with SGLT2 inhibitors in type 2 diabetes over a spectrum of cardiovascular and renal risk. *Cardiovasc Diabetol*, 19(1), 196.

Gorelick, D.A. et al., 2006, Aquaporin-11: a channel protein lacking apparent transport function expressed in brain. *BMC Biochem*, 7, 14.

Gorin, M.B. et al., 1984, The major intrinsic protein (MIP) of the bovine lens fiber membrane: characterization and structure based on cDNA cloning. *Cell*, 39, 49–59.

Han, B. et al., 2018, Aquaporin 11 rs2276415 variant and progression of chronic kidney disease. *Nephrol Dial Transpl*, 34(6), 970–973.

Hansen, M. et al., 2002, A single, bi-functional aquaglyceroporin in blood-stage *Plasmodium falciparum* malaria parasites. *J Biol Chem*, 277, 4874–4882.

Hartman, H.A., Lai, H.L., and Patterson, L.T., 2007, Cessation of renal morphogenesis in mice. *Dev Biol*, 310(2), 379–387.

Heller, K.B., Lin, E.C., and Wilson, T.H., 1980, Substrate specificity and transport properties of the glycerol facilitator of *Escherichia coli*. *J Bacteriol*, 144(1), 274–278.

Hoshino, Y. et al., 2019, Involvement of the NADPH oxidase 2 pathway in renal oxidative stress in *Aqp11*$^{-/-}$ mice. *Biochem Biophys Rep*, 17, 169–176.

Hosoi, K. et al., 2020, Dynamics of salivary gland AQP5 under normal and pathologic conditions. *Int J Mol Sci*, 21(4), 1182.

Hu, Q. et al., 2021, A novel aquaporin subfamily imports oxygen and contributes to *Pneumococcal* virulence by controlling the production and release of virulence factors. *MBIO*, e0130921.

Ikeda, M. et al., 2002, Characterization of aquaporin-6 as a nitrate channel in mammalian cells: requirement of pore-lining residue threonine 63. *J Biol Chem*, 277(42), 39873–39879.

Ikeda, M. et al., 2011, The NPC motif of aquaporin-11, unlike the NPA motif of known aquaporins, is essential for full expression of molecular function. *J Biol Chem*, 286(5), 3342–3350.

Inoue, Y. et al., 2014, Aberrant glycosylation and localization of polycystin-1 cause polycystic kidney in an AQP11 knockout model. *J Am Soc Nephrol*, 25(12), 2789–2799.

Ishibashi, K. et al., 1994, Molecular cloning and expression of a member of the aquaporin family with permeability to glycerol and urea in addition to water expressed at the basolateral membrane of kidney collecting duct cells. *Proc Natl Acad Sci U S A*, 91(14), 6269–6273.

Ishibashi, K. et al., 1997a, Cloning and functional expression of a new water channel abundantly expressed in the testis permeable to water, glycerol, and urea. *J Biol Chem*, 272(33), 20782–20786.

Ishibashi, K. et al., 1997b, Cloning and functional expression of a second new aquaporin abundantly expressed in testis. *Biochem Biophys Res Commun*, 237, 714–718.

Ishibashi, K. et al., 2000, Molecular cloning of a new aquaporin superfamily in mammals. In: Hohmann S, Nielsen S, editors. *Molecular biology and physiology of water and solute transport*. New York, Kluwer Academic/Plenum Publishers, p123–126.Ishibashi, K., Morishita, Y., and Tanaka, Y., 2017, The evolutionary aspects of aquaporin family. *Adv Exp Med Biol*, 969, 35–50.

Ishibashi, K., Tanaka, Y., and Morishita, Y., 2020, Perspectives on the evolution of aquaporin superfamily. *Vitam Horm*, 112, 1–27.

Ishibashi, K., Tanaka, Y., and Morishita, Y., 2021, The role of mammalian superaquaporins inside the cell: an update. *Biochim Biophys Acta Biomembr*, 1863(7), 183617.

Ishibashi, K. and Sasaki, S., 1998, The dichotomy of MIP family suggests two separate origins of water channels. *News Physiol Sci*, 13, 137–142.

Itoh, T. et al., 2005, Identification of a novel aquaporin, AQP12, expressed in pancreatic acinar cells. *Biochem Biophys Res Commun*, 330(3), 832–828.

Jennings, M.L., 2013, Transport of H2S and HS(-) across the human red blood cell membrane: rapid H2S diffusion and AE1-mediated Cl(-)/HS(-) exchange. *Am J Physiol Cell Physiol*, 305(9), C941–C950.

Jeremic, A., Cho, W.J., and Jena, B.P., 2005, Involvement of water channels in synaptic vesicle swelling. *Exp Bio Med*, 230, 674–680.

Johnson, K.D., Höfte, H., and Chrispeels, M.J., 1990, An intrinsic tonoplast protein of protein storage vacuoles in seeds is structurally related to a bacterial solute transporter (GIpF). *Plant Cell*, 6, 525–532.

Karner, C.M. et al., 2009, Wnt9b signaling regulates planar cell polarity and kidney tubule morphogenesis. *Nat Genet*, 41(7), 793–799.

Kishida, K. et al, 2000, Aquaporin adipose, a putative glycerol channel in adipocytes. *J Biol Chem*, 275, 20896–20902.

Kitchen, P. et al., 2020, Targeting aquaporin-4 subcellular localization to treat central nervous system edema. *Cell*, 181, 784–799.

Knepper, M.A., 1994, The aquaporin family of molecular water channels. *Proc Natl Acad Sci U S A*, 91(14), 6255–6258.

Koike, S. et al., 2016, Aquaporin-11 (AQP11) expression in the mouse brain. *Int J Mol Sci*, 17(6), 861.

Kozono, D. et al., 2003, Functional expression and characterization of an archaeal aquaporin AqpM from *Methanothermobacter marburgensis*. *J Biol Chem*, 278, 10649–10656.

Kreida, S. and Törnroth-Horsefield, S., 2015, Structural insights into aquaporin selectivity and regulation. *Curr Opin Struct Biol*, 33, 126–134.

Krüger, C. et al., 2021, The importance of aquaporin-8 for cytokine-mediated toxicity in rat insulin-producing cells. *Free Radic Biol Med*, 174, 135–143.

Laforenza, U., Scaffino, M.F., and Gastaldi, G., 2013, Aquaporin-10 represents an alternative pathway for glycerol efflux from human adipocytes. *PLoS ONE*, 8(1), e54474.

Lantinga-van Leeuwen, I.S. et al., 2007, Kidney-specific inactivation of the Pkd1 gene induces rapid cyst formation in developing kidneys and a slow onset of disease in adult mice. *Hum Mol Genet*, 16(24), 3188–3196.

Lee, J.K. et al., 2005, Structural basis for conductance by the archaeal aquaporin AqpM at 1.68 A. *Proc Natl Acad Sci USA*, 102, 18932–18937.

Li, H. et al., 2005, Expression and localization of two isoforms of AQP10 in human small intestine. *Biol Cell*, 97(11), 823–829.

Linhares Lino de Souza, D., Serrão, J.E., and Hansen, I.A., 2020, Aquaporin expression in the alimentary canal of the honey bee *Apis mellifera L.* (Hymenoptera: Apidae) and functional characterization of Am_Eglp 1. *PLoS ONE*, 15(9), e0236724.

Lismont, C., Revenco, I., and Fransen, M., 2019, Peroxisomal hydrogen peroxide metabolism and signaling in health and disease. *Int J Mol Sci*, 20, e3673.

Ma, T. et al., 1994, Expression, functional analysis, and in situ hybridization of a cloned rat kidney collecting duct water channel. *Am J Physiol*, 266(1 Pt 1), C189–C197.

Madeira, A. et al., 2014, Human aquaporin-11 is a water and glycerol channel and localizes in the vicinity of lipid droplets in human adipocytes. *Obesity (Silver Spring)*, 22(9), 2010–2017.

Malhotra, J.D. and Kaufman, R.J., 2007, Endoplasmic reticulum stress and oxidative stress: a vicious cycle or a double-edged sword? *Antioxid Redox Signal*, 9, 2277–2293.

Mathai, J.C. et al., 2009, No facilitator required for membrane transport of hydrogen sulfide. *Proc Natl Acad Sci U S A*, 106(39), 16633–16638.

Matsuzaki, T. et al., 2017, The distribution and function of aquaporins in the kidney: resolved and unresolved questions. *Anat Sci Int*, 92(2), 187–199.

Maurel, C. et al., 1994, Functional characterization of the *Escherichia coli* glycerol facilitator, GlpF, in *Xenopus* oocytes. *J Biol Chem*, 269(16), 11869–11872.

Medraño-Fernandez, I., et al., 2016, Stress regulates aquaporin-8 permeability to impact cell growth and survival. *Antioxid Redox Signal*, 24(18), 1031–1044.

Michenkova, M. et al., 2021, Carbon dioxide transport across membranes. *Interface Focus*, 11(2), 20200090.

Miller, E.W., Dickinson, B.C., and Chang, C.J., 2010, Aquaporin-3 mediates hydrogen peroxide uptake to regulate downstream intracellular signaling. *Proc Natl Acad Sci U S A*, 107(36), 15681–15686.

Mima, A., 2021, Sodium-glucose cotransporter 2 inhibitors in patients with non-diabetic chronic kidney disease. *Adv Ther*, 38(5), 2201–2212.

Mobasheri, A., Shakibaei, M., and Marples, D., 2004, Immunohistochemical localization of aquaporin 10 in the apical membranes of the human ileum: a potential pathway for luminal water and small solute absorption. *Histochem Cell Biol*, 121(6), 463–471.

Molinas, S.M., L. Trumper, L., and Marinelli, R.A., 2012, Mitochondrial aquaporin-8 in renal proximal tubule cells: evidence for a role in the response to metabolic acidosis. *Am J Physiol Renal Physiol*, 303, F458–F466.

Montiel, V. et al., 2020, Inhibition of aquaporin-1 prevents myocardial remodeling by blocking the transmembrane transport of hydrogen peroxide. *Sci Transl Med*, 12(564), eaay2176.

Morinaga, T., et al., 2002, Mouse aquaporin 10 gene (AQP10) is a pseudogene. *Biochem Biophys Res Commun*, 294(3), 630–634.

Morishita, Y., et al., 2005, Disruption of aquaporin-11 produces polycystic kidneys following vacuolization of the proximal tubule. *Mol Cell Biol*, 25(17), 7770–7779.

Moss, F.J. et al., 2020, Aquaporin-7: a dynamic aquaglyceroporin with greater water and glycerol permeability than its bacterial homolog GlpF. *Front Physiol*, 11, 728.

Muramatsu, S. and Mizuno, T., 1989. Nucleotide sequence of the region encompassing the *glpKF* operon and its upstream region containing a bent DNA sequence of *Escherichia coli*. *Nucleic Acids Res*, 17, 4378.

Németh-Cahalan, K.L. and Hall, J.E., 2000, pH and calcium regulate the water permeability of aquaporin 0. *J Biol Chem*, 275, 6777–6782.

Nordzieke, D.E. and Medrano-Fernandez, I., 2018, The plasma membrane: a platform for intra- and intercellular redox signaling. *Antioxidants (Basel)*, 7(11), 168.

Nozaki, K., Ishii, D., and Ishibashi, K., 2008, Intracellular aquaporins: clues for intracellular water transport? *Pflugers Archiv*, 456(4), 701–707.

Ohta, E. et al., 2009, Pancreas-specific aquaporin 12 null mice showed increased susceptibility to caerulein-induced acute pancreatitis. *Am J Physiol Cell Physiol*, 297(6), C1368–C1378.

Okada, S. et al., 2008, Aquaporin-11 knockout mice and polycystic kidney disease animals share a common mechanism of cyst formation. *FASEB J*, 22(10), 3672–3684.

Olesen, E.T. and Fenton, R.A., 2017, Aquaporin-2 membrane targeting: still a conundrum. *Am J Physiol Renal Physiol*, 312(4), F744–F747.

Ozu, M. et al., 2013, Human AQP1 is a constitutively open channel that closes by a membrane-tension-mediated mechanism. *Biophys J*, 104, 85–95.

Pao, G.M. et al., 1991, Evolution of the MIP family of integral membrane transport proteins. *Mol Microbiol*, 5, 33–37.

Park, J.I. et al., 2015, Genetic predisposition of donors affects the allograft outcome in kidney transplantation: single-nucleotide polymorphism of aquaporin-11. *Kidney Res Clin Pract*, 34(1), 47–52.

Piontek, K. et al., 2007, A critical developmental switch defines the kinetics of kidney cyst formation after loss of Pkd1. *Nat Med*, 13(12), 1490–1495.

Pizzino, G., Irrera, N. and Cucinotta, M., 2017, Oxidative stress: harms and benefits for human health. *Oxid Med Cell Longev*, 8416763.

Pluhackova, K. et al., 2022, Multiple pore lining residues modulate water permeability of GlpF. *Protein Sci*, 31(10), e4431.

Pommerrenig, B. et al., 2020, Functional evolution of nodulin 26-like intrinsic proteins: from bacterial arsenic detoxification to plant nutrient transport. *New Phytol*, 225, 1383–1396.

Prata, C., Hrelia, S. and Fiorentini, D., 2019, Peroxiporins in cancer. *Int J Mol Sci*, 20(6), 1371

Preston, G.M. and Agre, P., 1991, Isolation of the cDNA for erythrocyte integral membrane protein of 28 kD: member of an ancient channel family. *Proc Natl Acad Sci USA*, 88, 11110–11114.

Preston, G.M. et al., 1992, Appearance of water channels in *Xenopus* oocytes expressing red cell CHIP28 protein. *Science*, 256, 385–387.

Promeneur, D. et al., 2018, Aquaglyceroporin PbAQP is required for efficient progression through the liver stage of Plasmodium infection. *Sci Rep*, 8(1), 655.

Rao, Y., Jan, L.Y. and Jan, Y.N., 1990, Similarity of the product of the Drosophila neurogenic gene *big brain* to transmembrane channel proteins. *Nature*, 345, 163–167.

Renke, M. et al., 2008, The effect of N-acetylcysteine on proteinuria and markers of tubular injury in non-diabetic patients with chronic kidney disease. A placebo-controlled, randomized, open, cross-over study. *Kidney Blood Press Res*, 31, 404–410.

Rezende, F., Brandes, R.P. and Schröder, K., 2018, Detection of hydrogen peroxide with fluorescent dyes. *Antioxid Redox Signal*, 29(6), 585–602.

Rodrigues, C. et al., 2019, Human aquaporin-5 facilitates hydrogen peroxide permeation affecting adaption to oxidative stress and cancer cell migration. *Cancers (Basel)*, 11(7), 932.

Rodriguez, R.A. et al., 2020, Quantitative study of unsaturated transport of glycerol through aquaglyceroporin that has high affinity for glycerol. *RSC Adv*, 10(56), 34203–34214.

Rojek, A. et al., 2013, Liver-specific Aquaporin 11 knockout mice show rapid vacuolization of the rough endoplasmic reticulum in periportal

hepatocytes after amino acid feeding. *Am J Physiol Gastrointest Liver Physiol*, 304(5), G501–G515.
Roudier, N. et al., 2002, AQP3 deficiency in humans and the molecular basis of a novel blood group system, Gil. *J Biol Chem*, 277, 45854–45859.
Rü, W. et al., 2012, The formate channel *FocA* exports the products of mixed-acid fermentation. *Proc Natl Acad Sci U S A*, 109(33), 13254–13259.
Rützler, M. et al., 2017, Temporal deletion of *Aqp11* in mice is linked to the severity of cyst-like disease. *Am J Physiol Renal Physiol*, 312(2), F343–F351.
Sachse, A. and Wolf, G., 2007, Angiotensin II-induced reactive oxygen species and the kidney. *J Am Soc Nephrol*, 18, 2439–2446.
Saito, T. et al., 2017, Proteomic analysis of AQP11-null kidney: Proximal tubular type polycystic kidney disease. *Biochem Biophys Rep*, 13, 17–21.
Sakurai, F. et al., 2019, miR-27b-mediated suppression of aquaporin-11 expression in hepatocytes reduces HCV genomic RNA levels but not viral titers. *Virol J*, 16(1), 58.
Sandal, N.N. and Marcker, K.A., 1988, Soybean nodulin 26 is homologous to the major intrinsic protein of the bovine lens fiber membrane. *Nucleic Acids Res*, 16(19), 9347.
Shibuya, N. et al., 2013, A novel pathway for the production of hydrogen sulfide from D-cysteine in mammalian cells. *Nat Commun*, 4, 1366.
Sorrentino, I. et al., 2022, Transfer of H2O2 from Mitochondria to the endoplasmic reticulum via Aquaporin-11. *Redox Biol*, 55, 102410.
Stavang, J.A. et al., 2015, Phylogenomic and functional analyses of salmon lice aquaporins uncover the molecular diversity of the superfamily in Arthropoda. *BMC Genomics*, 16(1), 618.
Sudhakaran, S. et al., 2021, Significance of solute specificity, expression, and gating mechanism of tonoplast intrinsic protein during development and stress response in plants. *Physiol Plant*, 172(1), 258–274.
Tajkhorshid, E. et al., 2002, Control of the selectivity of the aquaporin water channel family by global orientational tuning. *Science*, 296, 525–530.
Tanaka, Y., Morishita, Y. and Ishibashi, K., 2015, Aquaporin10 is a pseudogene in cattle and their relatives. *Biochem Biophys Rep*, 1, 16–21.
Tanaka, Y. et al., 2016, Enhanced autophagy in polycystic kidneys of AQP11 null mice. *Int J Mol Sci*, 17(12), 1993.
Tchekneva, E.E. et al., 2008, Single amino acid substitution in aquaporin 11 causes renal failure. *J Am Soc Nephrol*, 19(10), 1955–1964.
Tesan, F.C. et al., 2021, AQPX-cluster aquaporins and aquaglyceroporins are asymmetrically distributed in trypanosomes. *Commun Biol*, 4(1), 953.
Tong, H. et al., 2019a, Prokaryotic aquaporins. *Cells*, 8(11), 1316.
Tong, H. et al., 2019b, A *Streptococcus* aquaporin acts as peroxiporin for efflux of cellular hydrogen peroxide and alleviation of oxidative stress. *J Biol Chem*, 294, 4583–4595.
Trott, J.F. et al., 2018, Arginine reprogramming in ADPKD results in arginine-dependent cystogenesis. *Am J Physiol Renal Physiol*, 315(6), F1855–F1868.
Uehlein, N. et al., 2003, The tobacco aquaporin NtAQP1 is a membrane CO2 pore with physiological functions. *Nature*, 425, 734–737.
Varadaraj, K. et al., 2019, C-terminal end of aquaporin 0 regulates lens gap junction channel function. *Invest Ophthalmol Vis Sci*, 60(7), 2525–2531.
Varadaraj, K. and Kumari, S.S., 2020, Lens aquaporins function as peroxiporins to facilitate membrane transport of hydrogen peroxide. *Biochem Biophys Res Commun*, 524(4), 1025–1029.
Verkman, A.S., 2005, Novel roles of aquaporins revealed by phenotype analysis of knockout mice. *Rev Physiol Biochem Pharmacol*, 155, 31–55.
Wang, Y. et al., 2009, Structure of the formate transporter *FocA* reveals a pentameric aquaporin-like channel. *Nature*, 462, 467–472.
Watanabe, S. et al., 2016, Aquaporin-9 facilitates membrane transport of hydrogen peroxide in mammalian cells. *Biochem Biophys Res Commun*, 471(1), 191–197.
Wree, D. et al., 2011, Requirement for asparagine in the aquaporin NPA sequence signature motifs for cation exclusion. *FEBS Journal*, 278 (5), 740–748.
Wu, B. et al., 2009, Concerted action of two cation filters in the aquaporin water channel. *EMBO Journal*, 28(15), 2188–2194.
Wudick, M.M. et al., 2015, Subcellular redistribution of root aquaporins induced by hydrogen peroxide. *Mol Plant*, 8(7), 1103–1114.
Xi, T. et al., 2018, miR-27a-3p protects against blood-brain barrier disruption and brain injury after intracerebral hemorrhage by targeting endothelial aquaporin-11. *J Biol Chem*, 293(52), 20041–20050.
Yakata, K., Tani, K. and Fujiyoshi, Y., 2011, Water permeability and characterization of aquaporin-11. *J Struct Biol*, 174(2), 315–320
Yang, B. and Verkman, A.S., 1997, Water and glycerol permeabilities of aquaporins 1–5 and MIP determined quantitatively by expression of epitope-tagged constructs in *Xenopus* oocytes. *J Biol Chem*, 272, 16140–16146.

Yang, B. et al., 2006a, Mouse model of inducible nephrogenic diabetes insipidus produced by floxed aquaporin-2 gene deletion. *Am J Physiol Renal Physiol*, 291(2), F465–F472.

Yang, B., Zhao, D. and Verkman, A.S., 2006b, Evidence against functionally significant aquaporin expression in mitochondria. *J Biol Chem*, 281, 16202–16206.

Yang, Y. et al., 2021, Amino acids in endoplasmic reticulum stress and redox signaling. *Adv Exp Med Biol*, 1332, 35–49.

Yasui, M. et al., 1999a, Aquaporin-6: An intracellular vesicle water channel protein in renal epithelia. *Proc Natl Acad Sci U S A*, 96, 5808–5813.

Yasui, M. et al., 1999b, Rapid gating and anion permeability of an intracellular aquaporin. *Nature*, 402(6758), 184–187.

Yool, A.J. and Campbell, E.M., 2012, Structure, function and translational relevance of aquaporin dual water and ion channels. *Mol Aspects Med*, 33(5–6), 553–561.

Zhang, J. et al., 2013, AQP1 expression alterations affect morphology and water transport in Schwann cells and hypoxia-induced up-regulation of AQP1 occurs in a HIF-1α-dependent manner. *Neuroscience*, 252, 68–79.

Zwiazek, J.J. et al, 2017, Significance of oxygen transport through aquaporins. *Sci Rep*, 7, 40411.

CHAPTER NINE

Peroxiporins in Plant-Microbial Interactions

Barbara Ludwig Navarro, Kathrin Blumenstein, and Daniela Elisabeth Nordzieke

ABBREVIATIONS

ABA	Abscisic acid
AM	Arbuscular mycorrhiza
AQGP	Aquaglyceroporin
AQP	Aquaporin
ar/R	Aromatic/arginine
DAB	3,3′-Diaminobenzidine
DNA	Deoxyribonucleic acid
DON	Deoxynivalenol
EM	Ectomycorrhiza
ETS	Effector-triggered susceptibility
F-actin	Filamentous actin
GLP	Glycerol-permeable aquaglyceroporin
H_2O_2	Hydrogen peroxide
$HgCl_2$	Mercuric chloride
hpi	Hours post infection
HR	Hypersensitive response
HyPer	Genetically encoded fluorescent sensor for the detection of intracellular H_2O_2
ICS	Isochorismate synthetase
MAMPs/PAMPs	Microbial- or pathogen-associated molecular patterns
MAPK	Mitogen-activated protein kinase
MIP	Major intrinsic proteins
NF	Nitrogen-fixing
Nox	NADPH oxidase
NB-LRR	Nucleotide-binding and leucine-rich repeat domains
NIP	NOD26-like intrinsic proteins
O_2^-	Superoxide radical anions
PAL	Phenylalanine ammonia-lyase
PCD	Programmed cell death
Phe	Phenylalanine
PIP	Plasma membrane intrinsic proteins
PO	Penetration organ
PP	Penetration peg
PR	Pathogenesis related
PRR	Pattern recognition receptors
PTI	PAMP-triggered immunity
Rboh	Respiratory burst oxidase homologs
ROS	Reactive oxygen species
SA	Salicylic acid
SAR	System-acquired resistance
SIP	Small basic intrinsic proteins
TIP	Tonoplast intrinsic proteins
XIP	X intrinsic proteins

9.1 INTRODUCTION

Peroxiporins are a sub-group of aquaporins responsible for the transport of hydrogen peroxide (H_2O_2), which is an important molecule for signaling plant defense responses and for pathogenesis (An et al., 2016; Li et al., 2020; Tian et al., 2016). Aquaporins are major intrinsic proteins (MIP) that mainly work as water (H_2O) channels (Wang et al., 2020); however, the transport of other uncharged substrates by aquaporins such as glycerol, urea, boric acid, arsenite, and CO_2 has been reported (Bienert et al., 2011; Katsuhara

DOI: 10.1201/9781003160649-9

et al., 2014; Uehlein et al., 2003). The functional mechanism of aquaporins is based on its conformation and consequently the electrostatic potential generated by the molecules inside the pore. A study regarding aquaporin conformation has been conducted with the spinach plasma membrane aquaporin SoPIP2;1. The open conformation of SoPIP2;1 blocks the entrance from the cytoplasm by a hydrophobic gate. This open conformation can be observed at 3.9 Å up to 16 Å resolution. At 2.1 Å resolution, a close conformation of loop D could block the pore gate (Törnroth-Horsefield et al., 2006). The water entrance is forced by a central water molecule, which works as a hydrogen bond donor to their neighboring molecules. This electrostatic potential is generated by two membrane loops, which dictate the opposite orientation of water molecules in the channel permitting rapid water diffusion (Tajkhorshid et al., 2002). Simulations for the *Escherichia coli* aquaporin GlpF indicate that seven to nine water molecules can be oriented inside the channel (Tajkhorshid et al., 2002).

The transport of H_2O_2 through water channels was postulated when experiments applying mercuric chloride ($HgCl_2$), an aquaporin activity inhibitor, blocked the membrane permeability of water and H_2O_2 (Henzler and Steudle, 2000). Comparisons between SoPIP2;1, *Arabidopsis thaliana* aquaporin AtPIP2;4, and human AQP1 indicates that these three aquaporins can transport H_2O and H_2O_2 but in different efficacies for each molecule. Different efficacies on H_2O and H_2O_2 cannot be explained by structural differences in the monomeric pore (Wang et al., 2020). Variances in transporting H_2O and H_2O_2 might be related to the flexible domains outside the membrane core of those channels (Wang et al., 2020), which can be responsible for changes in the electrostatic potential. In fact, other studies indicate that aromatic/arginine (ar/R) selectivity may be responsible for the substrate specificity of aquaporins (Dynowski et al., 2008). Curiously, the interaction between maize plasma membrane intrinsic proteins ZmPIP1;2 and ZmPIP2;1 has shown increased solute permeability. ZmPIP1 was localized in the endoplasmic reticulum, whereas ZmPIP2 was observed in the plasma membrane. However, co-expressed ZmPIP1 also has been found in the plasma membrane, which indicates that the PIP1-PIP2 interaction is a precondition for PIP1 to work transporting solutes in the plasma membrane (Zelazny et al., 2007). In fact, not all aquaporins may transport H_2O_2, as this molecule has higher polarity than water, and some may transport only H_2O_2, thus these channels are named peroxiporins (Černý et al., 2018; Hooijmaijers et al., 2012; Srivastava et al., 2016). In this chapter, peroxiporins refers to aquaporins that are able to transport H_2O_2.

9.2 PLANT AQUAPORINS

Aquaporin in plants can be divided in five subgroups according to their localization: plasma membrane intrinsic proteins (PIP), tonoplast intrinsic proteins (TIP), NOD26-like intrinsic proteins (NIP), small basic intrinsic proteins (SIP), and X intrinsic proteins (XIP; Bienert et al., 2011; Park et al., 2010). Previous phylogenetic analysis has shown that aquaporins are homologs between plant species and can be separated in four subfamilies (PIP, TIP, NIP, and SIP), as observed for maize and rice (Chaumount et al., 2001; Sakurai et al., 2005). A recent phylogenetic analysis including genes (some XIP genes) from tomato (Reuscher et al., 2013) and some other plants indicates that homology prevails (Table 9.1; Figure 9.1). In addition, PIP and TIP have been reported to transport H_2O_2 in plants (Bárzana et al., 2014; Bienert et al., 2014).

Hydrogen peroxide is a signaling molecule in several processes in plant growth, development, and defense mechanisms. Loss of viability in seeds are related to the accumulation of H_2O_2 during seed aging. During the quiescent stage enzymatic activity is reduced to a minimum level, however when there is rapid water intake, H_2O_2 is produced to stimulate germination process triggered by the increase of gibberellin biosynthesis, the decrease of abscisic acid levels and the activation of programmed cell death (PCD; Černý et al., 2018). Aquaporins are usually frequent on fine roots showing importance for the fast exchange of H_2O_2, water, and nutrients (Srivastava et al., 2016). The excess of light usually leads to the increase in the production of H_2O_2 within the thylakoid membrane, which is later diffused through chloroplast aquaporins (Mubarakshina Borisova et al., 2012). The gamethophyte fusion can also be influenced by H_2O_2 production, which activities PCD avoiding self-fertilization though self-incompatibility (Černý et al., 2018). The accumulation of reactive oxygen

TABLE 9.1
Plant Aquaporin Genes and Their Accession Numbers Used in the Phylogenetic Analysis

Gene	Plant	Scientific Name	Accession Number	Reference
PpPIP1;4	Peach	*Prunus persica*	XM_007218706.2	(Breia et al., 2020)
AtPIP1;4	*Arabidopsis*	*Arabidopsis thaliana*	NM_116268.4	(Breia et al., 2020)
VvPIP1;4	Grape	*Vitis vinifera*	DQ834696.1	(Breia et al., 2020)
OsPIP1;2	Rice	*Oryza sativa*	XM_015779980.1	(Breia et al., 2020)
NtAQP1	Tobacco	*Nicotiana tabacum*	AJ001416	(Chaumount et al., 2001)
ZmPIP1;1	Maize	*Zea mays*	X82633	(Chaumount et al., 2001)
ZmPIP1;2	Maize	*Zea mays*	AF131201	(Chaumount et al., 2001)
ZmPIP1;3	Maize	*Zea mays*	AF326487	(Chaumount et al., 2001)
ZmPIP1;4	Maize	*Zea mays*	AF326488	(Chaumount et al., 2001)
ZmPIP1;5	Maize	*Zea mays*	AF326489	(Chaumount et al., 2001)
ZmPIP1;6	Maize	*Zea mays*	AF326490	(Chaumount et al., 2001)
ZmPIP2;1	Maize	*Zea mays*	AF326491	(Chaumount et al., 2001)
ZmPIP2;2	Maize	*Zea mays*	AF326492	(Chaumount et al., 2001)
ZmPIP2;3	Maize	*Zea mays*	AF326493	(Chaumount et al., 2001)
ZmPIP2;4	Maize	*Zea mays*	AF326494	(Chaumount et al., 2001)
ZmPIP2;5	Maize	*Zea mays*	AF130975	(Chaumount et al., 2001)
ZmPIP2;6	Maize	*Zea mays*	AF326495	(Chaumount et al., 2001)
ZmPIP2;7	Maize	*Zea mays*	AF326496	(Chaumount et al., 2001)
ZmTIP1;1	Maize	*Zea mays*	AF037061	(Chaumount et al., 2001)
ZmTIP1;2	Maize	*Zea mays*	AF326500	(Chaumount et al., 2001)
ZmTIP2;1	Maize	*Zea mays*	AF326501	(Chaumount et al., 2001)
ZmTIP2;2	Maize	*Zea mays*	AF326502	(Chaumount et al., 2001)
ZmTIP2;3	Maize	*Zea mays*	AF326503	(Chaumount et al., 2001)
ZmTIP3;1	Maize	*Zea mays*	AF326504	(Chaumount et al., 2001)
ZmTIP4;1	Maize	*Zea mays*	AF326505	(Chaumount et al., 2001)
ZmTIP4;2	Maize	*Zea mays*	AF326506	(Chaumount et al., 2001)
ZmTIP4;3	Maize	*Zea mays*	AF326507	(Chaumount et al., 2001)
ZmTIP4;4	Maize	*Zea mays*	AF326508	(Chaumount et al., 2001)
ZmTIP5;1	Maize	*Zea mays*	AF326509	(Chaumount et al., 2001)
ZmSIP1;1	Maize	*Zea mays*	AF326497	(Chaumount et al., 2001)
ZmSIP1;2	Maize	*Zea mays*	AF326498	(Chaumount et al., 2001)
ZmSIP2;1	Maize	*Zea mays*	AF326499	(Chaumount et al., 2001)
ZmNIP1;1	Maize	*Zea mays*	AF326483	(Chaumount et al., 2001)
ZmNIP2;1	Maize	*Zea mays*	AF326484	(Chaumount et al., 2001)
ZmNIP2;2	Maize	*Zea mays*	AF326485	(Chaumount et al., 2001)
ZmNIP3;1	Maize	*Zea mays*	AF326486	(Chaumount et al., 2001)
OsPIP1;1	Rice	*Oryza sativa*	AK061769	(Sakurai et al., 2005)
OsPIP1;2	Rice	*Oryza sativa*	AK098849	(Sakurai et al., 2005)
OsPIP1;3	Rice	*Oryza sativa*	AK102174	(Sakurai et al., 2005)
OsPIP2;1	Rice	*Oryza sativa*	AK072519	(Sakurai et al., 2005)
OsPIP2;2	Rice	*Oryza sativa*	AK061782	(Sakurai et al., 2005)
OsPIP2;4	Rice	*Oryza sativa*	AK072632	(Sakurai et al., 2005)
OsPIP2;5	Rice	*Oryza sativa*	AK107700	(Sakurai et al., 2005)
OsPIP2;6	Rice	*Oryza sativa*	AK061312	(Sakurai et al., 2005)
OsPIP2;7	Rice	*Oryza sativa*	AK109439	(Sakurai et al., 2005)
OsPIP2;8	Rice	*Oryza sativa*	AK109024	(Sakurai et al., 2005)
OsTIP1;1	Rice	*Oryza sativa*	AK058322	(Sakurai et al., 2005)
OsTIP1;2	Rice	*Oryza sativa*	AK111747	(Sakurai et al., 2005)

(Continued)

TABLE 9.1
(Continued)

Gene	Plant	Scientific Name	Accession Number	Reference
OsTIP2;1	Rice	*Oryza sativa*	AK064728	(Sakurai et al., 2005)
OsTIP2;2	Rice	*Oryza sativa*	AK099141	(Sakurai et al., 2005)
OsTIP3;1	Rice	*Oryza sativa*	AK111931	(Sakurai et al., 2005)
OsTIP3;2	Rice	*Oryza sativa*	AK108116	(Sakurai et al., 2005)
OsTIP4;1	Rice	*Oryza sativa*	AK060193	(Sakurai et al., 2005)
OsTIP4;2	Rice	*Oryza sativa*	AK099190	(Sakurai et al., 2005)
OsTIP4;3	Rice	*Oryza sativa*	AK069192	(Sakurai et al., 2005)
OsTIP5;1	Rice	*Oryza sativa*	AK070602	(Sakurai et al., 2005)
OsNIP1;1	Rice	*Oryza sativa*	AK068806	(Sakurai et al., 2005)
OsNIP1;3	Rice	*Oryza sativa*	AK062320	(Sakurai et al., 2005)
OsNIP2;1	Rice	*Oryza sativa*	AK069842	(Sakurai et al., 2005)
OsNIP2;2	Rice	*Oryza sativa*	AK112022	(Sakurai et al., 2005)
OsNIP4;1	Rice	*Oryza sativa*	AK106825	(Sakurai et al., 2005)
OsSIP1;1	Rice	*Oryza sativa*	AK109424	(Sakurai et al., 2005)
OsSIP2;1	Rice	*Oryza sativa*	AK071190	(Sakurai et al., 2005)
SlPIP1;11	Tomato	*Solanum lycopersicum*	AB845604	(Reuscher et al., 2013)
SlPIP1;21	Tomato	*Solanum lycopersicum*	AB845605	(Reuscher et al., 2013)
SlPIP1;31	Tomato	*Solanum lycopersicum*	AB845606	(Reuscher et al., 2013)
SlPIP1;51	Tomato	*Solanum lycopersicum*	AB845607	(Reuscher et al., 2013)
SlPIP1;71	Tomato	*Solanum lycopersicum*	AB845608	(Reuscher et al., 2013)
SlPIP2;11	Tomato	*Solanum lycopersicum*	AB845609	(Reuscher et al., 2013)
SlPIP2;41	Tomato	*Solanum lycopersicum*	AB845610	(Reuscher et al., 2013)
SlPIP2;51	Tomato	*Solanum lycopersicum*	AB845611	(Reuscher et al., 2013)
SlPIP2;61	Tomato	*Solanum lycopersicum*	AB845612	(Reuscher et al., 2013)
SlPIP2;81	Tomato	*Solanum lycopersicum*	AB845613	(Reuscher et al., 2013)
SlPIP2;91	Tomato	*Solanum lycopersicum*	AB845614	(Reuscher et al., 2013)
SlPIP2;12	Tomato	*Solanum lycopersicum*	AB845615	(Reuscher et al., 2013)
SlTIP1;11	Tomato	*Solanum lycopersicum*	AB845616	(Reuscher et al., 2013)
SlTIP1;21	Tomato	*Solanum lycopersicum*	AB845617	(Reuscher et al., 2013)
SlTIP2;11	Tomato	*Solanum lycopersicum*	AB845618	(Reuscher et al., 2013)
SlTIP2;21	Tomato	*Solanum lycopersicum*	AB845619	(Reuscher et al., 2013)
SlTIP2;31	Tomato	*Solanum lycopersicum*	AB845620	(Reuscher et al., 2013)
SlTIP2;5	Tomato	*Solanum lycopersicum*	AB845621	(Reuscher et al., 2013)
SlTIP3;11	Tomato	*Solanum lycopersicum*	AB845622	(Reuscher et al., 2013)
SlTIP3;21	Tomato	*Solanum lycopersicum*	AB845623	(Reuscher et al., 2013)
SlTIP4;11	Tomato	*Solanum lycopersicum*	AB845624	(Reuscher et al., 2013)
SlNIP1;11	Tomato	*Solanum lycopersicum*	AB845625	(Reuscher et al., 2013)
SlNIP1;2	Tomato	*Solanum lycopersicum*	AB845626	(Reuscher et al., 2013)
SlNIP2;11	Tomato	*Solanum lycopersicum*	AB845627	(Reuscher et al., 2013)
SlNIP3;1	Tomato	*Solanum lycopersicum*	AB845628	(Reuscher et al., 2013)
SlNIP4;11	Tomato	*Solanum lycopersicum*	AB845629	(Reuscher et al., 2013)
SlNIP5;11	Tomato	*Solanum lycopersicum*	AB845630	(Reuscher et al., 2013)
SlNIP6;11	Tomato	*Solanum lycopersicum*	AB845631	(Reuscher et al., 2013)
SlNIP7;11	Tomato	*Solanum lycopersicum*	AB845632	(Reuscher et al., 2013)
SlSIP1;11	Tomato	*Solanum lycopersicum*	AB845633	(Reuscher et al., 2013)
SlSIP1;21	Tomato	*Solanum lycopersicum*	AB845634	(Reuscher et al., 2013)
SlSIP2;11	Tomato	*Solanum lycopersicum*	AB845635	(Reuscher et al., 2013)
SlXIP1;11	Tomato	*Solanum lycopersicum*	AB845636	(Reuscher et al., 2013)
SlXIP1;2	Tomato	*Solanum lycopersicum*	AB845637	(Reuscher et al., 2013)
SlXIP1;3	Tomato	*Solanum lycopersicum*	AB845638	(Reuscher et al., 2013)

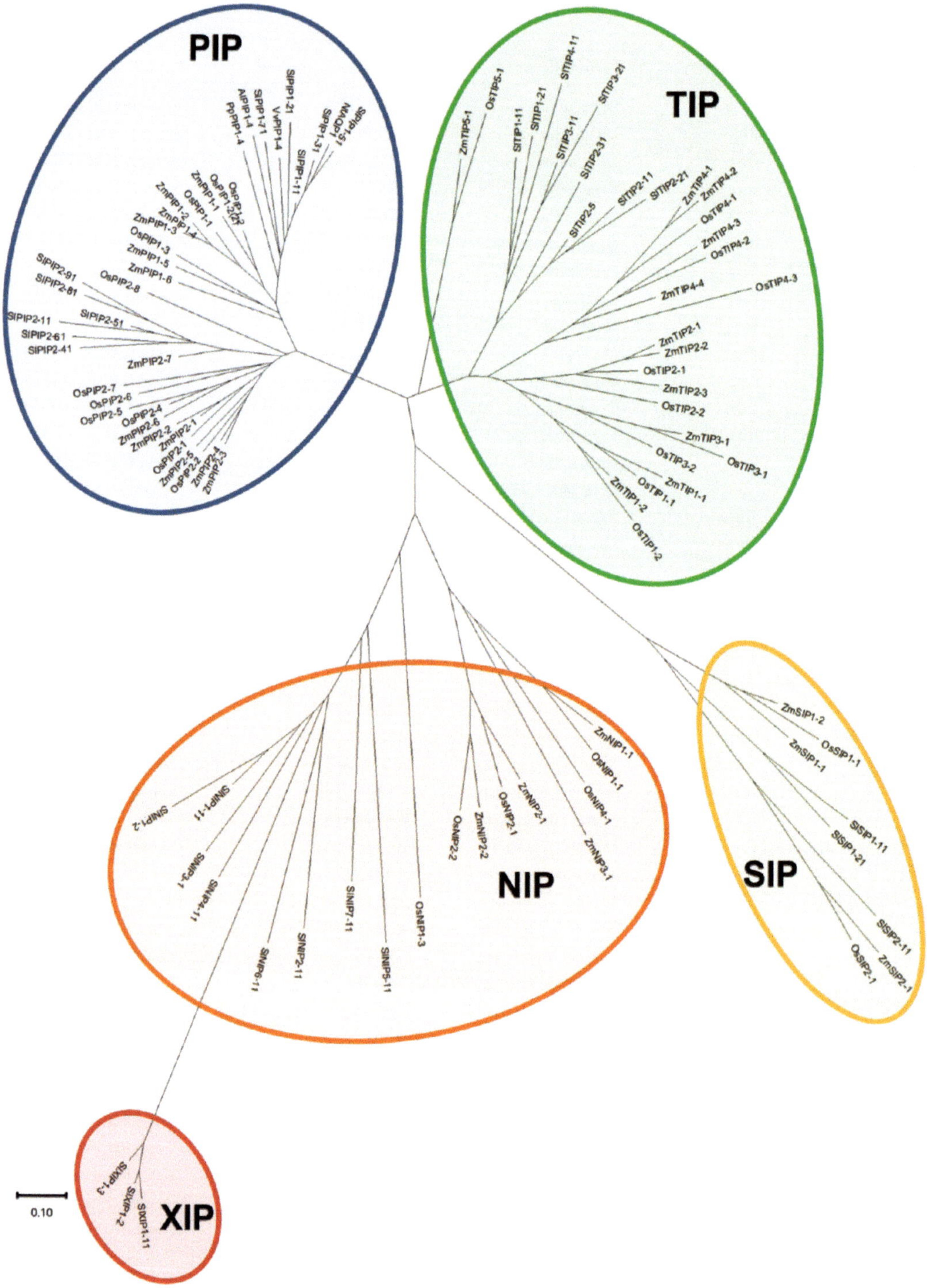

Figure 9.1 Phylogenetic analysis of DNA sequences of maize (*Zea mays*), rice (*Oryza sativa*), tomato (*Solanum lycopersicum*), and a few aquaporin genes from other plants. Alignment and phylogenetic analysis were performed in the software Mega X (version 11) applying Clustal W and neighbor joining methods. The scale bar of 0.1 is equal to 10% sequence divergence. NIP = NOD26-like intrinsic proteins; PIP = plasma membrane intrinsic proteins; SIP = small basic intrinsic proteins; TIP = tonoplast intrinsic proteins; XIP = X intrinsic proteins.

species (ROS) causes oxidative damage dictating fruit senescence. In the mitochondria membrane, ROS change the expression of mitochondrial proteins, the outer membrane transporters, leading to mitochondrial dysfunction and consequently fruit senescence (Tian et al., 2016). Curiously, peroxiporins have not be found in peroxisomal membranes despite the importance of H_2O_2 on peroxisome metabolism and cell functionality (Sandalio et al., 2021). Additionally, ROS may play important role on the induction of important plant defense mechanisms (9.1.).

9.2.1 Peroxiporins in Plant Defense

There are numerous induced plant defense mechanisms described for several pathosystems. These can be divided into structural mechanisms that physically block pathogen growth and biochemical mechanisms that inhibit pathogen development by the production of toxic substances. The build-up of vascular occlusions, named tyloses, impede the colonization of vascular pathogens (Eynck et al., 2009; Yadeta and Thomma, 2013). The reinforcement of cell wall by the deposition of lignin (Eynck et al., 2009; von Tiedemann et al., 2021), phenolics (Báidez et al., 2007; Eynck et al., 2009), and pectin compounds (Lionetti et al., 2012; Navarro et al., 2019) may block pathogen growth and spread to neighbor cells. Papillae consists in the accumulation of polysaccharides at the penetration site to protect cells from fungal invasion. The main polysaccharides accumulated consists in cellulose, arabinoxylan, and the most known, callose (Chowdhury et al., 2014; Voigt, 2014). Accumulation of ROS, such as H_2O_2 and superoxide radical anions (O_2^-), triggers hypersensitive response (HR) in the infection size leading to PCD, especially on resistant plants (Hückelhoven, 2007; Hückelhoven et al., 2000; Peters et al., 2017). The synthetization of phytoalexins, antimicrobial compounds after pathogen exposure, is another important resistant mechanism (Möbius and Hertweck, 2009). Pathogens can also synthesize toxins, which are usually important in pathogenesis, and one strategy of the plant is the degradation of these substances by the production of detoxifying enzymes (Anzai et al., 1989; Meeley and Walton, 1991).

Resistance mechanisms are triggered by the activation of the plant immune system. The host recognizes the plant invasion which leads to the activation of a signaling cascade described as a "zig-zag model" (Jones and Dangl, 2006). Transmembrane pattern recognition receptors (PRR proteins) recognize MAMPs/PAMPs (microbial- or pathogen-associated molecular patterns) in a process known as PAMP-triggered immunity (PTI). The releases of effectors (specific elicitors characterized as virulence factor) results in the process known as effector-triggered susceptibility (ETS). If pathogen recognition stops at ETS, the pathogen will be able to colonize the tissue. Conversely, if pathogen effectors are specifically recognized by PRR-proteins, the result is effector-triggered immunity (ETI). In several resistant hosts, the PRR-proteins are encoded in the genome at nucleotide binding and leucine-rich repeat domains (NB-LRR; Hurni et al., 2015). At ETI, the induction of a HR may occur (Jones and Dangl, 2006). The recognition of PAMPs by PRR proteins results in the accumulation of reactive oxygen species (ROS; Malinovsky et al., 2014). At PTI, ROS can activate protein kinase (MAPK) cascades and transcriptomes will be reprogrammed (Kachroo et al., 2017). The hypersensitive response consists of the accumulation of ROS in the apoplast and later in cell organelles, such as chloroplasts. The imbalance between ROS production and ROS detoxification increases ROS concentrations which may trigger PCD in the host tissue, avoiding pathogen colonization (Apel and Hirt, 2004). Many effectors target electron transport chains in an attempt to stop ROS production (Waszczak et al., 2018), especially biotrophic pathogens which require living host tissue to obtain nutrients (Glazebrook, 2005). In some pathosystems, necrotrophic pathogens may stimulate the production of H_2O_2, implying that ROS play an important role on the pathogenicity (van Baarlen et al., 2004; von Tiedemann, 1997). Just recently, the PIP TaPIP2;10 from wheat was shown to gate apoplastic H_2O_2 into the cytoplasm of plant cells upon pathogen attack by *Blumeria graminis* and *Fusarium graminearum*, the causal agents of powdery mildew and fusarium head blight, respectively. This H_2O_2 translocation activates plant defense responses and triggers PTI to prevent further pathogen invasion. As a consequence, the expression of TaPIP2;10 is enforced, increasing the resistance of wheat to both pathogens significantly (Wang et al., 2021).

TABLE 9.2

Plant Peroxiporins Related to the Defense Mechanisms against Microorganism Invasion

Peroxiporin	Plant	Location	Triggered Resistance Response	Reference
AtPIP1;2	*Arabidopsis thaliana*	Plasma membrane	Stomata closure	(Li et al., 2019)
AtPIP1;3	*Arabidopsis thaliana*	Plasma membrane	Transport of virulence factor	(Li et al., 2019)
AtPIP1;4	*Arabidopsis thaliana*	Plasma membrane	Regulation of SAR pathway	(Tian et al., 2016)
AtPIP2;1	*Arabidopsis thaliana*	Plasma membrane	Stomatal closure	(Rodrigues et al., 2017)
AtTIP1;1	*Arabidopsis thaliana*	Tonoplast	Intermediate signaling	(Maurel et al., 2009; Schüssler et al., 2008)
AtTIP1;2	*Arabidopsis thaliana*	Tonoplast	Intermediate signaling	(Maurel et al., 2009; Schüssler et al., 2008)
TaPIP2;10	*Triticum aestivum*	Plasma membrane	Regulation of pathogen-associated molecular pattern-triggered plant immunity	(Wang et al. 2021)

Among ROS, H_2O_2 is part of the system acquired resistance (SAR), which consists of the activation of broad-spectrum resistance mechanisms against secondary infections (Gao et al., 2015). This priming effect is triggered by a large transcriptomal reprogramming in cells distant to the first infection (Zeier, 2021), which will lead in an accumulation of salicylic acid (SA) in distal tissues. The SAR pathway is divided in two branches, one that involves ROS and a second related to SA (Wang et al., 2014). SA is synthesized by isochorismate synthetase (ICS) and phenylalanine ammonia-lyase (PAL) pathways. Pathogen infection can induce the conversion of phenylalanine (Phe) to trans-cinnamic acid, a reaction that is catalyzed by the enzyme PAL (Gao et al., 2015). The phenylalanine is a precursor of phenylpropanoid. This compound is related to the increase on phenolics promoting lignification and consequently cell wall fortification (Fortunato et al., 2017, 2015). In that sense, SAR activation and consequently the fast transport of H_2O_2 between cells will be determined the peroxiporin efficacy in transporting this molecule. In poplar hybrid tolerant to phytophthora species, an increase in the accumulation of aquaporin proteins can be correlated to the increase in the abundance of superoxide dismutase (Černý et al., 2022). The activation of some resistance mechanisms by H_2O_2 and the role of peroxiporins has been described above. Table 9.2 provides a list of peroxiporins reported in the literature and their roles/involvements in plant defense mechanisms.

9.2.1.1 Hypersensitive Response

Hypersensitive response is the PCD at the penetration site associated to disease resistance. Resistant hosts will increase ROS production, especially in the interaction with biotrophic pathogens (Balint-Kurti, 2019). The PCD can be triggered by two ways: ROS production can be activated where antioxidant compartments are low (e.g., in the apoplast) or the antioxidant capacity is insufficient because of high ROS production (e.g., in the chloroplast and peroxisome). Thus, the H_2O_2 produced in the apoplast can be rapidly transported by peroxiporins to be metabolized in the protoplast (Sedlárová et al., 2007). In normal conditions, there is a balance between H_2O_2 production and H_2O_2 scavenging, possibly due to the transport by peroxiporins. An increase in the gradient of H_2O_2 will increase membrane channel activity involving the calcium release and the increase in the number of peroxiporins (Smirnoff and Arnaud, 2019). For example, the treatment with SA has stimulated the accumulation of PIPs intracellularly on roots, as PIP accumulation is related to the increase in exogenous H_2O_2 (Maurel et al., 2009). The decrease in PIPs and in the activity of scavenging enzymes will lead to an unbalance and consequently in HR (Sedlárová et al., 2007). In the incompatible interaction between lettuce (*Lactuca sativa*) and *Bremia lactucae*, resistant plant showed H_2O_2 accumulation and high activity H_2O_2-scavenging enzymes at 6 hours post inoculation (hpi). A decrease in the H_2O_2 concentration could be observed at 12 hpi, indicating the role of H_2O_2 on avoiding pathogen penetration.

At 96 hpi, a peak in H_2O_2 and in the activity of H_2O_2-scavenging enzymes was correlated with HR expression (Sedlárová et al., 2007). Besides PIP functioning as a H_2O_2 scavenger, they can participate in the intermediate signaling process. *Arabidopsis* mutant plants for *tip1;1* and *tip1;2* showed high antocyanin content (+50%) under low light intensity (100 μmol photons m^{-2} s^{-1}), whereas the wild type has shown high antocyanin only under high light intensity (500 μmol photons m^{-2} s^{-1}). The inactivation of these peroxiporins provoked H_2O_2 accumulation and consequently activation of other ROS detoxification processes (Maurel et al., 2009).

The programmed cell death in plants can be divided in two types, apoptosis and autophagy. Apoptosis consist in the detachment of the plasma membrane from the cell wall in order to preserve organelles integrity (Veloso and van Kan, 2018). There is an engulfment of cytoplasmic material during this process; the chromatin is compensated, proteins are de novo synthesized, there is an activation of metacaspases (Coll et al., 2014; Dickman et al., 2017; Veloso and van Kan, 2018), whereas the tonoplast is intact (Greenberg and Yao, 2004). Autophagy consists of a mechanism of nutrient recycling to reuse basic elements such as amino acids, nucleotides, sugars, and lipids (Coll et al., 2014; Veloso and van Kan, 2018). PCD is effective for necrotrophic pathogens when the process of autophagy is overexpressed in comparison to apoptosis (Veloso and van Kan, 2018). These phenomenon has been reported for two pathogens, *Botrytis cinerea* and *Sclerotinia sclerotiorum* (Kabbage et al., 2013; Veloso and van Kan, 2018). *B. cinerea* induced the expression of *atg* genes, which are responsible for autophagy and autophagy (Lai et al., 2011). The oxalic acid produced by *S. sclerotiorum* suppressed autophagy, triggering fungal pathogenicity (Kabbage et al., 2013). The autophagic process is important because it regulates HR limiting cell death (Coll et al., 2014; Zhou et al., 2014) by decreasing the levels of ROS and consequently SA (Zhou et al., 2014).

9.2.1.2 Cell Wall Fortification

Several important strategies of plant defense mechanisms related to cell wall fortification. This can be a physical or chemical barriers such as the cell wall thickening by the accumulation of phenolics, pectin, callose, lignin, and other antimicrobial substances (Hückelhoven, 2007; Malinovsky et al., 2014). The accumulation of pectic compounds has observed in leaves of *Vitis vinifera* cv. Moscato infected with *Phakopsora euvitis*. A pectic thickening in sheath cells avoided pathogen invasion in the vascular bundles (Navarro et al., 2019). Pectic polysaccharides are cross-linked to hemicelluloses and phenolic compounds providing a structured and more functional complexity to the cell wall (Caffall and Mohnen, 2009), and the pectin accumulation can be associated with the binding of calcium ions to homogalacturans. These calcium pectate links confers greater rigidity to the cell wall (Peaucelle et al., 2012). The increase in calcium can be related to the increase in H_2O_2. DAB staining in lettuce leaves along xylem vessels indicates that H_2O_2 is involved in secondary cell wall thickening (Sedlárová et al., 2007).

Another strategy to avoid pathogen penetration is forming papillae formation. The papillae are mainly composed by callose, which is polysaccharide composed by (1,3)-β-glucan, and arabinoxylan and the outer layer contains arabinoxylan and cellulose. This tight structure of papillae makes the fungal penetration more difficult (Chowdhury et al., 2014; Piršelová and Matušíková, 2013). In grapevine, resistance to *Plasmopara viticola* is correlated to the increase in callose deposition. At 24 hpi, callose deposition was detected in the stomata close to germinating zoospores (Yu et al., 2016). In general, callose deposition is influenced by variations in ROS quantities (Luna et al., 2011). H_2O_2 can accumulate in forming papillae and can be used by peroxidases during cross-linking between phenolics and proteins for promoting cell wall reinforcement (Voigt, 2014). In *Arabidopsis*, the peroxiporin At1;4 is able to facilitate the transport of apoplastic H_2O_2 to the cytoplasm across the plasma membrane. The superoxide dismutase activity is constant, proving that the increase in cytoplasmatic H_2O_2 is promoted by the overexpression of AtPIP1;4 in yeast cells. The expression of At1;4 and consequently translocation of H_2O_2 induces callose formation in leaves infiltrated with flg22 and chitin being related to the PTI pathway. AtPIP1;4 was also regulates NPR1 and PR genes related to SAR (Tian et al., 2016).

Accumulation of phenolic compounds and lignin is related to the lignification process, a plant defense mechanism (Vance et al., 1980).

Deposition of phenolics compounds has been observed at the penetration sites (Hückelhoven, 2007) and close to papillae (Chowdhury et al., 2014). ROS can trigger the synthesis of phenolic compounds (Buckner et al., 2000). Lignin is a phenolic polymer (Malinovsky et al., 2014) and the process of lignin polymerization also requires H_2O_2 in the oxidation of monolignols and cinnamyl alcohols to free radicals by peroxidases (Buckner et al., 2000; Smirnoff and Arnaud, 2019; Vance et al., 1980). Respiratory burst oxidase homologs (Rboh, homologs for NOX oxidases in plants) can promote rapid extracellular dissipation of H_2O_2, which is triggered by a large calcium influx (Sagi and Fluhr, 2006). On the other hand, lignification may be inhibited by H_2O_2 scavenger potassium iodide indicating the importance of H_2O_2 and consequently peroxidase in the lignification (Smirnoff and Arnaud, 2019).

9.2.1.3 Stomata Closure

Stomata closure is a host response to avoid pathogen invasion, especially in the restriction of bacteria penetration (Melotto et al., 2006). In general, abscisic acid (ABA)-triggered stomatal closure may requires a high H_2O_2 membrane permeability (Grondin et al., 2015). The peroxiporin AtPIP1;2 facilitates the transport of H_2O_2 in the plasma membrane of guard cells triggering ABA hormone regulating stomatal closure (Li et al., 2020). Curiously, the PAMP flg22 requires AtPIP2;1 aquaporin to induce stomatal closure, similar to ABA. The flg22 activates AtPIP1;2 increasing osmotic water permeability. ABA and flg22 triggered the H_2O_2 accumulation in the wild type but not the *pip2;1* gene in the guard cells, which confirms that AtPIP2;1 is necessary for the accumulation of H_2O_2 in guard cells (Grondin et al., 2015; Rodrigues et al., 2017). The H_2O_2 scavenging process in the guard cells is taken by glutathione peroxidase, which can serve as a redox transducer (Černý et al., 2018). Other bacterial effectors such as harpin Hpa1 can interact with PIP1;3 promoting its translocation into the cytosol favoring *Xanthomonas oryzae* infection in rice (Li et al., 2019).

9.3 MAJOR INTRINSIC PROTEINS IN FUNGI

The kingdom of fungi covers worldwide distributed uni- and multicellular eukaryotic organisms, which depend on the degradation of organic matter. They are sessile organisms and therefore have to adapt to current environmental conditions to explore nutrients and propagate successfully. From a morphological point of view, fungi can grow as yeasts, molds, and mushrooms, however, these growth forms are not congruent with phylogenetic groups. Some fungi can also switch the growth form from yeast-like to filamentous, like it is described for the human pathogen *Candida albicans* and the maize pathogen *Ustilago maydis* (Banuett and Herskowitz, 1994; Snetselaar and Mims, 1992; Sudbery et al., 2004). Typical for multicellular fungi is the polar tip growth pattern of long filaments, the hyphae. There environmental signals (nutrients, pheromones, toxins) are sensed, and organic matter is degraded by lytic enzymes and taken up subapical. Depending on the external stimulus and intra-colony signals, the hyphae change their polarity, resulting in a growth redirection (e.g., to hosts or mating partners) or cellular differentiation (e.g., penetration structures, fruiting bodies; Turrà et al., 2016). For young germlings as well as inside a mature colony, polar growth along a signal gradient is crucial to allow for the formation of cellular fusions, which maintain efficient nutrient distribution and coordinated development of the colony (Fleissner and Herzog, 2016; Nordzieke, 2022). Despite these general features, fungi explore a lot of different lifestyles. They can live as saprophytes from dead organic matter, as symbionts or pathogens, and can also switch their lifestyle depending on the availability of the corresponding host or the status of its immune system.

Pettersson introduced the phylogenetic classification of the fungal MIP superfamily in 2005, which is also referred to as aquaporins (Pettersson et al., 2005). This system has been refined over the years, which was enabled by the growing availability of sequenced fungal genomes (Amira et al., 2021; Nehls and Dietz, 2014; Xu et al., 2013). Currently, it is estimated that four MIP clusters exist: (1) synonymous to orthodox aquaporins, (2) Fps-like aquaglyceroporins, (3) other aquaglyceroporins and homologs to yeast Yfl054, and (4) uncategorized (X)-intrinsic proteins (XIPs; Nehls and Dietz, 2014). Comparing the members of these clusters, it was possible to define two sister groups: the aquaporins (AQPs, clusters 1 and 4) and the aquaglyceroporins (AQGPs, clusters 2 and 3; Abascal et al., 2014; Amira et al., 2021; Nehls and Dietz, 2014). In general, it is considered that the AQPs are mainly involved in water transport, whereas the AQGPs can transport also other solutes like glycerol and urea (Amira et al., 2021). However, no direct conclusions from the classification to functionality can be drawn; in several investigations, AQGPs were shown to provide a much higher water transport capacity as "classical" AQPs (Dietz et al., 2011; Li et al., 2013; Nehls and Dietz, 2014; Xu et al., 2013). Additionally, MIPs were shown to form functional tetramers, containing multiple copies of the same (homotetramer) or different (heterotetramer) proteins, a fact that also affects the pore size and therefore the transport capacity (Fujiyoshi et al., 2002; Harvengt et al., 2000; Nehls and Dietz, 2014; Otto et al., 2010). These results emphasize the urgent need for functional characterization of the aquaporins regarding their transport capacities. So far, only limited data are available.

There is increasing evidence that plant and fungal H_2O_2–gating MIPs have substantial roles in the mutualistic (9.2.1) as well as the pathogenic interactions (9.2.2). Since in most cases the distinct molecular mechanisms and transport capacity of the participating MIPs is still in question, we focus here on the best-described interactions and indicate probable research direction for future investigations.

9.3.1 Mutual Plant-Fungi Interactions—Mycorrhiza

Symbiotic associations between soil-borne fungi belonging to the phyla Glomeromycota, Basidiomycota, or Ascomycota at the plant's rhizosphere are referred to as mycorrhiza (Prasad et al., 2017). In general, there are three types of mycorrhiza, all of them increasing the fitness of plants but showing fundamental differences in the morphological changes associated with the symbiotic interactions.

It is estimated that about 80% of all land plants, including flowering plants, bryophytes, and ferns, form highly specialized endosymbiotic associations with arbuscular mycorrhiza (AM) fungi (Bahadur et al., 2019; Begum et al., 2019; Varma, 2008). Typical for the AM symbiosis are rearrangements of the host cytoskeleton and the remodeling of organelles resulting in the formation of arbuscules inside root cells. Here, the metabolic exchange between the two interacting organisms takes place over the new formed periarbuscular membrane (Genre et al., 2005; Luginbuehl and Oldroyd, 2017). In contrast to endosymbiotic AM, the mutualistic interactions of land trees and shrubs with fungi takes place ectopically (EM) and is accompanied by the formation of a Hartig net, surrounding the plant root (Barea et al., 2011). Since the interaction is not that close, the fungi can grow without the presence of the host plant by feeding from other nutrient sources (Moore et al., 2015). In both AM and EM interactions, water and mineral nutrients like inorganic phosphate and ammonium are transferred from the fungus to the plant host and, in return, the plant provides assimilated carbohydrates to the symbiont (Harrison, 2005; Navarro García et al., 2011). As a consequence, the stress resistance of the interacting plants to abiotic and biotic stresses is increased (Augé, 2001; Kipfer et al., 2012; Ruiz-Lozano et al., 2016, 2012; Yi et al., 2008).

In the last decade, several studies investigated the impact of AM symbiosis on the capacity of the host plant to deal with drought stress (Begum et al., 2019). During drought, the plant faces a decreased soil water potential, which hinders the efficient water uptake (Ouziad et al., 2006). As one of the first drought effects, plant cells face a decreased turgor pressure, limiting the plant's ability to grow. Counteracting, the plant cells lower their osmotic potential by the accumulation of metabolites and ions like sugars and sugar alcohols, proline, and potassium, which allow for water influx and turgor pressure stabilization (Kishor and Sreenivasulu, 2014). Since roots are the first plant organs in contact with the drying soil, several measures are taken in roots to contain efficient water flux and to communicate the current stress situation to upper plant parts. For example, roots generate the plant hormone ABA, which allows for fast stomatal closure in leaves preventing water loss (Bárzana et al., 2015; Morgan, 1990; Taylor, 1991; Turner et al., 2001). However, stomatal closure has several harmful side effects. The reduced CO_2 uptake is followed by reduced carboxylation in leaves and an increase of ROS production. At the same time, the reduced water availability imbalances the processes of ROS generation and antioxidant defense in leaves and roots. Together, these processes result in a severe decrease of photosynthetic activity (Farooq et al., 2009). In the roots themselves, the water flow is redirected to cell-to-cell passages, allowing for efficient root hydraulic conductivity (Aroca et al., 2012). In several of these described short-time effects, MIPs take part (Javot et al., 2003; Luu and Maurel, 2005; Rodrigues et al., 2017; Sharma et al., 2021), therefore it is not surprising that changed expression patterns are reported for MIPs under drought stress conditions (Javot and Maurel, 2002). However, most studies focus on the water-transport capabilities of MIPs, ignoring the role of MIPs in the transport of other solutes of physiological importance. We will therefore concentrate here on the interaction of the AM fungus *Rhizophagus irregularis* with the economical important cereal *Zea mays*, since here the transport capacity for other solutes including H_2O_2 was examined (Bárzana et al., 2014).

In maize, 31 AQP isoforms were discovered, belonging to the four plant MIPs groups of PIPs, TIPs, NIPs, and SIPs (Chaumount et al., 2001). About the half of them showed differential expression patterns due to the associated AM fungus. Of these, ZmTIP1;1 and ZmTIP1;2 showed the highest H_2O_2 transport capacity in a yeast toxicity assay. Three further MIPs were shown to a moderate transport capacity: ZmPIP2;2, ZmNIP1;1, and ZmNIP2;1 (Bárzana et al., 2014). Interestingly, the 16 regulated AQPs were grouped into six response patterns to either short-term or sustainable drought stress. In terms of H_2O_2 transporting AQPs, the patterns 2, 3, 4, and 6 are the most relevant. For all peroxiporins, the symbiosis with *R. irregularis* lowers the expression of the corresponding genes compared to non-AM plants under well-watered conditions. Under short-term drought conditions, ZmTIP1;2 (pattern 2), ZmNIP2;1 (pattern 3), and ZmPIP2;2 and ZmTIP1;1 (both pattern 4) showed an increase in expression compared to non-AM plants, indicating a role in coping with short periods of water shortage. In line with this, sustainable drought conditions either did not alter the expression of these genes (pattern 3) or resulted in a decrease of expression (patterns 2 and 4). Only the expression of ZmNIP1;1 (pattern 6) was not altered under short-term drought conditions compared to non-AM plants. However, the expression of this gene increases strongly in AM plants under sustainable drought conditions, making a role in long-term adaptation probable (Bárzana et al., 2014). A recent study highlights further the H_2O_2-transporting ZmTPI1;1 and ZmPIP2;2 as probable key MIPs in drought stress resistance: both show a specific gene regulation by AM symbiosis with different maize cultivars depending on their sensitivity to drought (Quiroga et al., 2017). Nevertheless, in the past the mere usage of expression data to examine the impact of AQPs in stress responses has proven to be inconsistent, especially for PIP isoforms, and therefore have to be interpreted with care (Aroca et al., 2012; Quiroga et al., 2019).

But how do the H_2O_2-transporting MIPs influence the drought stress response? The answer to this question is difficult since these MIPs were shown to transport also other relevant solutes besides H_2O_2 (Figure 9.2; Bárzana et al., 2014). In Bárzana et al. (2014), the effect of AM symbiosis on different factors of water transport like the sap flow rate (transpiration flow as the ascent of sap within the xylem tissue) and the root hydraulic conductance (intrinsic ability of roots to conduct water across a water potential gradient) was investigated. For short-term stress it was shown that the sap flow rate and the root hydraulic conductance driven by osmotic forces was held constant by AM plants to comparable levels as in well-watered *Z. mays*, indicating a supporting function of the symbiont to cope with the stress. The same was true for the hydrostatic root hydraulic conductance in sustained drought conditions (Bárzana et al., 2014). For all three effects the water transport capability of the peroxiporins might be responsible; in the case of drought stress, water transport is shifted from the apoplastic path around the cells to the symplastic path, guiding the water flow through the plant cells mediated by plasmodesmata and water-transporting MIPs (Aroca et al., 2012; Steudle and Peterson, 1998). For this shift, especially the plasma-membrane standing MIPs are important (PIPs and NIPs). In support, the plant cell also has to increase its osmotic potential (e.g., by the supply of glycerol to the cytoplasm; Chen and Jiang, 2010). The two peroxiporins ZmNIP1;1 and ZmNIP1;2 were also shown to transport glycerol (Bárzana et al., 2014), so one of their function might be glycerol supply from organelles. Additionally, these MIPs could support the plant-fungus interaction during drought stress by providing more plant-derived glycerol to the fungal counterpart (Gustavsson et al., 2005; Quiroga et al., 2017). This in turn might lead to a better water uptake capacity of the fungal symbiont, which then would be reflected in increase water accessibility of the plant.

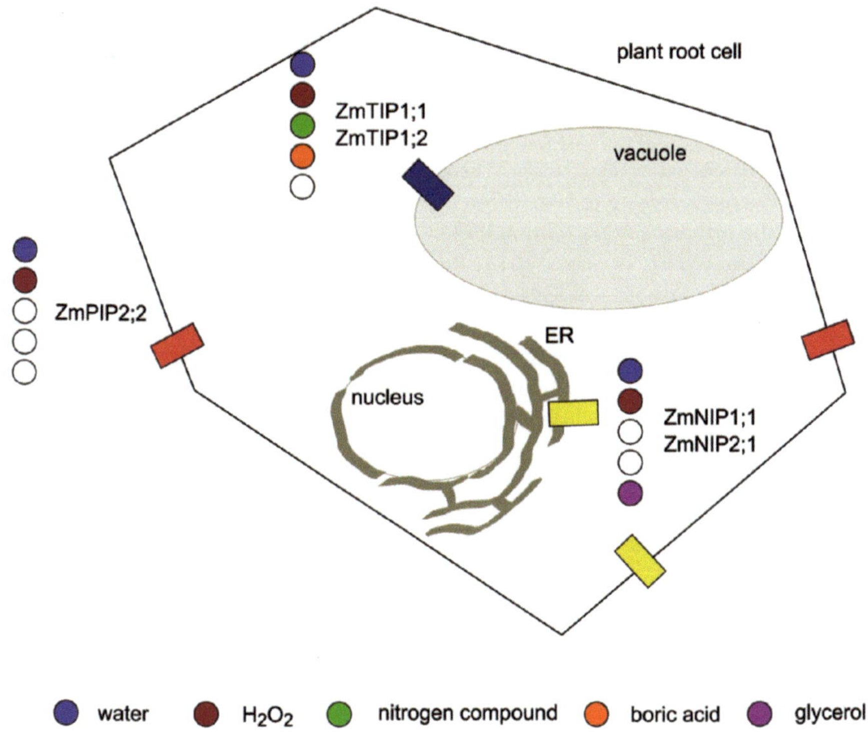

Figure 9.2 Maize peroxiporins regulated by AM symbiosis under drought stress. Schematic depiction of a plant root cell. Aquaporins with verified H_2O_2 transport capacities are indicated at their corresponding cellular sites. Circles indicate the specific transport capacities of the distinct peroxiporins depicted as rectangles (dark blue = tonoplast intrinsic proteins [TIP], red = membrane intrinsic proteins [PIP], yellow = Nod26-like intrinsic proteins [NIP]).

Source: **In Barzana et al 2014 no such figure is depicted. The figure here was drawn on our own, basing on the experimental findings described in Barzana et al 2014.**

Regarding H_2O_2 transport, the two main mechanisms discussed are either the detoxification of stress-induced H_2O_2 via MIPs or the involvement of the peroxiporins in redox signaling (Bienert et al., 2007, 2006; Quiroga et al., 2017; Ruiz-Lozano and Aroca, 2017). Since drought stress imbalances the redox status of the plant cells and results in a higher damage to molecules (DNA, protein, lipids) as well as cellular structures (membranes of organelles and cells; Miller et al., 2010), the discharge of H_2O_2 via MIPs is a reasonable plant defense mechanism. Indeed, when oxidative damage to lipids was accessed as a key figure of drought stress, the symbiosis of maize plants with *R. irregularis* resulted in significant less damage compared to non-AM plants (Bárzana et al., 2015). For this process, the most relevant peroxiporin might be the plasma membrane standing PIP, ZmPIP2;2. In line with these observations, the phosphorylation levels of ZmPIP2;2 are increased during drought, indicating an increase of solute transport (Quiroga et al., 2019). Additionally, data on stress responses in *Arabidopsis thaliana* point to a redox signaling-dependent mechanism. H_2O_2 accumulation in the plant cell is followed by a rapid internalization of plasma membrane–standing PIPs, resulting in a downregulated water uptake by root cells and a reduced hydraulic root conductivity (Boursiac et al., 2008). These results indicate a probable dual impact of AM symbiosis on H_2O_2 transporting MIPs: increased and stable expression of peroxiporins could on the one hand facilitate highly efficient redox signaling in response to drought-induced ROS stress. On the other hand, increased expression of water-transporting PIPs in the plasma membrane could provide an efficient way for ROS detoxification.

ABA is not only considered as the abiotic stress hormone, but is essential for an appropriate establishment and maintenance of AM symbiosis (Bahadur et al., 2019; Hong et al., 2013). As recent studies show, levels of ABA are increased in AM plants under drought stress compared to non-symbiotic plants and increases the expression levels of MIPs of the fungal counterpart *Rhizophagus intraradices*, GintAQPF1 and GintAQPF2 (Bahadur et al., 2019; Xu et al., 2018). These two membrane channels belong to two different fungal clusters, the orthodox AQPs (GintAQPF1) and the AQGPs (GintAQPF2; Li et al., 2013; Xu et al., 2013). Interestingly, the AQGPs is able to transport water more efficient and in higher quantities as the orthodox AQP, emphasizing the problematic deduction from MIP phylogeny to function in filamentous fungi (Nehls and Dietz, 2014). Concerning drought stress resistance, it is speculated that both channels are involved in water transport by mycorrhizal hyphae to the plant, a process that is increased during drought stress (Bahadur et al., 2019). However, as stands for most of the fungal MIPs, the transport capacity of these channels has not been analyzed for other solutes and therefore a probable role in redox processes in the mutualistic symbiosis is highly speculative. Besides the effect on fungal MIPs, increased levels of ABA do also affect redox-dependent processes in leaves cumulating in stomatal closure (Farooq et al., 2009). As it was shown for *A. thaliana*, binding of ABA to its cognate receptor induces a signaling cascade resulting in the activation of plant NADPH oxidase (Rboh) and calcium signaling (Balmant et al., 2016). At the same time, peroxiporin-mediated increase of H_2O_2 and water in guard cells by a PIP AQP is induced (Rodrigues et al., 2017). Together, these ABA-dependent mechanisms result in highly efficient stomatal closure. Whether these processes are altered in AM plants under drought stress has to be revealed in further investigations (Bahadur et al., 2019). Further, the question arises whether ABA could additionally induce redox processes in roots as well as in the symbiotic fungus, which could rely again on peroxiporin-gated transport.

9.3.2 Pathogenic Plant-Fungi Interactions

ROS are central to pathogenic plant-fungi interactions, playing out its dual personality ranging from a harmful reagent for defense or attack to a specific signaling molecule (Zhang et al., 2020). In filamentous fungi, signaling competent ROS is predominantly generated by NADPH oxidases (Nox) complexes. There are three isoforms of Nox complexes, NoxA/Nox1, NoxB/Nox2, homologous to the mammalian Nox2 and NoxC, a homolog of plant Rboh. Whereas NoxC is only encoded in some fungal genomes like *Podospora anserina* and *Magnaporthe oryzae*, the isoforms NoxA (Nox1) and NoxB (Nox2) are present in most fungal genomes and their functions and interaction partners are well characterized (Aguirre and Lambeth, 2010; Scott, 2015). The catalytic subunits of the NoxA and NoxB complexes require a distinct set of regulators for the generation of superoxide anions and are specific for cellular processes like sexual development (NoxA/Nox1), cellular fusions (NoxA/Nox1), penetration peg establishment (NoxB/Nox2), penetration peg elongation (NoxA/Nox1), germination of sexual spores (NoxB/Nox2), and chemotropic growth to host plant roots (NoxB/Nox2; Marschall and Tudzynski, 2016; Scott, 2015; Zhang et al., 2020). Two of these are crucial for proper plant disease development: (1) penetration of the plant surface outgoing from unicellular fungal penetration organs in which high osmotic pressure is established and (2) chemotropic growth of root pathogens to their host (Figure 9.3; Egan et al., 2007; Nordzieke et al., 2019; Segmüller et al., 2008; Zhao et al., 2016).

Despite the knowledge gained in the last decade about regulation of Nox complexes, the downstream processes of the generated superoxide anion are less defined. Increasing evidence indicates that superoxide dismutase is the first target of Nox-derived ROS, converting the superoxide anion to signaling competent H_2O_2 (Chung, 2012; Nordzieke et al., 2019). In analogy to the mammalian system, it is estimated that H_2O_2 is then subsequently gated by peroxiporins to their target structures (Nordzieke and Medraño-Fernandez, 2018; Zhang et al., 2020). Final target structures of redox signaling might be cytoskeleton components as septins and F-actin, as it was described for the plant penetration process of the rice pathogen *M. oryzae*. Here, remodeling of the F-actin cytoskeleton is dependent on NoxB, indicating a direct influence on the cytoskeleton and growth responses (Ryder et al., 2013). Since several other Nox-dependent processes rely on correct polarity establishment and directed growth (cellular fusions, chemotropic growth to host signals), the regulation the cytoskeleton might be an important function of

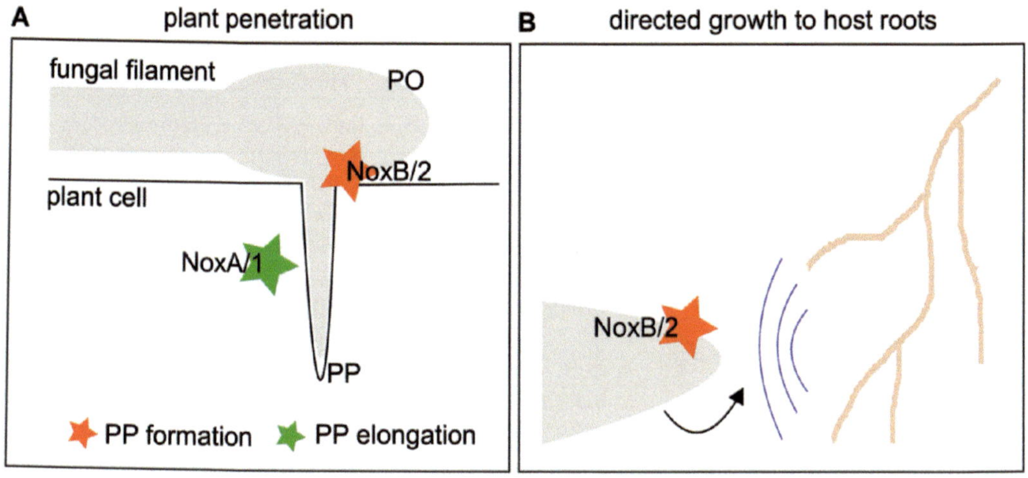

Figure 9.3 Nox-dependent redox signaling processes in fungal plant pathogenicity. (A) Penetration of plant surfaces by specific penetration organs (PO) is a two-step process in several fungal plant pathogens. First, NoxB/2-dependent generation of reactive oxygen species (ROS) induces the formation of the penetration peg (PP). Second, elongation of PP is mediated by a redox signal generated by NoxA/1. (B) Directed growth of the soil-borne fungus *Fusarium oxysporum* is mediated by a ROS signal coming from NoxB/2, which allows for host recognition.

fungal redox signaling per se. But how is H_2O_2 able to fulfill its signaling role and to reach its targets? So far, only one fungal MIP with a proven gating capacity for H_2O_2, AQP8 of the necrotrophic fungus *Botrytis cinerea*, is described (Figure 9.4; An et al., 2016). Deletion of the corresponding gene in *B. cinerea* revealed a reduced symptom development due to insufficient plant surface penetration (An et al., 2016). Follow-up microscopic investigations of Δ*AQP8* showed that the maturation process of the penetration organ is affected. In contrast, *B. cinerea* mutants of *noxB* and its regulator *pls1* generate mature penetration organs which fail to generate a proper penetration peg (Segmüller et al., 2008; Siegmund et al., 2013), indicating that AQP8 and NoxB are functional in different pathogenicity processes. Also, for AQP1 of *Fusarium graminearum* pathogenicity-relevant phenotypes were reported. *F. graminearum* is famous for the production of the mycotoxin deoxynivalenol (DON), an inhibitor of the eukaryotic protein biosynthesis, which spoils important grains as wheat, barley, and oat (D'Mello et al., 1999; Van De Walle et al., 2010). When analyzing an AQP1 deletion mutant, elevated levels of DON were detected. Additionally, Δ*AQP1* shows other phenotypes, like an aberrant stress response and abolished sexual development (Ding et al., 2018). Interestingly, these phenotypes overlap with previous reports from fungal Δ*noxA* mutants, making Δ*AQP1* a promising candidate to gate Nox-derived ROS (Cano-Domínguez et al., 2008; Dirschnabel et al., 2014; Wang et al., 2014).

In summary, the significance of H_2O_2 gating MIPs in fungal plant pathogens remains unclear. Whereas the genomes of phytopathogenic fungi encode several MIPs (Table 9.3; Figure 9.4), it is rather speculative whether or not these are involved in redox signaling. For future research it will be crucial to determine the transport capacity of each MIP present in a fungal species. These analyses should also take into account that some aquaporins of plants and bacteria were shown to explore different transport capacities when expressed together, indicating the formation of homotetramers or heterotetramers *in vivo* (Fetter et al., 2004; Luu and Maurel, 2005; Nehls and Dietz, 2014).

9.4 BACTERIAL AQUAPORINS

Prokaryotic AQPs are a rather unexplored field (Bienert et al., 2013; Borgnia and Agre, 2001; Calamita, 2000) and only few studies have investigated the roles of these transmembrane (integral membrane) proteins that are widely distributed in various organisms. More than 1700 integral

TABLE 9.3

Fungal Major Intrinsic Proteins of Phytopathogenic Species Used in Phylogenetic Analysis

Species	Accession Number	Locus Tag	MIP Clade	Reference
Botrytis cinerea (B05.10)	ATZ51251.1	BCIN_06g06680	classic AQP	(Amselem et al., 2011)
	ATZ45287.1	BCIN_01g00950	classic AQP	(Amselem et al., 2011)
	ATZ52480.1	BCIN_08g01870	Fps-like AQGP	(Amselem et al., 2011)
	ATZ46987.1	BCIN_02g03220	classic AQP	(Amselem et al., 2011)
	ATZ49537.1	BCIN_04g06710	classic AQP	(Amselem et al., 2011)
	ATZ45256.1	BCIN_01g00710	classic AQP	(Amselem et al., 2011)
	ATZ52240.1	BCIN_08g00120	classic AQP	(Amselem et al., 2011)
	ATZ53615.1	BCIN_09g04250	Fps-like AQGP	(Amselem et al., 2011)
Colletotrichum graminicola (M1.001)	XP_008096054.1	GLRG_07178	classic AQP	(O'Connell et al., 2012)
	XP_008094386.1	GLRG_05510	classic AQP	(O'Connell et al., 2012)
	XP_008094660.1	GLRG_05784	Fps-like AQGP	(O'Connell et al., 2012)
	XP_008098834.1	GLRG_09958	Fps-like AQGP	(O'Connell et al., 2012)
	XP_008095534.1	GLRG_06803	classic AQP	(O'Connell et al., 2012)
Fusarium oxysporum f.sp. *lycopersici* (4287)	XP_018247466.1	FOXG_10001	Fps-like AQGP	(Ma et al., 2010)
	XP_018257630.1	FOXG_16834	Fps-like AQGP	(Ma et al., 2010)
	XP_018247055.1	FOXG_09680	classic AQP	(Ma et al., 2010)
	XP_018242580.1	FOXG_06603	classic AQP	(Ma et al., 2010)
	XP_018251158.1	FOXG_12750	classic AQP	(Ma et al., 2010)
	XP_018250607.1	FOXG_12136	XIP	(Ma et al., 2010)
	XP_018233737.1	FOXG_01152	classic AQP	(Ma et al., 2010)
Magnaporthe oryzea (70–50)	XP_003717502.1	MGG_13615	Fps-like AQGP	(Dean et al., 2005)
	XP_003711708.1	MGG_05880	other AQGP	(Dean et al., 2005)
	XP_003717725.1	MGG_10783	classic AQP	(Dean et al., 2005)
	XP_003711197.1	MGG_13332	classic AQP	(Dean et al., 2005)
	XP_003719970.1	MGG_03904	classic AQP	(Dean et al., 2005)
	XP_003719667.1	MGG_04162	classic AQP	(Dean et al., 2005)
Ustilago maydis	XP_011387976.1	UMAG_10452	Fps-like AQGP	(Kämper et al., 2006)
	XP_011387547.1	UMAG_01508	Fps-like AQGP	(Kämper et al., 2006)
	XP_011388522.1	UMAG_02169	other AQGP	(Kämper et al., 2006)
	XP_011389539.1	UMAG_11481	classic AQP	(Kämper et al., 2006)
	XP_011386156.1	UMAG_00223	classic AQP	(Kämper et al., 2006)
Rhizoctonia solani	XP_043179795.1	RhiXN_00964	Fps-like AQGP	(Hane et al., 2014)
	XP_043179175.1	RhiXN_00344	classic AQP	(Hane et al., 2014)
	XP_043182831.1	RhiXN_07630	Fps-like AQGP	(Hane et al., 2014)
	XP_043181836.1	RhiXN_06588	other AQGP	(Hane et al., 2014)
	XP_043179334.1	RhiXN_00503	XIP	(Hane et al., 2014)
Trichoderma atroviride	XP_013945873.1	TRIATDRAFT_90169	other AQGP	(Kubicek et al., 2011)
	XP_013942882.1	TRIATDRAFT_39327	Fps-like AQGP	(Kubicek et al., 2011)
	XP_013944177.1	TRIATDRAFT_283564	Fps-like AQGP	(Kubicek et al., 2011)
	XP_013938926.1	TRIATDRAFT_31598	classic AQP	(Kubicek et al., 2011)
	XP_013944451.1	TRIATDRAFT_43816	classic AQP	(Kubicek et al., 2011)
	XP_013948951.1	TRIATDRAFT_6990	classic AQP	(Kubicek et al., 2011)
	XP_013940913.1	TRIATDRAFT_319992	XIP	(Kubicek et al., 2011)

AQGP = other aquaglyceroporin; classic AQP = classic aquaporins; Fps-like AQGP = Fps-like aquaglyceroporins; other XIP = uncategorized (X)-intrinsic proteins.

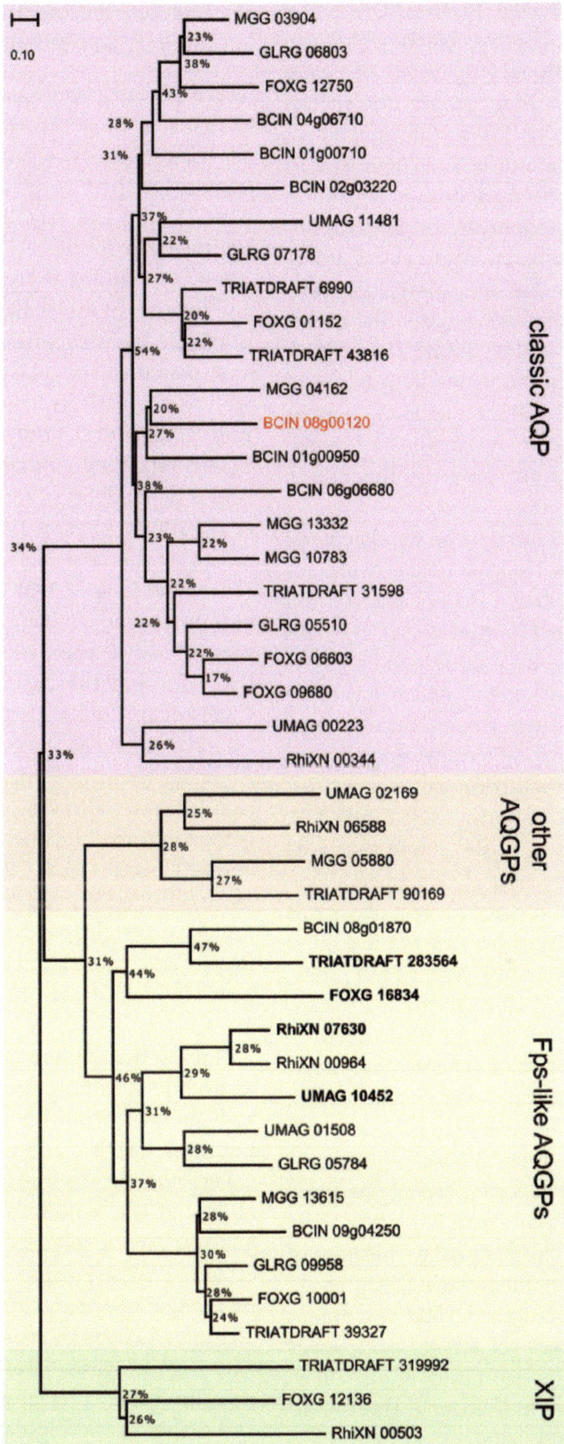

Figure 9.4 Phylogenetic analysis of fungal major intrinsic proteins (MIP) from selected phytopathogenic species. Alignment and phylogenetic analysis were performed using the software MEGA (version 11) using ClustalW and neighbor joining methods. MIPs group into four clades: the classic aquaporins (AQP), aquaglyceroporins (AQGPs) consisting of Fps-like and other AQGPs, and uncategorized intrinsic proteins (XIP). Locus tags are given for each MIP, bold letters indicate a conserved glycerol uptake protein domain, and verified fungal peroxiporins are indicated in red.

membrane proteins belong to the AQP family (Abascal et al., 2014; Zardoya, 2005) which can facilitate the bidirectional diffusion of water and uncharged solutes. Most prokaryotic species carry water-transporting MIPs, the aquaporins (Calamita, 2000; Hohmann et al., 2000; Tanghe et al., 2006), glycerol-transporting aquaporin homologues, the glyceroporins, or both multifunctional aquaglyceroporins (Agre et al., 1998). Also, some examples for the (probable) existence of bacterial peroxiporins exists. For example, Seaver and Imlay (2001) found that a H_2O_2 transmembrane concentration gradient exists in *Escherichia coli*, suggesting that the H_2O_2 permeability across the cellular membrane is limited, and the bacterial aquaporins may also function in facilitating H_2O_2 diffusion.

The only report of an intrinsic bacterial aquaporin that functions as a H_2O_2-inducible peroxiporin deals with the facultative anaerobic bacterium *Streptococcus oligofermentans* (Tong et al., 2019). This catalase-negative bacterium produces and tolerates the highest H_2O_2 levels known, which has to be exported to avoid oxidative stress, hence indicating they possess effective H_2O_2 excretion pathways (Tong et al., 2007). Streptococci carry genes that encode two clades of aquaporins, AQPs and GLPs. The studies by Tong et al. (2019) found increased expression of the AQP genes in H_2O_2-treated *S. oligofermentans*, implying that the AQPs might play a role in H_2O_2 excretion. Probably because of the similar electrochemical properties of H_2O and H_2O_2, some AQPs can transport H_2O_2 through the H_2O channel (Dynowski et al., 2008; Rodrigues et al., 2017). By imaging methodologies of genetic, physiological, biochemical, and single-molecule approaches, Tong et al. (2019) identified a streptococcal peroxiporin, So-AqpA. By using an intracellular-specific H_2O_2 fluorescence reporter, HyPer, the study demonstrated that So-AqpA, with Phe-40 as a key residue, facilitated the bidirectional permeation of H_2O_2 across the cellular membrane (Tong et al., 2019). The bacterial peroxiporin could be a potential target for suppression of Streptococci, especially the pathogenic species (Tong et al., 2019). This AQP facilitates the bidirectional transmembrane H_2O_2 transport and therefore has a key physiological role in H_2O_2 detoxification (Borgnia and Agre, 2001; Preston et al., 1992). Super-resolution imaging showed that H_2O_2 treatment increases So-AqpA protein molecules per cell 1.6- to 3-fold (Tong et al., 2019).

9.4.1 Mutualistic Redox Sensitive Bacteria-Plant Interactions

When soil nitrogen conditions are scarce, bacteria of the Rhizobiaceae family form a symbiosis with legumes, a family of plants especially known for their crop- and fruit-producing species. During the establishment of the symbiosis, nitrogen-fixing (NF) root nodules are built. These specialized organs help in the reduction and fixation of molecular nitrogen. The resulting organelle-like structure is called the symbiosome and contains the plant-derived membrane as well as the differentiated bacterium, the bacteroid (Chang et al., 2009; Guenther and Roberts, 2000). Establishment and function of the bacteroid is dependent on adequate ROS levels (Chang et al., 2009). Already 1 hour post infection, ROS efflux to the host *Medicago truncatula* is inhibited (Shaw and Long, 2003). This NF-dependent process is absent in incompatible hosts and thus may allow a compatible interaction between the plant and the bacteria (Bueno et al., 2001; Chang et al., 2009). Also, in a second phase of the interaction, alteration of ROS levels is observed. About 12 hours post inoculation, ROS accumulation increases allowing for root hair curling and infection thread formation (D'Haeze et al., 2003; Peleg-Grossman et al., 2007; Ramu et al., 2002). During bacteroid activity, however, ROS accumulation is tightly reduced to minimal amounts to prevent inactivation of the nitrogenase complex and diminished nitrogen fixation (Fischer, 1994).

Among the membrane proteins are MIPs such as nodule-specific proteins, nodulins. Guenther and Roberts (2000) identified two specialized nodulins, LIMP 1 and LIMP 2. The latter is also called nodulin 26 and mediates a high permeability of the symbiosome membrane to water and uncharged solutes (Dean et al., 1999; Rivers et al., 1997). Further, nodulin 26 may have osmoregulatory function, such as vacuolar water channels (Tyerman et al., 1999) and support the exchange of gases like CO_2, O_2 and NH_3. The nodulin LIMP 1 is a water-specific aquaporin and active in the regulation of the O_2-diffusion barrier in nodules (Guenther and Roberts, 2000). For both channels, studies for H_2O_2 transport capacity are missing, making an involvement in the tight redox regulation of the symbiosome speculative.

9.5 CONCLUSIONS AND FURTHER REMARKS

Peroxiporins are important H_2O_2 channels involved in several processes in plants and microorganisms. Some of these channels can also transport water but in different efficacies; therefore, experiments demonstrating their transport capacities need to be performed in the near future. Their role in response to biotic and abiotic stresses needs to be better characterized. The transport of ROS can influence plant defense mechanisms such as intermediate signaling, HR, stomata closure, cell wall reinforcement, and activation of SAR. Additionally, a peroxiporin has shown to transport a bacterial virulence factor leading to the induction of stomata closure. All these responses probably occur due to the increase in expression of peroxiporins and consequently fast redox signaling. In contrast, other PIPs may efficiently detoxify ROS. Thus, experiments supporting both hypotheses need to be conducted. Fungal peroxiporins can also play a role in plant pathogenesis as demonstrated for *Botrytis cinerea*, however future research must be conducted for the better understanding of peroxiporin role in other fungal pathogens. The symbiosis with mycorrhiza increases plant fitness against drought stress, involving plant peroxiporins regulation by the fungal counterpart. In rhizobacteria, MIPs can favor the permeability of water and uncharged solutes, however the transport of H_2O_2 remains to be elucidated. Peroxiporins may influence in many more processes than those already described. There is an urgent need for more research on its function and transport efficacy for the better understanding of plant-microbe interactions.

REFERENCES

Abascal, F., Irisarri, I., Zardoya, R., 2014. Diversity and evolution of membrane intrinsic proteins. *Biochimica et Biophysica Acta (BBA)—General Subjects, Aquaporins* 1840, 1468–1481. https://doi.org/10.1016/j.bbagen.2013.12.001

Agre, P., Bonhivers, M., Borgnia, M.J., 1998. The aquaporins, blueprints for cellular plumbing systems. *J Biol Chem* 273, 14659–14662. https://doi.org/10.1074/jbc.273.24.14659

Aguirre, J., Lambeth, J.D., 2010. Nox enzymes from fungus to fly to fish and what they tell us about Nox function in mammals. *Free Radic Biol Med* 49, 1342–1353. https://doi.org/10.1016/j.freeradbiomed.2010.07.027

Amira, M.B., Faize, M., Karlsson, M., Dubey, M., Frac, M., Panek, J., Fumanal, B., Gousset-Dupont, A., Julien, J.L., Chaar, H., Auguin, D., Mom, R., Label, P., Venisse, J.S., 2021. Fungal X-Intrinsic protein aquaporin from *Trichoderma atroviride*: structural and functional considerations. *Biomolecules* 11. https://doi.org/10.3390/biom11020338

Amselem, J., Cuomo, C.A., Kan, J.A.L. van, Viaud, M., Benito, E.P., Couloux, A., Coutinho, P.M., Vries, R.P. de, Dyer, P.S., Fillinger, S., Fournier, E., Gout, L., Hahn, M., Kohn, L., Lapalu, N., Plummer, K.M., Pradier, J.-M., Quévillon, E., Sharon, A., Simon, A., Have, A. ten, Tudzynski, B., Tudzynski, P., Wincker, P., Andrew, M., Anthouard, V., Beever, R.E., Beffa, R., Benoit, I., Bouzid, O., Brault, B., Chen, Z., Choquer, M., Collémare, J., Cotton, P., Danchin, E.G., Silva, C.D., Gautier, A., Giraud, C., Giraud, T., Gonzalez, C., Grossetete, S., Güldener, U., Henrissat, B., Howlett, B.J., Kodira, C., Kretschmer, M., Lappartient, A., Leroch, M., Levis, C., Mauceli, E., Neuvéglise, C., Oeser, B., Pearson, M., Poulain, J., Poussereau, N., Quesneville, H., Rascle, C., Schumacher, J., Ségurens, B., Sexton, A., Silva, E., Sirven, C., Soanes, D.M., Talbot, N.J., Templeton, M., Yandava, C., Yarden, O., Zeng, Q., Rollins, J.A., Lebrun, M.-H., Dickman, M., 2011. Genomic analysis of the necrotrophic fungal pathogens *Sclerotinia sclerotiorum* and *Botrytis cinerea*. *PLoS Genet* 7, e1002230. https://doi.org/10.1371/journal.pgen.1002230

An, B., Li, B., Li, H., Zhang, Z., Qin, G., Tian, S., 2016. Aquaporin8 regulates cellular development and reactive oxygen species production, a critical component of virulence in *Botrytis cinerea*. *New Phytol* 209, 1668–1680. https://doi.org/10.1111/nph.13721

Anzai, H., Yoneyama, K., Yamaguchi, I., 1989. Transgenic tobacco resistant to a bacterial disease by the detoxification of a pathogenic toxin. *Mol Genet Genom* 219, 492–494. https://doi.org/10.1007/BF00259626

Apel, K., Hirt, H., 2004. Reactive oxygen species: metabolism, oxidative stress, and signal transduction. *Annu Rev Plant Biol* 55, 373–399. https://doi.org/10.1146/annurev.arplant.55.031903.141701

Aroca, R., Porcel, R., Ruiz-Lozano, J.M., 2012. Regulation of root water uptake under abiotic stress conditions. *J Exp Bot* 63, 43–57. https://doi.org/10.1093/jxb/err266

Augé, R.M., 2001. Water relations, drought and vesicular-arbuscular mycorrhizal symbiosis. *Mycorrhiza* 11, 3–42. https://doi.org/10.1007/s005720100097

Bahadur, A., Batool, A., Nasir, F., Jiang, S., Mingsen, Q., Zhang, Q., Pan, J., Liu, Y., Feng, H., 2019. Mechanistic insights into arbuscular mycorrhizal fungi-mediated drought stress tolerance in plants. *Int J Mol Sci* 20. https://doi.org/10.3390/ijms20174199

Báidez, A.G., Gómez, P., Del Río, J.A., Ortuño, A., 2007. Dysfunctionality of the xylem in *Olea europaea* L. plants associated with the infection process by *Verticillium dahliae* Kleb. role of phenolic compounds in plant defense mechanism. *J Agric Food Chem* 55, 3373–3377. https://doi.org/10.1021/jf063166d

Balint-Kurti, P., 2019. The plant hypersensitive response: concepts, control and consequences. *Mol Plant Pathol* 20, 1163–1178. https://doi.org/10.1111/mpp.12821

Balmant, K.M., Zhang, T., Chen, S., 2016. Protein phosphorylation and redox modification in stomatal guard cells. *Front Physiol* 7, 26. https://doi.org/10.3389/fphys.2016.00026

Banuett, F., Herskowitz, I., 1994. Identification of fuz7, a *Ustilago maydis* MEK/MAPKK homolog required for a-locus-dependent and-independent steps in the fungal life cycle. *Genes Dev* 8, 1367–1378. https://doi.org/10.1101/gad.8.12.1367

Barea, J.M., Palenzuela, J., Cornejo, P., Sánchez-Castro, I., Navarro-Fernández, C., Lopéz-García, A., Estrada, B., Azcón, R., Ferrol, N., Azcón-Aguilar, C., 2011. Ecological and functional roles of mycorrhizas in semi-arid ecosystems of Southeast Spain. *J Arid Environ* 75, 1292–1301. https://doi.org/10.1016/j.jaridenv.2011.06.001

Bárzana, G., Aroca, R., Bienert, G.P., Chaumont, F., Ruiz-Lozano, J.M., 2014. New insights into the regulation of aquaporins by the arbuscular mycorrhizal symbiosis in maize plants under drought stress and possible implications for plant performance. *Mol Plant-Microbe Interact* 27, 349–363. https://doi.org/10.1094/MPMI-09-13-0268-R

Bárzana, G., Aroca, R., Ruiz-Lozano, J.M., 2015. Localized and non-localized effects of arbuscular mycorrhizal symbiosis on accumulation of osmolytes and aquaporins and on antioxidant systems in maize plants subjected to total or partial root drying. *Plant Cell Environ* 38, 1613–1627. https://doi.org/10.1111/pce.12507

Begum, N., Qin, C., Ahanger, M.A., Raza, S., Khan, M.I., Ashraf, M., Ahmed, N., Zhang, L., 2019. Role of arbuscular mycorrhizal fungi in plant growth regulation: implications in abiotic stress tolerance. *Front Plant Sci* 10, 1068. https://doi.org/10.3389/fpls.2019.01068

Bienert, G.P., Bienert, M.D., Jahn, T.P., Boutry, M., Chaumont, F., 2011. Solanaceae XIPs are plasma membrane aquaporins that facilitate the transport of many uncharged substrates. *Plant J Cell Mol Biol* 66, 306–317. https://doi.org/10.1111/j.1365-313X.2011.04496.x

Bienert, G.P., Desguin, B., Chaumont, F., Hols, P., 2013. Channel-mediated lactic acid transport: a novel function for aquaglyceroporins in bacteria. *Biochem J* 454, 559–570. https://doi.org/10.1042/BJ20130388

Bienert, G.P., Heinen, R.B., Berny, M.C., Chaumont, F., 2014. Maize plasma membrane aquaporin ZmPIP2;5, but not ZmPIP1;2, facilitates transmembrane diffusion of hydrogen peroxide. *Biochimica et Biophysica Acta* 1838, 216–222. https://doi.org/10.1016/j.bbamem.2013.08.011

Bienert, G.P., Møller, A.L.B., Kristiansen, K.A., Schulz, A., Møller, I.M., Schjoerring, J.K., Jahn, T.P., 2007. Specific aquaporins facilitate the diffusion of hydrogen peroxide across membranes. *J Biol Chem* 282, 1183–1192. https://doi.org/10.1074/jbc.M603761200

Bienert, G.P., Schjoerring, J.K., Jahn, T.P., 2006. Membrane transport of hydrogen peroxide. *Biochimica et Biophysica Acta (BBA)-Biomembranes* 1758, 994–1003. https://doi.org/10.1016/j.bbamem.2006.02.015

Borgnia, M.J., Agre, P., 2001. Reconstitution and functional comparison of purified GlpF and AqpZ, the glycerol and water channels from *Escherichia coli*. *PNAS* 98, 2888–2893. https://doi.org/10.1073/pnas.051628098

Boursiac, Y., Boudet, J., Postaire, O., Luu, D.T., Tournaire-Roux, C., Maurel, C., 2008. Stimulus-induced downregulation of root water transport involves reactive oxygen species-activated cell signalling and plasma membrane intrinsic protein internalization. *Plant J* 56, 207–218. https://doi.org/10.1111/j.1365-313X.2008.03594.x

Breia, R., Mósca, A.F., Conde, A., Correia, S., Conde, C., Noronha, H., Soveral, G., Gonçalves, B., Gerós, H., 2020. Sweet Cherry (*Prunus avium* L.) PaPIP1;4 Is a functional aquaporin upregulated by pre-harvest calcium treatments that prevent cracking. *Int J Mol Sci* 21, E3017. https://doi.org/10.3390/ijms21083017

Buckner, B., Johal, G.S., Janick-Buckner, D., 2000. Cell death in maize. *Physiol Plant* 108, 231–239. https://doi.org/10.1034/j.1399-3054.2000.108003231.x

Bueno, P., Soto, M.J., Rodríguez-Rosales, M.P., Sanjuan, J., Olivares, J., Donaire, J.P., 2001. Time-course of lipoxygenase, antioxidant enzyme activities and H_2O_2 accumulation during the early stages of Rhizobium legume symbiosis. *New Phytol* 152, 91–96. https://doi.org/10.1046/j.0028-646x.2001.00246.x

Caffall, K.H., Mohnen, D., 2009. The structure, function, and biosynthesis of plant cell wall pectic polysaccharides. *Carbohydr Res* 344, 1879–1900. https://doi.org/10.1016/j.carres.2009.05.021

Calamita, G., 2000. The *Escherichia coli* aquaporin-Z water channel. *Mol Microbiol* 37, 254–262. https://doi.org/10.1046/j.1365-2958.2000.02016.x

Cano-Domínguez, N., Álvarez-Delfín, K., Hansberg, W., Aguirre, J., 2008. NADPH oxidases NOX-1 and NOX-2 require the regulatory subunit NOR-1 to control cell differentiation and growth in *Neurospora crassa*. *Eukaryot Cell* 7, 1352–1361. https://doi.org/10.1128/ec.00137-08

Černý, M., Berka, M., Dvořák, M., Milenković, I., Saiz-Fernández, I., Brzobohatý, B., Ďurkovič, J., 2022. Defense mechanisms promoting tolerance to aggressive *Phytophthora* species in hybrid poplar. *Front Plant Sci* 13. https://doi.org/10.3389/fpls.2022.1018272

Černý, M., Habánová, H., Berka, M., Luklová, M., Brzobohatý, B., 2018. Hydrogen peroxide: its role in plant biology and crosstalk with signalling networks. *Int J Mol Sci* 19. https://doi.org/10.3390/ijms19092812

Chang, C., Damiani, I., Puppo, A., Frendo, P., 2009. Redox changes during the legume–Rhizobium symbiosis. *Mol Plant* 2, 370–377. https://doi.org/10.1093/mp/ssn090

Chaumount, F., Barrieu, F., Wojcik, E., Chrispeels, M.J., Jung, R., 2001. Aquaporins constitute a large and highly divergent protein family in maize. *Plant Physiol* 125, 1206–1215.

Chen, H., Jiang, J.-G., 2010. Osmotic adjustment and plant adaptation to environmental changes related to drought and salinity. *Environ Rev* 18, 309–319. https://doi.org/10.1139/A10-014

Chowdhury, J., Henderson, M., Schweizer, P., Burton, R.A., Fincher, G.B., Little, A., 2014. Differential accumulation of callose, arabinoxylan and cellulose in nonpenetrated versus penetrated papillae on leaves of barley infected with *Blumeria graminis f. sp. hordei*. *New Phytol* 204, 650–660. https://doi.org/10.1111/nph.12974

Chung, K.R., 2012. Stress response and pathogenicity of the necrotrophic fungal pathogen *Alternaria alternata*. *Scientifica (Cairo)* 2012, 635431. https://doi.org/10.6064/2012/635431

Coll, N.S., Smidler, A., Puigvert, M., Popa, C., Valls, M., Dangl, J.L., 2014. The plant metacaspase AtMC1 in pathogen-triggered programmed cell death and aging: functional linkage with autophagy. *Cell Death Differ* 21, 1399–1408. https://doi.org/10.1038/cdd.2014.50

Dean, R.A., Talbot, N.J., Ebbole, D.J., Farman, M.L., Mitchell, T.K., Orbach, M.J., Thon, M., Kulkarni, R., Xu, J.-R., Pan, H., Read, N.D., Lee, Y.-H., Carbone, I., Brown, D., Oh, Y.Y., Donofrio, N., Jeong, J.S., Soanes, D.M., Djonovic, S., Kolomiets, E., Rehmeyer, C., Li, W., Harding, M., Kim, S., Lebrun, M.-H., Bohnert, H., Coughlan, S., Butler, J., Calvo, S., Ma, L.-J., Nicol, R., Purcell, S., Nusbaum, C., Galagan, J.E., Birren, B.W. 2005. The genome sequence of the rice blast fungus *Magnaporthe grisea*. *Nature* 434, 980–986. https://doi.org/10.1038/nature03449

Dean, R.M., Rivers, R.L., Zeidel, M.L., Roberts, D.M., 1999. Purification and functional reconstitution of soybean nodulin 26. An aquaporin with water and glycerol transport properties. *Biochemistry* 38, 347–353. https://doi.org/10.1021/bi982110c

D'Haeze, W., Rycke, R.D., Mathis, R., Goormachtig, S., Pagnotta, S., Verplancke, C., Capoen, W., Holsters, M., 2003. Reactive oxygen species and ethylene play a positive role in lateral root base nodulation of a semiaquatic legume. *PNAS* 100, 11789–11794. https://doi.org/10.1073/pnas.1333899100

Dickman, M., Williams, B., Li, Y., de Figueiredo, P., Wolpert, T., 2017. Reassessing apoptosis in plants. *Nature Plants* 3, 773–779. https://doi.org/10.1038/s41477-017-0020-x

Dietz, S., von Bülow, J., Beitz, E., Nehls, U., 2011. The aquaporin gene family of the ectomycorrhizal fungus *Laccaria bicolor*: lessons for symbiotic functions. *New Phytol* 190, 927–940. https://doi.org/10.1111/j.1469-8137.2011.03651.x

Ding, M., Li, J., Fan, X., He, F., Yu, X., Chen, L., Zou, S., Liang, Y., Yu, J., 2018. Aquaporin1 regulates development, secondary metabolism and stress responses in *Fusarium graminearum*. *Curr Genet* 64, 1057–1069. https://doi.org/10.1007/s00294-018-0818-8

Dirschnabel, D.E., Nowrousian, M., Cano-Dominguez, N., Aguirre, J., Teichert, I., Kück, U., 2014. New insights into the roles of NADPH oxidases in sexual development and ascospore germination in *Sordaria macrospora*. *Genetics* 196, 729–744. https://doi.org/10.1534/genetics.113.159368

D'Mello, J.P.F., Placinta, C.M., Macdonald, A.M.C., 1999. *Fusarium* mycotoxins: a review of global implications for animal health, welfare and productivity. *Anim Feed Sci Technol* 80, 183–205. https://doi.org/10.1016/S0377-8401(99)00059-0

Dynowski, M., Schaaf, G., Loque, D., Moran, O., Ludewig, U., 2008. Plant plasma membrane water channels conduct the signalling molecule H_2O_2. *Biochem J* 414, 53–61. https://doi.org/10.1042/BJ20080287

Egan, M.J., Wang, Z.Y., Jones, M.A., Smirnoff, N., Talbot, N.J., 2007. Generation of reactive oxygen species by fungal NADPH oxidases is required for rice blast disease. *Proc Natl Acad Sci U S A* 104, 11772–11777. https://doi.org/10.1073/pnas.0700574104

Eynck, C., Koopmann, B., Karlovsky, P., von Tiedemann, A., 2009. Internal resistance in winter oilseed rape inhibits systemic spread of the vascular pathogen *Verticillium longisporum*. *Phytopathology* 99, 802–811.

Farooq, M., Wahid, A., Kobayashi, N., Fujita, D., Basra, S.M.A., 2009. Plant drought stress: effects, mechanisms and management. *Sustain Agric*, 153–188. https://doi.org/10.1051/agro:2008021

Fetter, K., Van Wilder, V., Moshelion, M., Chaumont, F., 2004. Interactions between plasma membrane aquaporins modulate their water channel activity. *Plant Cell* 16, 215–228. https://doi.org/10.1105/tpc.017194

Fischer, H.M., 1994. Genetic regulation of nitrogen fixation in rhizobia. *Microbiol Rev* 58, 352–386. https://doi.org/10.1128/mr.58.3.352-386.1994

Fleissner, A., Herzog, S., 2016. Signal exchange and integration during self-fusion in filamentous fungi. *Semin Cell Dev Biol* 57, 76–83. https://doi.org/10.1016/j.semcdb.2016.03.016

Fortunato, A.A., Araujo, L., Rodrigues, F.Á., 2017. Association of the production of phenylpropanoid compounds at the infection sites of *Corynespora cassiicola* with soybean resistance against target spot. *J Phytopathol* 165, 131–142. https://doi.org/10.1111/jph.12546

Fortunato, A.A., Debona, D., Bernardeli, A.M.A., Rodrigues, F.A., 2015. Defence-related enzymes in soybean resistance to target spot. *J Phytopathol* 163, 731–742. https://doi.org/10.1111/jph.12370

Fujiyoshi, Y., Mitsuoka, K., de Groot, B.L., Philippsen, A., Grubmüller, H., Agre, P., Engel, A., 2002. Structure and function of water channels. *Curr Opin Struct Biol* 12, 509–515. https://doi.org/10.1016/s0959-440x(02)00355-x

Gao, Q.-M., Zhu, S., Kachroo, P., Kachroo, A., 2015. Signal regulators of systemic acquired resistance. *Front Plant Sci* 6, 228. https://doi.org/10.3389/fpls.2015.00228

Genre, A., Chabaud, M., Timmers, T., Bonfante, P., Barker, D.G., 2005. Arbuscular mycorrhizal fungi elicit a novel intracellular apparatus in *Medicago truncatula* root epidermal cells before infection. *Plant Cell* 17, 3489–3499. https://doi.org/10.1105/tpc.105.035410

Glazebrook, J., 2005. Contrasting mechanisms of defense against biotrophic and necrotrophic pathogens. *Ann Rev Phytopathol* 43, 205–227. https://doi.org/10.1146/annurev.phyto.43.040204.135923

Greenberg, J.T., Yao, N., 2004. The role and regulation of programmed cell death in plant–pathogen interactions. *Cell Microbiol*, https://doi.org/10.1111/j.1462-5822.2004.00361.x

Grondin, A., Rodrigues, O., Verdoucq, L., Merlot, S., Leonhardt, N., Maurel, C., 2015. Aquaporins contribute to ABA-triggered stomatal closure through OST1-mediated phosphorylation. *Plant Cell* 27, 1945–1954. https://doi.org/10.1105/tpc.15.00421

Guenther, J.F., Roberts, D.M., 2000. Water-selective and multifunctional aquaporins from *Lotus japonicus* nodules. *Planta* 210, 741–748. https://doi.org/10.1007/s004250050675

Gustavsson, S., Lebrun, A.S., Norden, K., Chaumont, F., Johanson, U., 2005. A novel plant major intrinsic protein in *Physcomitrella patens* most similar to bacterial glycerol channels. *Plant Physiol* 139, 287–295. https://doi.org/10.1104/pp.105.063198

Hane, J.K., Anderson, J.P., Williams, A.H., Sperschneider, J., Singh, K.B., 2014. Genome sequencing and comparative genomics of the broad host-range pathogen *Rhizoctonia solani* AG8. *PLoS Genet* 10, e1004281. https://doi.org/10.1371/journal.pgen.1004281

Harrison, M.J., 2005. Signaling in the arbuscular mycorrhizal symbiosis. *Annu Rev Microbiol* 59, 19–42. https://doi.org/10.1146/annurev.micro.58.030603.123749

Harvengt, P., Vlerick, A., Fuks, B., Wattiez, R., Ruysschaert, J.M., Homble, F., 2000. Lentil seed aquaporins form a hetero-oligomer which is phosphorylated by a Mg^{2+}-dependent and Ca^{2+}-regulated kinase. *Biochem J* 352 (Pt 1), 183–190. https://doi.org/10.1042/bj3520183

Henzler, T., Steudle, E., 2000. Transport and metabolic degradation of hydrogen peroxide in *Chara corallina*: model calculations and measurements with the pressure probe suggest transport of H_2O_2 across water channels. *J Exp Bot* 51, 2053–2066.

Hohmann, S., Bill, R.M., Kayingo, G., Prior, B.A., Hohmann, S., Bill, R.M., Kayingo, G., Prior, B.A., Hohmann, S., Bill, R.M., Kayingo, G., Prior, B.A., 2000. Microbial MIP channels. *Trends Microbiol* 8, 33–38. https://doi.org/10.1016/S0966-842X(99)01645-5

Hong, J.H., Seah, S.W., Xu, J., 2013. The root of ABA action in environmental stress response. *Plant Cell Rep* 32, 971–983. https://doi.org/10.1007/s00299-013-1439-9

Hooijmaijers, C., Rhee, J.Y., Kwak, K.J., Chung, G.C., Horie, T., Katsuhara, M., Kang, H., 2012. Hydrogen peroxide permeability of plasma membrane aquaporins of *Arabidopsis thaliana*. *J Plant Res* 125, 147–153. https://doi.org/10.1007/s10265-011-0413-2

Hückelhoven, R., 2007. Cell wall-associated mechanisms of disease resistance and susceptibility. *Ann Revi Phytopathol* 45, 101–127. https://doi.org/10.1146/annurev.phyto.45.062806.094325

Hückelhoven, R., Fodor, J., Trujillo, M., Kogel, K.T., 2000. Barley Mla and Rar mutants compromised in the hypersensitive cell death response against *Blumeria graminis* f.sp. *hordei* are modified in their ability to accumulate reactive oxygen intermediated at sites of fungal invasion. *Planta* 212, 16–24.

Hurni, S., Scheuermann, D., Krattinger, S.G., Kessel, B., Wicker, T., Herren, G., Fitze, M.N., Breen, J., Presterl, T., Ouzunova, M., Keller, B., 2015. The maize disease resistance gene Htn1 against northern corn leaf blight encodes a wall-associated receptor-like kinase. *Proc Natl Acad Sci U S A* 112, 8780–8785. https://doi.org/10.1073/pnas.1502522112

Javot, H., Lauvergeat, V., Santoni, V., Martin-Laurent, F., Güçlü, J., Vinh, J., Heyes, J., Franck, K.I., Schaffner, A.R., Bouchez, D., 2003. Role of a single aquaporin isoform in root water uptake. *Plant Cell* 15, 509–522. https://doi.org/10.1105/tpc.008888

Javot, H., Maurel, C., 2002. The role of aquaporins in root water uptake. *Ann Bot* 90, 301–313. https://doi.org/10.1093/aob/mcf199

Jones, J.D.G., Dangl, J.L., 2006. The plant immune system. *Nature* 444, 323–329. https://doi.org/10.1038/nature05286

Kabbage, M., Williams, B., Dickman, M.B., 2013. Cell death control: the interplay of apoptosis and autophagy in the pathogenicity of *Sclerotinia sclerotiorum*. *PLoS Pathog* 9, e1003287. https://doi.org/10.1371/journal.ppat.1003287

Kachroo, A., Vincelli, P., Kachroo, P., 2017. Signaling mechanisms underlying resistance responses: what have we learned, and how is it being applied? *Phytopathology*, 107, 1452–1461. https://doi.org/10.1094/PHYTO-04-17-0130-RVW

Kämper, J., Kahmann, R., Bölker, M., Ma, L.-J., Brefort, T., Saville, B.J., Banuett, F., Kronstad, J.W., Gold, S.E., Müller, O., Perlin, M.H., Wösten, H.A.B., de Vries, R., Ruiz-Herrera, J., Reynaga-Peña, C.G., Snetselaar, K., McCann, M., Pérez-Martín, J., Feldbrügge, M., Basse, C.W., Steinberg, G., Ibeas, J.I., Holloman, W., Guzman, P., Farman, M., Stajich, J.E., Sentandreu, R., González-Prieto, J.M., Kennell, J.C., Molina, L., Schirawski, J., Mendoza-Mendoza, A., Greilinger, D., Münch, K., Rössel, N., Scherer, M., Vraneš, M., Ladendorf, O., Vincon, V., Fuchs, U., Sandrock, B., Meng, S., Ho, E.C.H., Cahill, M.J., Boyce, K.J., Klose, J., Klosterman, S.J., Deelstra, H.J., Ortiz-Castellanos, L., Li, W., Sanchez-Alonso, P., Schreier, P.H., Häuser-Hahn, I., Vaupel, M., Koopmann, E., Friedrich, G., Voss, H., Schlüter, T., Margolis, J., Platt, D., Swimmer, C., Gnirke, A., Chen, F., Vysotskaia, V., Mannhaupt, G., Güldener, U., Münsterkötter, M., Haase, D., Oesterheld, M., Mewes, H.-W., Mauceli, E.W., DeCaprio, D., Wade, C.M., Butler, J., Young, S., Jaffe, D.B., Calvo, S., Nusbaum, C., Galagan, J., Birren, B.W. 2006. Insights from the genome of the biotrophic fungal plant pathogen *Ustilago maydis*. *Nature* 444, 97–101. https://doi.org/10.1038/nature05248

Katsuhara, M., Sasano, S., Horie, T., Matsumoto, T., Rhee, J., Shibasaka, M., 2014. Functional and molecular characteristics of rice and barley NIP aquaporins transporting water, hydrogen peroxide and arsenite. *Plant Biotechnol* 31, 213–219. https://doi.org/10.5511/plantbiotechnology.14.0421a

Kipfer, T., Wohlgemuth, T., van der Heijden, M.G.A., Ghazoul, J., Egli, S., 2012. Growth response of drought-stressed *Pinus sylvestris* seedlings to single-and multi-species inoculation with ectomycorrhizal fungi. *PLoS ONE* 7, e35275. https://doi.org/10.1371/journal.pone.0035275

Kishor, K.P.B., Sreenivasulu, N., 2014. Is proline accumulation per se correlated with stress tolerance or is proline homeostasis a more critical issue? *Plant Cell Envir* 37, 300–311. https://doi.org/10.1111/pce.12157

Kubicek, C.P., Herrera-Estrella, A., Seidl-Seiboth, V., Martinez, D.A., Druzhinina, I.S., Thon, M., Zeilinger, S., Casas-Flores, S., Horwitz, B.A., Mukherjee, P.K., Mukherjee, M., Kredics, L., Alcaraz, L.D., Aerts, A., Antal, Z., Atanasova, L., Cervantes-Badillo, M.G., Challacombe, J., Chertkov, O., McCluskey, K., Coulpier, F., Deshpande, N., von Döhren, H., Ebbole, D.J., Esquivel-Naranjo, E.U., Fekete, E., Flipphi, M., Glaser, F., Gómez-Rodríguez, E.Y., Gruber, S., Han, C., Henrissat, B., Hermosa, R., Hernández-Oñate, M., Karaffa, L., Kosti, I., Le Crom, S., Lindquist, E., Lucas, S., Lübeck, M., Lübeck, P.S., Margeot, A., Metz, B., Misra, M., Nevalainen, H., Omann, M., Packer, N., Perrone, G., Uresti-Rivera, E.E., Salamov, A., Schmoll, M., Seiboth, B., Shapiro, H., Sukno, S., Tamayo-Ramos, J.A., Tisch, D., Wiest, A., Wilkinson, H.H., Zhang, M., Coutinho, P.M., Kenerley, C.M., Monte, E., Baker, S.E., Grigoriev, I.V. 2011. Comparative genome sequence analysis underscores mycoparasitism as the ancestral life style of *Trichoderma*. *Genome Biol* 12, R40. https://doi.org/10.1186/gb-2011-12-4-r40

Lai, Z., Wang, F., Zheng, Z., Fan, B., Chen, Z., 2011. A critical role of autophagy in plant resistance to necrotrophic fungal pathogens. *Plant J Cell Mol Biol* 66, 953–968. https://doi.org/10.1111/j.1365-313X.2011.04553.x

Li, G., Chen, T., Zhang, Z., Li, B., Tian, S., 2020. Roles of aquaporins in plant-pathogen interaction. *Plants (Basel, Switzerland)* 9. https://doi.org/10.3390/plants9091134

Li, P., Zhang, L., Mo, X., Ji, H., Bian, H., Hu, Y., Majid, T., Long, J., Pang, H., Tao, Y., Ma, J., Dong, H., 2019. Rice aquaporin PIP1;3 and harpin Hpa1 of bacterial blight pathogen cooperate in a type III effector translocation. *J Exp Bot* 70, 3057–3073. https://doi.org/10.1093/jxb/erz130

Li, T., Hu, Y.-J., Hao, Z.-P., Li, H., Wang, Y.-S., Chen, B.-D., 2013. First cloning and characterization of two functional aquaporin genes from an arbuscular mycorrhizal fungus *Glomus intraradices*. *New Phytol* 197, 617–630. https://doi.org/10.1111/nph.12011

Lionetti, V., Cervone, F., Bellincampi, D., 2012. Methyl esterification of pectin plays a role during plant-pathogen interactions and affects plant resistance to diseases. *J Plant Physiol* 169, 1623–1630. https://doi.org/10.1016/j.jplph.2012.05.006

Luginbuehl, L.H., Oldroyd, G.E.D., 2017. Understanding the arbuscule at the heart of endomycorrhizal symbioses in plants. *Curr Biol* 27, R952–R963. https://doi.org/10.1016/j.cub.2017.06.042

Luna, E., Pastor, V., Robert, J., Flora, V., Mauch-Mani, B., Ton, J., 2011. Callose deposition, a multifaceted plant defense response. *Mol Plant-Microbe Interact* 24, 183–193. https://doi.org/10.1094/MPMI-07-10-0149

Luu, D.-T., Maurel, C., 2005. Aquaporins in a challenging environment: molecular gears for adjusting plant water status. *Plant Cell Envir* 28, 85–96. https://doi.org/10.1111/j.1365-3040.2004.01295.x

Ma, L.-J., van der Does, H.C., Borkovich, K.A., Coleman, J.J., Daboussi, M.-J., Di Pietro, A., Dufresne, M., Freitag, M., Grabherr, M., Henrissat, B., Houterman, P.M., Kang, S., Shim, W.-B., Woloshuk, C., Xie, X., Xu, J.-R., Antoniw, J., Baker, S.E., Bluhm, B.H., Breakspear, A., Brown, D.W., Butchko, R.A.E., Chapman, S., Coulson, R., Coutinho, P.M., Danchin, E.G.J., Diener, A., Gale, L.R., Gardiner, D.M., Goff, S., Hammond-Kosack, K.E., Hilburn, K., Hua-Van, A., Jonkers, W., Kazan, K., Kodira, C.D., Koehrsen, M., Kumar, L., Lee, Y.-H., Li, L., Manners, J.M., Miranda-Saavedra, D., Mukherjee, M., Park, G., Park, J., Park, S.-Y., Proctor, R.H., Regev, A., Ruiz-Roldan, M.C., Sain, D., Sakthikumar, S., Sykes, S., Schwartz, D.C., Turgeon, B.G., Wapinski, I., Yoder, O., Young, S., Zeng, Q., Zhou, S., Galagan, J., Cuomo, C.A., Kistler, H.C., Rep, M. 2010. Comparative genomics reveals mobile pathogenicity chromosomes in *Fusarium*. *Nature* 464, 367–373. https://doi.org/10.1038/nature08850

Malinovsky, F.G., Fangel, J.U., Willats, W.G.T., 2014. The role of the cell wall in plant immunity. *Front Plant Sci* 5, 178. https://doi.org/10.3389/fpls.2014.00178

Marschall, R., Tudzynski, P., 2016. Reactive oxygen species in development and infection processes. *Semin Cell Dev Biol* 57, 138–146. https://doi.org/10.1016/j.semcdb.2016.03.020

Maurel, C., Santoni, V., Luu, D.-T., Wudick, M.M., Verdoucq, L., 2009. The cellular dynamics of plant aquaporin expression and functions. *Curr Opin Plant Biol* 12, 690–698. https://doi.org/10.1016/j.pbi.2009.09.002

Meeley, R.B., Walton, J.D., 1991. Enzymatic detoxification of HC-tonin, the host-selective cyclic peptide from *Cochliobolus carbonum*. *Plant Physiol* 97, 1989–1086.

Melotto, M., Underwood, W., Koczan, J., Nomura, K., He, S.Y., 2006. Plant stomata function in innate immunity against bacterial invasion. *Cell* 126, 969–980. https://doi.org/10.1016/j.cell.2006.06.054

Miller, G., Suzuki, N., Ciftci-Yilmaz, S., Mittler, R., 2010. Reactive oxygen species homeostasis and signalling during drought and salinity stresses. *Plant Cell Environ* 33, 453–467. https://doi.org/10.1111/j.1365-3040.2009.02041.x

Möbius, N., Hertweck, C., 2009. Fungal phytotoxins as mediators of virulence. *Curr Opin Plant Biol* 12, 390–398. https://doi.org/10.1016/j.pbi.2009.06.004

Moore, J.A.M., Jiang, J., Post, W.M., Classen, A.T., 2015. Decomposition by ectomycorrhizal fungi alters soil carbon storage in a simulation model. *Ecosphere* 6, 1–16. https://doi.org/10.1890/ES14-00301.1

Morgan, P.W., 1990. Effects of abiotic stresses on plant hormone systems. In: *Stress Responses in Plants: Adaptation and Acclimation Mechanisms*. Wiley-Liss, Inc., pp. 113–146.

Mubarakshina Borisova, M.M., Kozuleva, M.A., Rudenko, N.N., Naydov, I.A., Klenina, I.B., Ivanov, B.N., 2012. Photosynthetic electron flow to oxygen and diffusion of hydrogen peroxide through the chloroplast envelope via aquaporins. *Biochim Biophys Acta* 1817, 1314–1321. https://doi.org/10.1016/j.bbabio.2012.02.036

Navarro, B.L., Marques, J.P.R., Appezzato-da-Glória, B., Spósito, M.B., 2019. Histopathology of *Phakopsora euvitis* on *Vitis vinifera*. *Eur J Plant Pathol* 154, 1185–1193. https://doi.org/10.1007/s10658-019-01719-w

Navarro García, A., del Pilar Banon Árias, S., Morte, A., Sánchez-Blanco, M.J., 2011. Effects of nursery preconditioning through mycorrhizal inoculation and drought in *Arbutus unedo* L. plants. *Mycorrhiza* 21, 53–64. https://doi.org/10.1007/s00572-010-0310-x

Nehls, U., Dietz, S., 2014. Fungal aquaporins: cellular functions and ecophysiological perspectives. *Appl Microbiol Biotechnol* 98, 8835–8851. https://doi.org/10.1007/s00253-014-6049-0

Nordzieke, D.E., 2022. Hyphal fusions enable efficient nutrient distribution in *Colletotrichum graminicola* conidiation and symptom development on Maize. *Microorganisms* 10, 1146. https://doi.org/10.3390/microorganisms10061146

Nordzieke, D.E., Fernandes, T.R., El Ghalid, M., Turrà, D., Di Pietro, A., 2019. NADPH oxidase regulates chemotropic growth of the fungal pathogen *Fusarium oxysporum* towards the host plant. *New Phytol* 224, 1600–1612. https://doi.org/10.1111/nph.16085

Nordzieke, D.E., Medraño-Fernandez, I., 2018. The plasma membrane: a platform for intra- and

intercellular redox signaling. *Antioxidants (Basel, Switzerland)* 7. https://doi.org/10.3390/antiox7110168
O'Connell, R.J., Thon, M.R., Hacquard, S., Amyotte, S.G., Kleemann, J., Torres, M.F., Damm, U., Buiate, E.A., Epstein, L., Alkan, N., Altmüller, J., Alvarado-Balderrama, L., Bauser, C.A., Becker, C., Birren, B.W., Chen, Z., Choi, J., Crouch, J.A., Duvick, J.P., Farman, M.A., Gan, P., Heiman, D., Henrissat, B., Howard, R.J., Kabbage, M., Koch, C., Kracher, B., Kubo, Y., Law, A.D., Lebrun, M.-H., Lee, Y.-H., Miyara, I., Moore, N., Neumann, U., Nordström, K., Panaccione, D.G., Panstruga, R., Place, M., Proctor, R.H., Prusky, D., Rech, G., Reinhardt, R., Rollins, J.A., Rounsley, S., Schardl, C.L., Schwartz, D.C., Shenoy, N., Shirasu, K., Sikhakolli, U.R., Stüber, K., Sukno, S.A., Sweigard, J.A., Takano, Y., Takahara, H., Trail, F., van der Does, H.C., Voll, L.M., Will, I., Young, S., Zeng, Q., Zhang, J., Zhou, S., Dickman, M.B., Schulze-Lefert, P., Ver Loren van Themaat, E., Ma, L.-J., Vaillancourt, L.J. 2012. Lifestyle transitions in plant pathogenic *Colletotrichum* fungi deciphered by genome and transcriptome analyses. *Nat Genet* 44, 1060–1065. https://doi.org/10.1038/ng.2372
Otto, B., Uehlein, N., Sdorra, S., Fischer, M., Ayaz, M., Belastegui-Macadam, X., Heckwolf, M., Lachnit, M., Pede, N., Priem, N., Reinhard, A., Siegfart, S., Urban, M., Kaldenhoff, R., 2010. Aquaporin tetramer composition modifies the function of tobacco aquaporins. *J Biol Chem* 285, 31253–31260. https://doi.org/10.1074/jbc.M110.115881
Ouziad, F., Wilde, P., Schmelzer, E., Hildebrandt, U., Bothe, H., 2006. Analysis of expression of aquaporins and Na^+/H^+ transporters in tomato colonized by arbuscular mycorrhizal fungi and affected by salt stress. *Environ Exp Bot* 57, 177–186. https://doi.org/10.1016/j.envexpbot.2005.05.011
Park, W., Scheffler, B.E., Bauer, P.J., Campbell, B.T., 2010. Identification of the family of aquaporin genes and their expression in upland cotton (*Gossypium hirsutum* L.). *BMC Plant Biol* 10, 142.
Peaucelle, A., Braybrook, S., Höfte, H., 2012. Cell wall mechanics and growth control in plants: the role of pectins revisited. *Front Plant Sci* 3, 121. https://doi.org/10.3389/fpls.2012.00121
Peleg-Grossman, S., Volpin, H., Levine, A., 2007. Root hair curling and Rhizobium infection in *Medicago truncatula* are mediated by phosphatidylinositide-regulated endocytosis and reactive oxygen species. *J Exp Bot* 58, 1637–1649. https://doi.org/10.1093/jxb/erm013
Peters, L.P., Carvalho, G., Vilhena, M.B., Creste, S., Azevedo, R.A., Monteiro-Vitorello, C.B., 2017. Functional analysis of oxidative burst in sugarcane smut-resistant and –susceptible genotypes. *Planta* 245, 749–764. https://doi.org/10.1007/s00425-016-2642-z
Pettersson, N., Filipsson, C., Becit, E., Brive, L., Hohmann, S., 2005. Aquaporins in yeasts and filamentous fungi. *Biol Cell* 97, 487–500. https://doi.org/10.1042/BC20040144
Piršelová, B., Matušíková, I., 2013. Callose: the plant cell wall polysaccharide with multiple biological functions. *Acta Physiol Plant* 35, 635–644. https://doi.org/10.1007/s11738-012-1103-y
Prasad, R., Bhola, D., Akdi, K., Cruz, C., Sairam, K., Tuteja, N., Varma, A., 2017. Introduction to mycorrhiza: historical development. In: *Mycorrhiza-Function, Diversity, State of the Art*. Springer, pp. 1–7.
Preston, G.M., Carroll, T.P., Guggino, W.B., Agre, P., 1992. Appearance of water channels in *Xenopus oocytes* expressing red cell CHIP28 protein. *Science* 256, 385–387. https://doi.org/10.1126/science.256.5055.385
Quiroga, G., Erice, G., Aroca, R., Chaumont, F., Ruiz-Lozano, J.M., 2017. Enhanced drought stress tolerance by the arbuscular mycorrhizal symbiosis in a drought-sensitive maize cultivar is related to a broader and differential regulation of host plant aquaporins than in a drought-tolerant cultivar. *Front Plant Sci* 8, 1056. https://doi.org/10.3389/fpls.2017.01056
Quiroga, G., Erice, G., Ding, L., Chaumont, F., Aroca, R., Ruiz-Lozano, J.M., 2019. The arbuscular mycorrhizal symbiosis regulates aquaporins activity and improves root cell water permeability in maize plants subjected to water stress. *Plant Cell Environ* 42, 2274–2290. https://doi.org/10.1111/pce.13551
Ramu, S.K., Peng, H.-M., Cook, D.R., 2002. Nod factor induction of reactive oxygen species production is correlated with expression of the early nodulin gene rip1 in *Medicago truncatula*. *Mol Plant Microbe Interact* 15, 522–528. https://doi.org/10.1094/MPMI.2002.15.6.522
Reuscher, S., Akiyama, M., Mori, C., Aoki, K., Shibata, D., Shiratake, K., 2013. Genome-wide identification and expression analysis of aquaporins in tomato. *PLoS ONE* 8, e79052. https://doi.org/10.1371/journal.pone.0079052
Rivers, R.L., Dean, R.M., Chandy, G., Hall, J.E., Roberts, D.M., Zeidel, M.L., 1997. Functional analysis of nodulin 26, an aquaporin in soybean root nodule symbiosomes. *J Biol Chem* 272, 16256–16261. https://doi.org/10.1074/jbc.272.26.16256
Rodrigues, O., Reshetnyak, G., Grondin, A., Saijo, Y., Leonhardt, N., Maurel, C., Verdoucq, L., 2017.

Aquaporins facilitate hydrogen peroxide entry into guard cells to mediate ABA- and pathogen-triggered stomatal closure. *PNAS* 114, 9200–9205. https://doi.org/10.1073/pnas.1704754114

Ruiz-Lozano, J.M., Aroca, R., 2017. Plant aquaporins and mycorrhizae: their regulation and involvement in plant physiology and performance. In: Chaumont, F., Tyerman, S.D. (Eds.), *Plant Aquaporins, Signaling and Communication in Plants*. Springer International Publishing, Cham, pp. 333–353. https://doi.org/10.1007/978-3-319-49395-4_15

Ruiz-Lozano, J.M., Aroca, R., Zamarreño, Á.M., Molina, S., Andreo-Jiménez, B., Porcel, R., García-Mina, J.M., Ruyter-Spira, C., López-Ráez, J.A., 2016. Arbuscular mycorrhizal symbiosis induces strigolactone biosynthesis under drought and improves drought tolerance in lettuce and tomato. *Plant, Cell & Environment* 39, 441–452. https://doi.org/10.1111/pce.12631

Ruiz-Lozano, J.M., Porcel, R., Bárzana, G., Azcón, R., Aroca, R., 2012. Contribution of arbuscular mycorrhizal symbiosis to plant drought tolerance: state of the art. In: *Plant Responses to Drought Stress*. Springer, Berlin, Heidelberg, pp. 335–362. https://doi.org/10.1007/978-3-642-32653-0_13

Ryder, L.S., Dagdas, Y.F., Mentlak, T.A., Kershaw, M.J., Thornton, C.R., Schuster, M., Chen, J., Wang, Z., Talbot, N.J., 2013. NADPH oxidases regulate septin-mediated cytoskeletal remodeling during plant infection by the rice blast fungus. *Proc Natl Acad Sci U S A* 110, 3179–3184. https://doi.org/10.1073/pnas.1217470110

Sagi, M., Fluhr, R., 2006. Production of reactive oxygen species by plant NADPH oxidases. *Plant Physiol* 141, 336–340. https://doi.org/10.1104/pp.106.078089

Sakurai, J., Ishikawa, F., Yamaguchi, T., Uemura, M., Maeshima, M., 2005. Identification of 33 rice aquaporin genes and analysis of their expression and function. *Plant Cell Physiol* 46, 1568–1577. https://doi.org/10.1093/pcp/pci172

Sandalio, L.M., Peláez-Vico, M.A., Molina-Moya, E., Romero-Puertas, M.C., 2021. Peroxisomes as redox-signaling nodes in intracellular communication and stress responses. *Plant Physiol* 186, 22–35. https://doi.org/10.1093/plphys/kiab060

Schüssler, M.D., Alexandersson, E., Bienert, G.P., Kichey, T., Laursen, K.H., Johanson, U., Kjellbom, P., Schjoerring, J.K., Jahn, T.P., 2008. The effects of the loss of TIP1;1 and TIP1;2 aquaporins in *Arabidopsis thaliana*. *Plant J Cell Mol Biol* 56, 756–767. https://doi.org/10.1111/j.1365-313X.2008.03632.x

Scott, B., 2015. Conservation of fungal and animal nicotinamide adenine dinucleotide phosphate oxidase complexes. *Mol Microbiol* 95, 910–913. https://doi.org/10.1111/mmi.12946

Seaver, L.C., Imlay, J.A., 2001. Hydrogen peroxide fluxes and compartmentalization inside growing *Escherichia coli*. *J Bacteriol* 183, 7182–7189. https://doi.org/10.1128/JB.183.24.7182-7189.2001

Sedlářová, M., Luhová, L., Petrivalský, M., Lebeda, A., 2007. Localisation and metabolism of reactive oxygen species during *Bremia lactucae* pathogenesis in *Lactuca sativa* and wild *Lactuca* spp. *Plant Physiol Bioch* 45, 607–616. https://doi.org/10.1016/j.plaphy.2007.05.010

Segmüller, N., Kokkelink, L., Giesbert, S., Odinius, D., van Kan, J., Tudzynski, P., 2008. NADPH oxidases are involved in differentiation and pathogenicity in *Botrytis cinerea*. *Mol Plant Microbe Interact* 21, 808–819. https://doi.org/10.1094/MPMI-21-6-0808

Sharma, K., Gupta, S., Thokchom, S.D., Jangir, P., Kapoor, R., 2021. Arbuscular mycorrhiza-mediated regulation of polyamines and aquaporins during abiotic stress: deep insights on the recondite players. *Front Plant Sci* 12, 642101. https://doi.org/10.3389/fpls.2021.642101

Shaw, S.L., Long, S.R., 2003. Nod factor elicits two separable calcium responses in *Medicago truncatula* root hair cells. *Plant Physiol* 131, 976–984. https://doi.org/10.1104/pp.005546

Siegmund, U., Heller, J., van Kan, J.A., Tudzynski, P., 2013. The NADPH oxidase complexes in *Botrytis cinerea*: evidence for a close association with the ER and the tetraspanin Pls1. *PLoS ONE* 8. https://doi.org/10.1371/annotation/84c258cf-4f98-43d8-a4f0-7032cba36f22

Smirnoff, N., Arnaud, D., 2019. Hydrogen peroxide metabolism and functions in plants. *New Phytol* 221, 1197–1214. https://doi.org/10.1111/nph.15488

Snetselaar, K.M., Mims, C.W., 1992. Sporidial fusion and infection of maize seedlings by the smut fungus *Ustilago maydis*. *Mycologia* 84, 193–203. https://doi.org/10.1080/00275514.1992.12026126

Srivastava, A.K., Penna, S., van Nguyen, D., Tran, L.-S.P., 2016. Multifaceted roles of aquaporins as molecular conduits in plant responses to abiotic stresses. Critical Reviews in Biotechnology 36, 389–398. https://doi.org/10.3109/07388551.2014.973367

Steudle, E., Peterson, C.A., 1998. How does water get through roots? *J Exp Bot* 49, 775–788. https://doi.org/10.1093/jxb/49.322.775

Sudbery, P., Gow, N., Berman, J., 2004. The distinct morphogenic states of *Candida albicans*. *Trends Microbiol* 12, 317–324. https://doi.org/10.1016/j.tim.2004.05.008

Tajkhorshid, E., Nollert, P., Jensen, M.O., Miercke, L.J.W., O'Connell, J., Stroud, R.M., Schulten, K., 2002.

Control of the selectivity of the aquaporin water channel family by global orientational turning. *Science* 296, 595–530. https://doi.org/10.1029/2001GL013240

Tanghe, A., Dijck, P.V., Thevelein, J.M., 2006. Why do microorganisms have aquaporins? *Trends Microbiol* 14, 78–85. https://doi.org/10.1016/j.tim.2005.12.001

Taylor, I.B., 1991. Genetics of ABA synthesis. *Abscisic Acid Physiol Bioch*, 23–38.

Tian, S., Torres, R., Ballester, A.-R., Li, B., Vilanova, L., González-Candelas, L., 2016. Molecular aspects in pathogen-fruit interactions: virulence and resistance. *Postharvest Biol Technol* 122, 11–21. https://doi.org/10.1016/j.postharvbio.2016.04.018

Tong, H., Chen, W., Merritt, J., Qi, F., Shi, W., Dong, X., 2007. *Streptococcus oligofermentans* inhibits *Streptococcus mutans* through conversion of lactic acid into inhibitory H_2O_2: a possible counteroffensive strategy for interspecies competition. Mol Microbiol 63, 872–880. https://doi.org/10.1111/j.1365-2958.2006.05546.x

Tong, H., Wang, X., Dong, Y., Hu, Q., Zhao, Z., Zhu, Y., Dong, L., Bai, F., Dong, X., 2019. A *Streptococcus* aquaporin acts as peroxiporin for efflux of cellular hydrogen peroxide and alleviation of oxidative stress. *J Biol Chem* 294, 4583–4595. https://doi.org/10.1074/jbc.RA118.006877

Törnroth-Horsefield, S., Wang, Y., Hedfalk, K., Johanson, U., Karlsson, M., Tajkhorshid, E., Neutze, R., Kjellbom, P., 2006. Structural mechanism of plant aquaporin gating. *Nature* 439, 688–694. https://doi.org/10.1038/nature04316

Turner, N.C., Wright, G.C., Siddique, K.H.M., 2001. Adaptation of grain legumes (pulses) to water-limited environments. *Adv Agron* 71, 123–231. https://doi.org/10.1016/S0065-2113(01)71015-2

Turrà, D., Nordzieke, D., Vitale, S., El Ghalid, M., Di Pietro, A., 2016. Hyphal chemotropism in fungal pathogenicity. *Semin Cell Dev Biol* 57, 69–75. https://doi.org/10.1016/j.semcdb.2016.04.020

Tyerman, S.D., Bohnert, H.J., Maurel, C., Steudle, E., Smith, J.A.C., 1999. Plant aquaporins: their molecular biology, biophysics and significance for plant water relations. *J Exp Bot* 50, 1055–1071. https://doi.org/10.1093/jxb/50.Special_Issue.1055

Uehlein, N., Lovisolo, C., Siefritz, F., Kaldenhoff, R., 2003. The tobacco aquaporin NtAQP1 is a membrane CO_2 pore with physiological functions. *Nature* 425, 734–737. https://doi.org/10.1038/nature02027

van Baarlen, P., Staats, M., van Kan, J.A.L., 2004. Induction of programmed cell death in lily by the fungal pathogen *Botrytis elliptica*. *Mol Plant Pathol* 5, 559–574. https://doi.org/10.1111/j.1364-3703.2004.00253.x

Van De Walle, J., Sergent, Z., Piront, Z., Toussaint, O., Schneider, Y.-J., Larondelle, Y., 2010. Deoxynivalenol affects *in vitro* intestinal epithelial cell barrier integrity through inhibition of protein synthesis. *Toxicol Appl Pharmacol* 245, 291–298.

Vance, C.P., Kirk, T.K., Sherwood, R.T., 1980. Lignification as a mechanism of disease resistance. *Ann Revi Phytopathol* 18, 259–288. https://doi.org/10.1146/annurev.py.18.090180.001355

Varma, A., 2008. *Mycorrhiza: State of the art, genetics and molecular biology, eco-function, biotechnology, eco-physiology, structure and systematics*. Springer, Heidelberg.

Veloso, J., van Kan, J.A.L., 2018. Many shades of grey in *Botrytis*-host plant interactions. *Trends Plant Sci* 23, 613–622. https://doi.org/10.1016/j.tplants.2018.03.016

Voigt, C.A., 2014. Callose-mediated resistance to pathogenic intruders in plant defense-related papillae. *Front Plant Sci* 5, 168. https://doi.org/10.3389/fpls.2014.00168

von Tiedemann, A., 1997. Evidence for a primary role of active oxygen species in induction of host cell death during infection of bean leaves with *Botrytis cinerea*. *Physiol Mol Plant Pathol* 50, 151–166.

von Tiedemann, A., Koopmann, B., Hoech, K., 2021. Lignin composition and timing of cell wall lignification are involved in *Brassica napus* resistance to *Sclerotinia sclerotiorum* stem rot. *Phytopathology*, 111, 1438–1448. https://doi.org/10.1094/PHYTO-09-20-0425-R

Wang, C., El-Shetehy, M., Shine, M.B., Yu, K., Navarre, D., Wendehenne, D., Kachroo, A., Kachroo, P., 2014. Free radicals mediate systemic acquired resistance. *Cell Rep* 7, 348–355. https://doi.org/10.1016/j.celrep.2014.03.032

Wang, H., Schoebel, S., Schmitz, F., Dong, H., Hedfalk, K., 2020. Characterization of aquaporin-driven hydrogen peroxide transport. *Biochim Biophysica Acta Biomembr* 1862, 183065. https://doi.org/10.1016/j.bbamem.2019.183065

Wang, X., Lu, K., Yao, X., Zhang, L., Wang, F., Wu, D., Peng, J., Chen, X., Du, J., Wei, J., Ma, J., Chen, L., Zou, S., Zhang, C., Zhang, M., Dong, H., 2021. The aquaporin TaPIP2;10 confers resistance to two fungal diseases in wheat. *Phytopathology®* 111, 2317–2331. https://doi.org/10.1094/PHYTO-02-21-0048-R

Waszczak, C., Carmody, M., Kangasjärvi, J., 2018. Reactive oxygen species in plant signaling. *Ann Revi Plant Biol* 69, 209–236. https://doi.org/10.1146/annurev-arplant-042817-040322

Xu, H., Cooke, J.E.K., Zwiazek, J.J., 2013. Phylogenetic analysis of fungal aquaporins provides insight into their possible role in water transport of mycorrhizal associations. *Botany* 91, 495–504. https://doi.org/10.1139/cjb-2013-0041

Xu, L., Li, T., Wu, Z., Feng, H., Yu, M., Zhang, X., Chen, B., 2018. Arbuscular mycorrhiza enhances drought tolerance of tomato plants by regulating the 14-3-3 genes in the ABA signaling pathway. *Appl Soil Ecol* 125, 213–221. https://doi.org/10.1016/j.apsoil.2018.01.012

Yadeta, K.A., Thomma, B.P.H.J., 2013. The xylem as battleground for plant hosts and vascular wilt pathogens. *Front Plant Sci* 4, 97. https://doi.org/10.3389/fpls.2013.00097

Yi, H., Polanco, M.C., MacKinnon, M.D., Zwiazek, J.J., 2008. Responses of ectomycorrhizal *Populus tremuloides* and *Betula papyrifera* seedlings to salinity. *Environ Exp Bot* 62, 357–363. https://doi.org/10.1016/j.envexpbot.2007.10.008

Yu, Y., Li Jiao, Fu, S., Yin, L., Zhang, Y., Lu, J., 2016. Callose synthase family genes involved in the grapevine defense response to Downy mildew disease. *Phytopathology* 106, 56–64. https://doi.org/10.1094/PHYTO-07-15-0166-R

Zardoya, R., 2005. Phylogeny and evolution of the major intrinsic protein family. *Biol Cell* 97, 397–414. https://doi.org/10.1042/BC20040134

Zeier, J., 2021. Metabolic regulation of systemic acquired resistance. *Curr Opin Plant Biol* 62, 102050. https://doi.org/10.1016/j.pbi.2021.102050

Zelazny, E., Borst, J.W., Muylaert, M., Batoko, M., Hemminga, M.A., Chaumont, F., 2007. FRET imaging in living maize cells reveals that plasma membrane aquaporins interact to regulate their subcellular localization. *Proc Natl Acad Sci U S A* 104, 12359–12364. https://doi.org/10.1073/pnas.0701180104

Zhang, Z., Chen, Y., Li, B., Chen, T., Tian, S., 2020. Reactive oxygen species: a generalist in regulating development and pathogenicity of phytopathogenic fungi. *Comput Struct Biotechnol J* 18, 3344–3349. https://doi.org/10.1016/j.csbj.2020.10.024

Zhao, Y.L., Zhou, T.T., Guo, H.S., 2016. Hyphopodium-specific VdNoxB/VdPls1-dependent ROS-Ca^{2+} signaling is required for plant infection by *Verticillium dahliae*. *PLoS Pathog* 12, e1005793. https://doi.org/10.1371/journal.ppat.1005793

Zhou, J., Yu, J.-Q., Chen, Z., 2014. The perplexing role of autophagy in plant innate immune responses. *Mol Plant Pathol* 15, 637–645. https://doi.org/10.1111/mpp.12118

CHAPTER TEN

Role of Peroxiporins in Plant Stress Signaling and Development

Stephanie Frohn, Bernd H. Dreyer, Romy R. Schmidt-Schippers, and Jos H. M. Schippers

ABBREVIATIONS

AQP	Aquaporin
CO_2	Carbon dioxide
CZ	Central zone
ER	Endoplasmic reticulum
EZ	Elongation zone
GIP	GlpF-like intrinsic protein
H_2O_2	Hydrogen peroxide
HIP	Hybrid intrinsic protein
LIP	Large intrinsic protein
MZ	Meristematic zone
NIP	NOD26-like intrinsic protein
NOX	NADPH oxidase
OC	Organizing center
PIP	Plasma membrane intrinsic protein
PRX	Peroxidase
PZ	Peripheral zone
QC	Quiescent center
RBOH	Respiratory burst homolog
ROS	Reactive oxygen species
RZ	Rib zone
SAM	Shoot apical meristem
SIP	Small basic intrinsic protein
SOD	Superoxide dismutase
TIP	Tonoplast intrinsic protein
TZ	Transition zone
XIP	X intrinsic protein

10.1 INTRODUCTION

After oxygen became abundant in the atmosphere of our planet, life on earth had to evolve in the presence of reactive oxygen species (ROS). Hence, it is not surprising that today complex relationships exist between ROS and developmental-physiological processes. Among ROS, hydrogen peroxide (H_2O_2) was the first to be discovered around 200 years ago (Thénard, 1818). Its sterilizing nature made it a useful compound for the pharmaceutical industry, while its oxidizing features made it a valuable bleaching agent. The harsh nature of H_2O_2 gave rise to the idea that living organisms need to fight against oxidative stress, which was further encouraged by the discovery of the enzymes catalase and peroxidase in the 19th century (Sies, 2017). Until today, therefore, ROS has had a bad reputation as cell-killing agents. In past decades, however, compelling evidence accumulated in favor of a wide spectrum of essential regulatory functions. Indeed, ROS are now well established as universal signaling molecules in all kingdoms of life (Schippers et al., 2012; Lee et al., 2018; Sies and Jones, 2020). Moreover, in multicellular organisms, ROS gradients control stem cell fate, cell proliferation, and differentiation (Yang et al., 2018; Tan and Suda, 2018). Furthermore, the subcellular distribution of ROS is not even,

DOI: 10.1201/9781003160649-10

but each organelle and compartment has its own steady-state level. The cytosol has the lowest level of H_2O_2. Higher concentrations are found in organelles like chloroplast and mitochondria. The ER, the main site of oxidative protein folding, has even higher levels of H_2O_2. In plant cells the apoplast has the highest levels (Sies and Jones, 2020; Niemeyer et al., 2021). The distinction in H_2O_2 levels gives each compartment a specific signature which can readily be impaired by altered metabolic activity or due to external stresses, allowing for a stress read-out for the cell. Alterations in H_2O_2 levels in one cell, or one subcellular compartment will cause the formation of local ROS gradients that can entail diffusion and transport of H_2O_2, thereby potentially triggering or amplifying signal cascades. The active transport of H_2O_2 across biological membranes by water channels was first described for the multicellular freshwater algae *Chara corallina* (Henzler and Stuedle, 2000). This type of aquaporins (AQPs) were first designated "peroxoporins." Later on, this name was changed to "peroxiporins." The gradient-dependent transport of H_2O_2 across membranes might indicate for a role of peroxiporins in signaling between cellular compartments or cells. In addition, transport of H_2O_2 by peroxiporins might influence developmental processes as tissue gradients are known to modulate cell proliferation and differentiation.

AQPs form membrane channels to facilitate water transport across the plasma membrane or organellar membranes in nearly all living organisms. Although the pores are named after the molecule they transport the most, they also enable the movement of other neutral molecules, including H_2O_2, carbon dioxide (CO_2), glycerol, ammonium, urea, and metalloids like arsenic acid, boric acid, and silicic acid (Maurel et al., 2015; Ahmed et al., 2021). In contrast to mammals, which are known to encode for 13 AQP isoforms, plant genomes encode for more isoforms (>30 to >100) that vary in their tissue specific expression, subcellular localization, transport selectivity, and regulation properties (Laloux et al., 2018). Potentially, the sessile nature of plants and the uptake of many different solutes might explain the increased number and versatility of plant AQPs. Opening and closing of the pores is regulated at the posttranslational level through, for example, phosphorylation, methylation, and glycosylation (Santoni, 2017). In principle, two categories of AQPs exist, water-selective pores/channels and glycerol-facilitator channels. During the course of plant evolution, AQPs have diversified resulting in eight different sub-classes according to their sequence (Li et al., 2014; Hussain et al., 2020). AQP subclasses found in plants include plasma membrane intrinsic proteins (PIPs), tonoplast intrinsic proteins (TIPs), uncategorized X intrinsic proteins (XIPs), NOD26-like intrinsic proteins (NIPs), small basic intrinsic proteins (SIPs), algae-specific large intrinsic proteins (LIPs), and moss-specific (non-vascular plants) hybrid intrinsic proteins (HIPs) and GlpF-like intrinsic proteins (GIPs). Peroxiporins are reported among the different subclasses in the model plant *Arabidopsis thaliana*. The PIP family is subdivided into PIP1 and PIP2 clades, of which only members from the latter (AtPIP2;1, AtPIP2;2, AtPIP2;4, AtPIP2;5, and AtPIP2;7) are known to transport H_2O_2. Interestingly, their expression is downregulated upon H_2O_2 treatment (Hooijmaijers et al., 2012; Rodrigues et al., 2017; Wang et al., 2020). Among TIPs, AtTIP1;1; AtTIP1;2, and AtTIP2;3 act as peroxiporins as they were shown to be permeable for H_2O_2 (Bienert et al., 2007; Dynowski et al., 2008). Furthermore, the plasma-membrane localized AtNIP1;2 was found to increase the sensitivity of yeast toward H_2O_2 (Dynowski et al., 2008; Wang et al., 2017). The SIP-type AQPs are mainly localized to the ER (Ishikawa et al., 2005), which represents an oxidative environment with high levels of H_2O_2. So far, plant SIPs have not been reported to act as peroxiporins, however, human orthologues like AQP11 are also ER localized and were shown to release H_2O_2 from the ER (Bestetti et al., 2020). Therefore, plant SIPs might have a role in monitoring oxidative protein folding in the ER, or in downstream signaling events by locally releasing H_2O_2 in the cytosol. Also, among XIP AQPs, which mainly transport glycerol, peroxiporin activity has been demonstrated for the plasma membrane–located *Solanaceae* XIPs (Bienert et al., 2011). For the moss-specific HIP and GIP AQPs, thus far no information about possible H_2O_2 transport is known, although homologous bacterial GlpF proteins were shown to be H_2O_2 permeable (Bienert et al., 2013). Taken together, although a substantial number of peroxiporins have been identified in plants, the exact impact of H_2O_2 transport by these channels on physiological or developmental processes is poorly understood. Recently, it has become apparent that ROS gradients in plant tissues regulate stemness and cell differentiation (Tsukagoshi et al., 2010; Zeng et al., 2017; Yang et al., 2018). Here we review and discuss the current knowledge on these gradients in the root and shoot apical meristem of the plant and a possible role for peroxiporins in maintaining these gradients. In addition, ROS signaling is intimately related to stress Perception by plants, with a central role for H_2O_2 as secondary messenger. Within this chapter we will present several examples of stress signaling in plants with a link to peroxiporins.

10.2 ROOT DEVELOPMENT AND ROS GRADIENTS

The role for ROS as a developmental regulator in plants is well established for root development and growth. Plant roots show typical indeterminate growth that is maintained by an active meristematic zone that continuously produces new root cells resulting in an extensive underground system (Motte et al., 2019). Characteristically, ROS types are distinctly distributed along the root (Figure 10.1). The root tip contains the meristematic zone (MZ), which is typified by increased superoxide levels while it is largely devoid of H_2O_2 (Tsukagoshi et al., 2010). The meristematic zone is maintained by a small population of cells, the quiescent center (QC), which prevents differentiation of the stem cells but does not divide or does so only very slowly (Motte et al., 2019). The QC accumulates the phytohormone auxin, which promotes an oxidizing redox status by activation of ascorbate oxidase (Jiang et al., 2003; Dubrovsky and Ivanov, 2021). Mutants in ascorbate synthase display an increased number of QC cells and an

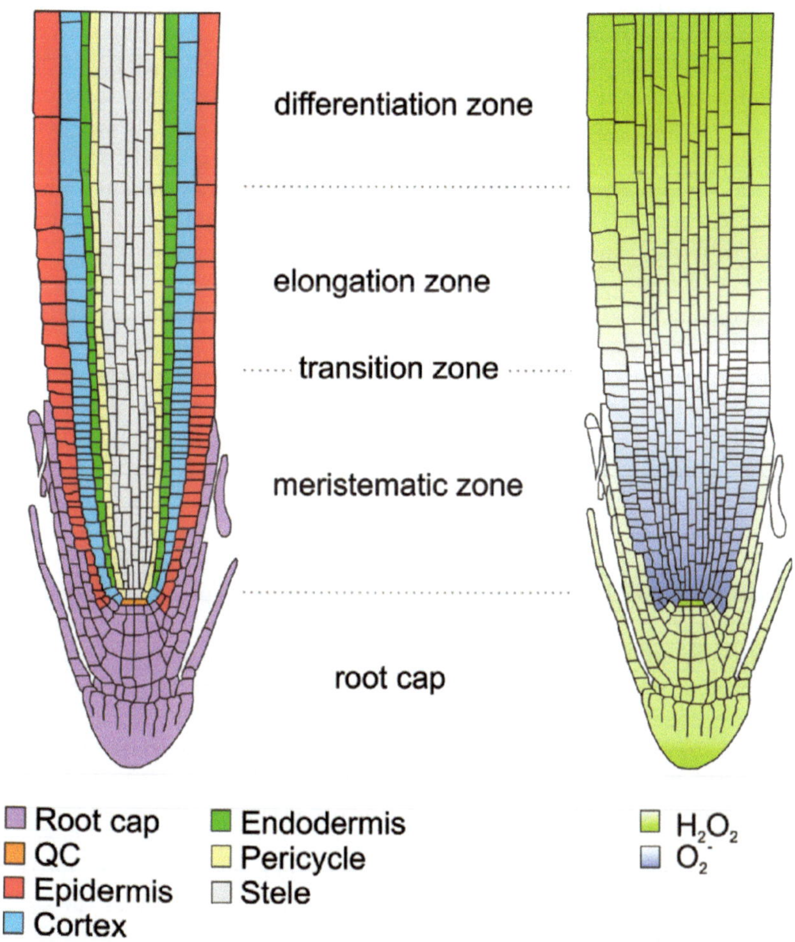

Figure 10.1 ROS gradients in the root. Left shows a schematic overview of the different root zones. Cell types are indicated as listed in the legend. On the right, the O_2^- and H_2O_2 gradients along the different root zones are shown.

elevated level of H_2O_2 (Kka et al., 2018). In addition, a P-loop NTPase is required to prevent QC cell division by maintaining an oxidized environment (Yu et al., 2016). In the wild-type situation, the QC-surrounding stem cells maintain a low level of H_2O_2, suggesting that transport of H_2O_2 from the QC into the surrounding cells is prevented.

For the root of the model plant *Arabidopsis*, several expression atlases have been established, allowing for the analysis of peroxiporin expression along the root. *PIP2* peroxiporins appear to be expressed mainly in the root zones that have higher levels of H_2O_2, including the QC, EZ, and DZ, but not in the MZ (Birnbaum et al., 2003; Nawy et al., 2005; Hachez et al., 2014). *PIP1* AQPs show a similar expression as the *PIP2* peroxiporins, except that several members already show a relative high expression in the MZ. The expression in the QC of peroxiporins is of interest as AQPs act bidirectionally and molecules move along a concentration gradient. Potentially the peroxiporins in the QC show a polar distribution allowing only for H_2O_2 transport between QC cells but not with the surrounding stem cells (Chevalier and Chaumont, 2015). The expression of *TIP* peroxiporins is similar to that of *PIP2*-type peroxiporins, indicating that their expression correlates with the H_2O_2 gradient in the root. To which extent peroxiporins maintain the ROS root gradient is thus far not known. For example, the *Arabidopsis* double mutant *tip1;1tip1;2* is vital (Schüssler et al., 2008), but the root meristem phenotype was not studied in these lines.

In general, growth and development of the root is maintained by constant cell divisions in the MZ (Motte et al., 2019). The ROS gradient along the root might affect growth by modifying redox sensitive proteins that influence cell proliferation and differentiation. However, very little is known about these proteins. Several regulators of the ROS gradient along the root have been identified. The high level of superoxide in the MZ is at least in part due to mitochondrial activity which requires the *Arabidopsis thaliana* P-loop NTPase APP1 (Yu et al., 2016). Considering that superoxide can spontaneously dismutate to H_2O_2, it is intriguing to understand how the proliferating root cells maintain low levels of H_2O_2. Recently, the root meristem growth factor 1 (RGF1) was found to promote an increase in root meristem size through the modulation of the root ROS gradient by the transcription factor GF1-INDUCIBLE TRANSCRIPTION FACTOR 1 (RITF1; (Yamada et al., 2020). One of the downstream targets of the RGF1-RITF1 pathway is the transcription factor PLETHORA2 (PLT2), which is a master regulator of root stem cell identity and appears to be stabilized by a post-transcriptional modification through ROS (Du and Scheres, 2017; Yamada et al., 2020). Another protein involved in the maintenance of the root ROS gradient is PROHIBITIN 3 (PHB3), which is required for maintaining cell division activity in the root meristem (Kong et al., 2018). Like APP1, PHB3 is mitochondrial localized; however, its loss causes an overaccumulation of ROS which strongly reduces root growth. The accumulation of H_2O_2 results in the upregulation of several ERF transcription factors that stimulate cellular differentiation of the meristematic cells and division of QC cells.

The transition zone (TZ) marks the switch from proliferation to elongation, which is noticeable by a shift from a high level of superoxide to a high level of H_2O_2. The transition between these two ROS types is regulated by the bHLH transcription factor UPBEAT1 (UPB1), which among other targets downregulates the expression of peroxidases in the transition zone (Tsukagoshi et al., 2010). In contrast, peroxiporins are not direct targets of UPB1, but they are targeted for instance by BRASSINAZOLE RESISTANT 1 (BZR1) and auxin response factor 6 (ARF6), suggesting they might act downstream of phytohormone signaling pathways (Sun et al., 2010; Oh et al., 2014). Subsequent cell elongation in the EZ dependents on the plasticity and composition of the primary cell wall (Schmidt et al., 2016). Cell expansion requires cell wall loosening and stiffening which is controlled in part by apoplastic ROS levels. Cell wall stiffening processes utilize apoplastic H_2O_2 through, for instance, peroxidases (PRX) that by a peroxidative cycle oxidize phenolic cell wall components (Dunand et al., 2007). On the other hand, a group of enzymes called NADPH oxidases (NOX) actively produce apoplastic superoxide, which is rapidly dismutated to H_2O_2. Through the hydroxylic cycle, PRXs form hydroxyl radicals utilizing H_2O_2 (Schopfer et al., 2002). The emerging hydroxyl radical catalyzes the oxidative cleavage of xyloglycan and other pectins resulting in cell wall loosening and cell expansion. This dual role of ROS in the determination of cell wall plasticity suggest that the balance of superoxide and H_2O_2 is a key factor in growth regulation (Schmidt et al., 2016).

The differentiation and growth of a root hair requires superoxide production by plasma membrane localized NOX proteins (Foreman et al., 2003). Furthermore, apoplastic H_2O_2 is utilized by peroxidases catalyzing the cross-linking of lignin monomers (Francoz et al., 2015; Tsukagoshi, 2016). Lignification is one characteristic of differentiation and supports the mechanical strength of plant cell walls. Additionally, in the endodermis, the formation of the so-called casparian strip, containing lignin and suberin, serves as a diffusion barrier in plant roots (Lee et al., 2013). The formation of the casparian strip depends on CASP proteins which together with peroxidases, NOX and superoxide dismutase (SOD) guide the deposition of lignin (Lee et al., 2013, Rojas-Murcia et al., 2020). Furthermore, ROS gradients are involved in lateral root development and emergence in the differentiation zone of the root. For instance, the size of the lateral root primordium is controlled in part by the transcription factor MYB36 through balancing ROS homeostasis (Fernández-Marcos et al., 2017). Additionally, UPB1 is active during early stages of lateral root formation, regulating ROS homeostasis by controlling *PRX* genes (Tsukagoshi et al., 2010). During lateral root formation most AQP genes are repressed by auxin (Péret et al., 2012). The expression of peroxiporin *PIP2;1* is progressively excluded from the site of the auxin response maximum in lateral root primordia, which is in line with the requirement of high superoxide levels in the newly established meristematic zone and exclusion of H_2O_2. Still a potential role for peroxiporins in these differentiation processes have not been reported. As peroxiporins transport molecules along a concentration gradient, it is likely that they allow for passage of apoplastic ROS to the cytosol for potential signaling and monitoring processes.

Next to the transfer of neutral ROS molecules by peroxiporins, it would be very interesting to understand if apoplastic superoxide is transported across the plasma membrane. So far, direct diffusion through chloride channels seems to be the best option for this (Moreland et al., 2006; Hawkins et al., 2007). The different mechanisms of ROS transport in plants need more investigation in the future, especially when considering the role of these channels during development.

10.3 ROS GRADIENTS AT THE SHOOT APICAL MERISTEM

The above-ground organs of a plant derive from stem cells that reside in the shoot apical meristem (SAM), except for cotyledons which like the SAM are established during embryogenesis. The current view on apical meristems is that they typically function as a signal integrator, coordinating cues from elsewhere in the plant and the environment to optimize growth and architecture of the plant shoot (Fouracre and Phoetig, 2020). In *Arabidopsis* the SAM is organized in distinct functional domains: the central zone (CZ), the peripheral zone (PZ), the organizing center (OC), and the rib zone (RZ; Figure 10.2A). The stem cells reside within the CZ. After a stem cell division, one daughter cell remains a stem cell while the second daughter cell will differentiate and becomes part of the surrounding PZ. Inside the PZ, the cells enter developmental pathways that lead to differentiation into lateral organs such as leaves, or to initiation of secondary meristems (e.g., floral meristems; Kitagawa and Jackson, 2019). CZ-derived cells that displace basally from the CZ to the RZ will differentiate into cells of the stem.

Proliferation of stem cells within the SAM is maintained by two distinct families of homeobox transcription factors, the WUSCHEL-LIKE (WOX) and KNOTTED-LIKE (KNOX). In *Arabidopsis* the loss of the KNOX gene *SHOOT MERISTEMLESS* (*STM*) results in premature differentiation of stem cells and the loss of the SAM during the seedling stage (Long et al., 1996). Stem cell proliferation by STM is promoted through the activation of cytokinin synthesis, a phytohormone that promotes cell division (Yanai et al., 2005). In addition, the organization of the SAM is established and maintained by the CLAVATA3(CLV3)/WUSCHEL(WUS)-feedback loop. In Arabidopsis *WUS* is expressed in the OC of the CZ to regulate the division of the surrounding stem cells (Zhou et al., 2015). WUS is a mobile transcription factor that moves to the CZ where one of its targets is *CLV3*. The activation of the CLAVATA signaling cascade in the CZ limits the expression of *WUS* to the OC. This feedback loop and the integration of other signaling

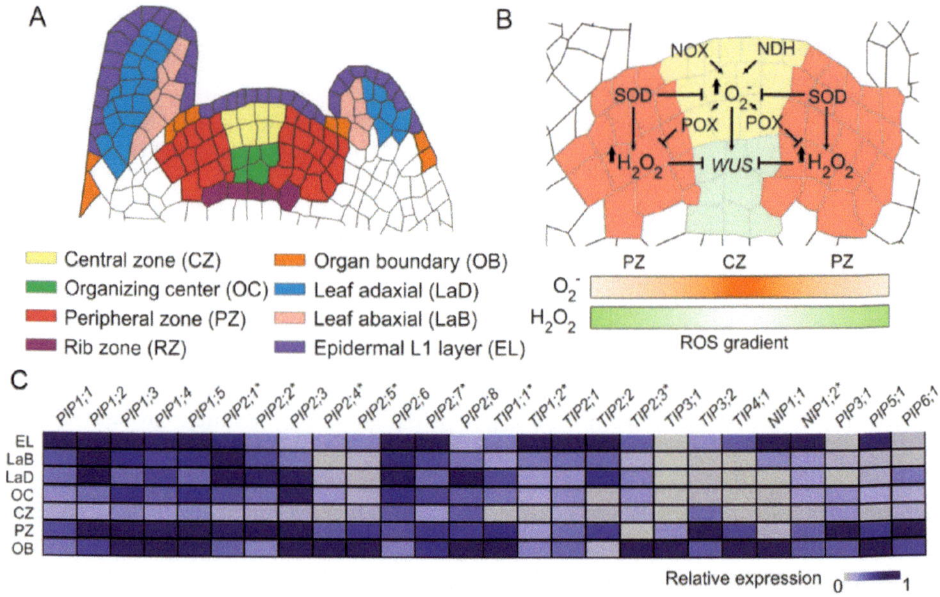

Figure 10.2 ROS gradients in the shoot apical meristem. (A) Schematic overview of the SAM. Tissue zones are indicated as listed in the legend. (B) The ROS gradient in the SAM contributes the maintenance of the stem cells. The accumulation of superoxide (O_2^-) in the central zone (CZ) activates the expression of WUS, which promotes maintenance of stem cells. NADPH oxidases (NOX) and NADH dehydrogenases (NDH) are responsible for the buildup of O_2^- in the CZ. In addition, SODs are not expressed in the CZ, but in the peripheral zone (PZ), where they convert O_2^- to hydrogen peroxide (H_2O_2), which promotes cell differentiation and restricts the expression of WUS. Within the CZ, H_2O_2 formation is counteracted by peroxidases (POX), which thereby prevent precocious differentiation. (C) Shown is the relative expression of AQPs within the SAM as based on (Tian et al., 2014). Peroxiporins (*) show a very similar expression as the non-H_2O_2 transporting AQPs.

molecules defines plant stem cell fate and was therefore considered as the heartbeat of SAM architecture (Weigel and Jürgens, 2002).

The seminal study of Zeng et al. (2017) adds the concept of ROS gradients as the heartbeat that defines the SAM architecture. The CZ of the SAM was shown to contain high levels of superoxide while the PZ is characterized by increased levels of H_2O_2 (Figure 10.2B). Superoxide induces the expression of *WUS*, thereby promoting stem cell maintenance. The accumulation of H_2O_2 in the PZ promotes the differentiation of stem cells by repressing *WUS*. The SAM ROS gradient is maintained through differential expression of ROS scavenging and producing enzymes. In the CZ, the expression of SODs is repressed to prevent enzymatic conversion of superoxide to H_2O_2 (Zeng et al., 2017). In addition, the inevitable formation of H_2O_2 in the CZ is counteracted through the specific upregulation of several peroxidase genes in stem cells. Moreover, the activity of mitochondrial NADH oxidases and plasma membrane–localized NOXs was found to be crucial for maintaining high superoxide levels in the CZ (Zeng et al., 2017). Within the PZ, the expression of SODs is upregulated, resulting in increased levels of H_2O_2 and the promotion of cell differentiation. Based on the primary and central role of ROS in the SAM, it will be interesting to explore if ROS are involved in specifying the SAM during embryogenesis and how ROS signals are integrated with other signaling molecules like the phytohormones cytokinin and auxin that play important roles in SAM maintenance.

ROS homeostasis inside the SAM is maintained by ROS scavenging and producing mechanisms. Although H_2O_2 is a mobile molecule, a sharp gradient exists between the PZ and the CZ, suggesting that transport of H_2O_2 between these two zones of the SAM is restricted. Peroxiporins might play a role in maintaining the observed SAM ROS gradient. Overexpression of the transcription factor TRANSLUCENT GREEN (TG) results in a reduced plant growth and the upregulation of several AQPs (Zhu et al., 2014). Although the impact on the SAM was not studied in detail, a clear growth defect was observed. It was found that TG activates the expression of several peroxiporins (AtTIP1;1, AtTIP2;3, and AtPIP2;2). These results suggest that misexpression of AQPs affects SAM size. That said, expression analysis of *AQP* genes within the SAM reveals a very similar expression pattern regarding the stem cell niche (Figure 10.2C). Irrespective of having peroxiporin activity, *AQP* gene expression within the central zone is very low. Among the peroxiporins, only *PIP2;2* shows high expression in the peripheral zone (PZ), while all others show maximum expression in either the epidermal layer or the organ boundary. The lower abundance of peroxiporins in the PZ might prevent unwanted H_2O_2 diffusion to the CZ. Considering that the PZ has high levels of H_2O_2 and that AQPs allow transport from a high to a low concentration, it is more likely that regulation of peroxiporins occurs at posttranslational and/or topological levels, instead of at the expression level. Potentially peroxiporins are inserted in a polar manner to facilitate the direction of H_2O_2 transport, and thereby might help to establish or maintain a ROS gradient. Therefore, we can expect that further research on the redox regulation of plant stem cell fate likely also include peroxiporins in *Arabidopsis*.

10.4 ROS TRANSPORT AND SYSTEMIC STRESS SIGNALING IN PLANTS

Plants continuously need to monitor their surroundings to be able to timely adjust their growth upon environmental constraints. Temperature, humidity, and light can change within seconds or minutes (Kollist et al., 2019), requiring that plants also sense and respond within these time frames. In multicellular organisms, systemic signaling pathways allow for priming of the whole organisms for an upcoming challenge that is initially only sensed by a few cells. Most stresses provoke a local ROS burst that acts as a warning signal, triggering cellular communication and organism-wide adaptation responses. The ROS wave describes a systemic cell-to-cell transmission of ROS signals (Fichman and Mittler, 2020). The initial ROS that is produced during this burst by respiratory burst homolog (RBOH or NOX) proteins was shown to diffuse from the site of production to adjacent cells (Miller et al., 2009), thereby activating in these cells also a ROS burst to amplify and transduce the signal throughout the plant (Figure 10.3). The apoplastic ROS, RBOH-derived H_2O_2 diffuses into the cytosol and will diffuse from cell to cell through plasmodesmata (Fichman et al., 2021). The cytosolic accumulation of the apoplastic H_2O_2 signal relies on peroxiporins. To determine

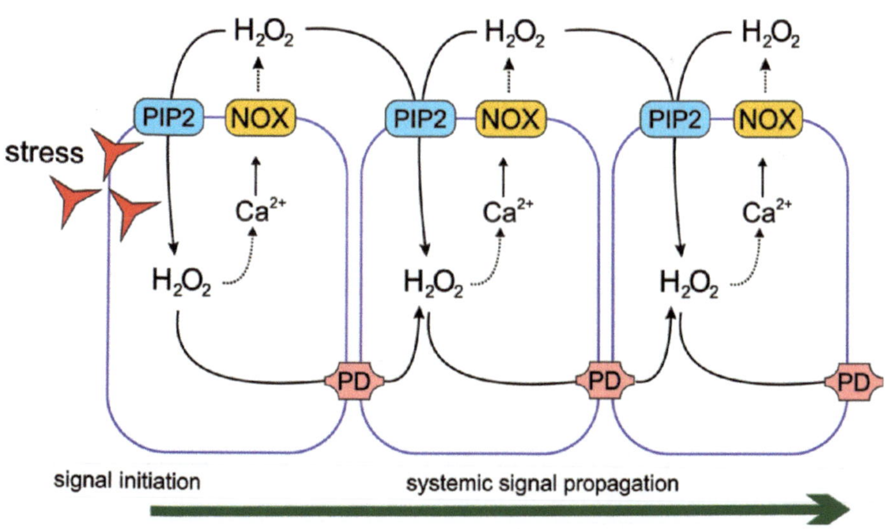

Figure 10.3 Peroxiporin-mediated systemic signaling in plants. In the model shown here, a stress is initially sensed by a single cell, resulting in the activation of a ROS burst through the action of NOX proteins. The apoplastic accumulated H_2O_2 enters the cytosol through peroxiporin PIP2;1, resulting in signal amplification through the activation of Ca^{2+} signals that further boost the ROS production by RBOHD. Subsequent systemic signal propagation relies on the transfer of the H_2O_2 signal to neighboring cells via either the apoplast or via plasmodesmata (PD). The neighboring cells will amplify the received signal through a ROS burst and transfer the signal further to the next cells.

whether systemic signaling in *Arabidopsis* requires an aquaporin-associated apoplastic route for ROS wave initiation and/or propagation, several *PIP* mutants were analyzed. This analysis revealed that the peroxiporin *PIP2;1* is required for initiation and propagation of the rapid systemic ROS signal (Rodrigues et al., 2017; Fichman et al., 2021). It was suggested that the RBOHD-produced ROS accumulate at the apoplast and enter cells through PIP2;1. The H_2O_2 influx will activate Ca^{2+} influx, which in turn further activates the RBOHD production of ROS to amplify the signal. In animal cells it has been reported that peroxiporins are required for signal amplification (Bertolotti et al., 2016), which based on the current data in plants appears to be an evolutionary conserved mechanism.

Acclimation to fluctuating light conditions in plants involves communication between the chloroplast and the nucleus. Recently it was shown that chloroplast directly relay a H_2O_2 signal to the nucleus through close association, avoiding the cytosol, thereby promoting transcriptional adjustment (Exposito-Rodriguez et al., 2017). The movement of H_2O_2 across the chloroplast membrane is considered to occur by diffusion down a concentration gradient facilitated by peroxiporins. Isolated chloroplasts exposed to high light intensities secrete H_2O_2, and this is blocked by the AQP inhibitor acetazolamide (Mubarakshina Borisova et al., 2012). Interestingly, the peroxiporins TIP1;1 and PIP2;1 have been detected in the chloroplast envelope by proteome studies (Simm et al., 2013). Whether these AQPs or others are responsible for the release of the ROS signal from chloroplast needs to be clarified.

Next to their active role in ROS signaling, it has been suggested that peroxiporins in the vacuolar membrane are involved in ROS homeostasis (Smirnoff and Arnaud, 2019). The vacuole represents the largest plant organelle that accumulates secondary compounds, inorganic ions, as well as sugars, amino acids and organic acids. In addition, the vacuole has a strong reducing capacity as it contains ascorbate and a whole battery of type III peroxidases. Therefore, it is likely that the vacuole assists in reducing cytosolic H_2O_2 levels by uptake through tonoplast peroxiporins and subsequent scavenging. This hypothesis is in line with the observation that RNA interference against tonoplast peroxiporins causes early cell death (Ma et al., 2004).

During leaf senescence, which represents a kind of programmed cell death, specialized vacuoles with high proteolytic capacity emerge (Otegui et al., 2005). Interestingly, these senescence-associated vacuoles lack TIP AQPs, which indicates that they are no longer capable of taking up H_2O_2 from the cytosol. A role for TIP peroxiporins in oxidative stress tolerance is also demonstrated by the overexpression of *TIP1;2* from *Thellungiella salsuginea* in *Arabidopsis* (Wang et al., 2014). TsTIP1;2 mediates H_2O_2 transport and was shown to improve oxidative stress tolerance when overexpressed. In addition, plants overexpressing *TsTIP1;2* are also more tolerant to abiotic stress that cause ROS accumulation, like salt and drought stress. These results underlie a role for the regulation of ROS homeostasis by TIP peroxiporins. That said, osmotic adjustment through water transport to maintain cell turgor pressure under drought or salt stress is also a vital aspect to which TIP AQPs contribute.

10.5 CONCLUSION

Research on H_2O_2 transport by AQPs in plants have resulted in the identification of numerous peroxiporins. This has been mainly achieved through heterologous expression experiments in yeast, while a detailed role for the H_2O_2 transporting function of these peroxiporins is still poorly explored. Only recently is has become apparent that specific peroxiporins are required for systematic signaling. However, much is still to be learned about the role of the other peroxiporins during development, stress signaling, and the maintenance of ROS homeostasis. The thus far uncovered roles of peroxiporins in plants indicates that many novel exciting mechanisms and functions await to be discovered.

10.6 ACKNOWLEDGMENTS

The authors would like to thank the Leibniz Institute of Plant Genetics and Crop Plant Research and the University of Bielefeld for their support.

REFERENCES

Ahmed S, Kouser S, Asgher M, Gandhi SG (2021) Plant aquaporins: A frontward to make crop plants drought resistant. *Physiol Plant* **172**: 1089–1105.

Bertolotti M, Farinelli G, Galli M, Aiuti A, Sitia R (2016) AQP8 transports NOX2-generated H_2O_2 across the plasma membrane to promote signaling in B cells. *J Leukoc Biol* **100**: 1071–1079.

Bestetti S, Galli M, Sorrentino I, Pinton P, Rimessi A, Sitia R, Medraño-Fernandez I (2020). Human aquaporin-11 guarantees efficient transport of H_2O_2 across the endoplasmic reticulum membrane. *Redox Biol* **28**: 101326.

Bienert GP, Bienert MD, Jahn TP, Boutry M, Chaumont F (2011) Solanaceae XIPs are plasma membrane aquaporins that facilitate the transport of many uncharged substrates. *Plant J* **66**: 306–317.

Bienert GP, Desguin B, Chaumont F, Hols P (2013) Channel-mediated lactic acid transport: a novel function for aquaglyceroporins in bacteria. *Biochem J* **454**: 559–570.

Bienert GP, Møller AL, Kristiansen KA, Schulz A, Møller IM, Schjoerring JK, Jahn TP (2007) Specific aquaporins facilitate the diffusion of hydrogen peroxide across membranes. *J Biol Chem* **282**: 1183–1192.

Birnbaum K, Shasha DE, Wang JY, Jung JW, Lambert GM, Galbraith DW, Benfey PN (2003) A gene expression map of the Arabidopsis root. *Science* **302**: 1956–1960.

Chevalier AS, Chaumont F (2015) Trafficking of plant plasma membrane aquaporins: multiple regulation levels and complex sorting signals. *Plant Cell Physiol* **56**: 819–829.

Du Y, Scheres B (2017) PLETHORA transcription factors orchestrate de novo organ patterning during Arabidopsis lateral root outgrowth. *Proc Natl Acad Sci U S A* **114**: 11709–11714.

Dubrovsky JG, Ivanov VB (2021) The quiescent centre of the root apical meristem: Conceptual developments from Clowes to modern times. *J Exp Bot* **72**: 6687–6707.

Dunand C, Crèvecoeur M, Penel C (2007) Distribution of superoxide and hydrogen peroxide in Arabidopsis root and their influence on root development: possible interaction with peroxidases. *New Phytol* **174**: 332–341.

Dynowski M, Schaaf G, Loque D, Moran O, Ludewig U (2008) Plant plasma membrane water channels conduct the signalling molecule H_2O_2. *Biochem J* **414**: 53–61.

Exposito-Rodriguez M, Laissue PP, Yvon-Durocher G, Smirnoff N, Mullineaux PM (2017) Photosynthesis-dependent H_2O_2 transfer from chloroplasts to nuclei provides a high-light signalling mechanism. *Nat Commun* **8**: 49.

Fernández-Marcos M, Desvoyes B, Manzano C, Liberman LM, Benfey PN, Del Pozo JC, Gutierrez C (2017) Control of Arabidopsis lateral root primordium boundaries by MYB36. *New Phytol* **213**: 105–112.

Fichman Y, Mittler R (2020) Rapid systemic signaling during abiotic and biotic stresses: is the ROS wave master of all trades? *Plant J* **102**: 887–896.

Fichman Y, Myers RJ Jr, Grant DG, Mittler R (2021) Plasmodesmata-localized proteins and ROS orchestrate light-induced rapid systemic signaling in *Arabidopsis*. *Sci Signal* **14**: eabf0322.

Foreman J, Demidchik V, Bothwell JH, Mylona P, Miedema H, Torres MA, Linstead P, Costa S, Brownlee C, Jones JD, Davies JM, Dolan L (2003) Reactive oxygen species produced by NADPH oxidase regulate plant cell growth. *Nature* **422**: 442–446.

Fouracre JP, Poethig RS (2020) Lonely at the top? Regulation of shoot apical meristem activity by intrinsic and extrinsic factors. *Curr Opin Plant Biol* **58**: 17–24.

Francoz E, Ranocha P, Nguyen-Kim H, Jamet E, Burlat V, Dunand C (2015) Roles of cell wall peroxidases in plant development. *Phytochemistry* **112**: 15–21.

Hachez C, Laloux T, Reinhardt H, Cavez D, Degand H, Grefen C, De Rycke R, Inzé D, Blatt MR, Russinova E, Chaumont F (2014) Arabidopsis SNAREs SYP61 and SYP121 coordinate the trafficking of plasma membrane aquaporin PIP2;7 to modulate the cell membrane water permeability. *Plant Cell* **26**: 3132–3147.

Hawkins BJ, Madesh M, Kirkpatrick CJ, Fisher AB (2007) Superoxide flux in endothelial cells via the chloride channel-3 mediates intracellular signaling. *Mol Biol Cell* **18**: 2002–2012.

Henzler T, Steudle E (2000) Transport and metabolic degradation of hydrogen peroxide in *Chara corallina*: model calculations and measurements with the pressure probe suggest transport of H_2O_2 across water channels. *J Exp Bot* **51**: 2053–2066.

Hooijmaijers C, Rhee JY, Kwak KJ, Chung GC, Horie T, Katsuhara M, Kang H (2012) Hydrogen peroxide permeability of plasma membrane aquaporins of *Arabidopsis thaliana*. *J Plant Res* **125**: 147–153.

Hussain A, Tanveer R, Mustafa G, Farooq M, Amin I, Mansoor S (2020) Comparative phylogenetic analysis of aquaporins provides insight into the gene family expansion and evolution in plants and their role in drought tolerant and susceptible chickpea cultivars. *Genomics* **112**: 263–275.

Ishikawa F, Suga S, Uemura T, Sato MH, Maeshima M (2005) Novel type aquaporin SIPs are mainly localized to the ER membrane and show cell-specific expression in *Arabidopsis thaliana*. *FEBS Lett* **579**: 5814–5820.

Jiang K, Meng YL, Feldman LJ (2003) Quiescent center formation in maize roots is associated with an auxin-regulated oxidizing environment. *Development* **130**: 1429–1438.

Kitagawa M, Jackson D (2019) Control of meristem size. *Annu Rev Plant Biol* **70**: 269–291.

Kka N, Rookes J, Cahill D (2018) The influence of ascorbic acid on root growth and the root apical meristem in *Arabidopsis thaliana*. *Plant Physiol Biochem* **129**: 323–330.

Kollist H, Zandalinas SI, Sengupta S, Nuhkat M, Kangasjärvi J, Mittler R (2019) Rapid responses to abiotic stress: priming the landscape for the signal transduction network. *Trends Plant Sci* **24**: 25–37.

Kong X, Tian H, Yu Q, Zhang F, Wang R, Gao S, Xu W, Liu J, Shani E, Fu C, Zhou G, Zhang L, Zhang X, Ding Z (2018) PHB3 maintains root stem cell niche identity through ROS-responsive AP2/ERF transcription factors in Arabidopsis. *Cell Rep* **22**: 1350–1363.

Laloux T, Junqueira B, Maistriaux LC, Ahmed J, Jurkiewicz A, Chaumont F (2018) Plant and mammal aquaporins: same but different. *Int J Mol Sci* **19**(2): 521.

Lee SJ, Kim DG, Lee KY, Koo JS, Lee BJ (2018) Regulatory mechanisms of thiol-based redox sensors: lessons learned from structural studies on prokaryotic redox sensors. *Arch Pharm Res* **41**: 583–593.

Lee Y, Rubio MC, Alassimone J, Geldner N (2013) A mechanism for localized lignin deposition in the endodermis. *Cell* **153**: 402–412.

Li G, Santoni V, Maurel C (2014) Plant aquaporins: roles in plant physiology. *Biochim Biophys Acta* **1840**: 1574–1582.

Long JA, Moan EI, Medford JI, Barton MK (1996) A member of the KNOTTED class of homeodomain proteins encoded by the STM gene of Arabidopsis. *Nature* **379**: 66–69.

Ma S, Quist TM, Ulanov A, Joly R, Bohnert HJ (2004) Loss of TIP1;1 aquaporin in Arabidopsis leads to cell and plant death. *Plant J* **40**: 845–859.

Maurel C, Boursiac Y, Luu DT, Santoni V, Shahzad Z, Verdoucq L (2015) Aquaporins in plants. *Physiol Rev* **95**: 1321–1358.

Miller G, Schlauch K, Tam R, Cortes D, Torres MA, Shulaev V, Dangl JL, Mittler R (2009) The plant NADPH oxidase RBOHD mediates rapid systemic signaling in response to diverse stimuli. *Sci Signal* **2**: ra45.

Moreland JG, Davis AP, Bailey G, Nauseef WM, Lamb FS (2006) Anion channels, including ClC-3, are required for normal neutrophil oxidative function, phagocytosis, and transendothelial migration. *J Biol Chem* **281**: 12277–12288.

Motte H, Vanneste S, Beeckman T (2019) Molecular and environmental regulation of root development. *Annu Rev Plant Biol* **70**: 465–488.

Mubarakshina Borisova MM, Kozuleva MA, Rudenko NN, Naydov IA, Klenina IB, Ivanov BN (2012) Photosynthetic electron flow to oxygen and diffusion of hydrogen peroxide through the chloroplast envelope via aquaporins. *Biochim Biophys Acta* **1817**: 1314–1321.

Nawy T, Lee JY, Colinas J, Wang JY, Thongrod SC, Malamy JE, Birnbaum K, Benfey PN (2005) Transcriptional profile of the Arabidopsis root quiescent center. *Plant Cell* **17**: 1908–1925.

Niemeyer J, Scheuring D, Oestreicher J, Morgan B, Schroda M (2021) Real-time monitoring of subcellular H_2O_2 distribution in *Chlamydomonas reinhardtii*. *Plant Cell* **33**: 2935–2949.

Oh E, Zhu JY, Bai MY, Arenhart RA, Sun Y, Wang ZY (2014) Cell elongation is regulated through a central circuit of interacting transcription factors in the Arabidopsis hypocotyl. *Elife* **3**: e03031.

Otegui MS, Noh YS, Martínez DE, Vila Petroff MG, Staehelin LA, Amasino RM, Guiamet JJ (2005) Senescence-associated vacuoles with intense proteolytic activity develop in leaves of Arabidopsis and soybean. *Plant J* **41**: 831–844.

Péret B, Li G, Zhao J, Band LR, Voß U, Postaire O, Luu DT, Da Ines O, Casimiro I, Lucas M, Wells DM, Lazzerini L, Nacry P, King JR, Jensen OE, Schäffner AR, Maurel C, Bennett MJ (2012) Auxin regulates aquaporin function to facilitate lateral root emergence. *Nat Cell Biol* **14**: 991–998.

Rodrigues O, Reshetnyak G, Grondin A, Saijo Y, Leonhardt N, Maurel C, Verdoucq L (2017) Aquaporins facilitate hydrogen peroxide entry into guard cells to mediate ABA- and pathogen-triggered stomatal closure. *Proc Natl Acad Sci U S A* **114**: 9200–9205.

Rojas-Murcia N, Hématy K, Lee Y, Emonet A, Ursache R, Fujita S, De Bellis D, Geldner N (2020) High-order mutants reveal an essential requirement for peroxidases but not laccases in Casparian strip lignification. *Proc Natl Acad Sci U S A* **117**: 29166–29177.

Santoni V (2017) Plant aquaporin posttranslational regulation. In *Plant Aquaporins*, ed. F. Chaumont and S. D. Tyerman, 83–105. Springer, Cham.

Schippers JHM, Nguyen HM, Lu D, Schmidt R, Mueller-Roeber B (2012) ROS homeostasis during development: an evolutionary conserved strategy. *Cell Mol Life Sci* **69**: 3245–3257.

Schmidt R, Kunkowska AB, Schippers JHM (2016) Role of reactive oxygen species during cell expansion in leaves. *Plant Physiol* **172**: 2098–2106.

Schopfer P, Liszkay A, Bechtold M, Frahry G, Wagner A (2002) Evidence that hydroxyl radicals mediate auxin-induced extension growth. *Planta* **214**: 821–828.

Schüssler MD, Alexandersson E, Bienert GP, Kichey T, Laursen KH, Johanson U, Kjellbom P, Schjoerring JK, Jahn TP (2008) The effects of the loss of TIP1;1 and TIP1;2 aquaporins in Arabidopsis thaliana. *Plant J* **56**: 756–767.

Sies H (2017) Hydrogen peroxide as a central redox signaling molecule in physiological oxidative stress: oxidative eustress. *Redox Biol* **11**: 613–619.

Sies H, Jones DP (2020) Reactive oxygen species (ROS) as pleiotropic physiological signalling agents. *Nat Rev Mol Cell Biol* **21**: 363–383.

Simm S, Papasotiriou DG, Ibrahim M, Leisegang MS, Müller B, Schorge T, Karas M, Mirus O, Sommer MS, Schleiff E (2013) Defining the core proteome of the chloroplast envelope membranes. *Front Plant Sci* **4**:11.

Smirnoff N, Arnaud D (2019) Hydrogen peroxide metabolism and functions in plants. *New Phytol* **221**: 1197–1214.

Sun Y, Fan XY, Cao DM, Tang W, He K, Zhu JY, He JX, Bai MY, Zhu S, Oh E, Patil S, Kim TW, Ji H, Wong WH, Rhee SY, Wang ZY (2010) Integration of brassinosteroid signal transduction with the transcription network for plant growth regulation in Arabidopsis. *Dev Cell* **19**: 765–777.

Tan DQ, Suda T (2018) Reactive oxygen species and mitochondrial homeostasis as regulators of stem cell fate and function. *Antioxid Redox Signal* **29**: 149–168.

Thénard MLJ (1818) Observations sur des nouvelles combinations entre l'oxigéne et divers acides. *Ann Chim Phys* **8**: 306–313.

Tian C, Zhang X, He J, Yu H, Wang Y, Shi B, Han Y, Wang G, Feng X, Zhang C, Wang J, Qi J, Yu R, Jiao Y (2014) An organ boundary-enriched gene regulatory network uncovers regulatory hierarchies underlying axillary meristem initiation. *Mol Syst Biol* **10**: 755.

Tsukagoshi H (2016) Control of root growth and development by reactive oxygen species. *Curr Opin Plant Biol* **29**: 57–63.

Tsukagoshi H, Busch W, Benfey PN (2010) Transcriptional regulation of ROS controls transition from proliferation to differentiation in the root. *Cell* **143**: 606–616.

Wang H, Schoebel S, Schmitz F, Dong H, Hedfalk K (2020) Characterization of aquaporin-driven hydrogen peroxide transport. *Biochim Biophys Acta Biomembr* **1862**: 183065.

Wang LL, Chen AP, Zhong NQ, Liu N, Wu XM, Wang F, Yang CL, Romero MF, Xia GX (2014) The *Thellungiella salsuginea* tonoplast aquaporin TsTIP1;2 functions in protection against multiple abiotic stresses. *Plant Cell Physiol* **55**: 148–161.

Wang Y, Li R, Li D, Jia X, Zhou D, Li J, Lyi SM, Hou S, Huang Y, Kochian LV, Liu J (2017) NIP1;2 is a plasma membrane-localized transporter mediating aluminum uptake, translocation, and tolerance in Arabidopsis. *Proc Natl Acad Sci U S A* **114**: 5047–5052.

Weigel D, Jürgens G (2002) Stem cells that make stems. *Nature* **415**: 751–754.

Yamada M, Han X, Benfey PN (2020) RGF1 controls root meristem size through ROS signalling. *Nature* **577**: 85–88.

Yanai O, Shani E, Dolezal K, Tarkowski P, Sablowski R, Sandberg G, Samach A, Ori N (2005) Arabidopsis KNOXI proteins activate cytokinin biosynthesis. *Curr Biol* **15**: 1566–1571.

Yang S, Yu Q, Zhang Y, Jia Y, Wan S, Kong X, Ding Z (2018) ROS: the fine-tuner of plant stem cell fate. *Trends Plant Sci* **23**: 850–853.

Yu Q, Tian H, Yue K, Liu J, Zhang B, Li X, Ding Z (2016) A P-Loop NTPase regulates quiescent center cell division and distal stem cell identity through the regulation of ROS homeostasis in Arabidopsis root. *PLoS Genet* **12**: e1006175.

Zeng J, Dong Z, Wu H, Tian Z, Zhao Z (2017) Redox regulation of plant stem cell fate. *EMBO J* **36**: 2844–2855.

Zhou Y, Liu X, Engstrom EM, Nimchuk ZL, Pruneda-Paz JL, Tarr PT, Yan A, Kay SA, Meyerowitz EM (2015) Control of plant stem cell function by conserved interacting transcriptional regulators. *Nature* **517**: 377–380.

Zhu D, Wu Z, Cao G, Li J, Wei J, Tsuge T, Gu H, Aoyama T, Qu LJ (2014) TRANSLUCENT GREEN, an ERF family transcription factor, controls water balance in Arabidopsis by activating the expression of aquaporin genes. *Mol Plant* **7**: 601–615.

CHAPTER ELEVEN

Evolution and Function of Peroxiporins in Piscine Spermatozoa

Joan Cerdà, François Chauvigné, and Roderick Nigel Finn

ABBREVIATIONS

AQP	Aquaporin
ATP	Adenosine triphosphate
ASK1	Apoptosis signal-regulating kinase 1
cAMP	Cyclic adenosine monophosphate
CASA	Computer-assisted semen analysis
$[Ca^{2+}]_i$	Intracellular calcium concentration
GSK3	Glycogen synthase kinase-3
JNK	Jun N-terminal kinase
MAPK	Mitogen-activated protein kinase
MKK4	Mitogen-activated protein kinase kinase 4
NADPH	Nicotinamide adenine dinucleotide phosphate
PKC	Protein kinase C
ROS	Reactive oxygen species

11.1 INTRODUCTION

Hydrogen peroxide (H_2O_2) is a relatively stable, non-radical form of reactive oxygen species (ROS) that may be beneficial or toxic to cell physiology depending upon the source, concentration, and location (Rhee, 2006; Sies, 2017; Herb et al., 2021). Benefits tend to occur at lower concentrations through paracrine or autocrine regulation of kinase-driven pathways affecting a range of homeostatic processes (Bienert et al., 2006; Gough and Cotter, 2011; Byon et al., 2016; Herb et al., 2021). At higher concentrations, the increased oxidative stress causes damage to biomolecules, which may be beneficial for phagocytic defense mechanisms, but otherwise negatively impacts cell physiology and is associated with disease and aging (Babior, 1987; Gough and Cotter, 2011; Sies et al., 2017; Prata et al., 2019; Purohit et al., 2019). Maintaining the right redox balance is thus crucial for healthy cell function.

In animal cells, H_2O_2 is typically generated as a by-product of mitochondrial oxidative phosphorylation or via a range of extracellular, intracellular, or cell surface enzyme systems (Rhee, 2006; Giorgio et al., 2007; Gough and Cotter, 2011; Sies, 2017; Herb et al., 2021). In some cells, such as muscle and sperm, osmotic stress is also known to induce excess ROS production (Burnaugh et al., 2010; McCarthy et al., 2010; Ørtenblad et al., 2003; Chauvigné et al., 2015a, 2021). To avoid redox imbalance cells engage a variety of catalytic and antioxidant defense systems (Berg et al., 2004). However, under high ROS load in metabolically active cells such as piscine sperm, or during paracrine signaling, it is now known that channels permeable to H_2O_2 are expressed in the plasma, mitochondrial or other intracellular organelle membranes to facilitate communication or efflux.

The first indirect evidence of such peroxiporins emerged more than 20 years ago from studies of isolated internodal cells of the green alga *Chara corallina* (Henzler and Steudle, 2000). Based upon modeling and non-specific inhibition

DOI: 10.1201/9781003160649-11

experiments with mercuric chloride ($HgCl_2$), it was suggested that transmembrane water channels (aquaporins) should be the most likely facilitators of H_2O_2 flux (Henzler and Steudle, 2000). At this point, it had already been established that different members of the aquaporin family of proteins are expressed in both prokaryotic and eukaryotic cell membranes, and conduct water, glycerol, or urea in a mercury-sensitive manner (Preston et al., 1992; Agre et al., 1993; Preston et al., 1993; Ishibashi et al., 1994; Ishibashi et al., 1998). Subsequent experimental evidence using yeast mutants showed that indeed selected plant and mammalian aquaporins conduct H_2O_2 across membranes (Bienert et al., 2007; Bienert and Chaumont, 2014). With further structural assessments of the manner in which H_2O_2 mimics H_2O to traverse the central pore of an aquaporin (i.e., one molecule at a time), it was originally suggested that all water-permeable aquaporins likely permeate H_2O_2 in varying capacities (Almasalmeh et al., 2014; Wragg et al., 2020). However, the fact that a given channel harbors the biophysical property of H_2O_2 permeation does not necessarily mean that it functions as a peroxiporin. Such a moniker should be given when the physiological function of the channel is H_2O_2 permeation, as for example revealed for AQP8 in mammalian B lymphoma and insulin-producing cells and piscine spermatozoa, and potentially for mammalian spermatozoa (Chauvigné et al., 2015a, 2021; Bertolotti et al., 2016; Laforenza et al., 2017; Krüger et al., 2021a, 2021b; Pellavio and Laforenza, 2021). The original name of the channel (aquaporin 8) and its gene symbol (AQP8) remain the same; consequently, to discuss the evolution of peroxiporins is to discuss the evolution of the broader superfamily of aquaporins.

In this review, we therefore provide a brief overview of the phylogenetic interrelationships of the aquaporin superfamily in jawed vertebrates (gnathostomes) and teleost fishes, with new findings for the H_2O_2 permeability properties of the subfamilies. We then review the expression of aquaporins in piscine spermatozoa and highlight how peroxiporin function is important for post-activated swimming performance. Finally, we review the evolution and function of signal transduction pathways regulating peroxiporin mitochondrial membrane trafficking in freshwater and marine teleost spermatozoa.

11.2 PHYLOGENETIC ASPECTS OF PEROXIPORINS IN TELEOSTS

Molecular phylogenetic studies of the aquaporin superfamily have to date uncovered 17 subfamilies (AQP0–AQP16) in gnathostomes (Figure 11.1A; Abascal et al., 2014; Di Giorgio et al., 2014; Finn et al., 2014; Chauvigné et al., 2015b; Finn and Cerdà, 2015; Finn and Cerdà, 2018; Chauvigné et al., 2019; Yilmaz et al., 2020; Ferré et al., 2023). These analyses further revealed an ancient origin of four separate grades that encompass the classical aquaporins (AQP0, 1, 2, 4, 5, 6, 14, and 15), AQP8-type aquaporins (AQP8 and 16), aquaglyceroporins (AQP3, 7, 9, 10, and 13), and unorthodox aquaporins (AQP11 and 12). All channels are considered to be H_2O transporters; however, some, such as the AQP8-type, the aquaglyceroporins, and the AQP14 channels, have broader permeability capacities due in part to wider cross-sectional areas of the selectivity filters at the aromatic-arginine (ar/R) constriction (Beitz et al., 2006; Engelund et al., 2013; Finn et al., 2015; Garneau et al., 2015; Gotfryd et al., 2018; Chauvigné et al., 2019; de Maré et al., 2020). The broader repertoires of molecules that can traverse the channels have been associated with the presentation of smaller non-aromatic residues or the arrangement of five rather than four residues at these pore junctions (Finn and Cerdà, 2018; Kirscht et al., 2018; Chauvigné et al., 2019). Despite these general classifications, it is becoming clear that solute selectivity is complex (Kitchen et al., 2019), with several exceptions including glycerol permeation of the classical aquaporin AQP6 (Holm et al., 2004) and the unorthodox aquaporin AQP11 (Madeira et al., 2014). It is thus important to experimentally test the permeability of a given channel with validation that it is expressed in the membrane to determine whether a solute can traverse the pore.

For H_2O_2, experiments have been conducted that demonstrate its permeation through mammalian or piscine orthologs of AQP0, 1, 3, 5, 7, 8, 9, 11, 12, and 14 (Figure 11.1A; Bienert et al., 2007; Miller et al., 2010; Gouleh et al., 2013; Almasalmeh et al., 2014; Bertolotti et al., 2013, 2016; Satooka and Hara-Chikuma, 2016; Watanabe et al., 2016; Chauvigné et al., 2015a, 2019, 2021; Rodrigues et al., 2019; Besetti et al., 2020; Varadaraj and Kumari, 2020; Krüger et al., 2021b; Chauvigné and Cerdà, unpublished data). The combined observations outlined in Figure 11.1 illustrate that the biophysical capacity for H_2O_2 permeation through gnathostome aquaporins is unrelated to their molecular phylogeny. It thus indeed seems likely that all aquaporin subfamilies possess the biophysical capacity to transport H_2O_2, with flux rates related to pore size and expression levels in the membrane.

For teleost fishes, molecular phylogenetic studies have revealed that they possess greater numbers of aquaporin gene duplicates compared to mammals (Figure 11.1B). This is in part due to a specific whole genome duplication at the root of the crown clade, but also due to tandem duplications preceding this event in the case of Aqp8 and 10 paralogs, or ensuing the event for Aqp1 paralogs (Tingaud-Sequeira et al., 2010; Finn et al., 2014; Finn and Cerdà, 2011; Chauvigné et al., 2015a, 2019; Cerdà et al., 2017; Yilmaz et al., 2020; Ferré et al., 2023). Thus the mitochondrial piscine peroxiporin previously termed Aqp8b (Chauvigné et al., 2015a) is now termed Aqp8bb (Chauvigné et al., 2021). The specific aquaporin gene copy number varies considerably between different teleost families due to additional lineage-specific tandem duplications and additional whole genome duplications, and differential gene loss (Finn et al., 2014; Yilmaz et al., 2020; Ferré et al., 2023). Despite such diversity, the orthologous subfamily composition and channel permeation preferences reflect those observed in tetrapods, with the exception of AQP2, 5, 6, 13, and 16, which have not been found in teleosts.

Experimental evidence for H_2O_2 permeation through piscine channels is more limited. To date, only five channels (Aqp1aa, 7, 8bb, 12, and 14), cloned from three species, the marine gilthead seabream (*Sparus aurata*) or the freshwater Atlantic salmon (*Salmo salar*) and zebrafish (*Danio rerio*), have been investigated in this regard (Figure 11.1B). Each channel tested belongs to the four ancient grades of aquaglyceroporins, classical, AQP8-type, and unorthodox aquaporins, and was found to permeate H_2O and H_2O_2 (Chauvigné et al., 2015a, 2019, 2021; Chauvigné and Cerdà, unpublished data). These observations reinforce the notion that H_2O_2 permeation is likely an ancient biophysical property of all H_2O-transporting aquaporins.

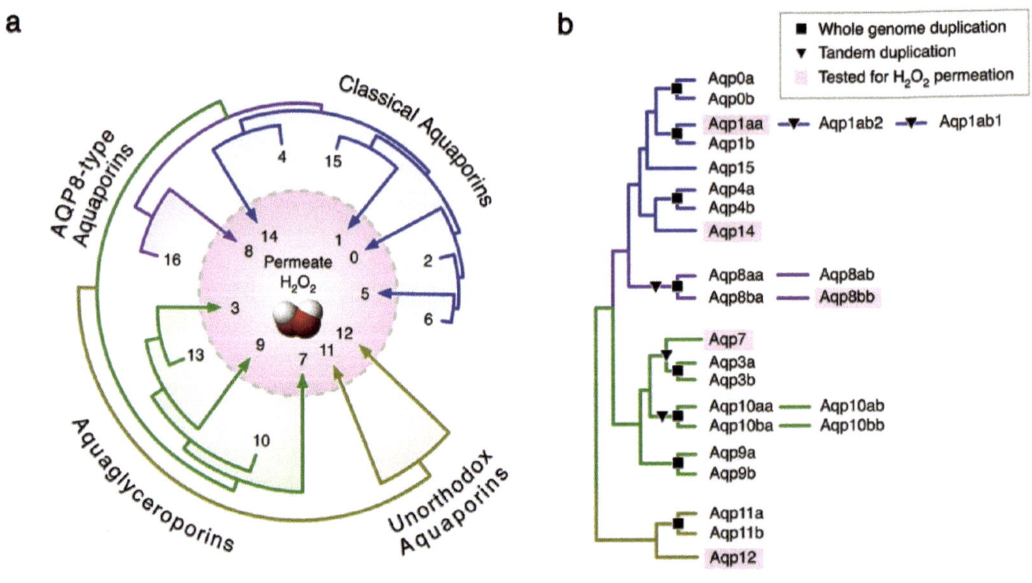

Figure 11.1 Phylogenetic interrelationships of water- and H_2O_2-transporting aquaporins. (A) Schematic view of the 17 aquaporin subfamilies identified in gnathostomes illustrating the four ancient grades of channel that retain H_2O_2 permeation properties. (B) Schematic view of the aquaporin subfamilies identified in teleost fishes. Tree branches for panels A and B are arranged according to the molecular phylogenetic studies performed by Finn and colleagues (2014) and Yilmaz and colleagues (2020). The newly discovered Aqp1b ohnolog derived from the fish-specific whole genome duplication and Aqp1ab1 and Aqp1ab2 tandem duplicates present in a teleost-specific aquaporin-1 cluster (TSA1C) are also included to illustrate current knowledge of the channel repertoires in the lineage (Ferré et al., 2023).

11.3 EXPRESSION OF AQUAPORINS IN FISH SPERMATOZOA

To understand the roles of aquaporins as potential peroxiporins in fish spermatozoa, we initially briefly review the modes of activation and subcellular locations of the channels. In both mammals and teleosts, osmotic changes associated with water and ion fluxes in spermatozoa are vital for the activation of motility and subsequent fertilization (Cooper and Yeung, 2003; Cosson et al., 2008a, 2008b). In freshwater fishes, such as salmonids, the external release of spermatozoa into the hypoosmotic medium induces K^+ efflux and plasma membrane hyperpolarization, which leads to a transient increase in intracellular Ca^{2+} ($[Ca^{2+}]_i$) by calcium channels (Morisawa and Suzuki, 1980; Morisawa et al., 1983; Cosson et al., 1989; Tanimoto et al., 1994; Alavi and Cosson, 2006). The increase in $[Ca^{2+}]_i$ stimulates the synthesis of cyclic AMP (cAMP) by the adenylyl cyclase, which activates protein kinases to phosphorylate axonemal proteins and initiate and maintain the movement of the flagellum. However, in some freshwater species, such as the cyprinid common carp (*Cyprinus carpio*), cAMP appears not to play a major role in the regulation of sperm motility (Krasznai et al., 2000). In contrast, in marine teleosts the intracellular Ca^{2+} signal activating the cAMP-dependent phosphorylation of flagellar proteins and sperm motility is triggered by the osmotic shock caused by exposure to the hyperosmotic seawater (Dzyuba and Cosson, 2014). The increment in $[Ca^{2+}]_i$, as well as of the intracellular concentration of other ions, is produced as a consequence of a reduction in cell volume, or through the stimulation of channel- or exchanger-mediated Ca^{2+} influx across the plasma membrane (Vines et al., 2002; Zilli et al., 2012; Boj et al., 2015a; Pérez et al., 2016).

The rapid water fluxes and changes in cell volume associated with the activation of motility of fish spermatozoa suggest that these processes may be regulated by aquaporins. In the marine teleost gilthead seabream, immunolocalization studies using paralog-specific antibodies have indeed shown that each spermatozoon expresses a surprisingly diverse and differentially localized repertoire of aquaporins, including Aqp1aa, 1ab1, 7, 8bb, and 10bb, as well as Aqp3a and 4a (Figure 11.2A; Zilli et al.,

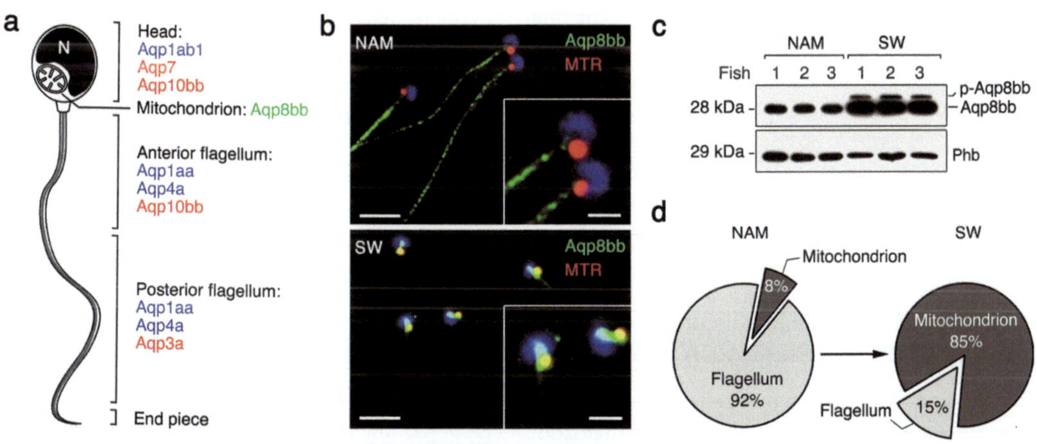

Figure 11.2 Aquaporins in the seabream spermatozoon. (A) Diagram showing the differential spatial localization of different aquaporin paralogs in seabream motile spermatozoa. The localization is based on data obtained from immunofluorescence microscopy using paralog-specific antibodies. Water-selective aquaporins are in blue, aquaglyceroporins in red, and the peroxiporin Aqp8bb in green. N = nucleus. (B) Representative immunostaining of Aqp8bb (green) in seabream immotile (diluted in non-activation medium [NAM]) and seawater (SW)-activated spermatozoa. The mitochondria are labeled with the mitochondrion-specific vital dye MitoTracker Red CMXRos (MTR, red), whereas the spermatozoa nuclei are counterstained with 4′,6-diamidino-2-phenylindole (DAPI, blue). Colocalized signals in mitochondria are yellow. Scale bars 5 μM (insets 1 μM). (C) Immunoblotting of Aqp8bb in mitochondrial extracts from non-activated and activated spermatozoa from three different males. The non-phosphorylated (Aqp8bb) and phosphorylated (p-Aqp8bb) forms of Aqp8bb are indicated. The inner mitochondrial membrane marker prohibitin (Phb) was employed as loading control. (D) Pie charts displaying the percentage of spermatozoa showing Aqp8bb mitochondrial localization after incubation in NAM or upon activation in SW.

2009; Chauvigné et al., 2013; Boj et al., 2015a; Chauvigné and Cerdà, unpublished data). In ejaculated but non-activated seabream spermatozoa, Aqp1aa and 4a are located along the anterior and posterior flagellum, whereas Aqp3a and 7 are restricted to the posterior flagellum or the head, respectively, and Aqp1ab1, 8bb, and 10bb are both in the head and the anterior flagellum. Upon seawater activation, Aqp1ab1 and 10bb are translocated to the plasma membrane of the head and the anterior flagellum, while Aqp8bb is rapidly phosphorylated and transported to the single spermatozoon mitochondrion (Figure 11.2B–D). In contrast, Aqp1aa, 3a, 4a, and 7 apparently remain unchanged. The expression of multiple aquaporins in fish sperm resembles the situation in mammalian spermatozoa, which also express AQP3, 7, and 8, as well as AQP11 (Boj et al., 2015b; Pellavio and Laforenza, 2021). However, with the exception of Aqp8bb, which functions as a mitochondrial peroxiporin in the post-activated spermatozoa of both marine and freshwater species (Chauvigné et al., 2021), information on aquaporin protein localization in the spermatozoa of fishes other than the seabream is currently lacking. The presence of multiple aquaporins in fish sperm cells is nevertheless plausible, since mercury, a typical non-specific inhibitor of aquaporin water conductance, can inhibit or reduce the duration of sperm motility in different species even at low concentrations (Cosson et al., 1999; Cosson et al., 2008b; Dzyuba and Cosson, 2014; Takei et al., 2015; Kollár et al., 2018; Pataki et al., 2021).

11.4 FUNCTION OF PEROXIPORINS IN PISCINE SPERMATOZOA

The specific and persistent localization of Aqp1aa, 3a, and 4a in the gilthead seabream spermatozoon flagellum upon seawater activation, as well as the fast regulation of the subcellular localization of Aqp1ab1, 8bb, and 10bb at this time, strongly suggest that these channels may play important roles during the activation or maintenance of sperm motility. However, only the physiological roles of Aqp1aa and 8bb have been uncovered so far.

In the seabream, as in other marine teleosts (Zilli et al., 2012), Ca^{2+} or K^+ channels seem not to be required for the initiation of flagellar motility (Zilli et al., 2008; Boj et al., 2015a). Rather, it is believed that the increase of $[Ca^{2+}]_i$ in spermatozoa during activation is due to the massive water efflux following the hyperosmotic shock in seawater and/or by the release of Ca^{2+} from intracellular organelles (Zilli et al., 2012; Boj et al., 2015a). This would lead to the activation of adenylyl cyclase and motility initiation by cAMP-dependent protein phosphorylation and dephosphorylation events (Zilli et al., 2011, 2012). However, the release of spermatozoa into seawater also leads to an osmotic stress-linked generation of ROS such as H_2O_2, possibly through the activation of membrane-bound NADPH oxidases (McCarthy et al., 2010; Burnaugh et al., 2010; Chauvigné et al., 2015a). The osmotically stress-induced H_2O_2 may subsequently diffuse into the mitochondrion, further augmenting the levels of ROS produced as a by-product of mitochondrial oxidative phosphorylation (Dreanno et al., 1999; Lahnsteiner and Caberlotto, 2012; Dzyuba and Cosson, 2014). Such an elevated ROS load in spermatozoa causes damage to the DNA and the plasma membrane, with further depolarization of the mitochondrial membrane potential, mitochondrial malfunctioning, and ATP depletion (Koppers et al., 2008; García et al., 2012; Chauvigné et al., 2015a; Gazo et al., 2015). This ultimately impairs the kinematic properties of sperm (motility, velocity) and may also alter intracellular signaling pathways through changes in tyrosine phosphorylation and acid phosphatase activity (Chauvigné et al., 2015a; Gazo et al., 2015).

To investigate the role of aquaporins during the osmotic stress-associated processes in spermatozoa, inhibition assays of sperm motility determined by computer-assisted sperm analysis (CASA) systems have been carried out in seabream using paralog-specific, affinity-purified antibodies. These experiments have shown that the immunological blockage of Aqp1aa H_2O and Aqp8bb H_2O_2 conductance, as well as steric inhibition of the Aqp8bb trafficking to the mitochondrion, induce a strong decline of flagellar movement and thus sperm motility and progression (Figure 11.3; Chauvigné et al., 2015a; Boj et al., 2015a). In contrast, inhibition of Aqp1ab1 and 7 do not affect sperm activation but latently compromise the trajectory and movement pattern of the spermatozoon (Boj et al., 2015a). Upon altering the function of Aqp1aa, but not of Aqp1ab1 and 7, the elevation of the $[Ca^{2+}]_i$ that normally occurs upon seawater activation is partially prevented. This

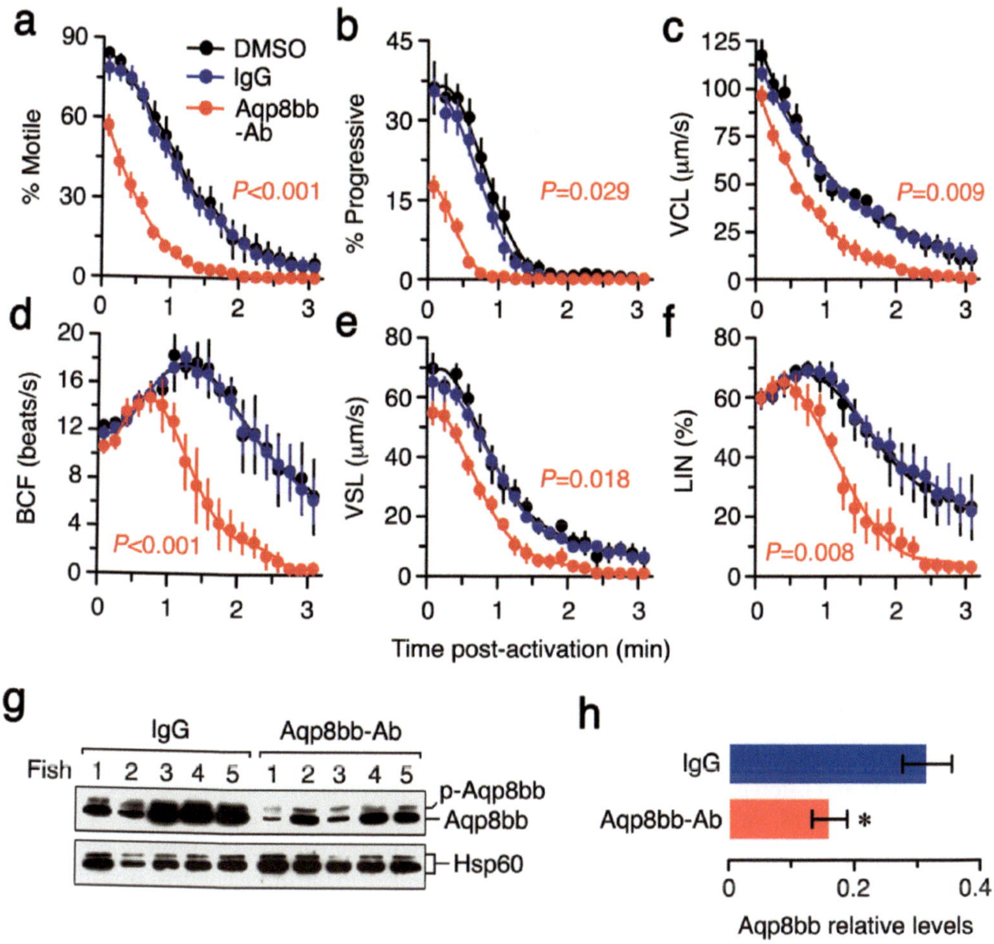

Figure 11.3 The mitochondrial peroxiporin Aqp8bb is required for seabream spermatozoa motility. (A–F) Kinetic parameters (mean ± SEM; n = 8 fish) of seawater-activated spermatozoa treated with immunoglobulin G (IgG) or an affinity-purified Aqp8bb rabbit antiserum (Aqp8bb-Ab). Values of the Aqp8bb-Ab-treated groups are significantly different (P is indicated in each panel). BCF = beat cross frequency; LIN = linearity; VCL = curvilinear velocity; VSL = straight-line velocity. (G) Immunoblot of mitochondrial Aqp8bb from seawater-activated sperm treated as above using the mitochondrial matrix marker heat shock protein 60 (Hsp60) as loading control. Phosphorylated (p-Aqp8bb) and dephosphorylated (Aqp8bb) peroxiporin are indicated. (H) Quantitation (mean ± SEM; n = 5 fish) of mitochondrial Aqp8bb normalized to Hsp60. The asterisk denotes statistically significant differences ($p < .05$) between IgG and Aqp8bb-Ab treated groups.

Source: **Chauvigné et al. (2015a).**

effect is reversed by exposure of spermatozoa to a Ca^{2+} ionophore, which restores $[Ca^{2+}]_i$ and motility (Boj et al., 2015a). The role of Aqp8bb is different since when the mitochondrial trafficking of this channel is immunologically inhibited in activated spermatozoa, ROS levels accumulate in the mitochondria leading to mitochondrial membrane depolarization, reduction of ATP synthesis, and the rapid arrest of sperm motility (Chauvigné et al., 2015a). These deleterious effects are fully reversed in the presence of a mitochondria-targeted antioxidant, which highlights the specific role of Aqp8bb as a mitochondrial peroxiporin (Chauvigné et al., 2015a).

The studies outlined above suggest that the major role of Aqp1aa in the seabream spermatozoon is to facilitate a rapid water efflux caused by the hyperosmotic shock, thus driving an intracellular Ca^{2+} wave to activate the flagellar movement. By contrast, Aqp8bb operates as a peroxiporin to allow the efflux of accumulated H_2O_2 in the mitochondria, thereby maintaining the production of ATP needed for flagellar motility. However, the inhibition of Aqp1aa function in spermatozoa partially blocks the trafficking of Aqp8bb to the mitochondrion, which can be rescued by a Ca^{2+} ionophore (Boj et al., 2015a). This indicates that the Aqp1aa-triggered intracellular Ca^{2+} surge plays a dual role by initiating flagellar motility through the stimulation of the axoneme as well as the mitochondrial trafficking of the Aqp8bb peroxiporin to mediate H_2O_2 detoxification for continuous ATP generation. These observations therefore suggest that both Aqp1aa-and Aqp8bb-mediated mechanisms are independent but coordinated events that regulate the activation and duration of sperm motility. This model implies a main function of Aqp1aa as a water transporter. However, it is not yet known if Aqp1aa also functions as a plasma membrane peroxiporin for the cellular efflux of H_2O_2 before or after ejaculation, as suggested for AQP3, 8, and/or 11 in mammalian spermatozoa (Laforenza et al., 2017; Pellavio and Laforenza, 2021).

In freshwater fishes, the role of Aqp1-like channels mediating the initiation of sperm motility may not be conserved, since in salmonids water permeability of the spermatozoon plasma membrane is low (Labbe and Maisse, 2001) and the volume changes of the spermatozoa following hypoosmotic activation seem not to be significant in some cases (Takei et al., 2015; Cosson, 2019). In addition, in species in which the spermatozoa do swell significantly upon freshwater activation, such as the common carp, treatment with mercurial compounds is ineffective at reducing volume increase (Perchec Poupard et al., 1997).

Nevertheless, the presence of Aqp8bb in the fish spermatozoa and its transport to the mitochondria during activation has recently been confirmed in freshwater teleosts, such as the zebrafish and Atlantic salmon (Chauvigné at al., 2021). This indicates that the expression of the Aqp8bb peroxiporin in the spermatozoon and its trafficking mechanism for mitochondrial ROS detoxification and the maintenance of ATP production are conserved traits in distantly related teleost lineages, regardless of whether the sperm are released into a hyper- or hypoosmotic medium. In some other teleost lineages, however, *aqp8bb* genes have been lost from the genome or are retained as partial copies unable to code for functional channels (pseudogenes; Finn et al., 2014). Therefore, whether other teleost Aqp8 paralogs, such as Aqp8aa, 8ab or 8ba, have co-opted the function of Aqp8bb remains to be investigated. Interestingly, in mammals ROS transport is involved in normal spermatozoa functioning (Baskaran et al., 2021), and AQP8 has been observed in the midpiece and likely in the mitochondria of the spermatozoon (Laforenza et al., 2017; Pellavio et al., 2020). However, it is not yet clear whether AQP8 functions as a mitochondrial peroxiporin in these cells. An AQP8-mediated mitochondrial detoxification mechanism might function during the hyperactivation process that occurs in the female oviduct under hypotonic conditions, which is reminiscent of the high motility of marine fish sperm (Cosson et al., 2008b) and requires increased intracellular pH and ATP production (Suarez, 2008).

11.5 MOLECULAR REGULATION OF THE AQP8BB PEROXIPORIN MITOCHONDRIAL TRAFFICKING

Recent studies of the spermatozoa of seabream, Atlantic salmon, and zebrafish have uncovered the signaling pathways regulating the trafficking of Aqp8bb to the mitochondria. By using different protein kinase inhibitors and activators, together with immunodetection of the kinase phosphorylation state, Chauvigné and colleagues (2021) determined that the cross-talk between protein kinase C (PKC) and apoptosis signal-regulating kinase 1 (ASK1)/mitogen-activated protein kinase kinase 4 (MKK4)/c-Jun N-terminal kinase (JNK) signaling pathways creates a multiplier cascade that triggers the glycogen synthase kinase (GSK3)-mediated trafficking of the Aqp8bb peroxiporin to the inner mitochondrial membrane of the seabream spermatozoon. In addition, a direct action of JNK, bypassing the GSK3 pathway, is also required for Aqp8bb mitochondrial transport, which likely further enhances the peroxiporin trafficking. This mechanism is partially inhibited by either blocking the Aqp1aa-mediated Ca^{2+} surge or the increase in ROS both occurring during the seawater-induced hyperosmotic shock (Chauvigné et al., 2021). This suggests that cytoplasmic Ca^{2+} and ROS are the upstream signals triggering the activation of the PKC and MAPK/JNK signaling pathways, respectively. Accordingly, inhibition of GSK3 or JNK catalytic activities impairs sperm motility, as observed when Aqp8bb trafficking, and hence mitochondrial function, are immunologically blocked (Chauvigné et al., 2015a, 2021).

Studies in zebrafish and Atlantic salmon spermatozoa showed that, in contrast to seabream inhibition of JNK does not affect mitochondrial Aqp8bb trafficking, indicating that the JNK signaling pathway is not involved in this mechanism (Chauvigné et al., 2021). However, both PKC and GSK3 inhibitors reduce Aqp8bb mitochondrial localization in salmon spermatozoa, whereas only the blockage of PKC elicited the same effect in zebrafish. This indicates a role of the PKC and PKC/GSK3 signaling pathways in the control of Aqp8bb transport in zebrafish and salmon spermatozoa, respectively. Interestingly, the hypoosmotic shock in freshwater does not increase measurable ROS levels in the zebrafish spermatozoon, unless the transport of Aqp8bb to the mitochondria is impaired by PKC inhibition (Chauvigné et al., 2021). These observations suggest that zebrafish spermatozoa have efficient detoxification pathways, possibly involving antioxidants, enzymes, and peroxisomes, to cope with ROS generated during the hypoosmotic shock (Morita et al., 2011; Lahnsteiner and Mansour, 2010; Hagedorn et al., 2012; Nynca et al., 2019). However, when the mitochondrial detoxification function of Aqp8bb is prevented, a situation of oxidative stress is induced and motility declines (Chauvigné et al., 2021). Therefore, in freshwater spermatozoa an osmotic shock-induced ROS/MAPK/JNK signaling pathway controlling Aqp8bb trafficking is probably absent, while the increase in $[Ca^{2+}]_i$ during activation is the major upstream signal that triggers the PKC or PKC/GSK3 cascade controlling Aqp8bb transport.

With the exception of the cross-talk between JNK and PKC that activate GSK3 in seabream, the regulatory pathways of Aqp8bb trafficking in teleost spermatozoa can be replicated in human hepatocarcinoma-derived HepG2 cells (Chauvigné et al., 2021). This cell line shows constitutive expression of the AQP8 ortholog in the inner mitochondrial membrane (Marchissio et al., 2012), and therefore the molecular machinery that transports AQP8 to the mitochondria must exist in these cells. The use of this heterologous cell model as a surrogate system for functional analyses employing wild-type and mutant Aqp8bb, as well as *in vitro* phosphorylation assays of recombinant channels, uncovered the role of Aqp8bb N-terminal phosphorylation on its trafficking to the mitochondria (Chauvigné et al., 2021). These experiments confirmed that Ca^{2+} and H_2O_2 are additive signals for seabream Aqp8bb mitochondrial transport, which trigger the PKC/GSK3 and JNK/GSK3 pathways, respectively, whereas only Ca^{2+} is able to traffic zebrafish and Atlantic salmon Aqp8bb to the mitochondria. More importantly, these studies revealed that GSK3-mediated phosphorylation of the seabream Aqp8bb N-terminal Ser^{16}, which is primed by the phosphorylation of downstream Ser residues (Ser^{19} and Ser^{20}) by yet unknown kinases, as well as the phosphorylation of Thr^{24} by JNK, are both necessary mechanisms for maximum accumulation of the channel in the mitochondria (Figure 11.4A–B). In the N-termini of zebrafish and salmon Aqp8bb, however, Ser residues at position 16 or 14, respectively, are the true target sites of PKC and GSK3. Phosphorylation of these residues directs the channel to the mitochondria, whereas a JNK phosphorylation site is missing in both channels (Figure 11.4B). These data thus provide further support for the role of PKC and GSK3 directly regulating the mitochondrial trafficking of Aqp8bb in the spermatozoa of zebrafish and salmon, respectively.

The molecular regulation of mitochondrial trafficking of the Aqp8bb peroxiporin in teleosts suggests that the signal transduction pathways controlling this mechanism evolved increasing levels of sophistication (Figure 11.4C; Chauvigné et al., 2021). In older lineages of freshwater teleosts, in which spermatozoon motility is activated through hypoosmotic shock, a direct Ca^{2+}-PKC-mediated phosphorylation of the Aqp8bb N-terminus is the only pathway involved. However, in freshwater salmonids, a PKC-regulated GSK3 cascade is added, and the channel is only phosphorylated by GSK3 to induce its trafficking. In contrast, in marine spermatozoa an additional ROS-triggered ASK1/MKK4/JNK pathway is recruited, which directly activates GSK3, but also cross-talks with Ca^{2+}- and ROS-activated PKC to trigger Aqp8bb N-terminal phosphorylation by GSK3, thus creating a multiplier pathway for the regulation of the peroxiporin trafficking. In addition, JNK bypasses the GSK3 pathway to directly phosphorylate the channel in a different N-terminal residue, further enhancing the peroxiporin trafficking. The conserved action of PKC in each of the pathways regulating Aqp8bb trafficking in freshwater and

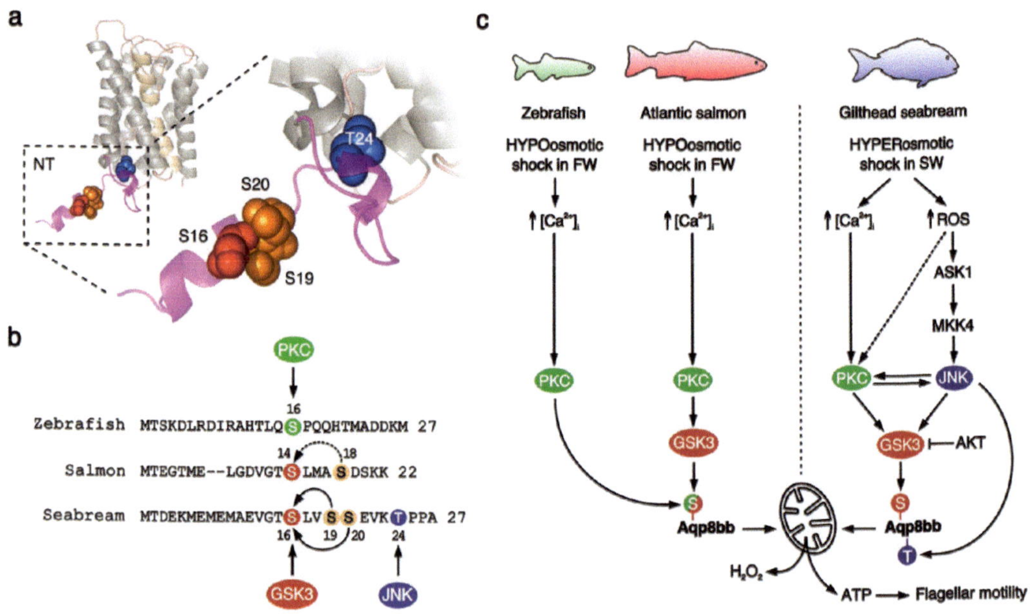

Figure 11.4 Molecular regulation of the peroxiporin Aqp8bb in teleost spermatozoa. (A) Mirrored lateral view of seabream Aqp8bb rendered by MacPymol using homologous *Pichia pastoris* Aqy1 (3ZOJ) as the structure mask. The enlarged render of the N-terminus (NT, magenta) shows the functional phosphorylation sites (spacefill) outlined in panel B. (B) Schematic representation of zebrafish, Atlantic salmon, and seabream Aqp8bb N-termini showing the phosphorylation sites by protein kinase C (PKC), glycogen synthase kinase 3 (GSK3) and c-Jun N-terminal kinase (JNK). The priming sites (Ser^{19} and Ser^{20}) for Ser^{16} phosphorylation by GSK3 in the seabream sequence are indicated by arrows, whereas a potential priming site for GSK3-mediated Ser^{14} phosphorylation in salmon Aqp8bb is indicated by a dashed arrow. (C) Proposed model by Chauvigné et al. (2021) for the evolution of the intracellular signal transduction pathways regulating Aqp8bb mitochondrial trafficking in piscine spermatozoa. In spermatozoa from more primitive freshwater teleost lineages, represented by zebrafish and salmonids, a rise of intracellular Ca^{2+} ($[Ca^{2+}]_i$) upon activation in hypotonic freshwater triggers the PKC-GSK3 signaling pathway, leading to the direct PKC (zebrafish) or GSK3 (Atlantic salmon) phosphorylation of N-terminal Aqp8bb residues, promoting rapid peroxiporin trafficking to the mitochondria. In modern marine teleosts, represented by the seabream, which release the sperm into the hypertonic seawater, cross-talk between $[Ca^{2+}]_i$-activated PKC and reactive oxygen species (ROS)-activated JNK and PKC signaling pathways converges to activate GSK3 with an additional JNK-mediated multiplier regulation of Aqp8bb mitochondrial trafficking to cope with a higher osmotic stress.

Source: Modified from Chauvigné et al., 2021.

seawater spermatozoa reveals how the multiplier mechanism evolved. Thus, in salmonids the addition of the GSK3 regulatory module can act as a multiplier if the expression levels of this kinase significantly exceed those of PKC, whereas in marine-activated spermatozoa the recruitment of the JNK pathway further enhances PKC activity, in addition to the activation of GSK3, which multiplies the action of GSK3 to phosphorylate Aqp8bb.

The evolution of a multiplier anti-apoptotic mechanism in activated marine spermatozoa, such as the JNK signaling pathway (Yue and López, 2020), to enhance the insertion of the Aqp8bb peroxiporin into the mitochondria, possibly reflects an adaptation of the sperm cells to maintain survival and swimming performance under a high osmotic and oxidative stress due to the hyperosmotic environment (García et al., 2012; Sagare-Patil et al., 2012). The importance of this stress pathway is clearly evident on the seabream spermatozoon kinematic properties, where JNK inhibition more rapidly suppresses the degree of motility, progressivity, and velocity of spermatozoa compared to GSK3 inhibition (Chauvigné et al., 2021). Consequently, it seems likely that positive selection of the JNK pathway with efficient catalytic kinetics would provide a selective advantage for competitive fertilization success of marine species, while those of the PKC and GSK3 pathways would be favorable for freshwater species (Chauvigné et al., 2021). Traits associated with the mitochondrial trafficking of the peroxiporin may thus underlie the mechanistic basis of sperm competition in fishes and should be investigated in the future.

11.6 CONCLUSIONS AND PERSPECTIVES

Current studies indicate that regardless of their phylogenetic affinities, most gnathostome water-transporting aquaporins likely retain an ancient biophysical property of H_2O_2 permeation. However, only channels that are expressed to facilitate H_2O_2 flux in a physiological context should receive the peroxiporin moniker. For fishes, aquaporin localization and peroxiporin function has mostly been characterized in the marine gilthead seabream, and therefore our understanding of the role of peroxiporins in teleost sperm physiology remains limited, particularly for freshwater species. In addition, although the role and regulation of the Aqp8bb peroxiporin have been uncovered, the potential peroxiporin functions of other aquaporins such as Aqp1aa, 1ab1, 3a, 4a, 7, and 10bb, which are expressed in spermatozoon membranes, remain to be investigated.

The discovery of the physiological function of the Aqp8bb peroxiporin and its trafficking mechanism in fish spermatozoa also raises many intriguing questions for future research. These include, for example, the existence of functional co-option mechanisms in spermatozoa among teleost Aqp8 paralogs, the involvement of direct protein interaction partners regulating Aqp8bb trafficking, or the presence of additional molecular mechanisms that specifically allow an efficient Aqp8bb-mediated efflux of ROS in the mitochondria. The elucidation of the Aqp8bb peroxiporin pathway may also offer new insight for the identification of polymorphisms or other genetic variations underpinning post-activated spermatozoon performance. Such traits may help to explain how sperm competition evolved in teleosts as well as to establish genetic markers of sperm quality that may be used in marker-assisted breeding programs in aquaculture.

In summary, a detailed understanding of the function and regulation of peroxiporins in piscine spermatozoa will provide new knowledge on the complex signaling network that optimizes motility under different environmental conditions, as well as insight into the molecular mechanisms involved in sperm evolution and postcopulatory sexual selection.

11.7 ACKNOWLEDGMENTS

The research carried out by the authors was financially supported by the Spanish Ministry of Science and Innovation (AGL2010-15597, AGL2013-41196, and AGL2016-76802), Generalitat de Catalunya (2014 SGR 1351 and 2017 SGR 1042), and the Research Council of Norway (Grants 254872/E40 and 294768/E40).

REFERENCES

Abascal, F., Irisarri, I., and R. Zardoya. 2014. Diversity and evolution of membrane intrinsic proteins. *Biochim Biophys Acta* 1840:1468–1481.

Agre, P., Sasaki, S., and M. J. Chrispeels. 1993. Aquaporins: a family of water channel proteins. *Am J Phys.* 265:F461.

Alavi, S. M., and J. Cosson. 2006. Sperm motility in fishes. (II) Effects of ions and osmolality: a review. *Cell Biol Int.* 30:1–14.

Almasalmeh, A., Krenc, D., Wu, B., and E. Beitz. 2014. Structural determinants of the hydrogen peroxide permeability of aquaporins. *FEBS J* 281:647–656.

Babior, B. M. 1987. The respiratory burst oxidase. *Trends Biochem Sci* 12:241–243.

Baskaran, S., Finelli, R., Agarwal, A., and R Henkel. 2021. Reactive oxygen species in male reproduction: a boon or a bane? *Andrologia* 53:e13577.

Beitz, E., Wu, B., Holm, L. M., Schultz, J. E., and T. Zeuthen. 2006. Point mutations in the aromatic/arginine region in aquaporin 1 allow passage of urea, glycerol, ammonia, and protons. *Proc Natl Acad sci USA* 103:269–274.

Berg, D., Youdin, M.B.H., and P. Riederer. 2004. Redox imbalance. *Cell Tissue Res* 318:201–213.

Bertolotti, M., Bestetti, S., García-Manteiga, J. M., Medraño-Fernandez, I., Dal Mas, A., Malosio, M. L., and R. Sitia. 2013. Tyrosine kinase signal modulation: a matter of H_2O_2 membrane permeability? *Antioxid Redox Signal* 19:1447–1451.

Bertolotti, M., Farinelli, G., Galli, M., Aiuti, A., and R. Sitia. 2016. AQP8 transports NOX2-generated H_2O_2 across the plasma membrane to promote signaling in B cells. *J Leukocyte Biol* 100:1071–1079.

Besetti, S., Galli, M., Sorrentino, I., Pinton, P., Rimessi, A., Sitia, R., and I. Medraño-Fernandez. 2020. Human aquaporin-11 guarantees efficient transport of H_2O_2 across the endoplasmic reticulum membrane. *Redox Biol* 28:101326.

Bienert, G. P., and F. Chaumont. 2014. Aquaporin-facilitated transmembrane diffusion of hydrogen peroxide. *Biochim Biophys Acta* 1840:1596–1604.

Bienert, G. P., Møller, A.L.B., Kristiansen, K. A., Schulz, A., Møller, I. M., Schjoerring, J. K., and T. P. Jahn. 2007. Specific aquaporins facilitate the diffusion of hydrogen peroxide across membranes. *J Biol Chem* 282:1183–1191.

Bienert, G. P., Schjoerring, D. K., and D. P. Jahn. 2006. Membrane transport of hydrogen peroxide. *Biochim Biophys Acta* 1758:994–1003.

Boj, M., Chauvigné, F., and J. Cerdà. 2015a. Coordinated action of aquaporins regulates sperm motility in a marine teleost. *Biol Reprod* 93(2):40.

Boj, M., Chauvigné, F., and J. Cerdà. 2015b. Aquaporin biology of spermatogenesis and sperm physiology in mammals and teleosts. *Biol Bull* 229:93–108.

Burnaugh, L., Ball, B. A., Sabeur, K., Thomas, A. D., and S. A. Meyers. 2010. Osmotic stress stimulates generation of superoxide anion by spermatozoa in horses. *Anim Reprod Sci* 117:249–260.

Byon, C. H., Heath, J. M., and Y. Chen. 2016. Redox signaling in cardiovascular pathophysiology: a focus on hydrogen peroxide and vascular smooth muscle cells. *Redox Biol* 9:244–253.

Cerdà, J., Chauvigné, F., and R. N. Finn. 2017. The physiological role and regulation of aquaporins in teleost germ cells. *Adv Exp Med Biol* 969:149–171.

Chauvigné, F., Boj, M., Finn, R. N., and J. Cerdà. 2015a. Mitochondrial aquaporin-8-mediated hydrogen peroxide transport is essential for teleost spermatozoon motility. *Sci Rep* 5:7789.

Chauvigné, F., Boj, M., Vilella, S., Finn, R. N., and J. Cerdà. 2013. Subcellular localization of selectively permeable aquaporins in the male germ line of a marine teleost reveals spatial redistribution in activated spermatozoa. *Biol Reprod* 89(2):37.

Chauvigné, F., Ducat, C., Ferré, A., Hansen, T., Carrascal, M., Abián, J., Finn, R. N., and J. Cerdà. 2021. A multiplier peroxiporin signal transduction pathway powers piscine spermatozoa. *Proc Natl Acad Sci USA* 118(10):e2019346118.

Chauvigné, F., Yilmaz, O., Ferré, A., Fjelldal, P. G., Finn, R. N., and J. Cerdà. 2019. The vertebrate Aqp14 water channel is a neuropeptide-regulated polytransporter. *Comm Biol* 2:462.

Chauvigné, F., Zapater, C., Stavang, J. A., Taranger, G. L., Cerdà, J., and R. N. Finn. 2015b. The pH sensitivity of Aqp0 channels in tetraploid and diploid teleosts. *FASEB J* 29:2172–2184.

Cooper, T. G., and C. H. Yeung. 2003. Acquisition of volume regulatory response of sperm upon maturation in the epididymis and the role of the cytoplasmic droplet. *Microsc Res Tech* 61:28–38.

Cosson, J. 2019. Fish sperm physiology: structure, factors regulating motility, and motility evaluation. In *Biological Research in Aquatic Science*, ed. Y. Bozkurt, Chapter 2, pp. 2–26. IntechOpen, Rijeka. http://doi.org/10.5772/intechopen.85139.

Cosson, M. P., Billard, R., and L. Letellier. 1989. Rise of internal Ca^{2+} accompanies the initiation of trout sperm motility. *Cell Motil Cytoskeleton* 14:3424–3434.

Cosson, J., Billard, R., Cibert, C., Dreanno, C., and M. Suquet. 1999. Regulation of axonemal wave parameters of fish spermatozoa by ionic factors. In *The Male Gamete: From Basic Knowledge to Clinical Applications*, ed. C. Gagnon, 161–186. Cache River Press, Vienna.

Cosson, J., Groison, A.-L., Suquet, M., Fauvel, C., Dreanno, C., and R. Billard. 2008a. Studying sperm motility in marine fish: an overview on the state of the art. *J Appl Ichthyol* 24:460–486.

Cosson, J., Groison, A.-L., Suquet, M., Fauvel, C., Dreanno, C., and R. Billard. 2008b. Marine fish spermatozoa: racing ephemeral swimmers. *Reproduction* 136:277–294.

De Maré, S. W., Venskutonytè, R., Eltschkner, S., de Groot, B., and K. Lindkvist. 2020. Structural basis for glycerol efflux and selectivity of human aquaporin 7. *Structure* 28:1–8.

Di Giorgio, J. P., Soto, G., Alleva, K., Jozefkowicz, C., Amodeo, G., Muschietti, J. P., and N. D. Ayub. 2014. Prediction of aquaporin function by integrating evolutionary and functional analyses. *J Membrane Biol* 247:107–125.

Dreanno, C., Cosson, J., Suquet, M., Seguin, F., Dorange, G., and R. Billard. 1999. Nucleotide content, oxidative phosphorylation, morphology, and fertilizing capacity of turbot (*Psetta maxima*) spermatozoa during the motility period. *Mol Reprod Dev* 53(2):230–243.

Dzyuba, V., and J. Cosson. 2014. Motility of fish spermatozoa: from external signaling to flagella response. *Reprod Biol* 14:165–175.

Engelund, M. B., Chauvigné, F., Christensen, B. M., Finn, R. N., Cerdà, J., and S. S. Madsen. 2013. Differential expression and novel permeability properties of three aquaporin 8 paralogs from seawater-challenged Atlantic salmon smolts. *J Exp Biol* 216:3873–3885.

Ferré, A., Chauvigné, F., Vlasova, A., Norberg, B., Bargelloni, L., Guigó, R., Finn, R. N., and J. Cerdà. 2023. Functional evolution of clustered aquaporin genes reveals insights into the oceanic success of teleost eggs. *Mol Biol Evol* 40(4): msad071.

Finn, R. N., and J. Cerdà. 2011. Aquaporin evolution in fishes. *Front Physiol* 2:44.

Finn, R. N., and J. Cerdà. 2015. Evolution and functional diversity of aquaporins. *Biol Bull* 229:6–23.

Finn, R. N., and J. Cerdà. 2018. Aquaporin. In *Encyclopedia of Signaling Molecules*, 2nd edition, ed. S. Choi, 1–18. Springer, New York.

Finn, R. N., Chauvigné, F., Hlidberg, J. B., Cutler, C. P., and J. Cerdà. 2014. The lineage-specific evolution of aquaporin gene clusters facilitated tetrapod terrestrial adaptation. *PLoS ONE* 9(11):e113686.

Finn, R. N., Chauvigné, F., Stavang, J. A., Belles, X., and J. Cerdà. 2015. Insect glycerol transporters evolved by functional co-option and gene replacement. *Nature Commun* 6:7814.

García, B. M., Moran, A. M., Fernández, L. G., Ferrusola, C. O., Rodriguez, A. M., Bolaños, J. M., da Silva, C. M., Martínez, H. R., Tapia, J. A., and F. J. Peña. 2012. The mitochondria of stallion spermatozoa are more sensitive than the plasmalemma to osmotic-induced stress: role of c-Jun N-terminal kinase (JNK) pathway. *J Androl* 33:105–113.

Garneau, A. P., Carpentier, G. A., Marcoux, A. A., Cotton, R. F., Simard, C. F., Rémus-Borel, W., Caron, L., Jacob-Wagner, M., Noël, M., Oowell. J. J., Bélanger, R., Côté, F., and P. Isenring. 2015. Aquaporins mediate silicon transport in humans. *PLoS ONE* 10(8):e0136149.

Gazo, I., Shaliutina-Kolešová, A., Dietrich, M. A., Linhartová, P., Shaliutina, O., and J. Cosson. 2015. The effect of reactive oxygen species on motility parameters, DNA integrity, tyrosine phosphorylation and phosphatase activity of common carp (*Cyprinus carpio* L.) spermatozoa. *Mol Reprod Dev* 82:48–57.

Giorgio, M., Trinei, M., Migliaccio, E., and P. G. Pelicci. 2007. Hydrogen peroxide: a metabolic by-product or a common mediator of ageing signals? *Nature Rev Mol Cell Biol* 8:722–727.

Gotfryd, K., Mósca, A. F., Missel, J. W., Truelsen, S. F., Wang, K., Spulber, M., Krabbe, S., Hélix-Nielsen, C., Laforenza, U., Soveral, G., Pedersen, P. A., and P. Gourdon. 2018. Human adipose glycerol flux is regulated by a pH gate in AQP10. *Nature Commun* 9:4749.

Gough, D. R., and D. G. Cotter. 2011. Hydrogen peroxide: a Jekyll and Hyde signaling molecule. *Cell Death and Disease* 2:e213.

Gouleh, I. A., Frazziano, G., Rodriguez, A. I., Csányi, G., Maniar, S., St Croix, C. M., Kelley, E. E., Egaña, L. A., Song, G. J., Bisello, A., Lee, Y.J., and P. J. Pagano. 2013. Aquaporin 1, Nox1, and Ask1 mediate oxidant-induced smooth muscle cell hypertrophy. *Cardiovascular Res* 97:134–142.

Hagedorn, M., McCarthy, M., Carter, V. L., and S. A. Meyers. 2012. Oxidative stress in zebrafish (*Danio rerio*) sperm. *PLoS ONE* 7(6):e39397.

Henzler, T., and E. Steudle. 2000. Transport and metabolic degradation of hydrogen peroxide in *Chara corallina*: model calculations and measurements with the pressure probe suggest transport of H_2O_2 across water channels. *J Exp Botany* 51:2053–2066.

Herb, M., Gluschko, A., and M. Schramm. 2021. Reactive oxygen species: not omnipresent but important in many locations. *Frontiers Cell Dev Biol* 9:716406.

Holm, L. M., Klaerke, D. A., and T. Zeuthen. 2004. Aquaporin 6 is permeable to glycerol and urea. *Pflugers Arch* 448:181–186.

Ishibashi, K., Kuwahara, M., Gu, Y., Tanaka, Y., Marumo, F., and S. Sasaki. 1998. Cloning and functional expression of a new aquaporin (AQP9) abundantly expressed in the peripheral leukocytes permeable to water and urea, but not to glycerol. *Biochem Biophys Res Comm* 244:268–274.

Ishibashi, K., Sasaki, S., Fushimi, K., Uchida, S., Kuwahara, M., Saito, H., Furukawa, T., Nakajima, K., Yamaguchi, Y., Gojobori, T., and F. Marumo. 1994. Molecular cloning and expression of a member of the aquaporin family with permeability to glycerol and urea in addition to water expressed at the basolateral membrane of kidney collecting duct cells. *Proc Natl Acad Sci USA* 91:6269–6273.

Kirscht, A., Sonntag, Y., Kjellbom, P., and U. Johanson. 2018. A structural preview of aquaporin 8 via homology modeling of seven vertebrate isoforms. *BMC Struct Biol* 18:2.

Kitchen, P., Salman, M. M., Pickel, S. U., Jennings, J., Törnroth-Horsefield, S., Conner, M. T., Bill, R. M., and A. C. Conner. 2019. Water channel pore size determines exclusion properties but not solute selectivity. *Sci Rep* 9:20369.

Kollár, T., Kása, E., Csorbai, B., Urbányi, B., Csenki-Bakos, Z., and A. Horváth. 2018. *In vitro* toxicology test system based on common carp (*Cyprinus carpio*) sperm analysis. *Fish Physiol Biochem* 44:1577–1589.

Koppers, A. J., De Iuliis, G. N., Finnie, J. M., McLaughlin, E. A., and R. J. Aitken. 2008. Significance of mitochondrial reactive oxygen species in the generation of oxidative stress in spermatozoa. *J Clin Endocrinol Metab* 93:3199–3207.

Krasznai, Z., Marian, T., Izumi, H., Damjanovich, S., Balkay, L., Tron, L., and M. Morisawa. 2000. Membrane hyperpolarization removes inactivation of Ca^{2+} channels, leading to Ca^{2+} influx and subsequent initiation of sperm motility in the common carp. *Proc Natl Acad Sci USA* 97:2052–2057.

Krüger, C., Jörns, A., Kaynert, J., Waldeck-Weiermair, M., Michel, T., Elsner, M. and S. Lenzen. 2021a. The importance of aquaporin-8 for cytokine-mediated toxicity in rat insulin-producing cells. *Free Radical Biol Med* 174:175–143.

Krüger, C., Waldeck-Weiermair, M., Kaynert, J., Pokrant, T., Komaragiri, Y., Otto, O., Michel, T., and M. Elsner. 2021b. AQP8 is a crucial H_2O_2 transporter in insulin-producing RINm5F cells. *Redox Biol* 43:101962.

Labbe, C., and G. Maisse. 2001. Characteristics and freezing tolerance of brown trout spermatozoa according to rearing water salinity. *Aquaculture* 201:287–299.

Laforenza, U., Pellavio, G., Marchetti, A. L., Omes, C., Todaro, F., and G. Gastaldi. 2017. Aquaporin-mediated water and hydrogen peroxide transport is involved in normal human spermatozoa functioning. *Int J Mol Sci* 18:66.

Lahnsteiner, F., and S. Caberlotto. 2012. Motility of gilthead seabream *Sparus aurata* spermatozoa and its relation to temperature, energy metabolism and oxidative stress. *Aquaculture* 370–371:76–83.

Lahnsteiner, F., and N. Mansour. 2010. A comparative study on antioxidant systems in semen of species of the Percidae, Salmonidae, Cyprinidae, and Lotidae for improving semen storage techniques. *Aquaculture* 307:130–140.

Madeira, A., Fernández-Valedo, S., Camps, M., Lorzano, A., Maura, T. F., Ceperuelo-Mallafre, V., Vendrell, J., and G. Soveral. 2014. Human aquaporin-11 is a water and glycerol channel and localizes in the vicinity of lipid droplets in human adipocytes. *Obesity* 22:2010–2017.

Marchissio, M. J., Francés, D. E., Carnovale, C. E., and R. A. Marinelli. 2012. Mitochondrial aquaporin-8 knockdown in human hepatoma HepG2 cells causes ROS-induced mitochondrial depolarization and loss of viability. *Toxicol Appl Pharmacol* 264:246–254.

McCarthy, M. J., Baumber, J., Kass, P. H., and S. A. Meyers. 2010. Osmotic stress induces oxidative cell damage to rhesus macaque spermatozoa. *Biol Reprod* 82:644–651.

Miller, E. W., Dickinson, B. C., and C. J. Chang. 2010. Aquaporin-3 mediates hydrogen peroxide uptake to regulate downstream intracellular signaling. *Proc Natl Acad Sci USA* 107:15681–15686.

Morisawa, M., and K. Suzuki. 1980. Osmolality and potassium ion: their roles in initiation of sperm motility in teleosts. *Science* 210:1145–1147.

Morisawa, M., Suzuki, K., and S. Morisawa. 1983. Effects of potassium and osmolality on spermatozoan motility of salmonid fishes. *J Exp Biol* 107:105–113.

Morita, M., Nakajima, A., Takemura, A., and M. Okuno. 2011. Involvement of redox- and phosphorylation-dependent pathways in osmotic adaptation in sperm cells of euryhaline tilapia. *J Exp Biol* 214:2096–2104.

Nynca, J., Słowińska, M., Judycka, S., and A. Ciereszko. 2019. Maladaptation of trout spermatozoa to fresh water is related to oxidative stress and proteome changes. *Reproduction* 157:485–499.

Ørtenblad, N., Young, J. F., Oksberg, N., Nielsen, J. H., and I. H. Lambert. 2003. Reactive oxygen species are important mediators of taurine release from skeletal muscle cells. *Am J Physiol Cell Physiol* 284:C1362–C1373.

Pataki, B., Roberta, B. I., Gazsi, G., Urbányi, B., Kollár, T., and Á. Horváth. 2021. Effect of age on the mercury sensitivity of zebrafish (*Danio rerio*) sperm. *Fish Physiol Biochem* 47:687–695.

Pellavio, G., and U. Laforenza. 2021. Human sperm functioning is related to the aquaporin-mediated water and hydrogen peroxide transport regulation. *Biochimie* 188:45–51.

Pellavio, G., Todaro, F., Alberizzi, P., Scotti, C., Gastaldi, G., Lolicato, M., Omes, C., Caliogna, L., Nappi, R. E., and U. Laforenza. 2020. HPV infection affects human sperm functionality by inhibition of aquaporin-8. *Cells* 9(5):1241.

Perchec Poupard, G., Gatti, J. L., Cosson, J., Jeulin, C., Fierville, F., and R. Billard. 1997. Effects of extracellular environment on the osmotic signal transduction involved in activation of motility of carp spermatozoa. *J Reprod Fertil* 110:315–327.

Pérez, L., Vílchez, M. C., Gallego, V., Morini, M., Peñaranda, D. S., and J. F. Asturiano. 2016. Role of calcium on the initiation of sperm motility in the European EEL. *Comp Biochem Physiol A Mol Integr Physiol* 191:98–106.

Prata, C., Hrelia, S., and D. Fiorentini. 2019. Peroxiporins in cancer. *Int J Mol Sci* 20:1371.

Preston, G. M., Carroll, T. P., Guggino, W. B., and P. Agre. 1992. Appearance of water channels in *Xenopus oocytes* expressing red cell CHIP28 protein. *Science* 256:385–387.

Preston, G. M., Jung, S. P., Guggino, W. B., and P. Agre. 1993. The mercury-sensitive residue at cysteine 189 in the CHIP28 water channel. *J Biol Chem* 268:17–20.

Purohit, V., Someone D. M., and C. A. Lyssiotis. 2019. Metabolic regulation of redox balance in cancer. *Cancers* 11:995.

Rhee, S. G. 2006. H_2O_2, a necessary evil for cell signaling. *Science* 312:1882–1883.

Rodrigues, C., Pimpão, C., Mósca, A. F., Coxixo, A. S., Lopes, D., da Silva, I. V., Pedersen, P. A., Antunes, F., and G. Soveral. 2019. Human aquaporin-5 facilitates hydrogen peroxide permeation affecting adaption to oxidative stress and cancer cell migration. *Cancers* 11:932.

Sagare-Patil, V., Galvankar, M., Satiya, M., Bhandari, B., Gupta, S. K., and D. Modi. 2012. Differential concentration and time dependent effects of progesterone on kinase activity, hyperactivation and acrosome reaction in human spermatozoa. *Int J Androl* 35:633–644.

Satooka, H., and M. Hara-chikuma. 2016. Aquaporin-3 controls breast cancer cell migration by regulating hydrogen peroxide transport and its downstream cell signaling. *Mol Cell Biol* 36:1206–1218.

Sies, H. 2017. Hydrogen peroxide as a central redox signaling molecule in physiological oxidative stress: oxidative stress. *Redox Biol* 11:613–619.

Sies, H., Berndt, C., and D. P. Jones. 2017. Oxidative stress. *Annu Rev Biochem* 86:715–748.

Suarez, S. S. 2008. Control of hyperactivation in sperm. *Hum Reprod Update* 14:647–657.

Takei, G. L., Mukai, C., and M. Okuno. 2015. Regulation of salmonid fish sperm motility by osmotic shock-induced water influx across the plasma membrane. *Comp Biochem Physiol A Mol Integr Physiol.* 182:84–92.

Tanimoto, S., Kudo, Y., Nakazawa, T., and M. Morisawa. 1994. Implication that potassium flux and increase in intracellular calcium are necessary for the initiation of sperm motility in salmonid fishes. *Mol Reprod Dev* 39:409–414.

Tingaud-Sequeira, A., Calusinska, M., Finn, R. N., Chauvigné, F., Lozano, J., and J. Cerdà. 2010. The zebrafish genome encodes the largest vertebrate repertoire of functional aquaporins with dual paralogy and substrate specificities similar to mammals. *BMC Evol Biol* 10:38.

Varadaraj, K., and S. S. Kumari. 2020. Lens aquaporins function as peroxiporins to facilitate membrane transport of hydrogen peroxide. *Biochim Biophys Acta* 524:1025–1029.

Vines, C. A., Yoshida, K., Griffin, F. J., Pillai, M. C., Morisawa, M., Yanagimachi, R., and G. N. Cherr. 2002. Motility initiation in herring sperm is regulated by reverse sodium-calcium exchange. *Proc Natl Acad Sci USA* 99:2026–2031.

Watanabe, S., Moniaga, C. S., Nielsen, S., and M. Hara-Chikuma. 2016. Aquaporin-9 facilitates membrane transport of hydrogen peroxide in mammalian cells. *Biochem Biophys Res Comm* 471:191–197.

Wragg, D., Leoni, S., and A. Casini. 2020. Aquaporin-driven hydrogen peroxide transport: a case of molecular mimicry? *RSC Chem Biol* 1:390–394.

Yilmaz, O., Ferré, A., Nilsen, F., Fjelldal, P. G., Cerdà, J., and R. N. Finn. 2020. Unravelling the complex duplication history of deuterostome glycerol transporters. *Cells* 9(7):1663.

Yue, J., and J. M. López. 2020. Understanding MAPK signaling pathways in apoptosis. *Int J Mol Sci* 21(7):2346.

Zilli, L., Beirão, J., Schiavone, R., Herraez, M. P., Cabrita, E., Storelli, C., and S. Vilella. 2011. Aquaporin inhibition changes protein phosphorylation pattern following sperm motility activation in fish. *Theriogenology* 76(4):737–744.

Zilli, L., Schiavone, R., Chauvigné, F., Cerdà, J., Storelli, C., and S. Vilella. 2009. Evidence for the involvement of aquaporins in sperm motility activation of the teleost gilthead sea bream (*Sparus aurata*). *Biol Reprod* 81(5):880–888.

Zilli, L., Schiavone, R., Storelli, C., and S. Vilella. 2008. Molecular mechanisms determining sperm motility initiation in two sparids (*Sparus aurata* and *Lithognathus mormyrus*). *Biol Reprod* 79(2):356–366.

Zilli, L., Schiavone, R., Storelli, C., and S. Vilella. 2012. Molecular mechanism regulating axoneme activation in marine fish: a review. *Int Aquat Res* 4:2.

CHAPTER TWELVE

Regulated Transport of H_2O_2 across the Plasma Membrane by Aquaporin-8

Milena Bertolotti,
José Manuel García-Manteiga, and Roberto Sitia

ABBREVIATIONS

AQP	Aquaporin
CBS	Cystathionine-beta-synthase
DTT	Dithiothreitol
ER	Endoplasmic reticulum
LPS	Lipopolysaccharide
MAMS	Mitochondria-associated ER membranes
ROS	Reactive oxygen species

12.1 INTRODUCTION

Our interest in peroxiporins started because of the senior author's ignorance. In the first decade of the millennium, most discussions in the lab were centered on the pathobiology of endoplasmic reticulum (ER). Simone Cenci, Niccolò Pengo, and Giada Bianchi were finding exciting connections between protein quality control in the ER and the sensitivity of myeloma cells to proteasome inhibitors (2006–2009). Noticing that the lumen of the ER tended to swell in the presence of tunicamycin or other stressors, we reasoned that dilution of the ER lumen would be an effective measure to acutely prevent protein aggregation. What about activating some aquaporins then? The idea sounded smart enough for the senior author of this chapter, who at the time viewed AQPs like hydrants in the hands of wise firemen, able to push a button and prompt water influx. It did not take much effort to learn that AQPs are channels, rather than pumps. Meanwhile, Gloria Bertoli and Thomas Simmen were playing with Ero1 mutants, which the Weissman, Kaiser, and Fass labs had shown to produce H_2O_2 in stoichiometric amounts to the disulfides made. And a single plasma cell, one of the favorite objects of study in the lab, can produce over 10^5 disulfides per second only to satisfy immunoglobulin biosynthesis, and presumably as many H_2O_2 molecules.

How do these cells cope with so many oxidants?

Are the latter important in limiting plasma cell life span?

Fascinated by these questions, Milena, then a master student in the lab, started reading about redox signaling and eventually traveled to Seoul to visit Sue Goo Rhee's lab. We reasoned that peroxiredoxins might play a role, especially Prdx4, an ER-resident isoform which increases about 40-fold during B to plasma cell differentiation. Much to our surprise, however, we found that Prd4^{KO} mice had only minor phenotypes in IgM secretion and plasma cell life span (Bertolotti et al. 2010). Thus, other mechanisms are at play. Our suspicions turned on AQPs. Key to the switch was the work of Seva Belousov and Chris Chang, who were able to generate excellent H_2O_2 sensors (see Chapter 5 in this volume).

In the fall of 2009, we tried to detect changes in intracellular H_2O_2 in I.29μ$^+$, a B cell lymphoma

DOI: 10.1201/9781003160649-12

which can be readily induced to differentiate toward IgM secretion with LPS. Our first attempts exploited the chemical sensor PG-1, which Chris Chang had generously provided to us. There were undoubtedly changes in the fluorescence of activated B cells. In the first setup of our experiments, cells were seeded in 96 or 24 well plates and placed onto a Typhoon detector, which was instructed to scan the samples at close-time intervals. Plotting the fluorescence data clearly revealed patterns which were consistent with a rapid entry of H_2O_2, followed by its slow clearance. In Italian, H_2O_2 is called *acqua ossigenata* ("oxygenated water"). Even an old hematologist could realize that its general shape and dipole are quite similar to H_2O. The finding that also water channels inhibitors like $HgCl_2$ inhibited the PG-1 fluorescence shifts induced by exogenously added H_2O_2 convinced us that indeed AQPs might be involved in the passage of H_2O_2 across membranes, as nicely anticipated in pioneering experiments (Bienert et al. 2007; Miller, Dickinson and Chang 2010). We thus selectively inhibited human AQP3, AQP4, AQP7, and AQP8, the isoforms expressed by our HeLa subline, which we had stably transfected to express ratiometric hydrogen peroxide sensors (HyPer1) in the cytosol, ER lumen, and mitochondrial matrix. Our observations that the silencing of AQP8, but not the other isoforms, prevented H_2O_2 entry into living cells generated great excitement.

Does the AQP8 conduit H_2O_2 also across the ER membrane? Answering this question required a trick, because without AQP8 in the plasma membrane, no H_2O_2 could reach the ER membrane, and a system to catalyze the production of H_2O_2 specifically inside the ER, like the D-amino acid oxidase (DAAO) system (Pollegioni et al. 1993; Tishkov and Khoronenkova 2005), was not available to us then. We thus used digitonin to selectively permeabilize the cholesterol-rich plasma membrane (Nerini-Molteni et al. 2004). AQP8 silencing had no detectable effect on mitochondrial H_2O_2 import. However, it inhibited in part activation of the luminal ER HyPer probe, consistent with an activity of AQP8 en route to the plasma membrane. Chapter 8 of this volume summarizes the roles of AQP11 in transporting H_2O_2 across the ER membrane (Bestetti et al. 2020; Sorrentino et al. 2022). The notion that devoted channels are needed also for interorganellar H_2O_2 transport delineates yet another possible level at which eukaryotes can control redoxstasis.

12.2 TISSUE DISTRIBUTION OF AQP8

AQP8 was originally identified in rat testis (Ishibashi, Kuwahara, Gu, et al. 1997). Its human amino acid sequence (hAQP8) was derived from testis and genomic cDNAs (Koyama et al. 1998) and found to be more similar to a plant water channel, the gamma-TIP (40%), than to hAQP2 and hAQP3 (30%). High AQP8 mRNA levels are detectable in rat and mouse hepatocytes (Ishibashi, Kuwahara, Kageyama, et al. 1997; Koyama et al. 1997; Elkjaer et al. 2001; Calamita et al. 2001; Ma, Yang and Verkman 1997; Ferri et al. 2003; Yang et al. 2005); exocrine (Koyama et al. 1997; Calamita et al. 2001; Ma, Yang and Verkman 1997) and endocrine pancreas (Kruger et al. 2021; Louchami et al. 2012); colon epithelium (Koyama et al. 1997; Elkjaer et al. 2001; Calamita et al. 2001; Ma, Yang and Verkman 1997; Hardin et al. 2004; Yang et al. 2005); and testis (Ishibashi, Kuwahara, Kageyama, et al. 1997; Elkjaer et al. 2001; Yang et al. 2005). AQP8 is abundant in the male reproductive tissues of fish (see Chapter 11 in this volume), and at lower mRNA levels in human testis and sperm (Yeung et al. 2010; Laforenza et al. 2016; Consortium 2013). In humans, consistent levels of AQP8 mRNA were detected only in exocrine pancreas (Koyama et al. 1998; Burghardt et al. 2003); colon mucosa (Koyama et al. 1998; Fischer et al. 2001); and a few areas of the brain (Figure 12.1). The paucity of AQP8 transcripts in human liver and other issues may reflect high stability of AQP8 polypeptides (Figure 12.1). Low levels are found in cardiomyocytes from elder individuals, lung resident macrophages, endocrine pancreas, mast cells, basophils, and hepatocytes (Gregoire et al. 2015). AQP8 can play critical roles also if expressed at low abundance. Considering that each AQP8 is thought to transport up to 10^9 molecules per second, few channels might conduit enough H_2O_2 to generate relevant signals, particularly if concentrated in select membrane regions. This notion answers another important question that often arises when discussing about peroxiporins. If indeed H_2O and H_2O_2 use the same channel, how can extracellular 50 μM H_2O_2 compete with 55 M H_2O? The solution likely lies in the existence of steep concentration H_2O_2 gradients across the

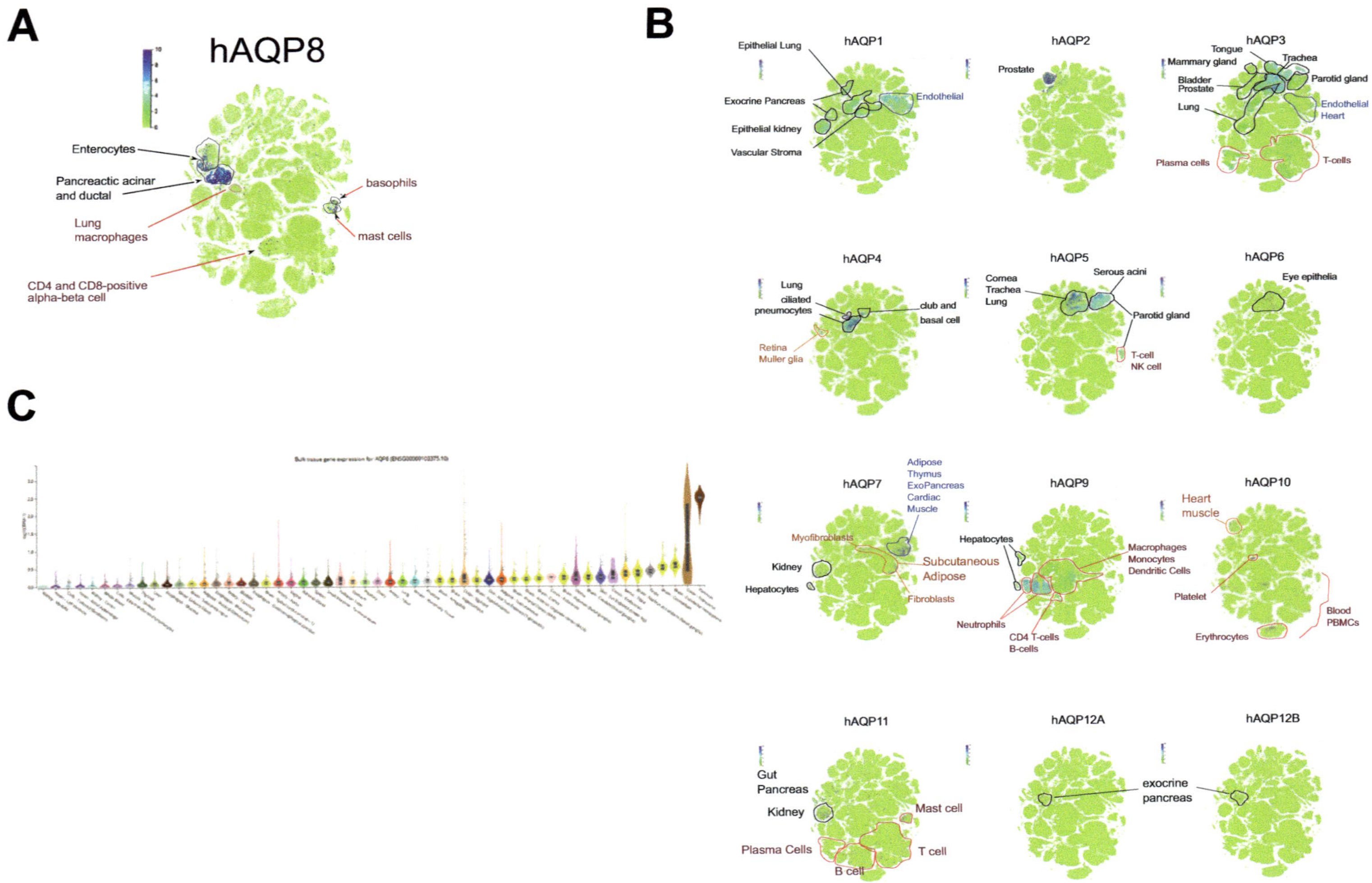

Figure 12.1 Tissue expression of AQP8. UMAP clustering showing the levels of expression of hAQP8 (A) and other AQPs (B) in single cells extracted from Tabula Sapiens datasets. Expression values are shown in transcripts per million (TPM). The clusters of cells belonging to specific cell types and tissues have been flagged based on cell assignments in the Tabula Sapiens CZ CELLxGENE. Cell type label colors are as follows: black, epithelial, blue, endothelial, orange, stromal; red, immune. (C) Bulk tissue RNA-Seq hAQP8 expression was obtained from the Gene Tissue Expression Portal (https://gtexportal.org/home/gene/AQP8).

membrane (Sorrentino and Medraño-Fernandez, Chapter 13 of this volume). Even if at a concentration of 1 millionth to that of H_2O, about 10^3 H_2O_2 molecules can transit through an AQP8 molecule each second. And the local concentration near the cytosolic side of a channel can reach high values.

As a common denominator among cell types and tissues, AQP8 expression is often associated with cells with high secretive potential (exocrine pancreas, colon epithelium, murine hepatocytes), susceptibility to redox signaling (insulin producing beta-cells), and/or high H_2O_2 detoxification needs (sperm cell flagellum). hAQP8 was detected in HepG2 hepatoma cells (Larocca et al. 2009; Marchissio et al. 2012) and other human tumors, including acute leukemias of erythroid phenotype (Vieceli Dalla Sega et al. 2014) and astrocytoma (Zhu et al. 2013). Increased AQP8 levels may correlate with tumor progression (see Section 12.8.1).

12.3 MODELS OF AQP8 STRUCTURE

AQP8 clearly displays the typical tetrameric arrangement of aquaporins, each monomer containing a functional pore formed by six transmembrane helices. Two additional half-transmembrane helices (helices B and E) each holding a conserved motif (Asn-Pro-Ala) and pointing toward each other, form a seventh pseudo-transmembrane helix. The NPA box poses a barrier for protons and most charged solutes. An aromatic/arginine region, situated in a narrow part of the pore, additionally contributes to exclude protons. This region is also referred to as the selectivity filter, as it determines substrate permeability.

So far, structural information for hAQP8 is limited to electron microscopy studies confirming homo-tetramerization (Agemark et al. 2012). However, the availability of crystal structures of other AQPs and the sequence conservation within the family (Ishibashi, Morishita and Tanaka 2017; Ishibashi, Tanaka and Morishita 2020) allowed to obtain a hAQP8 model that survived mutational tests. In fact, replacing residues predicted to serve key regulatory roles yielded the expected phenotypes (Bestetti et al. 2018).

Based on the similarities between H_2O and H_2O_2, it had been suggested that most if not all water-permeable AQPs have peroxiporin activity (Almasalmeh et al. 2014). Their pathophysiological role would vary among the family members and ultimately depend on the transport efficiency, expression levels, and tissue context. In AQP8 and other non-human aquaammoniaporins, the width of the pore is well suited for H_2O_2 transport (Almasalmeh et al. 2014; Prata, Hrelia and Fiorentini 2019; Bestetti et al. 2018).

The recently solved structure of the plant AtTIP2;1, which has structural and functional features in common with AQP8, and the development of AlphaFold analyses, confirmed the overall validity of homology models (Bestetti et al. 2018; Kirscht et al. 2018). Thus, vertebrate AQP8s are likely to have an AtTIP2;1-like selectivity filter. The only deviation is that they are slightly narrower and more hydrophobic, due to a phenylalanine replacing the histidine in loop C (Kirscht et al. 2018). Since vicinal arginines can impact the pKa of cysteines (Trost et al. 2017), R213 is likely to influence the susceptibility of C53 to oxidation (Bestetti et al. 2018), as it happens for tyrosine phosphatases (Meng, Fukada and Tonks 2002; Tonks 2005). Accessing C53, therefore, H_2O_2 may ultimately lead to transitory AQP8 inhibition. Accordingly, as we shall discuss in detail below, oxidative modifications of C53 gate human AQP8 during stress conditions (Bestetti et al. 2018; Medrano-Fernandez et al. 2016). Also the surface electrostatics of AQP8 (Kirscht et al. 2018), which hold clues to its orientation in membranes, may affect the pKa of Cys53. In AQP8, the surface near the selectivity filter and exposed to the extracellular space is more negative than its cytosolic face, where the N- and C-termini protrude. This may indicate an evolutionary selection based on stability at a lower pH as well as a mean to facilitate membrane insertion, as suggested by the positive-inside rule (Heijne 1986).

12.4 MECHANISMS OF AQP8 GATING

Having shown that AQP8 provides an essential conduit for amplifying tyrosine kinase signals (Bertolotti et al. 2013; Bertolotti et al. 2016), we surmised that cells evolved means to regulate its permeability. Accordingly, Iria Medraño-Fernandez and Stefano Bestetti showed that different kinds of cellular stress inhibited the entry of exogenous H_2O_2. As is often the case, a most unexpected result put us on the right track about the gating mechanisms. Paradoxically, in fact, the strong reductant DTT restored HyPer1 oxidation in heat-stressed cells, as long as H_2O_2 was present in the extracellular space (Medrano-Fernandez et al. 2016). This meant that AQP8 was gated by redox-dependent modification(s). Iria was then able to map the main target residue, C53. Molecular modeling revealed that this cysteine lies close to the narrowest part of the AQP8 channel (Bestetti et al. 2018). Initially, we assumed that sulfenylation might gate the channel, but this made little sense, since AQP8 is able to conduit exogenous H_2O_2 also when present at high concentrations. Luckily, Stefano Bestetti attended an EMBO redox meeting at San Feliu de Guixols, and besides swimming in the blue Mediterranean waters, he managed to talk at length with Peter Nagy, who suggested a role for persulfidation. Stefano and Iria had no difficulty in showing that addition of a H_2S donor (NaHS) to cells caused AQP8 gating even in the absence of stress. Cystathionine-beta-synthase (CBS) was the main H_2S source, as its silencing prevented gating in stressed cells. Importantly, NaHS restored closure, proving that H_2S is indeed the culprit. Gating is not immediate upon NaHS addition, because H_2S does not react with a thiol or a thiolate but requires prior sulfenylation. Hence a two-step gating mechanism of AQP8 became clear (Figure 12.2). Passage of H_2O_2 sulfenylates C53, which remains permeant until sufficient H_2S is produced by CBS, itself a redox-sensitive enzyme (Maclean et al. 2002), to cause persulfidation. Now the extra sulfur atom pushes the ring of a histidine (residue 72), ideally placed to gate the channel. Owing to the absence of reliable antibodies against human or murine AQP8, our results were obtained with tagged, overexpressed molecules. Their behavior was similar despite the different size and position of the tags. Aware of the fact that overexpression might cause artifacts, we concluded that AQP8 translocates cotranslationally into the ER, is N-glycosylated, and travels to the plasma membrane, where it accumulates in an Endo-H resistant configuration. Visual pulse chases coupled to H_2O_2 transport assays revealed that gated AQP8 molecules regain permeability in a few hours, also in the presence of inhibitors of protein synthesis (Medraño-Fernandez, Bestetti and Sitia, unpublished). The mechanisms that remove the persulfide moieties are unclear. Buried in the channel, C53 is inaccessible to bulky enzymes or charged reductants like cysteine or glutathione. A more detailed structural description of AQP8 and its main interactors in their physiological contexts would provide more

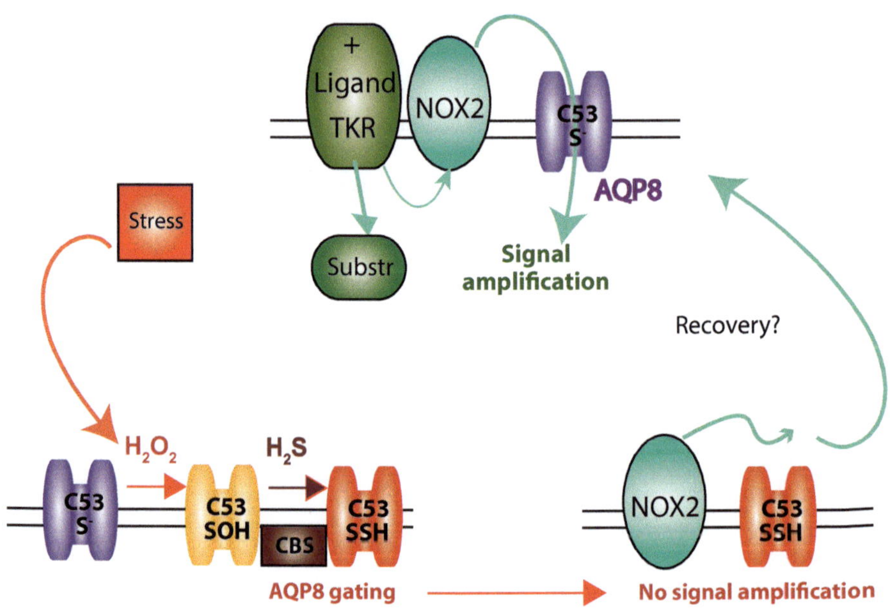

Figure 12.2 Two-step gating of AQP8 prevents H_2O_2 import into stressed cells. Located in the vicinity of the narrowest part of the AQP8 channel, C53 is first sulfenylated by H_2O_2 molecules in transit. This primes C53 for further modification by H_2S, produced by cystathionine beta synthase (CBS) an enzyme that is in turn activated in oxidative conditions. These sequential modifications gate AQP8, blocking further H_2O_2 entry into cells (Bestetti et al. 2018). Recovery from stress entails reopening of AQP8 conduits.

insights in mechanisms that mediate transport and gating. As previously shown for other AQPs, the outcoming tetrameric model of human AQP8 may be used to find drug binding sites or improve potential binders by interaction mapping.

12.5 SUBSTRATE SPECIFICITY

While all AQPs are thought to conduit water but not protons, differences exist in their permeability to other uncharged solutes. Of course, AQP8 is one of the first peroxiporins identified (Bienert et al. 2007), and it amplifies tyrosine kinase signals in lymphoid and non-lymphoid cells (Figure 12.3; see also Bertolotti et al. 2013; Bertolotti et al. 2016).

In addition to H_2O and H_2O_2, AQP8 was shown to be permeable to ammonium ions (NH_4^+), ammonia (NH_3), and urea, though the underlying transport mechanisms are not fully understood. Nitrogen is vital for all organisms. Most prokaryotes can take it up as NH_3 or NH_4^+. Diazotroph bacteria can also fix N_2 from the atmosphere. In mammals, NH_3 and NH_4^+ are obtained by import and digestion of amino acids in the intestine and liver. The cellular levels of these toxic species are tightly regulated: in hepatic failure, their increased blood levels may cause fatal brain edema.

Several studies showed that AQP8 transports NH_3 (Liu et al. 2006; Saparov et al. 2007), rescuing the growth of NH_4^+ uptake-defective yeast mutants (Saparov et al. 2007). Voltage clamp experiments suggested that AQP8 conducts NH_4^+ as well (Holm et al. 2005). However, comparative studies revealed only minor phenotypic differences in serum NH_3, colonic absorption, renal clearance, and liver accumulation (Yang et al. 2006) between control and AQP8 null mice, suggesting a marginal role of this peroxiporin as a facilitator of NH_3 transport. In human liver, AQP8 colocalizes with Rhesus, facilitating their activity as NH_3 and urea transporters (Marchissio et al. 2012). Despite mild evidences, in the kidney, AQP8 may facilitate mitochondrial NH_3 import (Soria et al. 2010, 2013) and adaptive responses of proximal tubule to acidosis (Molinas, Trumper and Marinelli 2012).

Unlike other NH_3-permeable AQPs (3, 7, 9, 10), AQP8 is impermeable to glycerol (Liu et al. 2006).

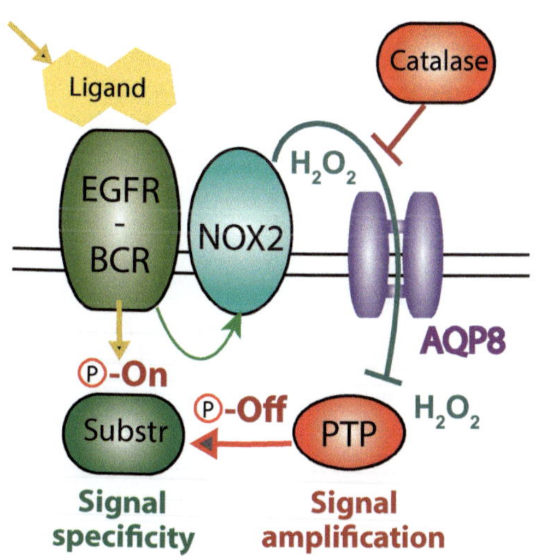

Figure 12.3 AQP8 allows amplification of TKR signals via import of H_2O_2. Schematic representation of the H_2O_2-dependent circuitry. EGFR or BCR triggering induces phosphorylation of key substrates and NOX2 activation. H_2O_2 reaches the cytosol via AQP8, amplifying signals via the transient inactivation of protein tyrosine phosphatases (PTP).

12.6 SUBCELLULAR LOCALIZATION OF AQP8

There is no doubt that AQP8 plays fundamental roles in the plasma membrane. Following the pioneering work of many laboratories (Ishibashi, Sasaki, et al. 1997; Koyama et al. 1998; Koyama et al. 1997; Ma, Yang and Verkman 1997; Garcia et al. 2001), we obtained several lines of evidence confirming its localization in the plasma membrane (Medrano-Fernandez et al. 2016; Bertolotti et al. 2013; Bertolotti et al. 2016; Bestetti et al. 2018). Thus, high-resolution imaging assays revealed that a tagged version of human AQP8, which efficiently rescues H_2O_2 (and H_2O) transport in silenced cells, decorated the plasma membrane (Medraño-Fernandez et al. 2016; Bestetti et al. 2018). Second, addition of extracellular catalase phenocopied the effects of AQP8 silencing in quenching tyrosine kinase signaling (see Bertolotti et al. 2013, 2016, and references therein). Unfortunately, so far, we could neither find in the market nor obtain trustworthy AQP8-specific antibodies to map the localization of endogenous molecules. Nonetheless, we are confident that AQP8 allows the entry into the cytosol of H_2O_2 molecules generated by NOX2 in the extracellular leaflet of the plasma membrane (Bertolotti et al. 2013, 2016).

Several laboratories reported the presence of functional AQP8 molecules in the inner mitochondrial membrane of murine 3T3-L1 adipocytes, liver, kidney, and neural stem cells (Gena et al.

2009; Marchissio et al. 2012; Molinas, Trumper and Marinelli 2012; Ikaga et al. 2015; Soria et al. 2013). In a recent study, a tagged AQP8-V5 was detected in the plasma and mitochondrial, but not in the peroxisomal, ER, Golgi, or secretory vesicle membranes (Kruger et al. 2021). To cope with the strong hypotonic or hypertonic shocks upon release into fresh or seawater, piscine sperm cells have been shown to translocate an AQP8 ortholog from the surface to the mitochondrial inner membrane (see Chapter 11 in this volume for details and references).

Further mechanistic studies are required to map these unorthodox transport pathways. Owing to the tight contacts between ER and mitochondria, which survive ultracentrifugation allowing MAM isolation, it may often be difficult to discriminate the precise localization of proteins that accumulate in this area. In a rat insulinoma-derived cell line, a band of about 28 kDa was observed in the inner mitochondrial membrane-enriched fraction by Western blotting and a slower one (≈34 kDa) in the plasma membrane, consistent with N-glycan processing during Golgi transport of the latter (Kruger et al. 2021). In certain cell types, the N-terminal transmembrane region of AQP8 might reroute part of the molecules from the ER sec61 to the mitochondrial TOM-TIM translocation systems (Cymer, von Heijne and White 2015). In our hands, chimeric AQP8s carrying different N-terminal tags were not detected in mitochondria. A small fraction displayed an electrophoretic mobility consistent with the absence of the single N-glycan added on most AQP8 molecules during their translocation into the ER. It remains to be seen whether these species reflect oligosaccharyl-transferase inefficiency or routing to mitochondria. The tags might prevent insertion of chimeric AQP8 into the inner mitochondrial membrane. A considerable fraction of AQP11 and AQP8 with unprocessed N-glycans accumulate in mitochondrial-ER contact sites (Bestetti et al. 2020) possibly facilitating inter-organellar H_2O_2 exchanges (Sorrentino et al. 2022) and at the same time mimicking mitochondrial localization. Another difficulty that must be reconciled with the dual localization of AQP8 in mammalian cells is the different lipid composition of mitochondrial and plasma membranes. Considering the high surface-to-volume ratio in mitochondria, passive diffusion could in principle mediate transport. In our opinion, however, channel protein(s) endowed with peroxiporin activity, not necessarily belonging to the AQP family, are likely to populate mitochondria. Clearly further investigations are required to decipher how a protein with six-transmembrane stretches abundantly synthesized in the ER and rapidly transported to the cell surface can be rerouted to the inner membrane of mitochondria. What makes their inner membrane permeant to ROS is a question of fundamental relevance for redox signaling and homeostasis.

12.7 PHYSIOLOGICAL ROLES OF AQP8

Elegant work from many labs demonstrated that H_2O_2 acts as a second messenger increasing the strength of tyrosine kinase signaling pathways (see Figure 12.3; Finkel 2011, and references therein). Accordingly, $HgCl_2$ or AQP8 silencing strongly inhibited the H_2O_2 fluxes that can be observed soon upon EGF stimulation of HeLa cells. Also the phosphorylation of many kinase targets, induced by EGF or LPS, was inhibited in cells lacking AQP8 (Bertolotti et al. 2013). AQP8 silencing was almost as efficient as the addition of catalase to the extracellular medium (Figure 12.3), confirming that most EGF-induced H_2O_2 molecules entered through AQP8 channels. Consistently with a role for AQP8 in amplifying growth signals, AQP8-overexpressing rat insulinoma cells grew faster than controls, while $AQP8^{KD}$ proliferated more slowly (Kruger et al. 2021). In B lymphocytes, we showed that AQP8 mediates the flux of H_2O_2 produced extracellularly by NOX2 upon B cell receptor (BCR) activation. Silencing AQP8 or NOX2, or adding extracellular catalase to intercept H_2O_2, had profound effects on key downstream events in B-cell physiology, such as IgM production and secretion (Bertolotti et al. 2016). These results extended elegant work by Capasso et al. (2010), demonstrating a key role for HVCN1, a proton channel important for the efficient production of extracellular H_2O_2 by NADPH oxidases, and ultimately for modulating the strength of BCR signals. HVCN1-deficient B cells, like $AQP8^{KD}$, yielded defective antibody responses (Bertolotti et al. 2016; Capasso et al. 2010). Taken together, these findings added fuel to the notion that the regulation of H_2O_2 entry allows cells to tune the intensity of signal transduction.

It is well established that the redox state of the cell controls ion channels/transporters, and, reciprocally, these transporters can regulate the redox environment. In higher organisms, different members of Ca^{2+}-conducting channels/transporters are known to be redox sensitive (Kozai, Ogawa and Mori 2014). Voltage-gated Ca^{2+} channels on the plasma membrane of mammalian cells are among them, their activity being regulated by glutathionylation of specific cysteine residues (Chandel and Bachhawat 2017). In rat hepatocytes, the reciprocal influence between redoxstasis and Ca^{2+} mobilization requires also the peroxiporin activity of AQP8 (Vilchis-Landeros et al. 2020). Moreover, chelating extracellular Ca^{2+} with EGTA does not prevent the H_2O_2-induced Ca^{2+} increases, arguing in favor of an intracellular origin of the cation, probably from the ER (Vilchis-Landeros et al. 2020). In all likelihood, the ER-mitochondrial interface is a central hub in this complex regulatory circuitry, integrating Ca^{2+} and redox- and ER stress–dependent signals (Yoboue et al. 2017). Accordingly, AQP11 and part of immature, endo-H sensitive, AQP8 accumulate in MAMs (Bestetti et al. 2020; Sorrentino et al. 2022) and mitochondrial dysfunction was observed in AQP8–knocked down preadipocytes (Ikaga et al. 2015).

In the male reproductive system, sperm motility is controlled by H_2O_2 (Laforenza et al. 2016) and correlates with the levels of the peroxiporins AQP3, 8, and 11. In fish, AQP8 is important in detoxifying spermatozoa from excess of H_2O_2 and in maintaining efficient ATP production and flagellar motility (Chauvigné et al. 2021).

12.8 ROLE IN PATHOLOGY

AQP8 is abundant in the digestive system and its dysregulation/abnormal expression is linked to inflammation and cancer (Wu et al. 2018; Zhang et al. 2020).

12.8.1 Cancer

HeLa cells expressing an AQP8 mutant resistant to stress-induced gating (C53S) proliferated more and displayed reduced levels of apoptosis when exposed to radiation or arsenite (Medrano-Fernandez et al. 2016). Whilst this mutation was not observed in primary human tumors histidine replacement of arginine 213, a residue that impacts thiolate formation in C53 of AQP8, has been found in some patients with gastrointestinal cancers (Figure 12.4A). However, these mutations were identified in few patients and no functional correlation demonstrated yet. More detailed studies are necessary to determine the impact of AQP8 mutants on tumorigenicity and cancer aggressiveness (White et al. 2017; Forbes et al. 2017; Cardone, Casavola and Reshkin 2005).

AQPs in general are abundantly expressed in cancer cells of different origins, particularly aggressive tumors. Their pro-tumorigenic activity might be linked also to H_2O transport. The demand for H_2O is higher for malignant than normal cells. Water intake at the front edge promotes migration. Accordingly, AQP5 depletion was recently shown to reduce confined migration of human adenocarcinoma cells (Bera et al. 2022). As discussed above, AQP8 amplifies EGFR signaling. In turn, EGF seems to induce AQP8 upregulation in esophageal cancer (Chang et al. 2014). This pathway was proposed to accelerate migration of a cancer line (Eca-109), enhancing plasma-membrane H_2O permeability, which in turn can boost cell mobility. In this case, the tumor promoting might be more related to the AQP8 role as classical H_2O channel, than as a peroxiporin. However, a contribution of redox signaling in affecting cell migration cannot be excluded, as ROS (and H_2O_2 in particular) relay signals that activate cell movements. Actin polymerization and restructuring (Moldovan et al. 2000; Ikeda et al. 2005) have been reported as well as transcriptional and protein phosphorylation changes (Harfouche et al. 2005; Abid et al. 2000), although the direct targets of H_2O_2 are not known (Hurd, DeGennaro and Lehmann 2012).

12.8.2 Inflammation

Several studies reported altered AQP7 and AQP8 expression in mucosal colonic tissue from patients with moderate to severe inflammatory bowel disease (Hardin et al. 2004; see Dicay et al. 2015, and references therein). Accordingly, the levels of AQP7, AQP8, and AQP4 decrease during progression of colitis in animal models. AQP8 was shown to be particularly relevant for maintaining efficient absorptive functions (Hardin et al. 2004). Decreased AQP8 mRNA levels were reported in ileal tissues from inflammatory bowel disease and ulcerative colitis patients, despite an increase in its protein levels (Zahn et al. 2007). AQP3 and AQP11 transcripts were

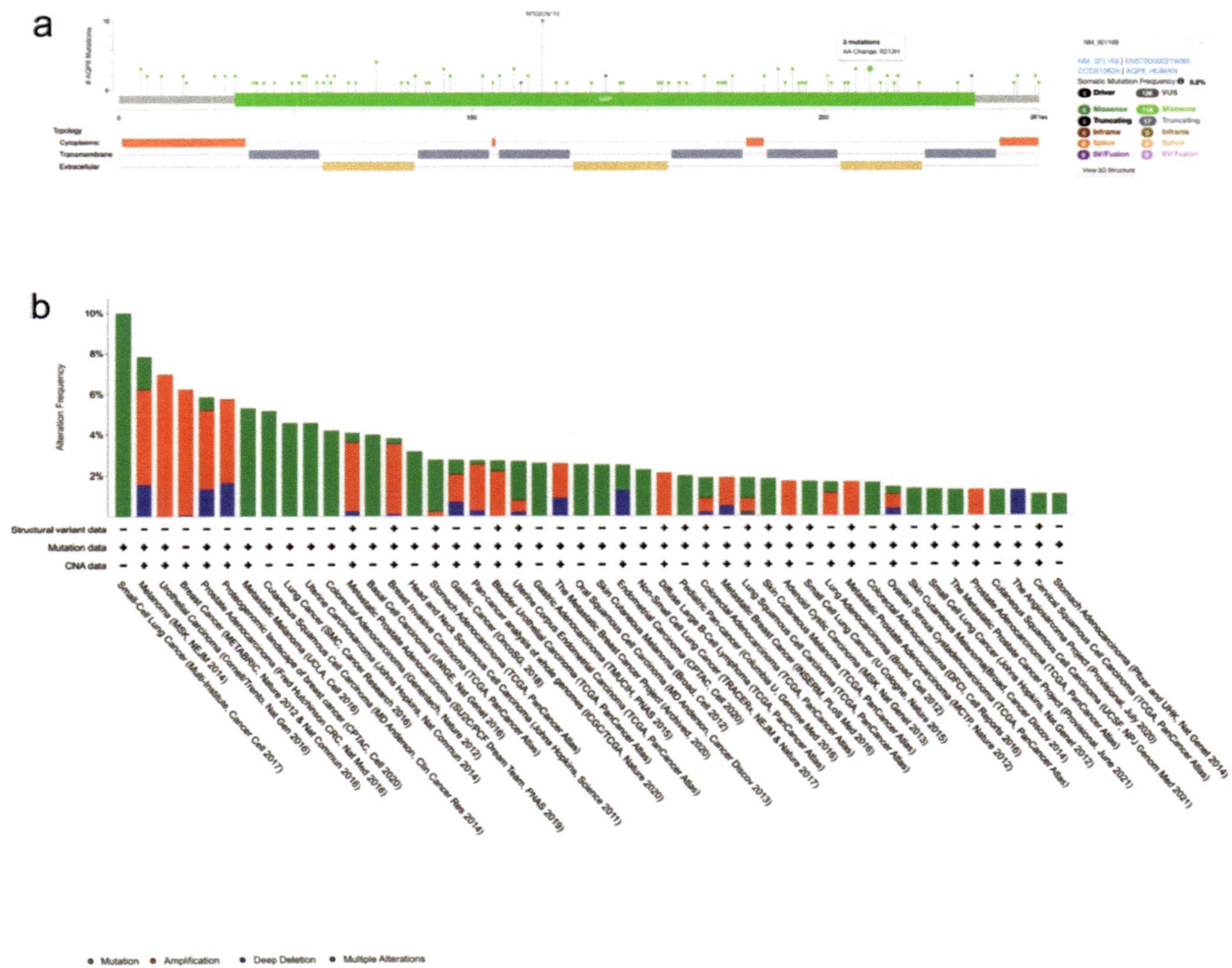

Figure 12.4 AQP8 mutations in human tumors. (A) Lollipop plot of mutations found within the hAQP8 gene. Tracks below the gene map the mutations to the correspondent protein topology based on UniProt predictions. Mutations found at R213 are flagged. (B) Alteration frequency in cancer studies with a minimum of 1% altered cases (46 out of 213). Red = amplification; blue = deletions; green = mutations. Analyses were performed using cBioPortal (Gao et al. 2013) across a combination of non-redundant studies comprising data from more than 60,000 patients.

either increased or unchanged in ileal but uniformly decreased in colonic tissues of patients. In addition, 5-fluorouracil (5-FU), a chemotherapeutic agent that often induces diarrhea, caused significant downregulation of AQP1, AQP4, and AQP8 in mouse intestine, and these changes correlated with an increase in inflammatory cytokine production (Sakai et al. 2013).

12.8.3 Metabolic Disease and Diabetes

H_2O_2 levels correlate with insulin resistance and their scavenging rescued sensitivity (Houstis, Rosen and Lander 2006). Some authors argue that insulin resistance, caused by nutrient overload and mitochondrial exhaustion, is the initiating event leading to mitochondrial dysfunction and increased H_2O_2 in metabolic disease (Holloszy 2013). However, insulin receptor deficiency does not generate oxidative stress in muscle (O'Neill et al. 2010), and most studies using environmental and nutritional factors to induce insulin resistance, such as a high-fat diet, report increased H_2O_2 and oxidative damage *prior to* the development of insulin resistance (Matsuzawa-Nagata et al. 2008; Barazzoni et al. 2012; Bonnard et al. 2008). The delicate and environmental-dependent friend/foe balance of H_2O_2 activity makes it difficult to draw firm conclusions as to whether increased H_2O_2 is a major player in decreased insulin signaling in fatty liver disease. Well-designed studies using tissue-specific and inducible genetic models (e.g., knockouts of either insulin receptors or their substrates) and/or more advanced tools that can detect changes in H_2O_2 production *in vivo* are needed to decipher its precise temporal links with insulin resistance and determine whether and how H_2O_2 levels can be manipulated in metabolic disease.

12.9 CONCLUDING REMARKS

Thousands of papers appear in PubMed when typing "cancer" and "H_2O_2" as key words. Yet, few if any drugs and treatments made it to the clinics. An intrinsic difficulty in successfully targeting redox-based signals is the fact that they can be friends or foes depending on the context in which they operate. A clearer view of the topology and regulation of these circuits is hence necessary, and the mechanisms that tune signal transmission acquire great relevance.

This chapter tries to summarize over a decade of studies on H_2O_2 transport and diffusion through peroxiporins like AQP8 in tuning inter- and intra-cellular redox signaling circuits. There remains little doubt that AQP8 is a key conductor at the plasma membrane. Can it work also at the ER membranes, where it folds and begin its journey to the plasma membrane? Considering that the ER lumen is constitutively rich in H_2O_2 molecules, one would reckon that cysteine 53 would be constantly sulfenylated. Since H_2S can freely diffuse across membranes (Riahi and Rowley 2014; Mathai et al. 2009), C53 persulfidation could keep AQP8 on its closed state. Our recent discovery that AQP11 is an ER-resident peroxiporin (Bestetti et al. 2020) may suggest that different organelles be equipped with specialized peroxiporins, with non-overlapping regulatory and gating mechanisms. In this respect, it is intriguing that in AQP11 the C53 homolog is part of a rather rare CCC sequence. It would be of great interest to determine whether also AQP11 is gated, and if so, to decipher the mechanisms.

Still missing is information on how the peroxiporin activity of AQP8 is restored during recovery from cell stress. Our preliminary data indicate that H_2O_2 permeability returns to normal also in the presence of protein synthesis inhibitors, despite slowly. Clearly no protein can get access to cysteine 53, buried in a channel that restricts transport to free cysteines.

Even more urgent is to understand how H_2O_2 molecules can permeate the inner mitochondrial membrane. Our data suggest that neither AQP8 nor AQP11 is required (Bestetti et al. 2018; Sorrentino et al. 2022). It is of paramount importance to establish whether and how AQP8 can be re-routed to mitochondria if other molecules display peroxiporin activity in the inner mitochondrial membrane, and if so, whether and how they are regulated.

Other open issues awaiting deep investigation relate to the interactomes of the proteins with established peroxiporin activity. Is AQP8 constitutively associated with NOX2? Are intracellular target proteins recruited in its vicinity? How can H_2O_2 molecules reach them in a cytosol rich in peroxidases and antioxidant

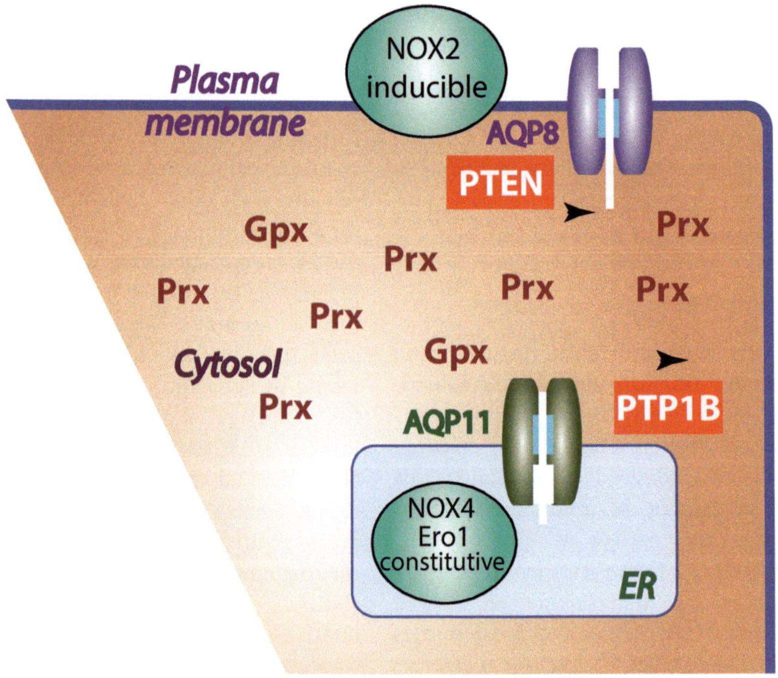

Figure 12.5 Toward a cartography of redox signaling. H_2O_2 molecules are thought to travel small distances in the crowded cytosol, rich in small and protein antioxidants (GSH, glutathione peroxidases, peroxiredoxins). Thus, localized sources and conduits are likely to mediate signaling in living cells, providing an extra layer of signal tuning. Accordingly, while surface AQP8 imports H_2O_2 generated by NADPH oxidases in the extracellular space upon acute TKR activation, AQP11 resides in the ER membrane, mediating a constitutive flux of the abundant H_2O_2, present in the lumen of this organelle (Bestetti et al. 2020) and possibly transmitting tonic signals that report cues on ER functionality.

systems? The observation that AQP8 and AQP11 exert their activity in different cellular districts argues in favor of a strictly controlled topological organization of the sources, conduits, and targets of redox signals (Figure 12.5). A precise cartography of such signaling paths is a formidable challenge that requires novel technologies to be tackled.

We look forward to soon reading the answers to these crucial questions in good scientific journals.

REFERENCES

Abid, M. R., Z. Kachra, K. C. Spokes, and W. C. Aird. 2000. "NADPH oxidase activity is required for endothelial cell proliferation and migration." *FEBS Lett* 486 (3):252–256. http://doi.org/10.1016/s0014-5793(00)02305-x.

Agemark, M., J. Kowal, W. Kukulski, K. Norden, N. Gustavsson, U. Johanson, A. Engel, and P. Kjellbom. 2012. "Reconstitution of water channel function and 2D-crystallization of human aquaporin 8." *Biochim Biophys Acta* 1818 (3):839–850. http://doi.org/10.1016/j.bbamem.2011.12.006.

Almasalmeh, A., D. Krenc, B. Wu, and E. Beitz. 2014. "Structural determinants of the hydrogen peroxide permeability of aquaporins." *FEBS J* 281 (3):647–656. http://doi.org/10.1111/febs.12653.

Barazzoni, R., M. Zanetti, G. Gortan Cappellari, A. Semolic, M. Boschelle, E. Codarin, A. Pirulli, L. Cattin, and G. Guarnieri. 2012. "Fatty acids acutely enhance insulin-induced oxidative stress and cause insulin resistance by increasing mitochondrial reactive oxygen species (ROS) generation and nuclear factor-kappaB inhibitor (IkappaB)-nuclear factor-kappaB (NFkappaB) activation in rat muscle, in the absence of mitochondrial dysfunction." *Diabetologia* 55 (3):773–782. http://doi.org/10.1007/s00125-011-2396-x.

Bera, K., A. Kiepas, I. Godet, Y. Li, P. Mehta, B. Ifemembi, C. D. Paul, A. Sen, S. A. Serra, K. Stoletov, J. Tao, G. Shatkin, S. J. Lee, Y. Zhang, A. Boen, P. Mistriotis, D. M. Gilkes, J. D. Lewis, C. M. Fan, A. P. Feinberg, M. A. Valverde, S. X. Sun, and K. Konstantopoulos. 2022. "Extracellular fluid viscosity enhances cell migration and cancer dissemination." *Nature* 611 (7935):365–373. http://doi.org/10.1038/s41586-022-05394-6.

Bertolotti, M., S. Bestetti, J. M. Garcia-Manteiga, I. Medrano-Fernandez, A. Dal Mas, M. L. Malosio, and R. Sitia. 2013. "Tyrosine kinase signal modulation: a matter of H2O2 membrane permeability?" *Antioxid Redox Signal* 19 (13):1447–1451. http://doi.org/10.1089/ars.2013.5330.

Bertolotti, M., G. Farinelli, M. Galli, A. Aiuti, and R. Sitia. 2016. "AQP8 transports NOX2-generated H2O2 across the plasma membrane to promote signaling in B cells." *J Leukoc Biol* 100 (5):1071–1079. http://doi.org/10.1189/jlb.2AB0116-045R.

Bertolotti, M., S. H. Yim, J. M. Garcia-Manteiga, S. Masciarelli, Y. J. Kim, M. H. Kang, Y. Iuchi, J. Fujii, R. Vené, A. Rubartelli, S. G. Rhee, and R. Sitia. 2010. "B- to plasma-cell terminal differentiation entails oxidative stress and profound reshaping of the antioxidant responses." *Antioxid Redox Signal* 13 (8):1133–1144.

Bestetti, S., M. Galli, I. Sorrentino, P. Pinton, A. Rimessi, R. Sitia, and I. Medrano-Fernandez. 2020. "Human aquaporin-11 guarantees efficient transport of H2O2 across the endoplasmic reticulum membrane." *Redox Biol* 28:101326. http://doi.org/10.1016/j.redox.2019.101326.

Bestetti, S., I. Medrano-Fernandez, M. Galli, M. Ghitti, G. P. Bienert, G. Musco, A. Orsi, A. Rubartelli, and R. Sitia. 2018. "A persulfidation-based mechanism controls aquaporin-8 conductance." *Sci Adv* 4 (5):eaar5770. http://doi.org/10.1126/sciadv.aar5770.

Bienert, G. P., A. L. Moller, K. A. Kristiansen, A. Schulz, I. M. Moller, J. K. Schjoerring, and T. P. Jahn. 2007. "Specific aquaporins facilitate the diffusion of hydrogen peroxide across membranes." *J Biol Chem* 282 (2):1183–1192. http://doi.org/10.1074/jbc.M603761200.

Bonnard, C., A. Durand, S. Peyrol, E. Chanseaume, M. A. Chauvin, B. Morio, H. Vidal, and J. Rieusset. 2008. "Mitochondrial dysfunction results from oxidative stress in the skeletal muscle of diet-induced insulin-resistant mice." *J Clin Invest* 118 (2):789–800. http://doi.org/10.1172/JCI32601.

Burghardt, B., M. L. Elkjaer, T. H. Kwon, G. Z. Racz, G. Varga, M. C. Steward, and S. Nielsen. 2003. "Distribution of aquaporin water channels AQP1 and AQP5 in the ductal system of the human pancreas." *Gut* 52 (7):1008–1016. http://doi.org/10.1136/gut.52.7.1008.

Calamita, G., A. Mazzone, A. Bizzoca, A. Cavalier, G. Cassano, D. Thomas, and M. Svelto. 2001. "Expression and immunolocalization of the aquaporin-8 water channel in rat gastrointestinal tract." *Eur J Cell Biol* 80 (11):711–719. http://doi.org/10.1078/0171-9335-00210.

Capasso, M., M. K. Bhamrah, T. Henley, R. S. Boyd, C. Langlais, K. Cain, D. Dinsdale, K. Pulford, M. Khan, B. Musset, V. V. Cherny, D. Morgan, R. D. Gascoyne, E. Vigorito, T. E. DeCoursey, I. C. MacLennan, and M. J. Dyer. 2010. "HVCN1 modulates BCR signal strength via regulation of BCR-dependent generation of reactive oxygen species." *Nat Immunol* 11 (3):265–272.

Cardone, R. A., V. Casavola, and S. J. Reshkin. 2005. "The role of disturbed pH dynamics and the Na+/H+ exchanger in metastasis." *Nat Rev Cancer* 5 (10):786–795. http://doi.org/10.1038/nrc1713.

Chandel, A., and A. K. Bachhawat. 2017. "Redox regulation of the yeast voltage-gated Ca(2+) channel homolog Cch1p by glutathionylation of specific cysteine residues." *J Cell Sci* 130 (14):2317–2328. http://doi.org/10.1242/jcs.202853.

Chang, H., Y. H. Shi, T. K. Talaf, and C. Lin. 2014. "Aquaporin-8 mediates human esophageal cancer Eca-109 cell migration via the EGFR-Erk1/2 pathway." *Int J Clin Exp Pathol* 7 (11):7663–7671.

Chauvigné, F., C. Ducat, A. Ferré, T. Hansen, M. Carrascal, J. Abián, R. N. Finn, and J. Cerda. 2021. "A multiplier peroxiporin signal transduction pathway powers piscine spermatozoa." *Proc Natl Acad Sci U S A* 118 (10). http://doi.org/10.1073/pnas.2019346118.

Consortium, G. TEx. 2013. "The genotype-tissue expression (GTEx) project." *Nat Genet* 45 (6):580–585. http://doi.org/10.1038/ng.2653.

Cymer, F., G. von Heijne, and S. H. White. 2015. "Mechanisms of integral membrane protein insertion and folding." *J Mol Biol* 427 (5):999–1022. http://doi.org/10.1016/j.jmb.2014.09.014.

Dicay, M. S., C. L. Hirota, N. J. Ronaghan, M. A. Peplowski, R. S. Zaheer, C. A. Carati, and W. K. MacNaughton. 2015. "Interferon-gamma suppresses intestinal epithelial aquaporin-1 expression via Janus kinase and STAT3 activation." *PLoS ONE* 10 (3):e0118713. http://doi.org/10.1371/journal.pone.0118713.

Elkjaer, M. L., L. N. Nejsum, V. Gresz, T. H. Kwon, U. B. Jensen, J. Frøkiaer, and S. Nielsen. 2001. "Immunolocalization of aquaporin-8 in rat kidney, gastrointestinal tract, testis, and airways." *Am J Physiol Renal Physiol* 281 (6):F1047–F1057. http://doi.org/10.1152/ajprenal.0158.2001.

Ferri, D., A. Mazzone, G. E. Liquori, G. Cassano, M. Svelto, and G. Calamita. 2003. "Ontogeny, distribution, and possible functional implications of an unusual aquaporin, AQP8, in mouse liver." *Hepatology* 38 (4):947–957. http://doi.org/10.1053/jhep.2003.50397.

Finkel, T. 2011. "Signal transduction by reactive oxygen species." *J Cell Biol* 194 (1):7–15. http://doi.org/10.1083/jcb.201102095.

Fischer, H., R. Stenling, C. Rubio, and A. Lindblom. 2001. "Differential expression of aquaporin 8 in human colonic epithelial cells and colorectal tumors." *BMC Physiol* 1:1. http://doi.org/10.1186/1472-6793-1-1.

Forbes, S. A., D. Beare, H. Boutselakis, S. Bamford, N. Bindal, J. Tate, C. G. Cole, S. Ward, E. Dawson, L. Ponting, R. Stefancsik, B. Harsha, C. Y. Kok, M. Jia, H. Jubb, Z. Sondka, S. Thompson, T. De, and P. J. Campbell. 2017. "COSMIC: somatic cancer genetics at high-resolution." *Nucleic Acids Res* 45 (D1):D777–D783. http://doi.org/10.1093/nar/gkw1121.

Gao, J., B. A. Aksoy, U. Dogrusoz, G. Dresdner, B. Gross, S. O. Sumer, Y. Sun, A. Jacobsen, R. Sinha, E. Larsson, E. Cerami, C. Sander, and N. Schultz. 2013. "Integrative analysis of complex cancer genomics and clinical profiles using the cBioPortal." *Sci Signal* 6 (269):pl1. http://doi.org/10.1126/scisignal.2004088.

Garcia, F., A. Kierbel, M. C. Larocca, S. A. Gradilone, P. Splinter, N. F. LaRusso, and R. A. Marinelli. 2001. "The water channel aquaporin-8 is mainly intracellular in rat hepatocytes, and its plasma membrane insertion is stimulated by cyclic AMP." *J Biol Chem* 276 (15):12147–12152. http://doi.org/10.1074/jbc.M009403200.

Gena, P., E. Fanelli, C. Brenner, M. Svelto, and G. Calamita. 2009. "News and views on mitochondrial water transport." *Front Biosci (Landmark Ed)* 14:4189–4198. http://doi.org/10.2741/3522.

Gregoire, F., V. Lucidi, A. Zerrad-Saadi, M. Virreira, N. Bolaky, V. Delforge, A. Lemmers, V. Donckier, J. Deviere, P. Demetter, J. Perret, and C. Delporte. 2015. "Analysis of aquaporin expression in liver with a focus on hepatocytes." *Histochem Cell Biol* 144 (4):347–363. http://doi.org/10.1007/s00418-015-1341-3.

Hardin, J. A., L. E. Wallace, J. F. Wong, E. V. O'Loughlin, S. J. Urbanski, D. G. Gall, W. K. MacNaughton, and P. L. Beck. 2004. "Aquaporin expression is downregulated in a murine model of colitis and in patients with ulcerative colitis, Crohn's disease and infectious colitis." *Cell Tissue Res* 318 (2):313–323. http://doi.org/10.1007/s00441-004-0932-4.

Harfouche, R., N. A. Malak, R. P. Brandes, A. Karsan, K. Irani, and S. N. Hussain. 2005. "Roles of reactive oxygen species in angiopoietin-1/tie-2 receptor signaling." *FASEB J* 19 (12):1728–1730. http://doi.org/10.1096/fj.04-3621fje.

Heijne, G. 1986. "The distribution of positively charged residues in bacterial inner membrane proteins correlates with the trans-membrane topology." *EMBO J* 5 (11):3021–3027. http://doi.org/10.1002/j.1460-2075.1986.tb04601.x.

Holloszy, J. O. 2013. "'Deficiency' of mitochondria in muscle does not cause insulin resistance." *Diabetes* 62 (4):1036–1040. http://doi.org/10.2337/db12-1107.

Holm, L. M., T. P. Jahn, A. L. Moller, J. K. Schjoerring, D. Ferri, D. A. Klaerke, and T. Zeuthen. 2005. "NH3 and NH4+ permeability in aquaporin-expressing Xenopus oocytes." *Pflugers Arch* 450 (6):415–428. http://doi.org/10.1007/s00424-005-1399-1.

Houstis, N., E. D. Rosen, and E. S. Lander. 2006. "Reactive oxygen species have a causal role in multiple forms of insulin resistance." *Nature* 440 (7086):944–948. http://doi.org/10.1038/nature04634.

Hurd, T. R., M. DeGennaro, and R. Lehmann. 2012. "Redox regulation of cell migration and adhesion." *Trends Cell Biol* 22 (2):107–115. http://doi.org/10.1016/j.tcb.2011.11.002.

Ikaga, R., I. Namekata, V. N. Kotiadis, H. Ogawa, M. R. Duchen, H. Tanaka, and N. Iida-Tanaka. 2015. "Knockdown of aquaporin-8 induces mitochondrial dysfunction in 3T3-L1 cells." *Biochem Biophys Rep* 4:187–195. http://doi.org/10.1016/j.bbrep.2015.09.009.

Ikeda, S., M. Yamaoka-Tojo, L. Hilenski, N. A. Patrushev, G. M. Anwar, M. T. Quinn, and M. Ushio-Fukai. 2005. "IQGAP1 regulates reactive oxygen species-dependent endothelial cell migration through interacting with Nox2." *Arterioscler Thromb Vasc Biol* 25 (11):2295–2300. http://doi.org/10.1161/01.ATV.0000187472.55437.af.

Ishibashi, K., M. Kuwahara, Y. Gu, Y. Kageyama, A. Tohsaka, F. Suzuki, F. Marumo, and S. Sasaki. 1997. "Cloning and functional expression of a new water channel abundantly expressed in the testis permeable to water, glycerol, and urea." *The Journal of biological chemistry* 272 (33):20782–20786.

Ishibashi, K., M. Kuwahara, Y. Kageyama, A. Tohsaka, F. Marumo, and S. Sasaki. 1997. "Cloning and functional expression of a second new aquaporin abundantly expressed in testis." *Biochem Biophys Res Commun* 237 (3):714–718. http://doi.org/10.1006/bbrc.1997.7219.

Ishibashi, K., Y. Morishita, and Y. Tanaka. 2017. "The evolutionary aspects of aquaporin family." *Adv Exp Med Biol* 969:35–50. http://doi.org/10.1007/978-94-024-1057-0_2.

Ishibashi, K., S. Sasaki, K. Fushimi, T. Yamamoto, M. Kuwahara, and F. Marumo. 1997. "Immunolocalization and effect of dehydration on AQP3, a basolateral water channel of kidney collecting ducts." *Am J Physiol* 272 (2 Pt 2):F235–F241.

Ishibashi, K., Y. Tanaka, and Y. Morishita. 2020. "Perspectives on the evolution of aquaporin superfamily." *Vitam Horm* 112:1–27. http://doi.org/10.1016/bs.vh.2019.08.001.

Kirscht, A., Y. Sonntag, P. Kjellbom, and U. Johanson. 2018. "A structural preview of aquaporin 8 via homology modeling of seven vertebrate isoforms." *BMC Struct Biol* 18 (1):2. http://doi.org/10.1186/s12900-018-0081-8.

Koyama, N., K. Ishibashi, M. Kuwahara, N. Inase, M. Ichioka, S. Sasaki, and F. Marumo. 1998. "Cloning and functional expression of human aquaporin8 cDNA and analysis of its gene." *Genomics* 54 (1):169–172. http://doi.org/10.1006/geno.1998.5552.

Koyama, Y., T. Yamamoto, D. Kondo, H. Funaki, E. Yaoita, K. Kawasaki, N. Sato, K. Hatakeyama, and I. Kihara. 1997. "Molecular cloning of a new aquaporin from rat pancreas and liver." *J Biol Chem* 272 (48):30329–30333. http://doi.org/10.1074/jbc.272.48.30329.

Kozai, D., N. Ogawa, and Y. Mori. 2014. "Redox regulation of transient receptor potential channels." *Antioxid Redox Signal* 21 (6):971–986. http://doi.org/10.1089/ars.2013.5616.

Kruger, C., M. Waldeck-Weiermair, J. Kaynert, T. Pokrant, Y. Komaragiri, O. Otto, T. Michel, and M. Elsner. 2021. "AQP8 is a crucial H2O2 transporter in insulin-producing RINm5F cells." *Redox Biol* 43:101962. http://doi.org/10.1016/j.redox.2021.101962.

Laforenza, U., G. Pellavio, A. L. Marchetti, C. Omes, F. Todaro, and G. Gastaldi. 2016. "Aquaporin-mediated water and hydrogen peroxide transport is involved in normal human spermatozoa functioning." *Int J Mol Sci* 18 (1). http://doi.org/10.3390/ijms18010066.

Larocca, M. C., L. R. Soria, M. V. Espelt, G. L. Lehmann, and R. A. Marinelli. 2009. "Knockdown of hepatocyte aquaporin-8 by RNA interference induces defective bile canalicular water transport." *Am J Physiol Gastrointest Liver Physiol* 296 (1):G93–100. http://doi.org/10.1152/ajpgi.90410.2008.

Liu, K., H. Nagase, C. G. Huang, G. Calamita, and P. Agre. 2006. "Purification and functional characterization of aquaporin-8." *Biol Cell* 98 (3):153–161. http://doi.org/10.1042/BC20050026.

Louchami, K., L. Best, P. Brown, M. Virreira, E. Hupkens, J. Perret, O. Devuyst, S. Uchida, C. Delporte, W. J. Malaisse, R. Beauwens, and A. Sener. 2012. "A new role for aquaporin 7 in insulin secretion." *Cell Physiol Biochem* 29 (1–2):65–74. http://doi.org/10.1159/000337588.

Ma, T., B. Yang, and A. S. Verkman. 1997. "Cloning of a novel water and urea-permeable aquaporin from mouse expressed strongly in colon, placenta, liver, and heart." *Biochem Biophys Res Commun* 240 (2):324–328. http://doi.org/10.1006/bbrc.1997.7664.

Maclean, K. N., M. Janosik, E. Kraus, V. Kozich, R. H. Allen, B. K. Raab, and J. P. Kraus. 2002. "Cystathionine beta-synthase is coordinately regulated with proliferation through a redox-sensitive mechanism in cultured human cells and Saccharomyces cerevisiae." *J Cell Physiol* 192 (1):81–92. http://doi.org/10.1002/jcp.10118.

Marchissio, M. J., D. E. Frances, C. E. Carnovale, and R. A. Marinelli. 2012. "Mitochondrial aquaporin-8 knockdown in human hepatoma HepG2 cells causes ROS-induced mitochondrial depolarization and loss of viability." *Toxicol Appl Pharmacol* 264 (2):246–254. http://doi.org/10.1016/j.taap.2012.08.005.

Mathai, J. C., A. Missner, P. Kugler, S. M. Saparov, M. L. Zeidel, J. K. Lee, and P. Pohl. 2009. "No facilitator required for membrane transport of hydrogen sulfide." *Proc Natl Acad Sci USA* 106 (39):16633-16638. http://doi.org/10.1073/pnas.0902952106.

Matsuzawa-Nagata, N., T. Takamura, H. Ando, S. Nakamura, S. Kurita, H. Misu, T. Ota, M. Yokoyama, M. Honda, K. Miyamoto, and S. Kaneko. 2008. "Increased oxidative stress precedes the onset of high-fat diet-induced insulin resistance and obesity." *Metabolism* 57 (8):1071–1077. http://doi.org/10.1016/j.metabol.2008.03.010.

Medrano-Fernandez, I., S. Bestetti, M. Bertolotti, G. P. Bienert, C. Bottino, U. Laforenza, A. Rubartelli, and R. Sitia. 2016. "Stress regulates aquaporin-8 permeability to impact cell growth and survival." *Antioxid Redox Signal* 24 (18):1031–1044. http://doi.org/10.1089/ars.2016.6636.

Meng, T. C., T. Fukada, and N. K. Tonks. 2002. "Reversible oxidation and inactivation of protein tyrosine phosphatases in vivo." *Molecular Cell* 9 (2):387–399.

Miller, E. W., B. C. Dickinson, and C. J. Chang. 2010. "Aquaporin-3 mediates hydrogen peroxide uptake to regulate downstream intracellular signaling." *Proc Natl Acad Sci U S A* 107 (36):15681–15686. http://doi.org/10.1073/pnas.1005776107.

Moldovan, L., N. I. Moldovan, R. H. Sohn, S. A. Parikh, and P. J. Goldschmidt-Clermont. 2000. "Redox changes of cultured endothelial cells and actin dynamics." *Circ Res* 86 (5):549–557. http://doi.org/10.1161/01.res.86.5.549.

Molinas, S. M., L. Trumper, and R. A. Marinelli. 2012. "Mitochondrial aquaporin-8 in renal proximal tubule cells: evidence for a role in the response to metabolic acidosis." *Am J Physiol Renal Physiol* 303 (3):F458–F466. http://doi.org/10.1152/ajprenal.00226.2012.

Molteni, S. N., A. Fassio, M. R. Ciriolo, G. Filomeni, E. Pasqualetto, C. Fagioli, and R. Sitia. 2004. "Glutathione limits Ero1-dependent oxidation in the endoplasmic reticulum." *J Biol Chem* 279 (31):32667–32673. http://doi.org/10.1074/jbc.M404992200.

O'Neill, E. D., J. P. Wilding, C. R. Kahn, H. Van Remmen, A. McArdle, M. J. Jackson, and G. L. Close. 2010. "Absence of insulin signalling in skeletal muscle is associated with reduced muscle mass and function: evidence for decreased protein synthesis and not increased degradation." *Age (Dordr)* 32 (2):209–222. http://doi.org/10.1007/s11357-009-9125-0.

Pollegioni, L., B. Langkau, W. Tischer, S. Ghisla, and M. S. Pilone. 1993. "Kinetic mechanism of D-amino acid oxidases from Rhodotorula gracilis and Trigonopsis variabilis." *J Biol Chem* 268 (19):13850–13857.

Prata, C., S. Hrelia, and D. Fiorentini. 2019. "Peroxiporins in cancer." *Int J Mol Sci* 20 (6). http://doi.org/10.3390/ijms20061371.

Riahi, S., and C. N. Rowley. 2014. "Why can hydrogen sulfide permeate cell membranes?" *J Am Chem Soc* 136 (43):15111–15113. http://doi.org/10.1021/ja508063s.

Sakai, H., A. Sagara, K. Matsumoto, S. Hasegawa, K. Sato, M. Nishizaki, T. Shoji, S. Horie, T. Nakagawa, S. Tokuyama, and M. Narita. 2013. "5-Fluorouracil induces diarrhea with changes in the expression of inflammatory cytokines and aquaporins in mouse intestines." *PLoS ONE* 8 (1):e54788. http://doi.org/10.1371/journal.pone.0054788.

Saparov, S. M., K. Liu, P. Agre, and P. Pohl. 2007. "Fast and selective ammonia transport by aquaporin-8." *J Biol Chem* 282 (8):5296–5301. http://doi.org/10.1074/jbc.M609343200.

Soria, L. R., E. Fanelli, N. Altamura, M. Svelto, R. A. Marinelli, and G. Calamita. 2010. "Aquaporin-8-facilitated mitochondrial ammonia transport." *Biochem Biophys Res Commun* 393 (2):217–221. http://doi.org/10.1016/j.bbrc.2010.01.104.

Soria, L. R., J. Marrone, G. Calamita, and R. A. Marinelli. 2013. "Ammonia detoxification via ureagenesis in rat hepatocytes involves mitochondrial aquaporin-8 channels." *Hepatology* 57 (5):2061–2071. http://doi.org/10.1002/hep.26236.

Sorrentino, I., M. Galli, I. Medrano-Fernandez, and R. Sitia. 2022. "Transfer of H2O2 from Mitochondria to the endoplasmic reticulum via Aquaporin-11." *Redox Biol* 55:102410. http://doi.org/10.1016/j.redox.2022.102410.

Tishkov, V. I., and S. V. Khoronenkova. 2005. "D-Amino acid oxidase: structure, catalytic mechanism, and practical application." *Biochemistry (Mosc)* 70 (1):40–54.

Tonks, N. K. 2005. "Redox redux: revisiting PTPs and the control of cell signaling." *Cell* 121 (5):667–670.

Trost, P., S. Fermani, M. Calvaresi, and M. Zaffagnini. 2017. "Biochemical basis of sulphenomics: how protein sulphenic acids may be stabilized by the protein microenvironment." *Plant Cell Environ* 40 (4):483–490. http://doi.org/10.1111/pce.12791.

Vieceli Dalla Sega, F., L. Zambonin, D. Fiorentini, B. Rizzo, C. Caliceti, L. Landi, S. Hrelia, and C. Prata. 2014. "Specific aquaporins facilitate Nox-produced hydrogen peroxide transport through plasma membrane in leukaemia cells." *Biochim Biophys Acta* 1843 (4):806–814. http://doi.org/10.1016/j.bbamcr.2014.01.011.

Vilchis-Landeros, M., R. Guinzberg, H. Riveros-Rosas, R. Villalobos-Molina, and E. Pina. 2020. "Aquaporin 8 is involved in H2O2-mediated differential regulation of metabolic signaling by alpha1—and beta-adrenoceptors in hepatocytes." *FEBS Lett* 594 (10):1564–1576. http://doi.org/10.1002/1873-3468.13763.

White, K. A., D. G. Ruiz, Z. A. Szpiech, N. B. Strauli, R. D. Hernandez, M. P. Jacobson, and D. L. Barber. 2017. "Cancer-associated arginine-to-histidine mutations confer a gain in pH sensing to mutant proteins." *Sci Signal* 10 (495). http://doi.org/10.1126/scisignal.aam9931.

Wu, Q., Z. F. Yang, K. J. Wang, X. Y. Feng, Z. J. Lv, Y. Li, and Z. X. Jian. 2018. "AQP8 inhibits colorectal cancer growth and metastasis by down-regulating PI3K/AKT signaling and PCDH7 expression." *Am J Cancer Res* 8 (2):266–279.

Yang, B., Y. Song, D. Zhao, and A. S. Verkman. 2005. "Phenotype analysis of aquaporin-8 null mice." *Am J Physiol Cell Physiol* 288 (5):C1161–C1170. http://doi.org/10.1152/ajpcell.00564.2004.

Yang, B., D. Zhao, E. Solenov, and A. S. Verkman. 2006. "Evidence from knockout mice against physiologically significant aquaporin 8-facilitated ammonia transport." *Am J Physiol Cell Physiol* 291 (3):C417–C423. http://doi.org/10.1152/ajpcell.00057.2006.

Yeung, C. H., C. Callies, F. Tüttelmann, S. Kliesch, and T. G. Cooper. 2010. "Aquaporins in the human testis and spermatozoa—identification, involvement in sperm volume regulation and clinical relevance." *Int J Androl* 33 (4):629–641. http://doi.org/10.1111/j.1365-2605.2009.00998.x.

Yoboue, E. D., A. Rimessi, T. Anelli, P. Pinton, and R. Sitia. 2017. "Regulation of calcium fluxes by GPX8, a type-II transmembrane peroxidase enriched at the mitochondria-associated endoplasmic reticulum membrane." *Antioxid Redox Signal* 27 (9):583–595. http://doi.org/10.1089/ars.2016.6866.

Zahn, A., C. Moehle, T. Langmann, R. Ehehalt, F. Autschbach, W. Stremmel, and G. Schmitz. 2007. "Aquaporin-8 expression is reduced in ileum and induced in colon of patients with ulcerative colitis." *World J Gastroenterol* 13 (11):1687–1695. http://doi.org/10.3748/wjg.v13.i11.1687.

Zhang, H., W. B. Du, X. M. Guo, L. K. Wang, J. M. Cheng, and L. J. Wei. 2020. "Identification of the AQP8-miR-92a network associated with the aggressive traits of colorectal cancer." *Biochem Biophys Res Commun* 527 (1):218–225. http://doi.org/10.1016/j.bbrc.2020.04.055.

Zhu, S. J., K. J. Wang, S. W. Gan, J. Xu, S. Y. Xu, and S. Q. Sun. 2013. "Expression of aquaporin8 in human astrocytomas: correlation with pathologic grade." *Biochem Biophys Res Commun* 440 (1):168–172. http://doi.org/10.1016/j.bbrc.2013.09.057.

CHAPTER THIRTEEN

Peroxiporins as Crucial Advocates of Redox Architecture

Ilaria Sorrentino, Angie Katherine Molina-Oviedo, Eduardo Arevalo-Nuñez de Arenas, and Iria Medraño-Fernandez

ABBREVIATIONS

AQP	Aquaporins
BCR	B-cell receptor
CBS	Cysthathionine-β-synthase
CSE	Cysthathionine-γ-lyase
EGFR	Epithelial growth factor receptor
eIF2α	Eukaryotic translation initiation factor 2 alpha
ER	Endoplasmic reticulum
Erk	Extracellular signal–regulated kinases
Ero1α	Endoplasmic reticulum oxidoreductase 1 alpha
ETC	Electron transport chain
GADD34	Growth arrest and DNA damage-inducible protein 34
Gpx	Glutaredoxins
GSH	Glutathione
HCO_4^-	Peroxy-monocarbonate
MAM	Mitochondrial-associated membranes
NOX	NADPH oxidases
PDI	Protein disulfide isomerase
PP1	Protein phosphatase 1
Prx	Peroxiredoxins
PTP1B	Protein tyrosine phosphatase 1B
ROS	Reactive oxygen species
SHP-2	Tyrosine-protein phosphatase non-receptor type 11
SOD	Superoxide dismutases
Trx	Thioredoxin
VDAC	Voltage-dependent anion channels

13.1 INTRODUCTION

Among the brain skills that allow humans to understand and learn from the elements that surround them, categorization arises as one of the most fundamental constituents of human cognition. Categorization is the ability to classify definite entities (e.g., tangible objects, ideas, experiences) into abstract groups based on their traits, characteristics, or comparable features. While this cognitive functioning capacitates us to recognize, differentiate, and process thoughts in a pragmatic way, there are certain risks associated. When a term or a concept is categorically classified into a deep penetrating class in the collective imaginary, it can be difficult to appreciate the nuances. This is precisely the case with reactive oxygen species (ROS). The gentler definition of ROS in textbooks portrays them as an unwanted by-product of biological processes, while the notion among the general public is even worse, especially by grace of some industries that have found a money spinner through the combined use of fear and poorly understood chemistry. And researchers are also humans categorizing their external world just in a specialized manner. Probably for this reason, the quantity of publications across the whole collection of scientific literature dedicated to defining, outlining, or expressing means to combat the toxicity of ROS outnumbers by a landslide the ones claiming beneficial effects.

DOI: 10.1201/9781003160649-13

Fortunately for us, humans are also individuals in continuous learning, a fact that in the long run corrects erroneous categorization even if it has remained static for years. Recent findings have turned the tables, recognizing that ROS are both deleterious and necessary molecules to sustain life (Sies and Jones 2020). Thus, an increasing number of reports recognize a versatile array of protein modifications induced by redox species, particularly H_2O_2, that ultimately influence cell responses (Rhee 1999; Finkel 2000; Sauer et al. 2000; Thannickal and Fanburg 2000; Sauer, Wartenberg and Hescheler 2001; Chiarugi et al. 2003; Foreman et al. 2003; Geiszt and Leto 2004; Cai 2005; Stone and Yang 2006; Ushio-Fukai 2006). Being caused by concentration-sensitive sequential oxidative reactions, these modifications not only trigger major outcomes such as proliferation, differentiation, or migration (Rhee 1999; Thannickal and Fanburg 2000), but in addition, fine-tune essential cell daily signaling routines (Meng, Fukada and Tonks 2002). Importantly, this dual perspective that redox biology aims to clarify is shedding light on some difficult-to-interpret evidence accumulated during the studies on human disorders in which ROS release their full potential as harmful elements. In cancer or chronic inflammation, diseases initiated or aggravated by ROS, antioxidant therapies have constantly failed (Steinhubl 2008; Forman and Zhang 2021; Luo et al. 2022), challenging all curative predictions. It is clear that we have missed some steps. Consequently, much attention in the field nowadays is directed toward delineating the precise mechanisms that govern the spatiotemporal specificity of these molecules and the facts that determine their behavior either as messengers or executioners. In other words, to comprehend how cellular organization defines, limits, and targets redox signaling in crowded cells.

On this conceptual frame, whose amplitude could be tackled by using the term "redox architecture" (Nordzieke and Medraño-Fernandez 2018), is where peroxiporins are now being revealed as crucial. Cellular membranes are an obligated traversing path for signaling-competent oxidative molecules (Henzler and Steudle 2000; Bienert et al. 2007; Miller, Dickinson and Chang 2010; Bertolotti et al. 2013). Moreover, inside the cell their effects are graded by their relative concentration (Sies and Jones 2020). Hence, it is perceptive that the expression, regulation, and rate of transport of these specialized H_2O_2 transporters would be major determinants of ROS outcomes. Thus, the concomitant analysis of positional production and quenching of ROS, together with the features of the gateways to the different cell compartments, may well have the key to disclose whether a particular input is perceived as a danger that must be decomposed or used to transduce signals.

13.2 ESTABLISHING THE FOUNDATIONS: THE PLASMA MEMBRANE

The outermost layer of the cell has been traditionally considered largely impermeant to ROS, a feature contemplated as a protective strategy. However, there is no innocuous substance when homeostasis must be maintained. Even quintessential molecules such as water are not inert, and their unrestricted net flow inside a cell would prevent normal function by dilution of molecular components. Not surprisingly, when cellular osmotic pressure is challenged, rapid re-equilibration of water content does not occur by passive diffusion, but rather using specialized aquaporin (AQP) channels (Preston et al. 1992). Following these lines, the plasma membrane results in an efficient barrier for most substances because it was carefully modeled by centuries of evolution to allow for selective passage of solutes in a controlled manner. And ROS are not an exception. So, taking a step back and changing the standpoint, the most exposed area of the cell also constitutes a privileged surface to sense and adequately trigger responses to the presence of molecules that either compromise or assist life. A place whose biochemical design must reconcile safeguarding the cell from the outside with communication with the inner side. To handle highly reactive molecules such as ROS, it is clear that the molecular organization of the plasma membrane in this respect should be exquisite.

Since ancient times, the arrangement of redox signaling components on the external lipid bilayer has been outlined by the pulse to survive to the introduction of oxygen in our atmosphere. Primordial organisms early evolved detoxifying pathways to cope with ROS (Ślesak, Ślesak and Kruk 2012), indicating that in the pre-aerobic world yet modest levels of these molecules may have exerted a strong evolutionary pressure (Dowling and Simmons 2009). Indeed, it has been suggested that dealing with abiotically produced exogenous

H_2O_2 may have been a decisive impetus for developing an oxygenic-based metabolism (McKay and Hartman 1991). Once mounted on extracellular membranes (Margulis 1992), the newly acquired capacity to extract energy from a water-splitting photosystem drove profound reshaping of both the extracellular and intracellular milieus of early aerobes. At one side, bimolecular oxygen was produced, ever changing the living conditions for all forms of life (Holland 2006). At the other side, ROS-generating leakage of electrons induced further adaptations to prevent genomic instability (Gacesa et al. 2016; Inupakutika et al. 2016), presumably contributing to the endosymbiotic events that led to the eukaryotic lineage (Sessions et al. 2009; J. Gross and Bhattacharya 2010). At the level of the plasma membrane, the structural barrier is reinforced against ROS by incorporating lipophilic antioxidants such as carotenoids and tocopherols (Hulbert et al. 2007), and adding sterols that make passive diffusion even more difficult (Mouritsen and Zuckermann 2004). As the metabolism of the latter is itself dependent on oxygen, a provocative hypothesis has proposed that the development of its enzymatic synthesis could represent an attempt to quench the potential menace resulting in the production of a novel metabolite that conferred other selective advantages (Galea and Brown 2009). Under the same lines, it is highly tempting to speculate that the advent of cytochromes may have responded to a similar fortunate event.

Whatever the true cause of their onset, refinement of cytochromes not only gave rise to one of the most breakthrough inventions to safely harness oxygen to generate energy. NADPH oxidases (NOX), seemingly the only protein family that produces ROS on purpose, are also supported by the presence of a cytochrome in their core (Kawahara, Quinn and Lambeth 2007; Sumimoto 2008). As little is known about the most distant part of the molecular evolutionary history of these enzymes, we can only make conjectures on the pulsion to conserve a potentially harmful ROS-generating protein cluster on the plasma membrane. Based on the present functions of the members of the NOX family in defense and redox signal transduction pathways, a plausible hypothesis would be that ROS were used to thrive among surrounding competitors, raising at the same time the necessity to communicate the presence of the autologous oxidative insult in the cell vicinity and prevent toxic side effects. In this regard, mounting the machinery on the external lipid bilayer perfectly pairs with economizing damage. Delimiting an outside from an inside compartment permits the localization different subsets of antioxidant enzymes with different reactivities and end products on either side. At the external leaflet, extracellular superoxide dismutases (Cu/Zn-SOD3) anchored to heparan sulfate proteoglycans convert the fast reactant $O_2^{\bullet-}$ produced by NOX into the more stable species H_2O_2 (Marklund 1985). In contrast, beneath the membrane, peroxiredoxins (particularly Prx1 and Prx2; Dalla Rizza et al. 2019) and glutathione peroxidases (headed by the abundant Gpx1; Brigelius-Flohé and Maiorino 2013) quench H_2O_2, maintaining an almost completely reduced scenario. The difference in concentration between both sides implies that a steep gradient of H_2O_2 is formed between the extracellular space and the cytosol upon NOX activation. Such a gradient constitutes the basis for redox signaling.

13.2.1 Traversing a Landmark Doorway: Plasma Membrane Peroxiporins

Exploitation of the gradient entails that privileged paths exist in the barrier so to internalize H_2O_2 molecules. The identity of the channel remained obscure for a long time since the very first observations that indicated that temporal concentrations of endogenously produced H_2O_2 (which should cause maladaptive oxidative stress) induced adaptive survival responses, therefore suggesting that NOX targets reside in the cytosol (Ramasarma 1982). Henzler and Steudle were pioneers proposing that some plasma membrane AQP may be functioning as peroxiporins rather than as pure water transporters (Henzler and Steudle 2000). Molecular genetic evidence followed, as Bienert and collaborators were able to characterize several efficient H_2O_2-transporting isoforms in an elegant wide screen among members of the family expressed in yeast membranes (Bienert et al. 2007). Thus, equally as with the water molecules that use these channels to balance the cell water content after an osmotic shock, transient rises in extracellular H_2O_2 are rapidly conducted to the more reducing interior of the cell by selected members of the AQP family.

Whether specific isoforms are endowed with peroxiporin activity relative to others is still a hard matter of debate. In principle, channel biophysical analyses have not supported a specialization on H_2O_2 transport that would not enable water passage and vice versa, suggesting that all family members in essence would be able to route H_2O_2 molecules (Almasalmeh et al. 2014). But biological signals work rarely on a yes/no basis, rather affinity and kinetics dominate inter- and intracellular redox dialogues. A major characteristic affecting the rate of permeation through a particular isoform would be pore diameter. Remarkably, comparison of the pore of the extensively validated peroxiporin AQP8 (Bienert et al. 2007; Miller, Dickinson and Chang 2010; Bertolotti et al. 2013; Medraño-Fernandez et al. 2016; Bestetti et al. 2018) with the conduit of other water-transporting AQP has revealed a wider diameter (Almasalmeh et al. 2014; Bestetti et al. 2018) that matches almost perfectly the slightly larger size of the molecule of H_2O_2 with respect to water (Henzler and Steudle 2000; Almasalmeh et al. 2014). This would predict that members of the aquaglyceroporin subfamily able to accommodate bulkier glycerol molecules could potentially also conduit H_2O_2. However, this is not the case, as only two members of this category have been shown to transport H_2O_2 (Miller, Dickinson and Chang 2010; Watanabe et al. 2016), indicating that other determinants of pore architecture should be taken into consideration. In that sense, a recent metadynamics analysis of H_2O_2 conductance by the aquaglyceroporin AQP3 points to the free energy required by the system around the proton exclusion region to thrive successful interactions with the residues lining the pore as one of these factors (Wragg, Leoni and Casini 2020). Thus, differences in internal energy barriers may sum to the structural context and the pore diameter and account for different H_2O_2 transport rates between isoforms, raising a question of undoubtful relevance: are all peroxiporin activities used indistinctly? Or more precisely, are there channels that preferentially overtake redox signaling while others are activated only under particular situations? In the ROS world, in which concentration matters, a panorama like the one drawn in the second option seems more likely to be true, and would justify apparently contradictory results obtained when analyzing the H_2O_2-conducting capacity of some of these transporters. For instance, this would apply to mammalian AQP1, an orthodox-classified AQP. The pore diameter of AQP1 measures in its narrowest area ~2.8 Å (Sui et al. 2001), defining a steric limit for the solutes allowed to traverse the protein but fit to water. Indeed, it did not show any peroxiporin activity even in the presence of relatively high concentrations of H_2O_2 in yeast survival assays (0.75 mM; Bienert et al. 2007), a fact later confirmed in mammalian cell lines under the same conditions that drove the passage through AQP8 (Miller, Dickinson and Chang 2010). However, when silencing the expression of AQP1 in aortic smooth muscle cells, which naturally lack any other previously characterized peroxiporin, a complete inhibition of H_2O_2 internalization upon an exogenous challenge was recorded, promoting its candidature for being included among the redox competent isoforms (al Ghouleh et al. 2013). Because the H_2O_2 uptake slope in these cells shows a much slower kinetics than in other cellular models endowed with AQP3 or AQP8 (Miller, Dickinson and Chang 2010; Bertolotti et al. 2013), we may consider whether these results truly categorize AQP1 as a relevant actor in more rapid redox-based signaling routes. Similarly, AQP5 has recently gained attention for its capacity to alleviate oxidative stress in yeast models, also being proposed as a functional peroxiporin (Rodrigues et al. 2016, 2019). However, the pore radius of AQP5 measures 1.02 Å at the ar/R exclusion filter, defining a region slightly more constricted than the one in AQP1 (Horsefield et al. 2008), and in the comparative study made by Bienert and collaborators this isoform did not show a relevant H_2O_2-conductive capacity, particularly when confronted with AQP8 (Bienert et al. 2007). Despite these putative limitations, AQP5 has several features that would be advantageous for a cell devoid of functional peroxiporins that needs to adapt its H_2O_2 permeability due to a stressful menace: first, it can be readily trafficked and inserted in cellular membranes after phosphorylation; and second, its activity is finely tuned by extracellular pH (Kitchen et al. 2015; Rodrigues et al. 2016). In the context of cancer, these characteristics turn rather interesting for transformed cells, as they would have the opportunity to exploit already enhanced malignant phosphorylation routes to insert a less efficient transporter in their membranes even if their own peroxiporins are safe-guardingly closed due to increasing intracellular ROS concentrations (Medraño-Fernandez et al. 2016), thus ensuring survival. Notably, most of the experiments showing that isoforms different from AQP3, 8, and 9 can also conduit H_2O_2 molecules have been done under close-to-pathological setups leading to oxidative stress (smooth muscle cell hypertrophy in the vascular wall for AQP1 (al Ghouleh et al. 2013), tumorigenesis and oxidative stress for AQP5 (Rodrigues et al. 2019), and cataractogenesis for AQP0 (Varadaraj and Kumari 2020). Hence, in an analogous situation to water in which AQP handle only 28% of cell basal transport but are the main responsible for the 800-fold water flow increase derived from an environmental osmotic stress (Kellen and Bassingthwaighte 2003), it is conceivable that distressful accumulation of oxidative species leads to a combined effort of all the available plasma membrane AQP repertoire to buffer the damage. In this outline, it may be worth to conceptually separate peroxiporins into signaling-associated isoforms, transmitters of redox information, and facultative channels, through which H_2O_2 is impelled by a strong redox gradient overcoming a more constricted pore.

Among the former (AQP3, AQP8, and AQP9), it would be equally important to carefully titrate their transporting rate in complex contexts in which more than one of these is expressed. As an example, selective expression exists in erythroid lineages, and primitive cells equipped with AQP3 and AQP8 peroxiporins preferentially accumulate H_2O_2 after exogenous exposure when compared with adult definitive erythroblasts bearing AQP1 and AQP9 channels (Kingsley et al. 2013), suggesting that AQP9 would be less competent at transporting redox equivalents and pointing to a functional necessity to adapt the transporting capacity of the redox routers to different native cell signaling requirements. Accordingly, AQP9 is the main peroxiporin expressed in leukocytes (Loitto et al. 2002), a cell type that if equipped with a high-rate transporter will perchance risk of dying from success while combating an infection using ROS. Instead, by expressing a "slower" plasma membrane peroxiporin that restricts the number of molecules entering the cytosol, these cells can turn this apparent disadvantage into an opportunity to tune their proliferation rates and eventually enhance the immune response (Moniaga et al. 2015) in the presence of concentrations of extracellular H_2O_2 that will be deleterious for an AQP8- or AQP3-expressing cell type. Other scenarios also highlight the relevance of determining which signals activate transport through one isoform rather than through another. HeLa cells, a model that mainly relies on AQP8 to internalize H_2O_2 molecules (Bertolotti et al. 2013), also expresses mRNA for AQP1, AQP3, and AQP9 (Pellavio et al. 2017), raising the possibility of alternative usage depending on the stimulus perceived and advancing yet to decipher modes of isoform-specific regulation depending perhaps on the scale of the input. However, the relative promiscuity of signals with respect to the isoform used predicts that this would not be a trivial affair: just to cite some cases, EGFR signaling has been shown to be mediated by AQP3 and AQP8 (Miller, Dickinson and Chang 2010; Bertolotti et al. 2013; Vieceli Dalla Sega et al. 2014; Hara-Chikuma, Watanabe and Satooka 2016); AQP3 and AQP9 participate on signal distribution upon LPS stimulation (da Silva et al. 2021; Tesse et al. 2021); and B-cells upregulate AQP3 and AQP8 after BCR activation (Moon et al. 2004; Bertolotti et al. 2016). Therefore, overcoming the complexity and overlapping nature of peroxiporins might only be possible by combinatorial advancing on our knowledge on the biophysical properties of the channels, their timed localization in restricted membrane microdomains, and the spatiotemporal redox status from cell to cell.

13.2.2 Corridors and Spillways: On the Organization of Redox Signaling Downstream Peroxiporins

Whatever the door selected to enter the cell and once inside, redox molecules have to defeat the very same ambient that praises them as signaling species. Cellular H_2O_2 levels are sequentially controlled by peroxidases, whose standard function is to protect the cytosol from oxidative challenges. However, the constant surveillance exerted by antioxidant enzymes results a necessary condition for categorizing H_2O_2 as a second messenger, as molecular signals must enter in a deprived compartment in order to be sensed. This is mechanistically well understood in the case of an old acquaintance within the signal transduction field, Ca^{2+}: its cytoplasmic concentration is kept in the low nanomolar range by active pumping, and signaling occurs when Ca^{2+} ions are released from intracellular stores or after regulated opening of plasma membrane ion channels (Clapham 2007). However, the redox cytoplasmic depletion strategy (based on in situ destruction rather than on species translocation) complicates our comprehension of how the H_2O_2 signal is effectively transmitted. Peroxiredoxins are thiol peroxidases that not only behave as ultrafast scavengers with reaction rates in the order of 10^7 M^{-1} s^{-1} (Peskin et al. 2007) but also fall among the most abundant proteins in mammalian cells, composing the 0.1%–0.8% of the total soluble protein pool (Chae et al. 1999). To illustrate the idea on how these enzymes should outcompete H_2O_2-derived signaling, the redox rate of PTP1B (a widely recognized redox target) is 10 million times slower than peroxiredoxins (Netto and Machado 2022). The ubiquitous presence of the redox buffer GSH, a low-weight thiol-bearing peptide that can be counted in millimolar units and reacts at comparable rates to phosphatases (Winterbourn and Hampton 2008), does not make the situation simpler. Notwithstanding and despite their lower reactivity toward H_2O_2, the list of cytosolic targets that have been found oxidized upon triggering of signaling cascades has grown exponentially in recent years. Key H_2O_2 targets are situated downstream from growth factor–controlled NOX activation and they must transmit their signals bypassing antioxidant defense systems rather than overwhelming them at risk of generalized toxicity. It is hard to envisage severely stressed cells to embark into proliferation, migration, and differential protein expression (Finkel 2000; Sauer et al. 2000; Chiarugi et al. 2003; Foreman et al. 2003; Geiszt and Leto 2004; Ushio-Fukai 2006), so redox signaling is actually happening under conditions in which omnipresent quenchers are presumably available to perform their primary job, raising the fundamental question of how it can be possible for a molecule of H_2O_2 immersed in a sea of antioxidants to find its targets while avoiding reduction and whether the cytosol is an amorphous or a highly ordered supramolecular entity.

Indeed, these types of conundrums are not unfamiliar to the field of signal transduction, as they are intimately ligated to the strict control of all biological pathways. Phosphatases themselves are roughly more abundant than kinases (Smoly et al. 2017), and their promiscuity toward substrates and fast kinetics with respect to their signaling counterparts (*Protein Phosphorylation : A Practical Approach/Edited by D. Grahame Hardie.* 1993) used to represent a challenge that puzzled out the understanding of how phosphorylation cascades can actually operate. A partial answer can be deduced from the very first paragraph of this section: cysteines in the catalytic site of phosphatases can be timely inactivated by a H_2O_2 wave induced by the same ligand that initiates phosphorylation (Lee et al. 2002; Leslie et al. 2004; den Hertog, Groen and van der Wijk 2005; Raman and Pervaiz 2019). The picture is further completed by adding a positional control by which kinases, targets, phosphatases, and their different proteinaceous regulators are separated or re-joined in distinct microdomains by using post-translational modifications or adaptor proteins (Virshup and Shenolikar 2009). In summary, the entire signaling network is governed by a highly coordinated spatiotemporal scheme that has been laboriously depicted by hundreds of indefatigable researchers over the last 70 years. We shall thus be comforted by the timeline, as in comparison redox biology is still in its infancy. However, some hints have already been unveiled. Similar to phosphatases, thiol peroxidases can be transiently inactivated in the presence of high concentrations of H_2O_2 due to the conversion of their catalytic cysteine to a sulfinic acid ($-SO_2H$) (K.-S. Yang et al. 2002), and some peroxiredoxin forms are even inhibited by phosphorylation (Woo et al. 2010). Perhaps more surprisingly, increasing evidence supports a signal intermediary activity of these enzymes much like the one that kinases perform, as they have been shown to be able to transduce the disulfide formed during their regular detoxification cycle to other proteins (Jarvis, Hughes and Ledgerwood 2012; Sobotta et al. 2015; Fernandez-Caggiano et al. 2016; Stöcker et al. 2018). As a result, two main models of possible redox signaling have been proposed: the floodgate model, in which H_2O_2-mediated inactivation of cytosolic peroxiredoxins allows redox signaling to occur; and the relay model, which sponsors a thiol peroxidase–based transfer mechanism of oxidation such as the one used in the ER to incorporate disulfide bonds into nascent proteins. Remarkably, although the latter has been lately gaining more and more followers due to the increasing list of targets using the mechanism, some degree of coexistence would reconcile apparently excluding results. Against overoxidation-dependent signaling, the process has not proven to be particularly physiologically relevant for worms (Thamsen et al. 2010) and flies (McGinnis et al. 2021) in deletion experiments, and kinetic modeling of the reactivity of peroxiredoxins demonstrates that relays are a far more efficient way to transmit redox signals deeper in the cell during normative signaling (Travasso et al. 2017). In favor, mathematical models do not completely discard the occurrence of overoxidative events in a highly discrete area in the vicinity of the source at the cost of a shallower cytoplasmic gradient (Travasso et al. 2017), and the facile hyperoxidation of only certain eukaryotic peroxiredoxins family members with respect to others indicates that this is a strongly selected feature, and so must be beneficial to the cell in one or more aspects (Karplus 2015). Moreover, particular signals such as those linked to inflammatory stimuli seem to require overoxidation (Jin et al. 1997; Kang et al. 1998; Rabilloud et al. 2002), suggesting that this mechanism could be used to transduce fire alarm cues to promote repair responses (Phalen et al. 2006). Indeed, the wide range of effects induced by H_2O_2 in cells are intrinsically linked to the concentration reached after entering the cytosol, and the necessity to grade it may plausibly demand different strategies to decode an adequate outcome. Crucial to completing the mechanistic map of the signal would be to topologically localize H_2O_2 reactivity, because only proteins situated within or recruited into high-H_2O_2 microdomains would be directly oxidized, while lower redox environments will preferentially use relays (Nordzieke and Medraño-Fernandez 2018). At this point it is inevitable bringing to the fore the H_2O_2-internalizing channels, as most of the H_2O_2 is restricted to a few tenths of micrometers from its supply sites in absence of oxidative stress (Travasso et al. 2017). As single points of entry across the plasma membrane, peroxiporins build up radial H_2O_2 gradients that have their highest expression around the cytoplasmic vestibule of the transporter. Therefore, not only putative direct targets but also both floodgate-sensitive and relaying peroxiredoxins should be positioned at convenient distances respect to the emanating flow to be operative. By virtue of their privileged situation at the origin of the signal, the channels may further serve as initial docking site driving the supramolecular clustering that organizes ROS-related proteins, as it will be possible to graduate specific combinations of peroxidases and target proteins starting from peroxiporin cytoplasmic tails and sequentially recruiting scaffolds. In such a manner, the positional control that similarly orchestrates phosphorylation pathways would be guaranteed, suggesting that H_2O_2 transporters are at the apex of the redox architecture of the cell (Figure 13.1). In agreement, imaging of H_2O_2

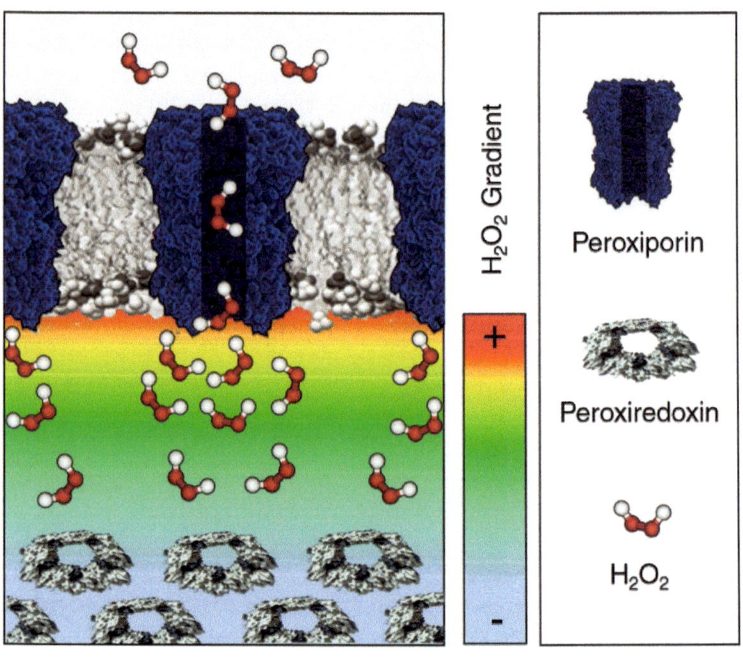

Figure 13.1 A hierarchical architecture is organized by strategical positioning of peroxiporins and thiol peroxidases such as peroxiredoxins. Acting as the customs of cellular membranes, the flux of molecules through peroxiporin pores presumably allow the achievement of high H_2O_2 buildups in the proximity of their cytoplasmic entrances, while signaling relays can be organized herein by strategical positioning of peroxiredoxins or even other intermediate protein scaffolds.

generation (Mishina et al. 2010) and protein oxidation (Paulsen et al. 2012) shows that accumulation of H_2O_2 and impact on redox-sensitive targets occurs in the vicinity of growth factor receptors, conceivably in the proximity of a peroxiporin and confined by a barrier of active peroxidases that limit diffusion. Thus, to determine whether and how redox signaling modalities are arranged along channel downstream aisles will probably constitute a fruitful path to follow in redox biology for years to come.

13.3 THE KITCHEN: THE ENDOPLASMIC RETICULUM AS A NOT-SO-ISOLATED REDOX ISLAND

Activated NOX complexes at the plasma membrane are frequently contemplated as the source of signaling-related H_2O_2, since they respond to the traditional input-signal-output scheme. However, this notion disregards the emerging concept of tonic or constitutive signaling, a collective denomination introduced to encompass the decoding of housekeeping molecular signals that internally assure cellular fitness. Importantly, particular membrane-delimited compartments within the cell house central redox-based routine processes that when dysregulated compromise survival. Enclosed generation entails the existence of conspicuous H_2O_2 gradients toward the cytosol (Malinouski et al. 2011) that could thus be potentially used to interpret the wealthiness of the metabolic status.

Among these organelles, the ER stands out by harboring the greatest internal H_2O_2 concentration (Gao et al. 2017; Melo et al. 2017). Oxidative folding of nascent proteins is a vigorous constitutive process mainly relying on the activity of the oxidoreductin Ero1α and its partner PDI (Tavender and Bulleid 2010). The reaction stoichiometrically yields one molecule of H_2O_2 per disulfide bond created, using molecular oxygen (O_2) as the terminal electron acceptor (E. Gross et al. 2006). Hence, an appealing hypothesis will be that the redox equivalents generated as by-product of folding might serve to tonically transmit whether the secretory pathway is proceeding normally. In turn, when reaching the cytosolic leaflet, oxidative modification of targets by H_2O_2 may further allow for tuning other cell machineries to the actual protein production demands by establishing inter-organellar communication pathways that offer a coordinated response in terms of energy supply, gene expression, or even cell behavior. For the premise to be true, a fundamental condition must be achieved: that H_2O_2 molecules produced inside the ER effectively reach the external interface with the cytosol. Therefore, and to trespass the otherwise rather impermeable lipid barrier, a peroxiporin would be needed (Appenzeller-Herzog et al. 2016). The voyage has been recently shown to be facilitated by the intracellularly expressed AQP11 (Bestetti et al. 2020). This isoform, belonging to the unorthodox AQP subfamily (Ishibashi et al. 2010), was indeed a promising candidate, as knockout mice for AQP11 expression die soon after being affected by a massive ER stress accompanied by abnormally elevated levels of intracellular H_2O_2 mainly manifested in the kidney (Yoshiyuki et al. 2005; Okada et al. 2008; Atochina-Vasserman et al. 2013), but also targeting other tissues such as the liver (Rojek et al. 2012). Moreover, protein expression databases such as the Human Protein Atlas (www.proteinatlas.org/ENSG00000178301-AQP11/summary/rna) support a wide tissue distribution that will be in accordance with a role as mediatory element of a signaling module that, if existing, would be functional in all cellular types. The study conducted by Bestetti and collaborators revealed yet more important evidence in favor of a H_2O_2-mediated tonic signal cause. When silencing the expression of AQP11 in HeLa cells, the basal redox state of the ER increased, concomitantly to a decrease in the oxidative status of the cytosol (Bestetti et al. 2020). The shift was indeed induced by H_2O_2 molecules, as it could be abrogated by expressing an ER-targeted catalase (Sorrentino et al. 2022). Accordingly, the interpretation of the data shown was that a constitutive flux pours through AQP11 following its concentration gradient and downregulating the channel derived in the accumulation of H_2O_2 inside the lumen of the organelle. However, one piece of the puzzle is still missing: which are the preferential targets of the discharged molecules?

Unfortunately, examples of redox signaling that can be attributable to the ER are still scarce, and mostly related to a source of H_2O_2 in the ER seemingly non-related to folding, the ER-resident NOX isoform NOX4 (Chen et al. 2008; Ru-Feng et al. 2010; Fazeli et al. 2012; Weyemi et al. 2012; Santos et al. 2016). Of note, NOX4 is the only member of its family known to directly build

H_2O_2 bypassing superoxide generation (Takac et al. 2011; Laurindo, Araujo and Abrahão 2014), and its membrane topology dictates that its products will spill into the ER lumen (Rousset et al. 2020). This implies that this oxidase will be associated with a peroxiporin so as to transmit is signal to the cytosol and hence could also putatively be the origin of the continuous flux permeating AQP11. However, although NOX4 was contemplated in the past as a constitutive enzyme (Geiszt et al. 2000; Nisimoto et al. 2010; von Löhneysen et al. 2012; Laurindo, Araujo and Abrahão 2014), this consideration was more likely due to the failure on identifying a trustworthy regulator than to a strict fact. Some reports suggest that NOX4 is, indeed, regulated and triggered by specific signaling pathways (Gorin et al. 2003; Kalyankar et al. 2004). Remarkably, the data shown in the paper by Sorrentino and colleagues reveal two interesting hints about the functioning of NOX4: first, that downregulation of this enzyme only minorly impacts the basal oxidative state of the ER, indicating that NOX4 is not a major contributor on maintaining the redox milieu inside the organelle (Sorrentino et al. 2022); and second, that silencing of Ero1α in combination with AQP11 does not lead to an increase in luminal H_2O_2, as would have been expected if NOX4 products are trapped inside the ER due to the absence of the conduit to the cytosol. These results suggest that either NOX4 is in an "off" state under resting conditions, exposing Ero1α as the most likely origin of the H_2O_2 traversing AQP11, or that Ero1α presence is somehow an activator of its oxidase activity. Importantly, both possibilities bring forth that AQP11-transported H_2O_2 is connected to folding, directly or indirectly. The fact that NOX4 upregulation has been constantly related to stress responses (Eric et al. 2004; Bedard and Krause 2007; M. Zhang et al. 2010; Lassègue, San Martín and Griendling 2012; Santos et al. 2016), a physiological context in which transmitting the prosperity of the secretory pathway has overstanding relevance (Harding et al. 2003; Hitomi et al. 2004; Høyer-Hansen et al. 2007; Vergés et al. 2007; Heath-Engel, Chang and Shore 2008), also weights in the same direction. For all these reasons, we foster that decoding the biological significance of an AQP11-mediated putative tonic signaling pathway will be a highly worthy path to follow for redox biology in the future.

Whatever their source, entering the cytosol implies that AQP11-driven spillways should be subjected to the same constrictions than the ones emanating from the plasma membrane. The best characterized target of ER-produced H_2O_2 is the phosphatase PTP1B, tethered to the external face of the organelle by a membrane localization domain (Haj et al. 2002; Eden et al. 2010; Yip, Saha and Chernoff 2010). As stated before, it is not clear this protein, which has a rate constant of reaction with H_2O_2 ranging from 10 to 40 $M^{-1} s^{-1}$ (Denu and Tanner 1998), competes effectively with the abundant cytosolic thiol peroxidases in the time frame required for signaling. A partial solution forcing a revision on how we conceptualize H_2O_2-driven modifications was found following the work started by Zhou and collaborators (Zhou et al. 2011). In their elegant study, they showed that the biological buffer, bicarbonate/CO_2, reacts with H_2O_2 yielding the highly reactive oxidant peroxy-monocarbonate (HCO_4^-), which in turn inactivates PTP1B by a two-electron oxidation mechanism. This redounds on an enhanced reaction rate of the phosphatase with H_2O_2 of one order of magnitude, even in presence of physiological concentrations of GSH. Still, the levels do not reach the rates reported for peroxiredoxins (Winterbourn 2008), and the very same authors suggest that conversion of H_2O_2 to HCO_4^- may act in tandem with topological compartmentalization of targets to increase specificity. Remarkably, the oxidative reaction mediated by HCO_4^- is also operative in other differently localized PTPs such as SHP-2, a protein tyrosine phosphatase that associates with most receptor tyrosine kinases to mediate complete ERK activation (Neel, Gu and Pao 2003), and bicarbonate-regulating enzymes that control intracellular levels are, as peroxiporins, membrane-bound proteins (Dorai et al. 2005; Svastova and Pastorekova 2013; Aalkjaer et al. 2014). Considering that H_2O_2 concentration at steady state decays steeply after the first ~0.3 mm adjacent to a membrane (Travasso et al. 2017), we can count on a profitable conjunction of all obligate components of this mechanism (H_2O_2, bicarbonate, and targets) in a defined area in the vicinity of the origin of the redox molecules. Moreover, in a space-limited scenario, it would be possible to achieve the conditions that predispose to faster HCO_4^--dependent inactivation of phosphatases respect to peroxiredoxins: lower Trx1/TrxR1 concentration or high bicarbonate levels (Dagnell et al. 2019) in concurrence with a fast flow of H_2O_2 molecules introduced in the cytosol by a peroxiporin channel (Nordzieke and Medraño-Fernandez 2018) and possibly thiol peroxidase inactivation, as HCO_4^--mediated peroxiredoxin hyperoxidation has been also demonstrated to be significantly swifter than the corresponding reaction with H_2O_2 (Peskin, Pace and Winterbourn 2019; Truzzi et al. 2019).

Spatial confinement of the redox-related proteins may be quite relevant in the context of the signals coming from the ER. The flux that runs through AQP11 continuously drives H_2O_2 molecules to the cytosol, suggesting that whether a protein is oxidatively modified or not will be defined by dynamic exclusions and inclusions from the spillway of the redox corridor downstream from the peroxiporin. For instance, this may well be the case of PP1, a ubiquitous serine/threonine phosphatase that acts on dozens of different substrates all around the cytosol (Ceulemans and Bollen 2004). During ER overload, protein synthesis attenuation is a conservative measure induced after phosphorylation of the translation initiation factor eIF2α (Walter and Ron 2011). Following relief of the stress, PP1 counteracts this pathway to promote the end of the adaptive signaling (Novoa et al. 2001). In the resolving phase, PP1 must be inhibited to permit eIF2α phosphorylation, an action ensued by the adaptor molecule GADD34 that directs the phosphatase to the ER surface and mediates contact with a NOX4-produced H_2O_2 flux (Santos et al. 2016). On the return to homeostasis, GADD34 interacts with eIF2a, allowing translation to resume (Brush, Weiser and Shenolikar 2003). Therefore, modulation of the stress response depends on an early targeted H_2O_2-mediated inhibition of PP1, and presumably on delocalizing the phosphatase from the redox influence at later stages. Another example can be found on the ER-vicinal Ras activation described in HUVEC cells exposed to an ER stressor virus peptide found in the blood of HIV-infected patients (Ru-Feng et al. 2010). Sustained co-culture of these cells with a HIV-Tat-producing cell line promoted translocation of RhoA to the ER membrane, NOX4-dependent H_2O_2 generation, and local Ras activation at the ER surface as a downstream consequence, suggesting that the membrane provides a platform to spatially organize ER redox signaling analogous to the one proposed for the plasma membrane (Nordzieke and Medraño-Fernandez 2018). Indeed, such cross-talk between the Ras and Nox proteins occurs in a similar manner in response to mitogenic signaling from lactosylceramide and angiotensin II, underneath the plasma membrane (Bhunia et al. 1997; Adachi et al. 2004). Thus, the capacity of acting over distinct pools of targets by using spatially specific signaling duos composed by a H_2O_2 producer and a transporting channel qualifies the message to reach the most appropriate outcome depending on the sensing region. In this scenario, the thiol peroxidase barrier-less compartment not only serves as ultimate constrain for H_2O_2 diffusion, limiting the amplitude of the signal, but also confines each flux to a particular area of interest (Figure 13.2). The same molecule (H_2O_2) can then be putatively used at more than one cell location with specifical meanings in each context. Real-time monitoring of H_2O_2 production upon growth factor activation seemingly confirms this case, as ligand binding triggers non-overlapping waves of H_2O_2 that arise with variable timing either from the plasma membrane or from the ER (Mishina et al. 2010).

Notably, the constitutive nature of the H_2O_2 flux emanating from the ER anticipates a novel redox signaling modality associated to variations in tonicity. The archetypical route that characterizes other H_2O_2-driven pathways is defined by the acute appearance and subsequent quenching of the second messenger in the cytosol. However, the flow of molecules that transit AQP11 is constant, implying that H_2O_2 translocation across the ER membrane can be increased to reach further targets, but also interrupted. Still to be determined is whether and how this peroxiporin isoform is regulated. The closure of AQP8, its cousin at the plasma membrane, can be induced by a redox-dependent two-step mechanism that involves first sulfenylation and then persulfidation of a critical cysteine, C53 (Bestetti et al. 2018). A homologous residue (C58) is also present within the pore of AQP11. Nevertheless, a sustained flux incessantly passing through the channel would entail progression to irreversible oxidative forms unless the cysteine is protected. Therefore, the gating mechanism, if existing, cannot be based by definition in the same principles than the one governing AQP8. Yet, evolution conserved a redox-sensitive amino acid inside a H_2O_2 conduit, suggesting that it is necessary for some purpose (Marino and Gladyshev 2010). In addition to C58,

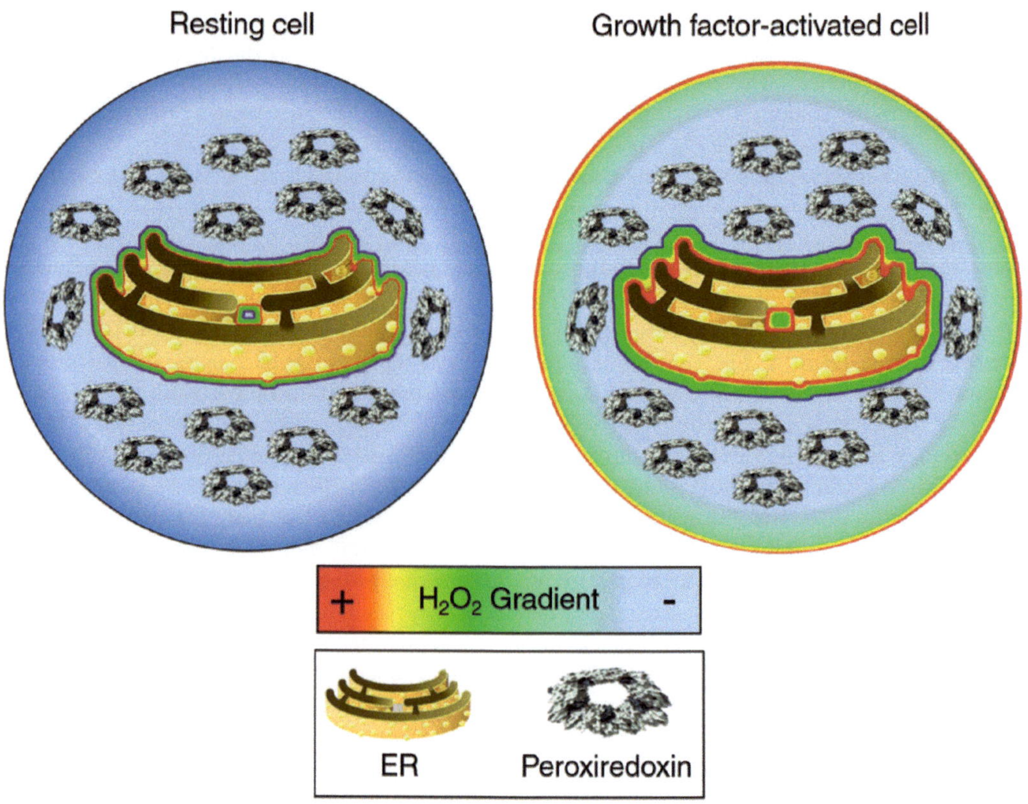

Figure 13.2 Thiol peroxidases as gatekeepers of the borders of redox signaling. The isolating component provided by the rapid quenching catalyzed by elements of the antioxidant defense permits the function of different redox compartments separately or in synchrony. Thus, in the resting cell (left), the constitutive flux pouring through AQP11 reaches targets vicinal to the ER membrane and is efficiently contained by peroxiredoxins. In contrast, when a stimulus activates a growth factor in the plasma membrane (right), oxidative influence is recorded near to the plasma membrane affecting resident protein pools. Note that the same stimulus can be transduced by a signaling cascade into the cell interior simultaneously, inducing a second wave that alters the flux that surrounds the ER in a local manner due to action of the very same peroxiredoxin shield.

other intriguing features decorate AQP11 amino acid sequence: first, this residue forms part of a rare tri-cysteine motif, too close for a disulfide bond, but infrequent enough in the proteome not to be noticed; second, situated at the cytoplasmic side, cysteine 227 mutation produces a gain-of-function channel with enhanced water permeability (Takahashi et al. 2014) that causes renal failure (Tchekneva et al. 2008); and last, methionines have been lately gaining attention as previously unadvertised redox sensors (Fricke et al. 2019; Kato et al. 2019), and AQP11 is highly enriched in these residues, most of them structurally grouped and exposed to the cytoplasm (our unpublished observations).

Indeed, it is in the interface of the ER with other organelles in the cell where peroxiporin regulation may reach its highest levels of complexity by virtue of juxtaposed balances in redox homeostasis. As stated above, a reductive shield composed by thiol peroxidases and other antioxidant molecules will guarantee redox signaling specificity by confinement of signals. However, cell compartments do not respond anymore to a stand-alone definition and crosstalk is revealing more and more necessary to offer highly interconnected responses to daily redox variations (Yoboue, Sitia and Simmen 2018). As proof of concept, it has been recently reported that a mitochondrially originated H_2O_2 flux enters into the ER via AQP11 upon depletion of Ero1a (Sorrentino et al. 2022), resembling the mechanism for store-operated Ca^{2+} entry orchestrated between the ER and the plasma membrane (Chenlu Li et al.

2021). Though the ultimate significance of this ER refilling is still obscure, it cannot be overlooked that the most probable aim of the flux will be the maintenance of the redox status quo, whether it singularly or complementary involves the intra-luminal ER oxidative ambient or the gradient toward the cytosol. In any instance, it is clear that this type of inter-organellar communications will not be exceptional cases and most likely will compete the entire cell for the sake of cooperation. Remarkably, the exchange of redox molecules has a spatial component requiring the de novo formation of mitochondrial-associated membranes (MAM; Sorrentino et al. 2022). Thus, it is possible that these structures function as isolating synapses dynamically armed to prevent peroxiredoxins from reducing the strength of the signal. At the same time, MAM could concentrate the means for gathering transporting channels in both organelles, directly by locally supporting rejoining, or indirectly by trapping the molecules that will warrant coordinated opening of communicating conduits. Can a similar conjunction be operating in the crossroads of the ER and the plasma membrane? Even if the conceptual answer would be tentatively positive, the fact that PTP1B and other ER-linked phosphatases are found active on these contact sites to gain access to plasma membrane-bound substrates (Foti, Audhya and Emr 2001; Snyder et al. 2005; Y. Zhang et al. 2008; Eden et al. 2010; Monteleone et al. 2012) argues on the contrary, suggesting that they would constitute redox excluding domains if any. Notwithstanding, it is still possible that the mechanisms of regulation of plasma-membrane and ER-resident peroxiporins are shared at any time, responding to the general cell policy of economizing signals when integrating modulatory routes. Cystathionine-β-synthase (CBS), the critical redox sensor that ultimately persulfidates and closes AQP8 at the plasma membrane (Bestetti et al. 2018), is mainly a cytosolic protein (Zuhra et al. 2020). Moreover, its product H_2S is one of the scarce molecules able to fit into a peroxiporin pore, apart from their specific solutes (Mathai et al. 2009; Cuevasanta et al. 2012; Bestetti et al. 2018). Therefore, a possibility yet to be explored is that CBS yields opposing effects on peroxiporins transiting the ER: by one side, it could be the responsible for the putative protection of C58 from terminal oxidation (Zivanovic et al. 2019), maintaining AQP11 in an open state; by the other, it could assure that other peroxiporins such as AQP8, folded and assembled in the ER (Hohmann, Nielsen and Nielsen 2000; Turnbull et al. 2000; Chunling Li and Wang 2017), are not fully functional until their arrival to their final destination. Alternatively, cysthationine-g-lyase (CSE), another cytosolic H_2S-releasing enzyme in the transsulfuration pathway (Sbodio, Snyder and Paul 2019), could complement or substitute CBS in this role.

Whatsoever results true, the crossover expression between different compartments of H_2O_2 transporters and potential regulators physically links the redox status of each individual cellular structure with the one of its neighboring organelles and precludes the notion of redox-rich islands floating isolated in the reduced environment of the cytosol. Thus, although much effort will be still necessary to understand the complete picture, defeating this misconception seems an optimal stone from which start building.

13.4 A CENTRAL STOVE: MITOCHONDRIA AS SINKS AND SOURCES

Probably influenced by the isolation procedures that allow for obtaining purified and "alive" mitochondria, redox signaling from or directed by this compartment has been even more often neglected than the one emanating from the ER. Under these circumstances, it comes as no surprise that ROS produced in the respiratory electron transport chain (ETC) were considered in the past just as troublesome by-products with pernicious consequences (Boveris and Chance 1973). However, the global health of the cell is highly dependent on mitochondrial fitness, manifesting the establishment of tight interconnections with the rest of the cellular components. Thus, the pulsion to conserve respiratory complexes which occasionally leak electrons to form ROS during endosymbiont adaptation may well reflect an internal necessity of communication. Indeed, much evidence can be lately traced across the literature that relate mitochondrial ROS production to measurable outcomes inside and outside the organelle (as examples from a growing body of reports, see Livigni et al. 2005; Moreno-Loshuertos et al. 2006; West et al. 2011; Kil et al. 2012; Kelly et al. 2015; and Sorrentino et al. 2022; for more general reviews on the subject, see Chandel 2014; Brand 2020; and Sies and Jones 2020).

The major producers of ROS within mitochondria are complex I and complex III of the ETC (McBride, Neuspiel and Wasiak 2006; Quinlan et al. 2012, 2013; Goncalves et al. 2015). Both embedded in the inner membrane of the mitochondria, the topology of their leaking sites determines the side in which the redox species are produced (Quinlan et al. 2013). In complex I, site I_Q releases superoxide and H_2O_2 inside the matrix, where an enclosed superoxide dismutase (MnSOD or SOD2) catalyzes the former into the latter (Murphy 2008). In complex III, site III_{Q0} produces superoxide into either side of the inner membrane (St-Pierre et al. 2002; Muller, Liu and van Remmen 2004). At the intermembrane space, the cytosolic superoxide dismutase (Cu/ZnSOD or SOD1) transforms the radical to H_2O_2. Other less wealthy sources also fill the matrix with redox equivalents, including the flavin I_F site of complex I, certain dehydrogenases and the reactions that culminate in the β-oxidation of fatty acids among others (Brand 2016). Being the most prominent destination of redox species, it is little wonder that the mitochondrial matrix is populated by residing peroxidases (Prx3, GPx1, Prx5, and GPx4) that together account for 99.9% of the H_2O_2 consumed (Cox, Winterbourn and Hampton 2009). Just as its cytosolic counterparts, Prx3 is a highly abundant protein, representing approximately 1.6% of the total soluble mitochondrial protein pool (Watabe et al. 1994). Therefore, the mitochondrial matrix presumably might be as reduced as the cytosol, predicting that mitochondria will behave more as a sink for rather than as a source of H_2O_2 (Zoccarato, Cavallini and Alexandre 2004; Drechsel and Patel 2010). However, the net rate of production is what will determine whether a particular compartment exports or imports H_2O_2 (Mueller et al. 2002; Munro and Treberg 2017), and recently detection of mitochondria-released H_2O_2 directly at the surface of the organelle and in the bulk cytosol has been achieved using a new generation of exquisitely sensitive genetically encoded sensors (Hoehne et al. 2022). Remarkably, these observations match to a hypothesis proposed more than ten years ago that claimed that mitochondria can dynamically switch between being a sink or a source by adjusting its steady-state levels of H_2O_2 on demand of parameters such as mitochondria abundance, changes in metabolic workload, availability of oxidative substrates, and external ROS pressure (Zoccarato, Cavallini and Alexandre 2004; Starkov 2008).

If temporal gradients of H_2O_2 can arise to or from mitochondria, a pertaining question will be whether a peroxiporin is needed to traverse their membranes. The general consensus is that a specialized channel will not be necessary in the outer membrane, where abundant porins/VDAC feasibly allow for undampened diffusion (Zalman, Nikaido and Kagawa 1980; Mihara, Blobel and Sato 1982; Benz 1994; Colombini, Blachly-Dyson and Forte 1996; Benz 2021). Thus, when assessing the efflux of H_2O_2 molecules from complex III_{Q0} site, the most straightforward conclusion will be that free permeation rules in such location. However, it is not clear how a totally unrestrained conductance may be an efficient way to signal outside the organelle. Undirected diffusion through the entire outer membrane sheet will likely hamper concentration buildups, rendering peroxidase-based relays as the only means for transduction (de Cubas et al. 2021). Moreover, delocalized exiting and spreading of H_2O_2 will only hit a few molecules at each time, expanding the risks for message dilution. In this context, the capacity of VDAC proteins to constitute mitochondrial anchoring platforms for cytosolic proteins (Fiek et al. 1982; Brdiczka 1990; Adams et al. 1991; de Pinto et al. 2003) suggests that the strategy of gathering signaling elements in the vicinity of a H_2O_2-transporting conduit may be a common feature of peroxiporins, independently from belonging to one or a different family of channels. Furthermore, it has been recently shown that VDAC can be gated by a redox-dependent mechanism in which an oxidative-sensitive protein disrupts the flow of metabolites traversing the pore (Lipper et al. 2019), matching the redoxstat control of H_2O_2 fluxes first proposed for AQP8 conductance at the plasma membrane (Bestetti et al. 2018). In addition, the spatial component may well also play a role, especially when considering the structural requirements for establishing functional redox liaisons with the ER (Sorrentino et al. 2022). In accordance, VDAC is highly enriched in MAM (Shoshan-Barmatz et al. 2004; Bononi et al. 2012), predicting a direct communicating path between both organelles in which their respective channels result in close contact in a synaptic-like manner.

With respect to the mitochondrial inner membrane, the matter of whether a H_2O_2 transporter exists is frequently the center of passionate debates in redox-specialized meetings. By one side, the curves of probe oxidation obtained during experiments either using matrix-targeted sensors and exogenous bolus of H_2O_2 (e.g., Bertolotti et al. 2013; Mishina et al. 2019; Bestetti et al. 2020; Hoehne et al. 2022) or cytosolic readouts and matrix-expressed chemogenetic substrate-controlled H_2O_2 generators (e.g., Pak et al. 2020) show slopes compatible with a non-impeded diffusion, the opposite to what would have been expected when a lipid barrier is present. However, cardiolipin-based membranes are less packaged (Unsay et al. 2013), and due to their increased surface-to-volume ratio mitochondria are expected to equilibrate unbalanced inside-outside polar concentrations more rapidly than other cell compartments (B. Yang, Zhao and Verkman 2006), challenging the need for a dedicated channel to cross. Controversy arises as well on the identity of a putative transporter and its functionality. While a substantial party defends that a portion of AQP8 resides at the mitochondrial inner membrane (Ferri et al. 2003; Calamita et al. 2005; Ikaga et al. 2015) and is a functional peroxiporin (Marchissio et al. 2012; Krüger et al. 2021), others have found no trace of the channel at this location (B. Yang, Zhao and Verkman 2006) or were not able to record any measurable impact on mitochondrial H_2O_2 import upon AQP8 silencing (Bertolotti et al. 2013). One possible explanation reconciling these seemingly contrasting results would be that AQP8 trafficking is cell type– or even moment type–dependent. In this last sense, a recent manuscript reports on the mitochondrial insertion of a peroxiporin driven by post-traductional modifications of its N-terminal (Chauvigné et al. 2021). The isoform, a piscine ortholog of AQP8 expressed in spermatozoa, executed a species-conserved phosphorylation program that culminated on the translocation of the channel into the inner membrane to facilitate H_2O_2 efflux under oxidative stress. Although these data require further substantiation to disclose the biophysical components that allow protein transfer between compartments, the described mechanism sums to the recently accumulating evidence documenting unorthodox trafficking phenomena triggered by stress-controlled pathways (Gee et al. 2011; Jung et al. 2016; Sitia and Rubartelli 2018; Cortés Sanchón et al. 2021). Importantly, the inside story behind could represent the starting point that explains how an ER-folded protein can be incorporated to mitochondrial membranes, whose residing proteins are generally encoded in the nucleus and synthesized on cytoplasmic ribosomes (Freitag, Janes and Neupert 1982; Mihara, Blobel and Sato 1982; Grevel, Pfanner and Becker 2020).

In any case, with respiratory complex contributions to the total H_2O_2 pool approaching 30% in a wide range of cell types (Fang, Wong and Brand 2020; Brand 2020), determining the paths that the redox molecules emitted follow to reach the rest of the cell promises another good choice for future lines of investigation, just as was true for the ER.

13.5 CONCLUDING REMARKS

We are just only realizing that redox signaling may be a vague term to embrace a plethora of different mechanistic strategies to transmit information while safeguarding the integrity of the molecular components of the cell from potentially harming molecules. Even the variety recalls to the versatility of these processes, it also claims for the difficulties on applying reductionist approaches to study redox biology. However, that all redox fluxes use one kind or another transporting channel constitutes an irrefutable fact. Therefore, there is a huge opportunity for ascertain the different structural and temporal requirements that spatially localize each modality of redox signaling by using peroxiporins. The way to the house has been opened; now it is time to unveil the redox architecture of the cell.

REFERENCES

Aalkjaer, Christian, Ebbe Boedtkjer, Inyeong Choi, and Soojung Lee. 2014. "Cation-Coupled Bicarbonate Transporters". *Comprehensive Physiology*, 1605–1637. https://doi.org/10.1002/cphy.c130005.

Adachi, Takeshi, David R Pimentel, Tyler Heibeck, Xiuyun Hou, Yong J Lee, Bingbing Jiang, Yasuo Ido, and Richard A Cohen. 2004. "S-Glutathiolation of Ras Mediates Redox-Sensitive Signaling by Angiotensin II in Vascular Smooth Muscle Cells * [Boxs]". *Journal of Biological Chemistry* 279 (28). Elsevier: 29857–29862. https://doi.org/10.1074/jbc.M313320200.

Adams, V, L Griffin, J Towbin, B Gelb, K Worley, and E R McCabe. 1991. "Porin Interaction with Hexokinase and Glycerol Kinase: Metabolic Microcompartmentation at the Outer Mitochondrial Membrane". *Biochemical*

Medicine and Metabolic Biology 45 (3): 271–291. https://doi.org/10.1016/0885-4505(91)90032-g.

Almasalmeh, Abdulnasser, Dawid Krenc, Binghua Wu, and Eric Beitz. 2014. "Structural Determinants of the Hydrogen Peroxide Permeability of Aquaporins". *The FEBS Journal* 281 (3). John Wiley & Sons, Ltd: 647–656. https://doi.org/10.1111/febs.12653.

Appenzeller-Herzog, Christian, Gabor Bánhegyi, Ivan Bogeski, Kelvin J A Davies, Agnès Delaunay-Moisan, Henry Jay Forman, Agnes Görlach, et al. 2016. "Transit of H2O2 Across the Endoplasmic Reticulum Membrane Is Not Sluggish". *Free Radical Biology and Medicine* 94: 157–160. https://doi.org/10.1016/j.freeradbiomed.2016.02.030.

Atochina-Vasserman, Elena N, Asel Biktasova, Elena Abramova, Dong-Sheng Cheng, Vasiliy v Polosukhin, Harikrishna Tanjore, Saki Takahashi, et al. 2013. "Aquaporin 11 Insufficiency Modulates Kidney Susceptibility to Oxidative Stress". *American Journal of Physiology-Renal Physiology* 304 (10). American Physiological Society: F1295–F1307. https://doi.org/10.1152/ajprenal.00344.2012.

Bedard, Karen, and Karl-Heinz Krause. 2007. "The NOX Family of ROS-Generating NADPH Oxidases: Physiology and Pathophysiology". *Physiological Reviews* 87 (1). American Physiological Society: 245–313. https://doi.org/10.1152/physrev.00044.2005.

Benz, Roland. 1994. "Permeation of Hydrophilic Solutes through Mitochondrial Outer Membranes: Review on Mitochondrial Porins". *Biochimica et Biophysica Acta (BBA)—Reviews on Biomembranes* 1197 (2): 167–196. https://doi.org/10.1016/0304-4157(94)90004-3.

Benz, Roland. 2021. "Historical Perspective of Pore-Forming Activity Studies of Voltage-Dependent Anion Channel (Eukaryotic or Mitochondrial Porin) Since Its Discovery in the 70th of the Last Century". *Frontiers in Physiology* 12. https://doi.org/10.3389/fphys.2021.734226.

Bertolotti, Milena, Stefano Bestetti, Jose M García-Manteiga, Iria Medraño-Fernandez, Andrea Dal Mas, Maria Luisa Malosio, and Roberto Sitia. 2013. "Tyrosine Kinase Signal Modulation: A Matter of H2O2 Membrane Permeability?" *Antioxidants & Redox Signaling* 19 (13): 1447–1451. https://doi.org/10.1089/ars.2013.5330.

Bertolotti, Milena, Giada Farinelli, Mauro Galli, Alessandro Aiuti, and Roberto Sitia. 2016. "AQP8 Transports NOX2-Generated H2O2 across the Plasma Membrane to Promote Signaling in B Cells". *Journal of Leukocyte Biology* 100 (5). United States: 1071–1079. https://doi.org/10.1189/jlb.2AB0116-045R.

Bestetti, Stefano, Mauro Galli, Ilaria Sorrentino, Paolo Pinton, Alessandro Rimessi, Roberto Sitia, and Iria Medraño-Fernandez. 2020. "Human Aquaporin-11 Guarantees Efficient Transport of H(2)O(2) across the Endoplasmic Reticulum Membrane". *Redox Biology* 28 (January): 101326. https://doi.org/10.1016/j.redox.2019.101326.

Bestetti, Stefano, Iria Medraño-Fernandez, Mauro Galli, Michela Ghitti, Gerd P Bienert, Giovanna Musco, Andrea Orsi, Anna Rubartelli, and Roberto Sitia. 2018. "A Persulfidation-Based Mechanism Controls Aquaporin-8 Conductance". *Science Advances* 4 (5): eaar5770. https://doi.org/10.1126/sciadv.aar5770.

Bhunia, Anil Kumar, Hui Han, Ann Snowden, and Subroto Chatterjee. 1997. "Redox-Regulated Signaling by Lactosylceramide in the Proliferation of Human Aortic Smooth Muscle Cells *". *Journal of Biological Chemistry* 272 (25). Elsevier: 15642–15649. https://doi.org/10.1074/jbc.272.25.15642.

Bienert, Gerd P, Anders L B Møller, Kim A Kristiansen, Alexander Schulz, Ian M Møller, Jan K Schjoerring, and Thomas P Jahn. 2007. "Specific Aquaporins Facilitate the Diffusion of Hydrogen Peroxide across Membranes". *The Journal of Biological Chemistry* 282 (2). United States: 1183–1192. https://doi.org/10.1074/jbc.M603761200.

Bononi, Angela, Sonia Missiroli, Federica Poletti, Jan M Suski, Chiara Agnoletto, Massimo Bonora, Elena de Marchi, et al. 2012. "Mitochondria-Associated Membranes (MAMs) as Hotspot Ca2+ Signaling Units". In *Calcium Signaling*, edited by Md. Shahidul Islam, 411–437. Dordrecht: Springer Netherlands. https://doi.org/10.1007/978-94-007-2888-2_17.

Boveris, Alberto, and Britton Chance. 1973. "The Mitochondrial Generation of Hydrogen Peroxide. General Properties and Effect of Hyperbaric Oxygen". *Biochemical Journal* 134 (3): 707–716. https://doi.org/10.1042/bj1340707.

Brand, Martin D. 2016. "Mitochondrial Generation of Superoxide and Hydrogen Peroxide as the Source of Mitochondrial Redox Signaling". *Free Radical Biology & Medicine* 100 (November). United States: 14–31. https://doi.org/10.1016/j.freeradbiomed.2016.04.001.

Brand, Martin D. 2020. "Riding the Tiger—Physiological and Pathological Effects of Superoxide and Hydrogen Peroxide Generated in the Mitochondrial Matrix". *Critical Reviews in Biochemistry and Molecular Biology* 55 (6). Taylor & Francis: 592–661. https://doi.org/10.1080/10409238.2020.1828258.

Brdiczka, D. 1990. "Interaction of Mitochondrial Porin with Cytosolic Proteins". *Experientia* 46 (2): 161–166. https://doi.org/10.1007/BF02027312.

Brigelius-Flohé, Regina, and Matilde Maiorino. 2013. "Glutathione Peroxidases". *Biochimica et Biophysica Acta (BBA)—General Subjects* 1830 (5): 3289–3303. https://doi.org/10.1016/j.bbagen.2012.11.020.

Brush, Matthew H, Douglas C Weiser, and Shirish Shenolikar. 2003. "Growth Arrest and DNA Damage-Inducible Protein GADD34 Targets Protein Phosphatase 1α to the Endoplasmic Reticulum and Promotes Dephosphorylation of the α Subunit of Eukaryotic Translation Initiation Factor 2". *Molecular and Cellular Biology* 23 (4). American Society for Microbiology: 1292–1303. https://doi.org/10.1128/MCB.23.4.1292-1303.2003.

Cai, Hua. 2005. "Hydrogen Peroxide Regulation of Endothelial Function: Origins, Mechanisms, and Consequences". *Cardiovascular Research* 68 (1). England: 26–36. https://doi.org/10.1016/j.cardiores.2005.06.021.

Calamita, Giuseppe, Domenico Ferri, Patrizia Gena, Giuseppa E Liquori, Annie Cavalier, Daniel Thomas, and Maria Svelto. 2005. "The Inner Mitochondrial Membrane Has Aquaporin-8 Water Channels and Is Highly Permeable to Water *". *Journal of Biological Chemistry* 280 (17). Elsevier: 17149–17153. https://doi.org/10.1074/jbc.C400595200.

Ceulemans, Hugo, and Mathieu Bollen. 2004. "Functional Diversity of Protein Phosphatase-1, a Cellular Economizer and Reset Button". *Physiological Reviews* 84 (1). American Physiological Society: 1–39. https://doi.org/10.1152/physrev.00013.2003.

Chae, Ho Zoon, Hyung Jung Kim, Sang Won Kang, and Sue Goo Rhee. 1999. "Characterization of Three Isoforms of Mammalian Peroxiredoxin That Reduce Peroxides in the Presence of Thioredoxin". *Diabetes Research and Clinical Practice* 45 (2). Elsevier: 101–112. https://doi.org/10.1016/S0168-8227(99)00037-6.

Chandel, Navdeep S. 2014. "Mitochondria as Signaling Organelles". *BMC Biology* 12 (1): 34. https://doi.org/10.1186/1741-7007-12-34.

Chauvigné, François, Carla Ducat, Alba Ferré, Tom Hansen, Montserrat Carrascal, Joaquín Abián, Roderick Nigel Finn, and Joan Cerdà. 2021. "A Multiplier Peroxiporin Signal Transduction Pathway Powers Piscine Spermatozoa". *Proceedings of the National Academy of Sciences* 118 (10). Proceedings of the National Academy of Sciences: e2019346118. https://doi.org/10.1073/pnas.2019346118.

Chen, Kai, Michael T Kirber, Hui Xiao, Yu Yang, and John F Keaney Jr. 2008. "Regulation of ROS Signal Transduction by NADPH Oxidase 4 Localization". *Journal of Cell Biology* 181 (7): 1129–1139. https://doi.org/10.1083/jcb.200709049.

Chiarugi, Paola, Giovambattista Pani, Elisa Giannoni, Letizia Taddei, Renata Colavitti, Giovanni Raugei, Mark Symons, Silvia Borrello, Tommaso Galeotti, and Giampietro Ramponi. 2003. "Reactive Oxygen Species as Essential Mediators of Cell Adhesion: The Oxidative Inhibition of a FAK Tyrosine Phosphatase Is Required for Cell Adhesion". *The Journal of Cell Biology* 161 (5): 933–944. https://doi.org/10.1083/jcb.200211118.

Clapham, David E. 2007. "Calcium Signaling". *Cell* 131 (6). Elsevier: 1047–1058. https://doi.org/10.1016/j.cell.2007.11.028.

Colombini, Marco, Elizabeth Blachly-Dyson, and Michael Forte. 1996. "VDAC, a Channel in the Outer Mitochondrial Membrane". In *Ion Channels*, edited by Toshio Narahashi, 169–202. Boston, MA: Springer US. https://doi.org/10.1007/978-1-4899-1775-1_5.

Cortés Sanchón, Adrián, Harshitha Santhosh Kumar, Matilde Mantovani, Ivan Osinnii, José María Mateos, Andres Kaech, Dimitri Shcherbakov, Rashid Akbergenov, and Erik C Böttger. 2021. "ER-Misfolded Proteins Become Sequestered with Mitochondria and Impair Mitochondrial Function". *Communications Biology* 4 (1): 1350. https://doi.org/10.1038/s42003-021-02873-w.

Cox, Andrew G, Christine C Winterbourn, and Mark B Hampton. 2009. "Mitochondrial Peroxiredoxin Involvement in Antioxidant Defence and Redox Signalling". *Biochemical Journal* 425 (2): 313–325. https://doi.org/10.1042/BJ20091541.

Cubas, Laura de, Valeriy v Pak, Vsevolod v Belousov, José Ayté, and Elena Hidalgo. 2021. "The Mitochondria-to-Cytosol H2O2 Gradient Is Caused by Peroxiredoxin-Dependent Cytosolic Scavenging". *Antioxidants* 10 (5). https://doi.org/10.3390/antiox10050731.

Cuevasanta, Ernesto, Ana Denicola, Beatriz Alvarez, and Matías N Möller. 2012. "Solubility and Permeation of Hydrogen Sulfide in Lipid Membranes". *PLoS ONE* 7 (4). Public Library of Science: e34562. https://doi.org/10.1371/journal.pone.0034562.

Dagnell, Markus, Qing Cheng, Syed Husain Mustafa Rizvi, Paul E Pace, Benoit Boivin, Christine C Winterbourn, and Elias S J Arnér. 2019. "Bicarbonate Is Essential for Protein-Tyrosine Phosphatase 1B (PTP1B) Oxidation and Cellular Signaling through EGF-Triggered Phosphorylation Cascades". *Journal of Biological Chemistry* 294 (33). Elsevier: 12330–12338. https://doi.org/10.1074/jbc.RA119.009001.

Dalla Rizza, Joaquín, Lía M Randall, Javier Santos, Gerardo Ferrer-Sueta, and Ana Denicola. 2019. "Differential Parameters between Cytosolic 2-Cys Peroxiredoxins, PRDX1 and PRDX2". *Protein Science*

28 (1). John Wiley & Sons, Ltd: 191–201. https://doi.org/10.1002/pro.3520.

Denu, John M, and Kirk G Tanner. 1998. "Specific and Reversible Inactivation of Protein Tyrosine Phosphatases by Hydrogen Peroxide: Evidence for a Sulfenic Acid Intermediate and Implications for Redox Regulation". *Biochemistry* 37 (16). American Chemical Society: 5633–5642. https://doi.org/10.1021/bi973035t.

Dorai, Thambi, Ihor S Sawczuk, Jaromir Pastorek, Peter H Wiernik, and Janice P Dutcher. 2005. "The Role of Carbonic Anhydrase IX Overexpression in Kidney Cancer". *European Journal of Cancer* 41 (18). Elsevier: 2935–2947. https://doi.org/10.1016/j.ejca.2005.09.011.

Dowling, Damian K, and Leigh W Simmons. 2009. "Reactive Oxygen Species as Universal Constraints in Life-History Evolution". *Proceedings of the Royal Society B: Biological Sciences* 276 (1663). Royal Society: 1737–1745. https://doi.org/10.1098/rspb.2008.1791.

Drechsel, Derek A, and Manisha Patel. 2010. "Respiration-Dependent H_2O_2 Removal in Brain Mitochondria via the Thioredoxin/Peroxiredoxin System*". *Journal of Biological Chemistry* 285 (36). Elsevier: 27850–27858. https://doi.org/10.1074/jbc.M110.101196.

Eden, Emily R, Ian J White, Anna Tsapara, and Clare E Futter. 2010. "Membrane Contacts between Endosomes and ER Provide Sites for PTP1B–Epidermal Growth Factor Receptor Interaction". *Nature Cell Biology* 12 (3): 267–272. https://doi.org/10.1038/ncb2026.

Eric, Pedruzzi, Guichard Cécile, Ollivier Véronique, Driss Fathi, Fay Michèle, Prunet Céline, Marie Jean-Claude, et al. 2004. "NAD(P)H Oxidase Nox-4 Mediates 7-Ketocholesterol-Induced Endoplasmic Reticulum Stress and Apoptosis in Human Aortic Smooth Muscle Cells". *Molecular and Cellular Biology* 24 (24). American Society for Microbiology: 10703–10717. https://doi.org/10.1128/MCB.24.24.10703-10717.2004.

Fang, Jingqi, Hoi-Shan Wong, and Martin D Brand. 2020. "Production of Superoxide and Hydrogen Peroxide in the Mitochondrial Matrix is Dominated by Site IQ of Complex I in Diverse Cell Lines". *Redox Biology* 37: 101722. https://doi.org/10.1016/j.redox.2020.101722.

Fazeli, Gholamreza, Helga Stopper, Reinhard Schinzel, Chih-Wen Ni, Hanjoong Jo, and Nicole Schupp. 2012. "Angiotensin II Induces DNA Damage via AT1 Receptor and NADPH Oxidase Isoform Nox4". *Mutagenesis* 27 (6): 673–681. https://doi.org/10.1093/mutage/ges033.

Fernandez-Caggiano, Mariana, Ewald Schröder, Hyun-Ju Cho, Joseph Burgoyne, Javier Barallobre-Barreiro, Manuel Mayr, and Philip Eaton. 2016. "Oxidant-Induced Interprotein Disulfide Formation in Cardiac Protein DJ-1 Occurs via an Interaction with Peroxiredoxin 2". *The Journal of Biological Chemistry* 291 (19): 10399–10410. https://doi.org/10.1074/jbc.m115.699850.

Ferri, Domenico, Amelia Mazzone, Giuseppa Esterina Liquori, Grazia Cassano, Maria Svelto, and Giuseppe Calamita. 2003. "Ontogeny, Distribution, and Possible Functional Implications of an Unusual Aquaporin, AQP8, in Mouse Liver". *Hepatology* 38 (4). John Wiley & Sons, Ltd: 947–957. https://doi.org/10.1002/hep.1840380421.

Fiek, Christian, Roland Benz, Norbert Roos, and Dieter Brdiczka. 1982. "Evidence for Identity between the Hexokinase-Binding Protein and the Mitochondrial Porin in the Outer Membrane of Rat Liver Mitochondria". *Biochimica et Biophysica Acta (BBA)—Biomembranes* 688 (2): 429–440. https://doi.org/10.1016/0005-2736(82)90354-6.

Finkel, T. 2000. "Redox-Dependent Signal Transduction". *FEBS Letters* 476 (1–2). England: 52–54. https://doi.org/10.1016/s0014-5793(00)01669-0.

Foreman, Julia, Vadim Demidchik, John H F Bothwell, Panagiota Mylona, Henk Miedema, Miguel Angel Torres, Paul Linstead, et al. 2003. "Reactive Oxygen Species Produced by NADPH Oxidase Regulate Plant Cell Growth". *Nature* 422 (6930). England: 442–446. https://doi.org/10.1038/nature01485.

Forman, Henry Jay, and Hongqiao Zhang. 2021. "Targeting Oxidative Stress in Disease: Promise and Limitations of Antioxidant Therapy". *Nature Reviews Drug Discovery* 20 (9): 689–709. https://doi.org/10.1038/s41573-021-00233-1.

Foti, Michelangelo, Anjon Audhya, and Scott D Emr. 2001. "Sac1 Lipid Phosphatase and Stt4 Phosphatidylinositol 4-Kinase Regulate a Pool of Phosphatidylinositol 4-Phosphate That Functions in the Control of the Actin Cytoskeleton and Vacuole Morphology". *Molecular Biology of the Cell* 12 (8). American Society for Cell Biology (mboc): 2396–2411. https://doi.org/10.1091/mbc.12.8.2396.

Freitag, Helmut, Michael Janes, and Walter Neupert. 1982. "Biosynthesis of Mitochondrial Porin and Insertion into the Outer Mitochondrial Membrane of Neuruspora Crassa". *European Journal of Biochemistry* 126 (1). John Wiley & Sons, Ltd: 197–202. https://doi.org/10.1111/j.1432-1033.1982.tb06766.x.

Fricke, Tabea C, Frank Echtermeyer, Johannes Zielke, Jeanne de la Roche, Milos R Filipovic,

Stéphane Claverol, Christine Herzog, et al. 2019. "Oxidation of Methionine Residues Activates the High-Threshold Heat-Sensitive Ion Channel TRPV2". *Proceedings of the National Academy of Sciences* 116 (48). Proceedings of the National Academy of Sciences: 24359–24365. https://doi.org/10.1073/pnas.1904332116.

Gacesa, Ranko, Walter C Dunlap, David J Barlow, Roman A Laskowski, and Paul F Long. 2016. "Rising Levels of Atmospheric Oxygen and Evolution of Nrf2". *Scientific Reports* 6 (1): 27740. https://doi.org/10.1038/srep27740.

Galea, Anne M, and Andrew J Brown. 2009. "Special Relationship between Sterols and Oxygen: Were Sterols an Adaptation to Aerobic Life?". *Free Radical Biology and Medicine* 47 (6): 880–889. https://doi.org/10.1016/j.freeradbiomed.2009.06.027.

Gao, Congcong, Yong Tian, Rubo Zhang, Jing Jing, and Xiaoling Zhang. 2017. "Endoplasmic Reticulum-Directed Ratiometric Fluorescent Probe for Quantitive Detection of Basal H(2)O(2)". *Analytical Chemistry* 89 (23). United States: 12945–12950. https://doi.org/10.1021/acs.analchem.7b03809.

Gee, Heon Yung, Shin Hye Noh, Bor Luen Tang, Kyung Hwan Kim, and Min Goo Lee. 2011. "Rescue of ΔF508-CFTR Trafficking via a GRASP-Dependent Unconventional Secretion Pathway". *Cell* 146 (5). Elsevier: 746–760. https://doi.org/10.1016/j.cell.2011.07.021.

Geiszt, Miklós, Jeffrey B Kopp, Péter Várnai, and Thomas L Leto. 2000. "Identification of Renox, an NAD(P)H Oxidase in Kidney". *Proceedings of the National Academy of Sciences* 97 (14). Proceedings of the National Academy of Sciences: 8010–8014. https://doi.org/10.1073/pnas.130135897.

Geiszt, Miklós, and Thomas L Leto. 2004. "The Nox Family of NAD(P)H Oxidases: Host Defense and Beyond". *The Journal of Biological Chemistry* 279 (50). United States: 51715–51718. https://doi.org/10.1074/jbc.R400024200.

Ghouleh, Imad al, Giovanna Frazziano, Andres I Rodriguez, Gábor Csányi, Salony Maniar, Claudette M St Croix, Eric E Kelley, et al. 2013. "Aquaporin 1, Nox1, and Ask1 Mediate Oxidant-Induced Smooth Muscle Cell Hypertrophy". *Cardiovascular Research* 97 (1): 134–142. https://doi.org/10.1093/cvr/cvs295.

Goncalves, Renata L S, Casey L Quinlan, Irina V Perevoshchikova, Martin Hey-Mogensen, and Martin D Brand. 2015. "Sites of Superoxide and Hydrogen Peroxide Production by Muscle Mitochondria Assessed <em>ex Vivo</Em> under Conditions Mimicking Rest and Exercise *". *Journal of Biological Chemistry* 290 (1). Elsevier: 209–227. https://doi.org/10.1074/jbc.M114.619072.

Gorin, Yves, Jill M Ricono, Nam-Ho Kim, Basant Bhandari, Goutam Ghosh Choudhury, and Hanna E Abboud. 2003. "Nox4 Mediates Angiotensin II-Induced Activation of Akt/Protein Kinase B in Mesangial Cells". *American Journal of Physiology-Renal Physiology* 285 (2). American Physiological Society: F219–F229. https://doi.org/10.1152/ajprenal.00414.2002.

Grevel, Alexander, Nikolaus Pfanner, and Thomas Becker. 2020. "Coupling of Import and Assembly Pathways in Mitochondrial Protein Biogenesis". *Biological Chemistry* 401 (1): 117–129. https://doi.org/10.1515/hsz-2019-0310.

Gross, Einav, Carolyn S Sevier, Nimrod Heldman, Elvira Vitu, Moran Bentzur, Chris A Kaiser, Colin Thorpe, and Deborah Fass. 2006. "Generating Disulfides Enzymatically: Reaction Products and Electron Acceptors of the Endoplasmic Reticulum Thiol Oxidase Ero1p". *Proceedings of the National Academy of Sciences* 103 (2). Proceedings of the National Academy of Sciences: 299–304. https://doi.org/10.1073/pnas.0506448103.

Gross, Jeferson, and Debashish Bhattacharya. 2010. "Uniting Sex and Eukaryote Origins in an Emerging Oxygenic World". *Biology Direct* 5 (1): 53. https://doi.org/10.1186/1745-6150-5-53.

Haj, Fawaz G, Peter J Verveer, Anthony Squire, Benjamin G Neel, and Philippe I H Bastiaens. 2002. "Imaging Sites of Receptor Dephosphorylation by PTP1B on the Surface of the Endoplasmic Reticulum". *Science* 295 (5560). American Association for the Advancement of Science: 1708–1711. https://doi.org/10.1126/science.1067566.

Hara-Chikuma, Mariko, Sachiko Watanabe, and Hiroki Satooka. 2016. "Involvement of Aquaporin-3 in Epidermal Growth Factor Receptor Signaling via Hydrogen Peroxide Transport in Cancer Cells". *Biochemical and Biophysical Research Communications* 471 (4): 603–609. https://doi.org/10.1016/j.bbrc.2016.02.010.

Harding, Heather P, Yuhong Zhang, Huiquing Zeng, Isabel Novoa, Phoebe D Lu, Marcella Calfon, Navid Sadri, et al. 2003. "An Integrated Stress Response Regulates Amino Acid Metabolism and Resistance to Oxidative Stress". *Molecular Cell* 11 (3). Elsevier: 619–633. https://doi.org/10.1016/S1097-2765(03)00105-9.

Heath-Engel, H M, N C Chang, and G C Shore. 2008. "The Endoplasmic Reticulum in Apoptosis and Autophagy: Role of the BCL-2 Protein Family". *Oncogene* 27 (50): 6419–6433. https://doi.org/10.1038/onc.2008.309.

Henzler, T, and E Steudle. 2000. "Transport and Metabolic Degradation of Hydrogen Peroxide in Chara Corallina: Model Calculations and Measurements with the Pressure Probe Suggest Transport of H(2)O(2) across Water Channels". *Journal of Experimental Botany* 51 (353). England: 2053–2066. https://doi.org/10.1093/jexbot/51.353.2053.

Hertog, Jeroen den, Arnoud Groen, and Thea van der Wijk. 2005. "Redox Regulation of Protein-Tyrosine Phosphatases". *Archives of Biochemistry and Biophysics* 434 (1): 11–15. https://doi.org/10.1016/j.abb.2004.05.024.

Hitomi, Junichi, Taiichi Katayama, Yutaka Eguchi, Takashi Kudo, Manabu Taniguchi, Yoshihisa Koyama, Takayuki Manabe, et al. 2004. "Involvement of Caspase-4 in Endoplasmic Reticulum Stress-Induced Apoptosis and Aβ-Induced Cell Death". *Journal of Cell Biology* 165 (3): 347–356. https://doi.org/10.1083/jcb.200310015.

Hoehne, Michaela Nicole, Lianne J H C Jacobs, Kim Jasmin Lapacz, Gaetano Calabrese, Lena Maria Murschall, Teresa Marker, Harshita Kaul, et al. 2022. "Spatial and Temporal Control of Mitochondrial H2O2 Release in Intact Human Cells". *The EMBO Journal* 41 (7). John Wiley & Sons, Ltd: e109169. https://doi.org/10.15252/embj.2021109169.

Hohmann, Stefan, Søren Nors Nielsen, and Søren Nielsen. 2000. *Molecular Biology and Physiology of Water and Solute Transport*. Springer Science & Business Media.

Holland, Heinrich D. 2006. "The Oxygenation of the Atmosphere and Oceans". *Philosophical Transactions of the Royal Society B: Biological Sciences* 361 (1470). Royal Society: 903–915. https://doi.org/10.1098/rstb.2006.1838.

Horsefield, Rob, Kristina Nordén, Maria Fellert, Anna Backmark, Susanna Törnroth-Horsefield, Anke C Terwisscha van Scheltinga, Jan Kvassman, Per Kjellbom, Urban Johanson, and Richard Neutze. 2008. "High-Resolution x-Ray Structure of Human Aquaporin 5". *Proceedings of the National Academy of Sciences* 105 (36). Proceedings of the National Academy of Sciences: 13327–13332. https://doi.org/10.1073/pnas.0801466105.

Høyer-Hansen, Maria, Lone Bastholm, Piotr Szyniarowski, Michelangelo Campanella, György Szabadkai, Thomas Farkas, Katiuscia Bianchi, et al. 2007. "Control of Macroautophagy by Calcium, Calmodulin-Dependent Kinase Kinase-β, and Bcl-2". *Molecular Cell* 25 (2). Elsevier: 193–205. https://doi.org/10.1016/j.molcel.2006.12.009.

Hulbert, A J, Reinald Pamplona, Rochelle Buffenstein, and W A Buttemer. 2007. "Life and Death: Metabolic Rate, Membrane Composition, and Life Span of Animals". *Physiological Reviews* 87 (4). American Physiological Society: 1175–1213. https://doi.org/10.1152/physrev.00047.2006.

Ikaga, Reina, Iyuki Namekata, Vassilios N Kotiadis, Haruko Ogawa, Michael R Duchen, Hikaru Tanaka, and Naoko Iida-Tanaka. 2015. "Knockdown of Aquaporin-8 Induces Mitochondrial Dysfunction in 3T3-L1 Cells". *Biochemistry and Biophysics Reports* 4: 187–195. https://doi.org/10.1016/j.bbrep.2015.09.009.

Inupakutika, Madhuri A, Soham Sengupta, Amith R Devireddy, Rajeev K Azad, and Ron Mittler. 2016. "The Evolution of Reactive Oxygen Species Metabolism". *Journal of Experimental Botany* 67 (21): 5933–5943. https://doi.org/10.1093/jxb/erw382.

Ishibashi, Kenichi, Shintaro Kondo, Shigeki Hara, and Yoshiyuki Morishita. 2010. "The Evolutionary Aspects of Aquaporin Family". *American Journal of Physiology-Regulatory, Integrative and Comparative Physiology* 300 (3). American Physiological Society: R566–R576. https://doi.org/10.1152/ajpregu.90464.2008.

Jarvis, Reagan M, Stephanie M Hughes, and Elizabeth C Ledgerwood. 2012. "Peroxiredoxin 1 Functions as a Signal Peroxidase to Receive, Transduce, and Transmit Peroxide Signals in Mammalian Cells". *Free Radical Biology and Medicine* 53 (7): 1522–1530. https://doi.org/10.1016/j.freeradbiomed.2012.08.001.

Jin, Dong-Yan, Ho Zoon Chae, Sue Goo Rhee, and Kuan-Teh Jeang. 1997. "Regulatory Role for a Novel Human Thioredoxin Peroxidase in NF-κB Activation*". *Journal of Biological Chemistry* 272 (49). Elsevier: 30952–30961. https://doi.org/10.1074/jbc.272.49.30952.

Jung, Jinsei, Jiyoon Kim, Shin Hye Roh, Ikhyun Jun, Robert D Sampson, Heon Yung Gee, Jae Young Choi, and Min Goo Lee. 2016. "The HSP70 Co-Chaperone DNAJC14 Targets Misfolded Pendrin for Unconventional Protein Secretion". *Nature Communications* 7 (1): 11386. https://doi.org/10.1038/ncomms11386.

Kalyankar, Mahadev, Motoshima Hiroyuki, Wu Xiangdong, Ruddy Jean Marie, Arnold Rebecca S, Cheng Guangjie, Lambeth J David, and Goldstein Barry J. 2004. "The NAD(P)H Oxidase Homolog Nox4 Modulates Insulin-Stimulated Generation of H2O2 and Plays an Integral Role in Insulin Signal Transduction". *Molecular and Cellular Biology* 24 (5). American Society for Microbiology: 1844–1854. https://doi.org/10.1128/MCB.24.5.1844-1854.2004.

Kang, Sang Won, Ho Zoon Chae, Min Seok Seo, Kanghwa Kim, Ivan C Baines, and Sue Goo Rhee. 1998. "Mammalian Peroxiredoxin Isoforms Can Reduce Hydrogen Peroxide Generated in Response to Growth Factors and Tumor Necrosis Factor-A*". *Journal of Biological Chemistry* 273 (11): 6297–6302. https://doi.org/10.1074/jbc.273.11.6297.

Karplus, P Andrew. 2015. "A Primer on Peroxiredoxin Biochemistry". *Free Radical Biology and Medicine* 80: 183–190. https://doi.org/10.1016/j.freeradbiomed.2014.10.009.

Kato, Masato, Yu-San Yang, Benjamin M Sutter, Yun Wang, Steven L McKnight, and Benjamin P Tu. 2019. "Redox State Controls Phase Separation of the Yeast Ataxin-2 Protein via Reversible Oxidation of Its Methionine-Rich Low-Complexity Domain". *Cell* 177 (3). Elsevier: 711–721.e8. https://doi.org/10.1016/j.cell.2019.02.044.

Kawahara, Tsukasa, Mark T Quinn, and J David Lambeth. 2007. "Molecular Evolution of the Reactive Oxygen-Generating NADPH Oxidase (Nox/Duox) Family of Enzymes". *BMC Evolutionary Biology* 7 (1): 109. https://doi.org/10.1186/1471-2148-7-109.

Kellen, Michael R, and James B Bassingthwaighte. 2003. "Transient Transcapillary Exchange of Water Driven by Osmotic Forces in the Heart". *American Journal of Physiology-Heart and Circulatory Physiology* 285 (3). American Physiological Society: H1317–H1331. https://doi.org/10.1152/ajpheart.00587.2002.

Kelly, Beth, Gillian M Tannahill, Michael P Murphy, and Luke A J O'Neill. 2015. "Metformin Inhibits the Production of Reactive Oxygen Species from NADH:Ubiquinone Oxidoreductase to Limit Induction of Interleukin-1β (IL-1β) and Boosts Interleukin-10 (IL-10) in Lipopoly saccharide (LPS)-Activated Macrophages*". *Journal of Biological Chemistry* 290 (33). Elsevier: 20348–20359. https://doi.org/10.1074/jbc.M115.662114.

Kil, In Sup, Se Kyoung Lee, Keun Woo Ryu, Hyun Ae Woo, Meng-Chun Hu, Soo Han Bae, and Sue Goo Rhee. 2012. "Feedback Control of Adrenal Steroidogenesis via H2O2-Dependent, Reversible Inactivation of Peroxiredoxin III in Mitochondria". *Molecular Cell* 46 (5): 584–594. https://doi.org/10.1016/j.molcel.2012.05.030.

Kingsley, Paul D, Emily Greenfest-Allen, Jenna M Frame, Timothy P Bushnell, Jeffrey Malik, Kathleen E McGrath, Christian J Stoeckert, and James Palis. 2013. "Ontogeny of Erythroid Gene Expression". *Blood* 121 (6): e5–13. https://doi.org/10.1182/blood-2012-04-422394.

Kitchen, Philip, Fredrik Öberg, Jennie Sjöhamn, Kristina Hedfalk, Roslyn M Bill, Alex C Conner, Matthew T Conner, and Susanna Törnroth-Horsefield. 2015. "Plasma Membrane Abundance of Human Aquaporin 5 Is Dynamically Regulated by Multiple Pathways". *PLoS ONE* 10 (11). Public Library of Science: e0143027. https://doi.org/10.1371/journal.pone.0143027.

Krüger, Christina, Markus Waldeck-Weiermair, Jonas Kaynert, Thomas Pokrant, Yesaswini Komaragiri, Oliver Otto, Thomas Michel, and Matthias Elsner. 2021. "AQP8 Is a Crucial H2O2 Transporter in Insulin-Producing RINm5F Cells". *Redox Biology* 43: 101962. https://doi.org/10.1016/j.redox.2021.101962.

Lassègue, Bernard, Alejandra San Martín, and Kathy K Griendling. 2012. "Biochemistry, Physiology, and Pathophysiology of NADPH Oxidases in the Cardiovascular System". *Circulation Research* 110 (10). American Heart Association: 1364–1390. https://doi.org/10.1161/CIRCRESAHA.111.243972.

Laurindo, Francisco R M, Thaís L S Araujo, and Thalita B Abrahão. 2014. "Nox NADPH Oxidases and the Endoplasmic Reticulum". *Antioxidants & Redox Signaling* 20 (17): 2755–2775. https://doi.org/10.1089/ars.2013.5605.

Lee, Seung-Rock, Kap-Seok Yang, Jaeyul Kwon, Chunghee Lee, Woojin Jeong, and Sue Goo Rhee. 2002. "Reversible Inactivation of the Tumor Suppressor PTEN by H_2O_2*". *Journal of Biological Chemistry* 277 (23). Elsevier: 20336–20342. https://doi.org/10.1074/jbc.M111899200.

Leslie, N R, Y Lindsay, S H Ross, and C P Downes. 2004. "Redox Regulation of Phosphatase Function". *Biochemical Society Transactions* 32 (6): 1018–1020. https://doi.org/10.1042/BST0321018.

Li, Chenlu, Tiantian Qian, Ruyue He, Chun Wan, Yinghui Liu, and Haijia Yu. 2021. "Endoplasmic Reticulum–Plasma Membrane Contact Sites: Regulators, Mechanisms, and Physiological Functions". *Frontiers in Cell and Developmental Biology* 9. https://doi.org/10.3389/fcell.2021.627700.

Li, Chunling, and Weidong Wang. 2017. "Molecular Biology of Aquaporins". *Advances in Experimental Medicine and Biology* 969. United States: 1–34. https://doi.org/10.1007/978-94-024-1057-0_1.

Lipper, Colin H, Jason T Stofleth, Fang Bai, Yang-Sung Sohn, Susmita Roy, Ron Mittler, Rachel Nechushtai, José N Onuchic, and Patricia A Jennings. 2019. "Redox-Dependent Gating of VDAC by MitoNEET". *Proceedings of the National Academy of Sciences* 116 (40). Proceedings of the National Academy of

Sciences: 19924–19929. https://doi.org/10.1073/pnas.1908271116.

Livigni, Alessandra, Antonella Scorziello, Savina Agnese, Annagrazia Adornetto, Annalisa Carlucci, Corrado Garbi, Imma Castaldo, Lucio Annunziato, Enrico v Avvedimento, and Antonio Feliciello. 2005. "Mitochondrial AKAP121 Links CAMP and Src Signaling to Oxidative Metabolism". *Molecular Biology of the Cell* 17 (1). American Society for Cell Biology (mboc): 263–271. https://doi.org/10.1091/mbc.e05-09-0827.

Löhneysen, Katharina von, Deborah Noack, Patti Hayes, Jeffrey S Friedman, and Ulla G Knaus. 2012. "Constitutive NADPH Oxidase 4 Activity Resides in the Composition of the B-Loop and the Penultimate C Terminus". *The Journal of Biological Chemistry* 287 (12). American Society for Biochemistry and Molecular Biology: 8737–8745. https://doi.org/10.1074/jbc.M111.332494.

Loitto, Vesa-Matti, Tony Forslund, Tommy Sundqvist, Karl-Eric Magnusson, and Mikael Gustafsson. 2002. "Neutrophil Leukocyte Motility Requires Directed Water Influx". *Journal of Leukocyte Biology* 71 (2). John Wiley & Sons, Ltd: 212–222. https://doi.org/10.1189/jlb.71.2.212.

Luo, Maochao, Li Zhou, Zhao Huang, Bowen Li, Edouard C Nice, Jia Xu, and Canhua Huang. 2022. "Antioxidant Therapy in Cancer: Rationale and Progress". *Antioxidants* 11 (6). https://doi.org/10.3390/antiox11061128.

Malinouski, Mikalai, You Zhou, Vsevolod v Belousov, Dolph L Hatfield, and Vadim N Gladyshev. 2011. "Hydrogen Peroxide Probes Directed to Different Cellular Compartments". *PLoS ONE* 6 (1): e14564. https://doi.org/10.1371/journal.pone.0014564.

Marchissio, Maria Julia, Daniel Eleazar Antonio Francés, Cristina Ester Carnovale, and Raúl Alberto Marinelli. 2012. "Mitochondrial Aquaporin-8 Knockdown in Human Hepatoma HepG2 Cells Causes ROS-Induced Mitochondrial Depolarization and Loss of Viability". *Toxicology and Applied Pharmacology* 264 (2): 246–254. https://doi.org/10.1016/j.taap.2012.08.005.

Margulis, L S. 1992. *Symbiosis in Cell Evolution: Microbial Communities in the Archean and Proterozoic Eons*. W.H. Freeman.

Marino, Stefano M, and Vadim N Gladyshev. 2010. "Cysteine Function Governs Its Conservation and Degeneration and Restricts Its Utilization on Protein Surfaces". *Journal of Molecular Biology* 404 (5): 902–916. https://doi.org/10.1016/j.jmb.2010.09.027.

Marklund, Stefan L. 1985. "Product of Extracellular-Superoxide Dismutase Catalysis". *FEBS Letters* 184 (2). John Wiley & Sons, Ltd: 237–239. https://doi.org/10.1016/0014-5793(85)80613-X.

Mathai, John C, Andreas Missner, Philipp Kügler, Sapar M Saparov, Mark L Zeidel, John K Lee, and Peter Pohl. 2009. "No Facilitator Required for Membrane Transport of Hydrogen Sulfide". *Proceedings of the National Academy of Sciences* 106 (39). Proceedings of the National Academy of Sciences: 16633–16638. https://doi.org/10.1073/pnas.0902952106.

McBride, Heidi M, Margaret Neuspiel, and Sylwia Wasiak. 2006. "Mitochondria: More Than Just a Powerhouse". *Current Biology* 16 (14). Elsevier: R551–R560. https://doi.org/10.1016/j.cub.2006.06.054.

McGinnis, Austin, Vladimir I Klichko, William C Orr, and Svetlana N Radyuk. 2021. "Hyperoxidation of Peroxiredoxins and Effects on Physiology of Drosophila". *Antioxidants* 10 (4). https://doi.org/10.3390/antiox10040606.

McKay, C P, and H Hartman. 1991. "Hydrogen Peroxide and the Evolution of Oxygenic Photosynthesis". *Origins of Life and Evolution of the Biosphere* 21 (3): 157–163. https://doi.org/10.1007/BF01809444.

Medraño-Fernandez, Iria, Stefano Bestetti, Milena Bertolotti, Gerd P Bienert, Cinzia Bottino, Umberto Laforenza, Anna Rubartelli, and Roberto Sitia. 2016. "Stress Regulates Aquaporin-8 Permeability to Impact Cell Growth and Survival". *Antioxidants & Redox Signaling* 24 (18): 1031–1044. https://doi.org/10.1089/ars.2016.6636.

Melo, Eduardo Pinho, Carlos Lopes, Peter Gollwitzer, Stephan Lortz, Sigurd Lenzen, Ilir Mehmeti, Clemens F Kaminski, David Ron, and Edward Avezov. 2017. "TriPer, an Optical Probe Tuned to the Endoplasmic Reticulum Tracks Changes in Luminal H2O2". *BMC Biology* 15 (1): 24. https://doi.org/10.1186/s12915-017-0367-5.

Meng, Tzu-Ching, Toshiyuki Fukada, and Nicholas K Tonks. 2002. "Reversible Oxidation and Inactivation of Protein Tyrosine Phosphatases in Vivo". *Molecular Cell* 9 (2). United States: 387–399. https://doi.org/10.1016/s1097-2765(02)00445-8.

Mihara, K, G Blobel, and R Sato. 1982. "In Vitro Synthesis and Integration into Mitochondria of Porin, a Major Protein of the Outer Mitochondrial Membrane of Saccharomyces Cerevisiae". *Proceedings of the National Academy of Sciences* 79 (23). Proceedings of the National Academy of Sciences: 7102–7106. https://doi.org/10.1073/pnas.79.23.7102.

Miller, Evan W, Bryan C Dickinson, and Christopher J Chang. 2010. "Aquaporin-3 Mediates Hydrogen Peroxide Uptake to Regulate Downstream

Intracellular Signaling". *Proceedings of the National Academy of Sciences of the United States of America* 107 (36): 15681–15686. https://doi.org/10.1073/pnas.1005776107.

Mishina, Natalia M, Pyotr A Tyurin-Kuzmin, Kseniya N Markvicheva, Alexander v Vorotnikov, Vsevolod A Tkachuk, Vibor Laketa, Carsten Schultz, Sergey Lukyanov, and Vsevolod v Belousov. 2010. "Does Cellular Hydrogen Peroxide Diffuse or Act Locally?". *Antioxidants & Redox Signaling* 14 (1). Mary Ann Liebert, Inc., publishers: 1–7. https://doi.org/10.1089/ars.2010.3539.

Mishina, Natalie M, Yulia A Bogdanova, Yulia G Ermakova, Anastasiya S Panova, Daria A Kotova, Dmitry S Bilan, Benjamin Steinhorn, Elias S J Arnér, Thomas Michel, and Vsevolod V Belousov. 2019. "Which Antioxidant System Shapes Intracellular H_2O_2 Gradients?" *Antioxidants & Redox Signaling* 31 (9). Mary Ann Liebert, Inc.: 664–670. https://doi.org/doi:10.1089/ars.2018.7697.

Moniaga, Catharina Sagita, Sachiko Watanabe, Tetsuya Honda, Søren Nielsen, and Mariko Hara-Chikuma. 2015. "Aquaporin-9-Expressing Neutrophils Are Required for the Establishment of Contact Hypersensitivity". *Scientific Reports* 5 (1): 15319. https://doi.org/10.1038/srep15319.

Monteleone, Melisa C, Ana E González Wusener, Juan E Burdisso, Cecilia Conde, Alfredo Cáceres, and Carlos O Arregui. 2012. "ER-Bound Protein Tyrosine Phosphatase PTP1B Interacts with Src at the Plasma Membrane/Substrate Interface". *PLoS ONE* 7 (6). Public Library of Science: e38948. https://doi.org/10.1371/journal.pone.0038948.

Moon, Chulso, Raphael Rousseau, Jean-Charles Soria, Mohammad Obaidul Hoque, Juna Lee, Se Jin Jang, Barry Trink, David Sidransky, and Li Mao. 2004. "Aquaporin Expression in Human Lymphocytes and Dendritic Cells". *American Journal of Hematology* 75 (3). John Wiley & Sons, Ltd: 128–133. https://doi.org/10.1002/ajh.10476.

Moreno-Loshuertos, Raquel, Rebeca Acín-Pérez, Patricio Fernández-Silva, Nieves Movilla, Acisclo Pérez-Martos, Santiago Rodriguez de Cordoba, M Esther Gallardo, and José Antonio Enríquez. 2006. "Differences in Reactive Oxygen Species Production Explain the Phenotypes Associated with Common Mouse Mitochondrial DNA Variants". *Nature Genetics* 38 (11): 1261–1268. https://doi.org/10.1038/ng1897.

Mouritsen, Ole G, and Martin J Zuckermann. 2004. "What's so Special about Cholesterol?". *Lipids* 39 (11): 1101–1113. https://doi.org/10.1007/s11745-004-1336-x.

Mueller, Sebastian, Angelika Weber, Reiner Fritz, Sabine Mütze, Daniel Rost, Henning Walczak, Alfred Völkl, and Wolfgang Stremmel. 2002. "Sensitive and Real-Time Determination of H2O2 Release from Intact Peroxisomes". *Biochemical Journal* 363 (3): 483–491. https://doi.org/10.1042/bj3630483.

Muller, Florian L, Yuhong Liu, and Holly van Remmen. 2004. "Complex III Releases Superoxide to Both Sides of the Inner Mitochondrial Membrane". *The Journal of Biological Chemistry* 279 (47). United States: 49064–49073. https://doi.org/10.1074/jbc.M407715200.

Munro, Daniel, and Jason R Treberg. 2017. "A Radical Shift in Perspective: Mitochondria as Regulators of Reactive Oxygen Species". *Journal of Experimental Biology* 220 (7): 1170–1180. https://doi.org/10.1242/jeb.132142.

Murphy, Michael P. 2008. "How Mitochondria Produce Reactive Oxygen Species". *Biochemical Journal* 417 (1): 1–13. https://doi.org/10.1042/BJ20081386.

Neel, Benjamin G, Haihua Gu, and Lily Pao. 2003. "The "Shp'ing News: SH2 Domain-Containing Tyrosine Phosphatases in Cell Signaling". *Trends in Biochemical Sciences* 28 (6): 284–293. https://doi.org/10.1016/S0968-0004(03)00091-4.

Netto, Luís Eduardo S, and Luciana Elena S F Machado. 2022. "Preferential Redox Regulation of Cysteine-Based Protein Tyrosine Phosphatases: Structural and Biochemical Diversity". *The FEBS Journal* 289 (18). John Wiley & Sons, Ltd: 5480–5504. https://doi.org/10.1111/febs.16466.

Nisimoto, Y, H M Jackson, H Ogawa, T Kawahara, and J D Lambeth. 2010. "Constitutive NADPH-Dependent Electron Transferase Activity of the Nox4 Dehydrogenase Domain". *Biochemistry* 49: 2433.

Nordzieke, Daniela E, and Iria Medraño-Fernandez. 2018. "The Plasma Membrane: A Platform for Intra- and Intercellular Redox Signaling". *Antioxidants (Basel, Switzerland)* 7 (11). https://doi.org/10.3390/antiox7110168.

Novoa, Isabel, Huiquing Zeng, Heather P Harding, and David Ron. 2001. "Feedback Inhibition of the Unfolded Protein Response by GADD34-Mediated Dephosphorylation of EIF2α". *Journal of Cell Biology* 153 (5): 1011–1022. https://doi.org/10.1083/jcb.153.5.1011.

Okada, Shinji, Takumi Misaka, Yasuko Tanaka, Ichiro Matsumoto, Kenichi Ishibashi, Sei Sasaki, and Keiko Abe. 2008. "Aquaporin-11 Knockout Mice and Polycystic Kidney Disease Animals Share a Common Mechanism of Cyst Formation". *The FASEB Journal* 22 (10). John Wiley & Sons, Ltd: 3672–3684. https://doi.org/10.1096/fj.08-111872.

Pak, Valeriy v, Daria Ezeriņa, Olga G Lyublinskaya, Brandán Pedre, Pyotr A Tyurin-Kuzmin, Natalie

M Mishina, Marion Thauvin, et al. 2020. "Ultrasensitive Genetically Encoded Indicator for Hydrogen Peroxide Identifies Roles for the Oxidant in Cell Migration and Mitochondrial Function". *Cell Metabolism* 31 (3): 642–653.e6. https://doi.org/10.1016/j.cmet.2020.02.003.

Paulsen, Candice E, Thu H Truong, Francisco J Garcia, Arne Homann, Vinayak Gupta, Stephen E Leonard, and Kate S Carroll. 2012. "Peroxide-Dependent Sulfenylation of the EGFR Catalytic Site Enhances Kinase Activity". *Nature Chemical Biology* 8 (1): 57–64. https://doi.org/10.1038/nchembio.736.

Pellavio, Giorgia, Marta Rui, Laura Caliogna, Emanuela Martino, Giulia Gastaldi, Simona Collina, and Umberto Laforenza. 2017. "Regulation of Aquaporin Functional Properties Mediated by the Antioxidant Effects of Natural Compounds". *International Journal of Molecular Sciences* 18 (12). https://doi.org/10.3390/ijms18122665.

Peskin, Alexander V, Felicia M Low, Louise N Paton, Ghassan J Maghzal, Mark B Hampton, and Christine C Winterbourn. 2007. "The High Reactivity of Peroxiredoxin 2 with H(2)O(2) Is Not Reflected in Its Reaction with Other Oxidants and Thiol Reagents". *The Journal of Biological Chemistry* 282 (16). United States: 11885–11892. https://doi.org/10.1074/jbc.M700339200.

Peskin, Alexander V, Paul E Pace, and Christine C Winterbourn. 2019. "Enhanced Hyperoxidation of Peroxiredoxin 2 and Peroxiredoxin 3 in the Presence of Bicarbonate/CO2". *Free Radical Biology and Medicine* 145: 1–7. https://doi.org/10.1016/j.freeradbiomed.2019.09.010.

Phalen, Timothy J, Kelly Weirather, Paula B Deming, Vikas Anathy, Alan K Howe, Albert van der Vliet, Thomas J Jönsson, Leslie B Poole, and Nicholas H Heintz. 2006. "Oxidation State Governs Structural Transitions in Peroxiredoxin II That Correlate with Cell Cycle Arrest and Recovery". *Journal of Cell Biology* 175 (5): 779–789. https://doi.org/10.1083/jcb.200606005.

Pinto, Vito de, Angela Messina, Rosita Accardi, Rita Aiello, Francesca Guarino, Flora Tomasello, Massimo Tommasino, et al. 2003. "New Functions of an Old Protein: The Eukaryotic Porin or Voltage Dependent Anion Selective Channel (VDAC)". *The Italian Journal of Biochemistry* 52 (January): 17–24.

Preston, G M, T P Carroll, W B Guggino, and P Agre. 1992. "Appearance of Water Channels in Xenopus Oocytes Expressing Red Cell CHIP28 Protein". *Science (New York, N.Y.)* 256 (5055). United States: 385–387. https://doi.org/10.1126/science.256.5055.385.

Protein Phosphorylation : A Practical Approach/Edited by D. Grahame Hardie. 1993. *Protein Phosphorylation : A Practical Approach*. Practical Approach Series. Oxford: Oxford University Press.

Quinlan, Casey L, Jason R Treberg, Irina V Perevoshchikova, Adam L Orr, and Martin D Brand. 2012. "Native Rates of Superoxide Production from Multiple Sites in Isolated Mitochondria Measured Using Endogenous Reporters". *Free Radical Biology and Medicine* 53 (9): 1807–1817. https://doi.org/10.1016/j.freeradbiomed.2012.08.015.

Quinlan, Casey L, Irina v Perevoshchikova, Martin Hey-Mogensen, Adam L Orr, and Martin D Brand. 2013. "Sites of Reactive Oxygen Species Generation by Mitochondria Oxidizing Different Substrates". *Redox Biology* 1 (1): 304–312. https://doi.org/10.1016/j.redox.2013.04.005.

Rabilloud, Thierry, Manfred Heller, Françoise Gasnier, Sylvie Luche, Catherine Rey, Ruedi Aebersold, Mohamed Benahmed, Pierre Louisot, and Joël Lunardi. 2002. "Proteomics Analysis of Cellular Response to Oxidative Stress: EVIDENCE FOR in Vivo Overoxidation of Peroxiredoxins at Their Active Site*". *Journal of Biological Chemistry* 277 (22). Elsevier: 19396–19401. https://doi.org/10.1074/jbc.M106585200.

Raman, Deepika, and Shazib Pervaiz. 2019. "Redox Inhibition of Protein Phosphatase PP2A: Potential Implications in Oncogenesis and Its Progression". *Redox Biology* 27: 101105. https://doi.org/10.1016/j.redox.2019.101105.

Ramasarma, T. 1982. "Generation of H2O2 in Biomembranes". *Biochimica et Biophysica Acta (BBA)—Reviews on Biomembranes* 694 (1): 69–93. https://doi.org/10.1016/0304-4157(82)90014-4.

Rhee, S G. 1999. "Redox Signaling: Hydrogen Peroxide as Intracellular Messenger". *Experimental & Molecular Medicine* 31 (2). United States: 53–59. http://doi.org/10.1038/emm.1999.9.

Rodrigues, Claudia, Andreia Filipa Mósca, Ana Paula Martins, Tatiana Nobre, Catarina Prista, Fernando Antunes, Ana Cipak Gasparovic, and Graça Soveral. 2016. "Rat Aquaporin-5 Is PH-Gated Induced by Phosphorylation and Is Implicated in Oxidative Stress". *International Journal of Molecular Sciences* 17 (12). http://doi.org/10.3390/ijms17122090.

Rodrigues, Claudia, Catarina Pimpão, Andreia F Mósca, Ana S Coxixo, Duarte Lopes, Inês Vieira da Silva, Per Amstrup Pedersen, Fernando Antunes, and Graça Soveral. 2019. "Human Aquaporin-5 Facilitates Hydrogen Peroxide Permeation Affecting Adaption to Oxidative Stress and Cancer Cell

Migration". *Cancers* 11 (7). http://doi.org/10.3390/cancers11070932.

Rojek, Aleksandra, Ernst-Martin Füchtbauer, Annette Füchtbauer, Sabina Jelen, Anders Malmendal, Robert A Fenton, and Søren Nielsen. 2012. "Liver-Specific Aquaporin 11 Knockout Mice Show Rapid Vacuolization of the Rough Endoplasmic Reticulum in Periportal Hepatocytes after Amino Acid Feeding". *American Journal of Physiology-Gastrointestinal and Liver Physiology* 304 (5). American Physiological Society: G501–G515. http://doi.org/10.1152/ajpgi.00208.2012.

Rousset, Francis, Leilei Zhang, Bernard Lardy, Françoise Morel, and Minh Vu Chuong Nguyen. 2020. "Transmembrane Nox4 Topology Revealed by Topological Determination by Ubiquitin Fusion Assay, a Novel Method to Uncover Membrane Protein Topology". *Biochemical and Biophysical Research Communications* 521 (2): 383–388. https://doi.org/10.1016/j.bbrc.2019.10.098.

Ru-Feng, Wu, Ma Zhenyi, Liu Zhe, and Terada Lance S. 2010. "Nox4-Derived H2O2 Mediates Endoplasmic Reticulum Signaling through Local Ras Activation". *Molecular and Cellular Biology* 30 (14). American Society for Microbiology: 3553–3568. http://doi.org/10.1128/MCB.01445-09.

Santos, Celio X C, Anne D Hafstad, Matteo Beretta, Min Zhang, Chris Molenaar, Jola Kopec, Dina Fotinou, et al. 2016. "Targeted Redox Inhibition of Protein Phosphatase 1 by Nox4 Regulates EIF2α-Mediated Stress Signaling". *The EMBO Journal* 35 (3): 319–334. http://doi.org/10.15252/embj.201592394.

Sauer, H, G Rahimi, J Hescheler, and M Wartenberg. 2000. "Role of Reactive Oxygen Species and Phosphatidylinositol 3-Kinase in Cardiomyocyte Differentiation of Embryonic Stem Cells". *FEBS Letters* 476 (3). England: 218–223. http://doi.org/10.1016/s0014-5793(00)01747-6.

Sauer, H, M Wartenberg, and J Hescheler. 2001. "Reactive Oxygen Species as Intracellular Messengers During Cell Growth and Differentiation". *Cellular Physiology and Biochemistry* 11 (4): 173–186. http://doi.org/10.1159/000047804.

Sbodio, Juan I, Solomon H Snyder, and Bindu D Paul. 2019. "Regulators of the Transsulfuration Pathway". *British Journal of Pharmacology* 176 (4). John Wiley & Sons, Ltd: 583–593. https://doi.org/10.1111/bph.14446.

Sessions, Alex L, David M Doughty, Paula v Welander, Roger E Summons, and Dianne K Newman. 2009. "The Continuing Puzzle of the Great Oxidation Event". *Current Biology* 19 (14): R567–R574. https://doi.org/10.1016/j.cub.2009.05.054.

Shoshan-Barmatz, Varda, Ran Zalk, Dan Gincel, and Noga Vardi. 2004. "Subcellular Localization of VDAC in Mitochondria and ER in the Cerebellum". *Biochimica et Biophysica Acta (BBA)—Bioenergetics* 1657 (2): 105–114. https://doi.org/10.1016/j.bbabio.2004.02.009.

Sies, Helmut, and Dean P Jones. 2020. "Reactive Oxygen Species (ROS) as Pleiotropic Physiological Signalling Agents". *Nature Reviews Molecular Cell Biology* 21 (7): 363–383. http://doi.org/10.1038/s41580-020-0230-3.

Silva, Inês Vieira da, Carlos Cardoso, Helios Martínez-Banaclocha, Angela Casini, Pablo Pelegrín, and Graça Soveral. 2021. "Aquaporin-3 Is Involved in NLRP3-Inflammasome Activation Contributing to the Setting of Inflammatory Response". *Cellular and Molecular Life Sciences* 78 (6): 3073–3085. http://doi.org/10.1007/s00018-020-03708-3.

Sitia, Roberto, and Anna Rubartelli. 2018. "The Unconventional Secretion of IL-1β: Handling a Dangerous Weapon to Optimize Inflammatory Responses". *Seminars in Cell & Developmental Biology* 83: 12–21. https://doi.org/10.1016/j.semcdb.2018.03.011.

Ślesak, Ireneusz, Halina Ślesak, and Jerzy Kruk. 2012. "Oxygen and Hydrogen Peroxide in the Early Evolution of Life on Earth: In Silico Comparative Analysis of Biochemical Pathways". *Astrobiology* 12 (8). Mary Ann Liebert, Inc., publishers: 775–784. http://doi.org/10.1089/ast.2011.0704.

Smoly, Ilan, Netta Shemesh, Michal Ziv-Ukelson, Anat Ben-Zvi, and Esti Yeger-Lotem. 2017. "An Asymmetrically Balanced Organization of Kinases versus Phosphatases across Eukaryotes Determines Their Distinct Impacts". *PLOS Computational Biology* 13 (1). Public Library of Science: e1005221. https://doi.org/10.1371/journal.pcbi.1005221.

Snyder, Eric M, Yi Nong, Claudia G Almeida, Surojit Paul, Timothy Moran, Eun Young Choi, Angus C Nairn, et al. 2005. "Regulation of NMDA Receptor Trafficking by Amyloid-β". *Nature Neuroscience* 8 (8): 1051–1058. http://doi.org/10.1038/nn1503.

Sobotta, Mirko C, Willy Liou, Sarah Stöcker, Deepti Talwar, Michael Oehler, Thomas Ruppert, Annette N D Scharf, and Tobias P Dick. 2015. "Peroxiredoxin-2 and STAT3 Form a Redox Relay for H2O2 Signaling". *Nature Chemical Biology* 11 (1): 64–70. http://doi.org/10.1038/nchembio.1695.

Sorrentino, Ilaria, Mauro Galli, Iria Medraño-Fernandez, and Roberto Sitia. 2022. "Transfer of H2O2 from Mitochondria to the Endoplasmic Reticulum via Aquaporin-11". *Redox Biology* 55: 102410. https://doi.org/10.1016/j.redox.2022.102410.

Starkov, Anatoly A. 2008. "The Role of Mitochondria in Reactive Oxygen Species Metabolism and Signaling". *Annals of the New York Academy of Sciences* 1147 (1). John Wiley & Sons, Ltd: 37–52. https://doi.org/10.1196/annals.1427.015.

Steinhubl, Steven R. 2008. "Why Have Antioxidants Failed in Clinical Trials?". *American Journal of Cardiology* 101 (10). Elsevier: S14–S19. http://doi.org/10.1016/j.amjcard.2008.02.003.

Stöcker, Sarah, Michael Maurer, Thomas Ruppert, and Tobias P Dick. 2018. "A Role for 2-Cys Peroxiredoxins in Facilitating Cytosolic Protein Thiol Oxidation". *Nature Chemical Biology* 14 (2): 148–155. http://doi.org/10.1038/nchembio.2536.

Stone, James R, and Suping Yang. 2006. "Hydrogen Peroxide: A Signaling Messenger". *Antioxidants & Redox Signaling* 8 (3–4). Mary Ann Liebert, Inc., publishers: 243–270. http://doi.org/10.1089/ars.2006.8.243.

St-Pierre, Julie, Julie A Buckingham, Stephen J Roebuck, and Martin D Brand. 2002. "Topology of Superoxide Production from Different Sites in the Mitochondrial Electron Transport Chain*". *Journal of Biological Chemistry* 277 (47). Elsevier: 44784–44790. http://doi.org/10.1074/jbc.M207217200.

Sui, Haixin, Bong-Gyoon Han, John K Lee, Peter Walian, and Bing K Jap. 2001. "Structural Basis of Water-Specific Transport through the AQP1 Water Channel". *Nature* 414 (6866): 872–878. http://doi.org/10.1038/414872a.

Sumimoto, Hideki. 2008. "Structure, Regulation and Evolution of Nox-Family NADPH Oxidases That Produce Reactive Oxygen Species". *The FEBS Journal* 275 (13). John Wiley & Sons, Ltd: 3249–3277. https://doi.org/10.1111/j.1742-4658.2008.06488.x.

Svastova, Eliska, and Silvia Pastorekova. 2013. "Carbonic Anhydrase IX". *Cell Adhesion & Migration* 7 (2). Taylor & Francis: 226–231. http://doi.org/10.4161/cam.23257.

Takac, Ina, Katrin Schröder, Leilei Zhang, Bernard Lardy, Narayana Anilkumar, J David Lambeth, Ajay M Shah, Francoise Morel, and Ralf P Brandes. 2011. "The E-Loop Is Involved in Hydrogen Peroxide Formation by the NADPH Oxidase Nox4*". *Journal of Biological Chemistry* 286 (15). Elsevier: 13304–13313. http://doi.org/10.1074/jbc.M110.192138.

Takahashi, Saki, Kanako Muta, Hiroko Sonoda, Ayaka Kato, Ahmed Abdeen, and Masahiro Ikeda. 2014. "The Role of Cysteine 227 in Subcellular Localization, Water Permeability, and Multimerization of Aquaporin-11". *FEBS Open Bio* 4 (1). John Wiley & Sons, Ltd: 315–320. https://doi.org/10.1016/j.fob.2014.03.005.

Tavender, Timothy J, and Neil J Bulleid. 2010. "Molecular Mechanisms Regulating Oxidative Activity of the Ero1 Family in the Endoplasmic Reticulum". *Antioxidants & Redox Signaling* 13 (8). United States: 1177–1187. doi:10.1089/ars.2010.3230.

Tchekneva, Elena E, Zaza Khuchua, Linda S Davis, Veronika Kadkina, Stephen R Dunn, Sebastian Bachman, Kenichi Ishibashi, et al. 2008. "Single Amino Acid Substitution in Aquaporin 11 Causes Renal Failure". *Journal of the American Society of Nephrology* 19 (10). American Society of Nephrology: 1955–1964. http://doi.org/10.1681/ASN.2008030296.

Tesse, Angela, Patrizia Gena, Michael Rützler, and Giuseppe Calamita. 2021. "Ablation of Aquaporin-9 Ameliorates the Systemic Inflammatory Response of LPS-Induced Endotoxic Shock in Mouse". *Cells* 10 (2). http://doi.org/10.3390/cells10020435.

Thamsen, Maike, Caroline Kumsta, Fei Li, and Ursula Jakob. 2010. "Is Overoxidation of Peroxiredoxin Physiologically Significant?". *Antioxidants & Redox Signaling* 14 (4). Mary Ann Liebert, Inc., publishers: 725–730. http://doi.org/10.1089/ars.2010.3717.

Thannickal, Victor J, and Barry L Fanburg. 2000. "Reactive Oxygen Species in Cell Signaling". *American Journal of Physiology-Lung Cellular and Molecular Physiology* 279 (6). American Physiological Society: L1005–L1028. http://doi.org/10.1152/ajplung.2000.279.6.L1005.

Travasso, Rui D M, Fernando Sampaio dos Aidos, Anahita Bayani, Pedro Abranches, and Armindo Salvador. 2017. "Localized Redox Relays as a Privileged Mode of Cytoplasmic Hydrogen Peroxide Signaling". *Redox Biology* 12: 233–245. https://doi.org/10.1016/j.redox.2017.01.003.

Truzzi, Daniela R, Fernando R Coelho, Veronica Paviani, Simone v Alves, Luis E S Netto, and Ohara Augusto. 2019. "The Bicarbonate/Carbon Dioxide Pair Increases Hydrogen Peroxide-Mediated Hyperoxidation of Human Peroxiredoxin 1". *Journal of Biological Chemistry* 294 (38). Elsevier: 14055–14067. http://doi.org/10.1074/jbc.RA119.008825.

Turnbull, Isaiah, Yun Lu, Alan S Verkman, and William R Skach. 2000. "Biogenesis and Folding of Aquaporin Water Channels in the Endoplasmic Reticulum". In *Molecular Biology and Physiology of Water and Solute Transport*, edited by Stefan Hohmann and Søren Nielsen, 49–55. Boston, MA: Springer US. http://doi.org/10.1007/978-1-4615-1203-5_7.

Unsay, Joseph D, Katia Cosentino, Yamunadevi Subburaj, and Ana J García-Sáez. 2013. "Cardiolipin Effects on Membrane Structure and Dynamics". *Langmuir* 29 (51). American Chemical

Society: 15878–15887. http://doi.org/10.1021/la402669z.

Ushio-Fukai, Masuko. 2006. "Localizing NADPH Oxidase-Derived ROS". *Science's STKE: Signal Transduction Knowledge Environment* 2006 (349). United States: re8. http://doi.org/10.1126/stke.3492006re8.

Varadaraj, Kulandaiappan, and S Sindhu Kumari. 2020. "Lens Aquaporins Function as Peroxiporins to Facilitate Membrane Transport of Hydrogen Peroxide". *Biochemical and Biophysical Research Communications* 524 (4): 1025–1029. https://doi.org/10.1016/j.bbrc.2020.02.031.

Vergés, Emili, Neus Colomina, Eloi Garí, Carme Gallego, and Martí Aldea. 2007. "Cyclin Cln3 Is Retained at the ER and Released by the J Chaperone Ydj1 in Late G1 to Trigger Cell Cycle Entry". *Molecular Cell* 26 (5). Elsevier: 649–662. http://doi.org/10.1016/j.molcel.2007.04.023.

Vieceli Dalla Sega, Francesco, Laura Zambonin, Diana Fiorentini, Benedetta Rizzo, Cristiana Caliceti, Laura Landi, Silvana Hrelia, and Cecilia Prata. 2014. "Specific Aquaporins Facilitate Nox-Produced Hydrogen Peroxide Transport through Plasma Membrane in Leukaemia Cells". *Biochimica et Biophysica Acta (BBA)—Molecular Cell Research* 1843 (4): 806–814. https://doi.org/10.1016/j.bbamcr.2014.01.011.

Virshup, David M, and Shirish Shenolikar. 2009. "From Promiscuity to Precision: Protein Phosphatases Get a Makeover". *Molecular Cell* 33 (5). Elsevier: 537–545. http://doi.org/10.1016/j.molcel.2009.02.015.

Walter, Peter, and David Ron. 2011. "The Unfolded Protein Response: From Stress Pathway to Homeostatic Regulation". *Science* 334 (6059). American Association for the Advancement of Science: 1081–1086. http://doi.org/10.1126/science.1209038.

Watabe, Shoji, Hiroyuki Kohno, Hiroshi Kouyama, Tomoko Hiroi, Nagasumi Yago, and Tohru Nakazawa. 1994. "Purification and Characterization of a Substrate Protein for Mitochondrial ATP-Dependent Protease in Bovine Adrenal Cortex". *The Journal of Biochemistry* 115 (4): 648–654. doi:10.1093/oxfordjournals.jbchem.a124390.

Watanabe, Sachiko, Catharina Sagita Moniaga, Søren Nielsen, and Mariko Hara-Chikuma. 2016. "Aquaporin-9 Facilitates Membrane Transport of Hydrogen Peroxide in Mammalian Cells". *Biochemical and Biophysical Research Communications* 471 (1). United States: 191–197. http://doi.org/10.1016/j.bbrc.2016.01.153.

West, A Phillip, Igor E Brodsky, Christoph Rahner, Dong Kyun Woo, Hediye Erdjument-Bromage, Paul Tempst, Matthew C Walsh, Yongwon Choi, Gerald S Shadel, and Sankar Ghosh. 2011. "TLR Signalling Augments Macrophage Bactericidal Activity through Mitochondrial ROS". *Nature* 472 (7344): 476–480. http://doi.org/10.1038/nature09973.

Weyemi, U, O Lagente-Chevallier, M Boufraqech, F Prenois, F Courtin, B Caillou, M Talbot, et al. 2012. "ROS-Generating NADPH Oxidase NOX4 Is a Critical Mediator in Oncogenic H-Ras-Induced DNA Damage and Subsequent Senescence". *Oncogene* 31 (9): 1117–1129. http://doi.org/10.1038/onc.2011.327.

Winterbourn, Christine C. 2008. "Reconciling the Chemistry and Biology of Reactive Oxygen Species". *Nature Chemical Biology* 4 (5): 278–286. http://doi.org/10.1038/nchembio.85.

Winterbourn, Christine C, and Mark B Hampton. 2008. "Thiol Chemistry and Specificity in Redox Signaling". *Free Radical Biology & Medicine* 45 (5). United States: 549–561. http://doi.org/10.1016/j.freeradbiomed.2008.05.004.

Woo, Hyun Ae, Sun Hee Yim, Dong Hae Shin, Dongmin Kang, Dae-Yeul Yu, and Sue Goo Rhee. 2010. "Inactivation of Peroxiredoxin I by Phosphorylation Allows Localized H(2)O(2) Accumulation for Cell Signaling". *Cell* 140 (4). United States: 517–528. http://doi.org/10.1016/j.cell.2010.01.009.

Wragg, Darren, Stefano Leoni, and Angela Casini. 2020. "Aquaporin-Driven Hydrogen Peroxide Transport: A Case of Molecular Mimicry?". *RSC Chemical Biology* 1 (5). RSC: 390–394. http://doi.org/10.1039/D0CB00160K.

Yang, Baoxue, Dan Zhao, and A S Verkman. 2006. "Evidence against Functionally Significant Aquaporin Expression in Mitochondria*". *Journal of Biological Chemistry* 281 (24). Elsevier: 16202–16206. http://doi.org/10.1074/jbc.M601864200.

Yang, Kap-Seok, Sang Won Kang, Hyun Ae Woo, Sung Chul Hwang, Ho Zoon Chae, Kanghwa Kim, and Sue Goo Rhee. 2002. "Inactivation of Human Peroxiredoxin I during Catalysis as the Result of the Oxidation of the Catalytic Site Cysteine to Cysteine-Sulfinic Acid*". *Journal of Biological Chemistry* 277 (41). Elsevier: 38029–38036. http://doi.org/10.1074/jbc.M206626200.

Yip, Shu-Chin, Sayanti Saha, and Jonathan Chernoff. 2010. "PTP1B: A Double Agent in Metabolism and Oncogenesis". *Trends in Biochemical Sciences* 35 (8). Elsevier: 442–449. http://doi.org/10.1016/j.tibs.2010.03.004.

Yoboue, Edgar Djaha, Roberto Sitia, and Thomas Simmen. 2018. "Redox Crosstalk at Endoplasmic Reticulum (ER) Membrane Contact Sites (MCS) Uses Toxic Waste to Deliver Messages". *Cell Death & Disease* 9 (3): 331. http://doi.org/10.1038/s41419-017-0033-4.

Yoshiyuki, Morishita, Matsuzaki Toshiyuki, Harachikuma Mariko, Andoo Ayaka, Shimono Mariko, Matsuki Asako, Kobayashi Katsuki, et al. 2005. "Disruption of Aquaporin-11 Produces Polycystic Kidneys Following Vacuolization of the Proximal Tubule". *Molecular and Cellular Biology* 25 (17). American Society for Microbiology: 7770–7779. http://doi.org/10.1128/MCB.25.17.7770-7779.2005.

Zalman, L S, H Nikaido, and Y Kagawa. 1980. "Mitochondrial Outer Membrane Contains a Protein Producing Nonspecific Diffusion Channels". *Journal of Biological Chemistry* 255 (5): 1771–1774. https://doi.org/10.1016/S0021-9258(19)85942-2.

Zhang, Min, Alison C Brewer, Katrin Schröder, Celio X C Santos, David J Grieve, Minshu Wang, Narayana Anilkumar, et al. 2010. "NADPH Oxidase-4 Mediates Protection against Chronic Load-Induced Stress in Mouse Hearts by Enhancing Angiogenesis". *Proceedings of the National Academy of Sciences* 107 (42). Proceedings of the National Academy of Sciences: 18121–18126. http://doi.org/10.1073/pnas.1009700107.

Zhang, Yang, Deepa v Venkitaramani, Clare M Gladding, Yongfang Zhang, Pradeep Kurup, Elek Molnar, Graham L Collingridge, and Paul J Lombroso. 2008. "The Tyrosine Phosphatase STEP Mediates AMPA Receptor Endocytosis after Metabotropic Glutamate Receptor Stimulation". *Journal of Neuroscience* 28 (42). Society for Neuroscience: 10561–10566. http://doi.org/10.1523/JNEUROSCI.2666-08.2008.

Zhou, Haiying, Harkewal Singh, Zachary D Parsons, Sarah M Lewis, Sanjib Bhattacharya, Derrick R Seiner, Jason N LaButti, Thomas J Reilly, John J Tanner, and Kent S Gates. 2011. "The Biological Buffer Bicarbonate/CO2 Potentiates H2O2-Mediated Inactivation of Protein Tyrosine Phosphatases". *Journal of the American Chemical Society* 133 (40). American Chemical Society: 15803–15805. http://doi.org/10.1021/ja2077137.

Zivanovic, Jasmina, Emilia Kouroussis, Joshua B Kohl, Bikash Adhikari, Biljana Bursac, Sonia Schott-Roux, Dunja Petrovic, et al. 2019. "Selective Persulfide Detection Reveals Evolutionarily Conserved Antiaging Effects of S-Sulfhydration". *Cell Metabolism* 30 (6): 1152–1170.e13. https://doi.org/10.1016/j.cmet.2019.10.007.

Zoccarato, Franco, Lucia Cavallini, and Adolfo Alexandre. 2004. "Respiration-Dependent Removal of Exogenous H_2O_2 in Brain Mitochondria: INHIBITION BY Ca^{2+}*". *Journal of Biological Chemistry* 279 (6). Elsevier: 4166–4174. http://doi.org/10.1074/jbc.M308143200.

Zuhra, Karim, Fiona Augsburger, Tomas Majtan, and Csaba Szabo. 2020. "Cystathionine-β-Synthase: Molecular Regulation and Pharmacological Inhibition". *Biomolecules* 10 (5). http://doi.org/10.3390/biom10050697.

CHAPTER FOURTEEN

The History of the Discovery of Peroxiporins

Tobias Henzler, Michael Broberg Palmgren, and Gerd Patrick Bienert

ABBREVIATIONS

AQP	Aquaporin
CPP	Cell pressure probe
ER	Endoplasmic reticulum
H_2O	Water
H_2O_2	Hydrogen peroxide
HDO	Deuterated water
MAPK	Mitogen-activated protein kinase
MIP	Major intrinsic protein
NO	Nitric oxide
*OH	Hydroxyl radical
PIP	Plasma membrane intrinsic protein
RNS	Reactive nitrogen species
ROS	Reactive oxygen species
TIP	Tonoplast intrinsic protein

14.1 THE HISTORY ON THE DISCOVERY OF PEROXIPORINS

Major intrinsic proteins (MIPs) shape a family of transmembrane channel proteins that have been found widespread in bacteria, protozoa, fungi, insects, plants, and vertebrates, thus covering all kingdoms of life (Chaumont and Tyerman, 2017). The initial functional characterization of MIP proteins is associated with the groundbreaking work of Agre's laboratory (Preston et al., 1992), demonstrating that MIPs are as functional water channels when heterologously expressed in *Xenopus laevis* oocytes and that MIPs form hydrophilic pores in membranes. This identification of proteins capable of facilitating the transport of water was explaining the high membrane water permeability of various cell types, which could not solely be explained by simple passive diffusion of water molecules across lipid bilayers. In the years following this discovery, the important role of MIPs in water transport processes of microorganisms, plants, and vertebrates became obvious, and the name "aquaporin" (AQP; from the Latin *aqua*, meaning "water; the sea; rain") was rapidly adopted for MIPs transporting water and, subsequently, for the whole MIP family (Chaumont and Tyerman, 2017).

Thereafter, numerous studies demonstrated that MIPs are not only essential for water transport homeostasis in various organisms, but additionally fulfill physiologically relevant uptake, translocation, sequestration, or extrusion transport functions by facilitating the bidirectional membrane diffusion of a variety of metabolically important small solutes. AQP-facilitated transport of glycerol (Fu et al., 2000; Stroud et al., 2003); urea (Gerbeau et al., 1999; Liu et al., 2003); carbon dioxide (Uehlein et al., 2003); nitric oxide (Herrera et al., 2006); lactic acid (Tsukaguchi et al., 1998; Choi and Roberts, 2007; Bienert et al., 2013); acetic acid (Mollapour and Piper, 2007); ammonia (Jahn et al., 2004; Loque et al., 2005); arsenous acid (Ma et al., 2006; Bienert et al., 2008); boric acid (Takano

DOI: 10.1201/9781003160649-14

et al., 2006); silicic acid (Ma et al., 2006); antimonous acid (Bienert et al., 2008; Kamiya and Fujiwara, 2009); selenous acid (Zhao et al., 2010) and in the focus of this book, hydrogen peroxide (H_2O_2; Bienert et al., 2007; Dynowski et al., 2008) have been identified and further studied in detail.

Based on their different transport properties AQPs conducting water- and glycerol were designated as functional "aquaglyceroporins" (Maurel et al., 1994; Fu et al., 2000), AQPs conducting ammonia and water as "aquaammoniaporins" (Jahn et al., 2004; Kirscht et al., 2016) and AQPs conducting the undissociated (non-charged) oxyacids of metalloids as "metalloidoporins" (Pommerrenig et al., 2015).

Dependent on its concentration of H_2O_2, a redox oxygen species (ROS) is an essential second messenger (<100 nM), capable of promoting the growth and stress resistance of living cells or of causing their death (>100 nM; Sies and Jones, 2020). H_2O_2 regulates a large set of developmental and physiological processes and stress responses within cells by either directly triggering chemical reactions and influence responsive targets, such as metabolites or proteins, or by providing the impetus for signaling pathways (e.g., involving calcium channels, kinases, phosphatases and other proteins; Vissers et al., 2017). H_2O_2 acts either as a paracrine (intercellular) or an autocrine (intracellular) signal (Bienert et al., 2006; Sies and Jones, 2020). An intercellular signaling function means that the signal molecule needs to be transported across at least one membrane barrier. When H_2O_2 acts as an intracellular signal, it often serves as an interorganellar signal: while H_2O_2 is mostly produced in membrane-surrounded organelles, such as mitochondria, chloroplasts, peroxisomes, nuclei, and the endoplasmic reticulum (ER), its signal molecule action is observed outside these organelles (Vissers et al., 2017). Various cellular and subcellular membranes represent a significant resistance for intercellular and interorganellar passive H_2O_2 diffusion. Consequently, spatio-temporally and quantitatively regulated H_2O_2 membrane permeation does depend on proteinaceous H_2O_2 transport facilitators. Today we know that the physiological regulation of the transmembrane diffusion of water is mediated by the water channeling "aquaporins" (Maurel et al., 2015), and that the control of the physiological membrane passage of H_2O_2 is accomplished by another subgroup of the MIP transport protein superfamily. AQPs capable of transporting H_2O_2 were designated first as "peroxoporins" (Henzler and Steudle, 2000) and nowadays as "peroxiporins" (Marinelli and Marchissio, 2013).

The story of peroxiporins began in 2000, when Henzler and Steudle hypothesized and described for the first time that AQPs are capable of transporting H_2O_2 in a study investigating H_2O_2 transport in the freshwater algae *Chara corallina* (Henzler and Steudle, 2000). The authors ended their abstract of this groundbreaking study with the following statement:

> The absolute values of P_s and σ_s of H_2O_2 indicated that hydrogen peroxide, a molecule with chemical properties close to that of water, uses water channels (aquaporins) to cross the cell membrane rapidly. When water channels were inhibited with the blocker mercuric chloride ($HgCl_2$), the permeabilities of both water and H_2O_2, were substantially reduced. In fact, for the latter, it was not measurable. It is suggested that some of the water channels in Chara (and, perhaps, in other species) serve as "peroxoporins" rather than as "aquaporins."
>
> (Henzler and Steudle, 2000)

This statement, which proved to be true, was based on calculations and observations from analyzing the transmembrane diffusion of water and H_2O_2 in parallel in internodal cells of *C. corallina*. The permeability coefficient of H_2O_2, which is a measure of the speed with which a compound crosses a given membrane, was found to be almost identical to that of water, and treatment with mercury, an AQP blocker, significantly repressed the influx of both water and H_2O_2 into the cells. A relatively high refection coefficient for H_2O_2 was leading to the hypothesis of Henzler and Steudle (2000), that only some, but not all, AQPs serve as "peroxoporins."

14.2 THE ERA OF THE DISCOVERY OF PEROXIPORINS IN THE ERNST STEUDLE LABORATORY—A PERSONAL PERSPECTIVE

The following lines will be more of a historical nature delivering firsthand lab-bench insights into the very first developments and discoveries in the field of peroxiporins rather than a rigid scientific summary.

The last decade of the 20th century was the decade of the discovery of water channel proteins, also known as AQPs, in mammals and plants. During this period, Ernst Steudle, former Professor at the chair of Plant Ecology at the University of Bayreuth, Germany, was one of the acknowledged experts of water and solute transport in plants. His team contributed to some remarkable first discoveries around the functions and roles of AQPs for membrane transport of plant cells and whole organs such as roots (Zhu and Steudle, 1991; Clarkson et al., 2000; Zimmermann et al., 2000). As a young physicist, Tobias Henzler had the chance and the privilege to be part of the Steudle team. At that time, Ernst Steudle delegated a lot of the practical research around water channels to his new undergraduate and then Ph.D. student. Moreover, Steudle enjoyed engaging all of his team members in hours of discussions and reflections on experimental setups, results, and the implications for the greater picture. Unfortunately, Steudle passed away much too early in the beginning of his retirement (†June 22, 2011). When Henzler started in his lab, Ernst Steudle had established a very useful and efficient experimental model system to study water and solute transport phenomena across plant plasma membranes. The model system was an internodal cell of the green algae *Chara corallina* belonging to the class of *Charophyceae*. *C. corallina* has a single internodal cell of several centimeters in length and 1 to 3 mm in diameter and displays a hydraulic cell pressure of 5 to 10 bars. This cell was very easy to handle and could be fixed in a stable way in a tube where cells, not solutes could be circumfused with solutions, which could be exchanged within one or two seconds. Moreover, this internodal cell was suitable to be probed by means of the so-called cell pressure probe (CPP), which allowed to continuously monitor the cell's turgor pressure for several hours. The CPP is a micromanometer device that Ernst Steudle had developed, tested, and established during his own Ph.D. as a method to measure a set of parameters that characterize water and solute transport across plasma membranes (Hüsken et al., 1978). Water fluxes across a plant membrane can be induced either by generating a hydrostatic pressure difference across the membrane, for which a small, defined volume of cell liquid is injected into or pulled out of the cell by means of the CPP, or alternatively, an osmotic pressure gradient can be induced by quickly exchanging the external solution flowing around the cell by another solution containing an additional chemical compound that can or cannot permeate the membrane. The latter will result in an exponential decrease of the measured cell pressure from which the water permeability of the membrane can be determined. The former will result in a biphasic pressure relaxation: the first phase is characterized by the water permeation out of the cell and the second phase by the solution permeating into the cell. From these biphasic curves, the solute permeation and reflection coefficients can be determined.

The description of the transport phenomena being observed in the CPP experiments was founded on the well-established physical theory of thermodynamics of irreversible processes. Based on the findings of Nobel Prize laureate in physics, Lars Onsager (1931), the great biophysicist Jack Dainty adapted the theory to membrane transport and defined the three above-mentioned transport parameters (the water permeability of the membrane, the solute permeation coefficient, and the reflection coefficient) that describe the interaction between water and solute flow across membranes (Dainty and Ginzburg, 1963). Accordingly, one of the greatest honors in the scientific career of Henzler was that Jack Dainty chaired a session in a scientific conference where he presented for the first time the findings of the reversible closing of water channels in *Chara*.

Undergraduate students who took part in a practical course about water and solute transport in Ernst Steudle's lab had typically been introduced to the *Chara* model system and to the CPP technique in order to perform various hydraulic and osmotic experiments to learn about osmosis and membrane transport. With his educational background as a physico-chemist, Ernst Steudle suggested a plenitude of well-known and less well-known chemical substances that should be used to probe their transport across the *Chara* plasma membrane: salts, sugars, long- and short-chained alcohols such as mannitol (non-permeable), and acetone (very permeable through the lipid bilayer). One of his favorite experiments was to determine the sugar and alcohol content of a de-alcoholized beer from a local Franconian brewery; the *Chara*-CPP system was so sensitive that ethanol contents of as little as 0.5% vol were possible to be determined.

When Henzler did his first practical courses his lab, one day, Ernst Steudle excitedly rushed into the laboratory waving in his hands a new paper from Christophe Maurel (Maurel et al., 1993) and said: "We need to test this with *Chara*, now!" Christophe Maurel and colleagues showed in this paper that when a tonoplast intrinsic protein, the γ-TIP from *Arabidopsis thaliana*, belonging to the AQP water channel family, was heterologously expressed in *Xenopus laevis* oocytes, those frog eggs swelled much faster than control eggs and even exploded when subjected to a hypo-osmotic shock. Furthermore, the faster swelling was inhibited by treatment of mercuric chloride, which blocked the protein channel, and the observed swelling could be reversed by mercaptoethanol, which removed the mercury inhibition from the protein. The conclusion was that γ-TIP forms a channel in the oocyte membrane that facilitates the water permeability and therefore represents a functional water channel or aquaporin. Ernst Steudle's hypothesis was that *Chara* also might contain water channel proteins as they showed an a priori high water conductance, and that these proteins could be blocked by mercury. When first tests were performed in the Steudle lab using a treatment of mercury on the *Chara* internodes, his hypothesis proved to be correct: the hydrostatic and osmotic water permeability was significantly reduced upon treatment with mercury and could be restored to nearly 100% by treatment with mercaptoethanol. Thus, all members of the Steudle lab became convinced that mercury-sensitive water channel proteins are present in *Chara*'s plasma membrane.

This observation triggered a series of experiments with *Chara* and different solutes in order to prove the presence and investigate the properties of the water channels in this model system. Ernst Steudle's idea was to use a series of small molecules that are similar in size and shape to water and to see whether their transport characteristics could be influenced by mercury and, thus, the closure of the "water" channel proteins. The first osmotic substance which was tested was deuterated water (HDO). Although extremely similar to water, this molecule had a measurable osmotic effect when used in higher concentrations (3–4 M); biphasic pressure relaxation was induced and the permeability of HDO determined. The permeability of internodal *Chara* cells to HDO was significantly reduced upon treatment with mercury showing that also HDO crossed the membrane preferably via water channel proteins. The same tests were performed in the following with larger, lipophilic molecules such as acetone and ethanol. However, no reduction of acetone and ethanol permeabilities upon mercury treatment had been quantified suggesting that these molecules crossed the membrane essentially via the lipid bilayer. However, a strange phenomenon appeared, namely a negative osmosis; an osmotic gradient of acetone, for example, would not result in an outflow of water and a subsequent pressure decrease but rather an intake of water against the pressure gradient. This led Henzler and Steudle to the proposition of a composite transport model of the membrane, where the membrane is composed of arrays of lipidic bilayer structures and of arrays of water channels, both exhibiting their own distinct transport coefficients (Henzler and Steudle, 1995). The negative osmosis in the case of acetone and ethanol was explained by the fact that, upon treatment of mercury, all transport of water and of the solute is mediated via the lipidic bilayer array, the latter having a negative reflection coefficient. A negative reflection coefficient is the mathematical description of the fact that the fast influx of the solute molecules will drag with it also a significant flow of water in the same direction.

The last molecule left over between water (H_2O or HDO) on one side, and the small lipophilic molecules such as acetone and ethanol on the other side was, as suggested again by the physicochemist Ernst Steudle, H_2O_2. Tobias Henzler performed various experiments using the *Chara*-CPP system and proved the chemical prediction of Ernst Steudle to be correct. It was not much surprising to observe that the high permeability of H_2O_2 was nearly reduced to zero upon closure of water channels by the use of mercury. This was a very strong indication that also H_2O_2 permeates through water channels to cross the membrane. This led Henzler and Steudle then to the proposition that one of the functions of water channels in *Chara* was the facilitation of H_2O_2 transport through the plasma membrane; consequently, Henzler and Steudle suggested that AQPs being permeable to H_2O_2 should be called peroxoporins (Henzler and Steudle, 2000). The discovery that AQP seem to transport also other molecules such as H_2O_2 was also a kind of "relief" for the team of Steudle, continuously wondering why a plant living submerged in water, as *C. corallina*, would need water channels at all. There is no apparent need to facilitate the transport of water in an aquatic environment where even the permeability of the lipid bilayer would be sufficient to react on slow changes in the environment. To rapidly transport (or to transport at all) H_2O_2 across the plasma membrane made more physiological sense for these channel proteins than the transport of water, especially as H_2O_2 was known to be a signaling messenger molecule for plants in response to adverse environmental effects such as abiotic and biotic stresses. *Peroxoporins* would help plants to mediate the signal much faster across membranes than without their presence.

When the Steudle team thought that all possible molecules had been tested in the *Chara* system, Ernst Steudle came yet with another very small molecule he wanted his team to test transport of with their *Chara*-CPP model system, something even smaller than water. His idea was to test whether hydroxyl radicals (*OH), as generated by the well-known Fenton reaction, would also use water channels to cross the membrane. However, in this case, insufficient osmotic pressure had been generated to measure transport characteristics of *OH itself because the radical nature of the molecule was so toxic that the cells did not survive. However, another unexpected action of *OH was discovered when lowering its concentration to a not immediately toxic level (Henzler et al., 2004). Since hydrostatic pressure relaxations were continuously performed during the treatments in order to see when the effect on water conductance takes place, it was observed that, in the presence of *OH, the conductance did quickly and significantly decrease. Even a negative osmosis had been observed with acetone and the same effect as described before was seen when the channels had been blocked with mercury. The effects of *OH were as reversible as the effects of mercury by simply removing the Fenton reactants from the medium and, thus, the *OH, too. Based on these results it was suggested that *OH may play an important role in gating AQPs and, thus, control the permeation and signaling via H_2O_2 (Henzler et al., 2004). Further investigations conducted and published by one of the last Ph.D. students of Ernst Steudle, Qing Ye, showed that this gating is also present in cells of whole tissues such as maize roots (Ye and Steudle, 2006). Years later, molecular and genetic evidence for the reversible oxidative AQP gating mechanisms—initially described in the Steudle lab—have been provided by the Maurel and Sitia laboratories in 2015 and 2018, respectively (Wudick et al., 2015; Bestetti et al., 2018).

Looking from a retrospective to the research performed by Henzler and Steudle and colleagues, everything seems very straightforward. But as with every research and experiment, the team had to cope with various difficulties in their trials. Although the Steudle team was lucky to have a stable model system that allowed for transport experiments to run for hours and dozens of hydraulic and osmotic pressure relaxations to be made, one should bear in mind that the various chemical treatments were not only a real challenge for the viability of *Chara* cells but also for the scientists' aiming at a substantial number of measurements for proper statistical analysis of their observations. Not only were mercury, mercaptoethanol, acetone, H_2O_2, and *OH alone toxic for their physiology, but often cell had been treated for hours with combinations of these compounds. This often led to a sudden decrease of turgor pressure (= cell death) in the middle of a promising experiment, and data could not be used for statistical analysis. Thus, the Steudle team had to perform lots of experiments and often got frustrated by the difficulty of increasing the number of successful experiments and replicates above the level of significancy of the effects. Often, Ernst Steudle silently appeared behind the backs of his scientists and looked over their shoulders, his thin-lensed reading glasses on the very tip of his nose, and suddenly asked, "What did you find?" But he also comforted his team when something went wrong and helped with good advice about the experimental procedure and an eventual reduction of the dose of the toxic compounds. He never lost his calm and often relaxed the tense situation in the lab with one of his humorous remarks. Also, the expert technical help and knowledge of the long-term lab assistant Burkard Stumpf cannot be sufficiently valued in the context of pressure probes measurements on cells and roots as well as the dear C. *corallina*. Looking back today, almost 30 years into the past, it must be said that it has been a great chance and privilege to have worked in Ernst Steudle's lab in these exciting times and to have had the occasion of a small contribution to the full picture of AQPs, which might indeed be the plants' *peroxoporins*. Tobias Henzler left academic research in 2003, the year the Nobel Prize in Chemistry was awarded to Peter Agre and Roderick MacKinnon for their contribution to the discovery of AQPs and ion channels.

14.3 YOU'LL NEVER KNOW UNTIL YOU TRY! HOW RESEARCH OBJECTIVES MAY SHIFT

The experimental evidence for the hypothesis of Henzler and Steudle (i.e., that individual AQPs facilitate the transport of H_2O_2) (Henzler and Steudle, 2000) was established seven years later, in 2007, by Gerd Patrick Bienert, Thomas Jahn, and colleagues in the Plant Nutrition laboratory of Jan Kofod Schjoerring in Copenhagen, Denmark (Bienert et al., 2007). Bienert et al. (2007) demonstrated that certain AQPs act as functional H_2O_2 channels when heterologously expressed in *Saccharomyces cerevisiae* yeast cells differing in their sensitivity toward oxidative stress and, consequently, to externally supplied elevated levels of H_2O_2. This finding was the outcome of a Ph.D. project (2005–2008) that was originally addressing other AQP substrate selectivities than that for H_2O_2 (Figure 14.1).

One focus of the Ph.D. work of Bienert was on the identification and molecular characterization of AQPs facilitating the diffusion of nitrogenous molecules. Thomas Jahn and colleagues in the lab of Schjoerring had demonstrated in 2004 that specific AQPs are permeable to ammonia in addition to water and function as *aquaammoniaporins* (Jahn et al., 2004). In line with this finding, the thesis of Bienert aimed (1) at analyzing the quantitative occurrence of ammonia

Figure 14.1 The tale of the fortunate discovery of peroxiporins.

permeability in different AQP isoforms from different organisms, to qualitatively understand protein features necessary for the ammonia specificity, and (2) at identifying whether AQPs are also permeable to other nitrogenous molecules than ammonia. To that aim, different bacterial, fungal, protozoan, plant, and mammalian AQP isoforms with varying, but for different AQP subgroup representative selectivity filters, had been selected and compiled for a comprehensive screening targeting the identification of novel nitrogenous AQP substrates. Potential novel AQP substrates were selected based on physico-chemical properties which are shared with known AQP substrates such as water and glycerol and which have been known to be crucial for the permeation through an AQP. Based on such physico-chemical characteristics, and on the observation that AQPs have been shown to transport small gaseous molecules (e.g., carbon dioxide, ammonia) and some AQPs are highly expressed in cells and tissues that produce the small gaseous signaling molecule, nitric oxide (NO), NO was one of the targets hypothesized to permeate through AQPs. Nitric oxide, a reactive nitrogen species (RNS), proved to be a signaling molecule in many physiological and pathological processes in mammals, including humans and plants. The discovery of its role as a cardiovascular signaling molecule was awarded 1998 with the Nobel Prize in Physiology or Medicine. It was unknown at the time whether NO exits cells where it is produced and enters cells where it acts by free passive diffusion through the lipid bilayer of the cell membrane without need for a specific transport protein, or whether specific transport proteins are involved. In 2002, Wong and colleagues had published a study evaluating whether yeast Tsa1p and Tsa2p, two peroxiredoxins, can protect cells from RNS (Wong et al., 2002). To address this question, the authors tested the sensitivity of the Δ*tsa1,2* yeast double mutant strain toward two RNS compounds, peroxynitrite and sodium nitroprusside, causing nitrosative stress. Notably, the Δ*tsa1,2* double mutant was hypersensitive to these compounds compared to wild-type yeast, demonstrating a pivotal role for peroxiredoxins in the protection against nitrosative stress which can also be induced by NO. The Δ*tsa1,2* double mutant represented therefore a tool to test if certain transport proteins facilitate the transport of NO across the yeast membrane simply by assessing cell viability and performance in response to the exposure to elevated NO amounts in the growth environment of this yeast strain. Non-enzymatic mechanisms such as conversion of nitrite to NO under acidic pH conditions can easily produce significant amounts of NO. Such NO-generating conditions in a closed Petri dish cultivation system can therefore be used to assess the sensitivity of yeast strains or yeast transformants to NO in typical yeast growth–based chemical sensitivity assays (Petropavlovskiy et al., 2020). Consequently, Bienert grew the NO hypersensitive Δ*tsa1,2* double mutant and its corresponding wild-type yeast, on a high nitrite and low pH medium in order to expose the empty vector transformed Δ*tsa1,2* double mutant or the Δ*tsa1,2* double mutant transformed with the different AQP isoforms to toxic NO stress conditions. Yeasts being transformed with an AQP channel protein being putatively permeable to NO were hypothesized to display an increased sensitivity (= reduced growth rate) under high NO conditions in this yeast toxicity growth assay compared to non-transformed Δ*tsa1,2* mutants. Having in mind the hypothesis of Henzler and Steudle (i.e., that AQPs are suggested to facilitate the transmembrane diffusion of H_2O_2; Henzler and Steudle, 2000), the physico-chemical similarity of H_2O_2 and H_2O (Ardon, 1965; Bienert et al., 2006), as well as the observations of Wong and colleagues (Wong et al., 2002)—that is, that Δ*tsa1,2* mutants exhibited also a hypersensitivity to externally elevated H_2O_2 levels—Bienert and Jahn decided during an intense scientific discussion at Blågårds Plads to additionally test the growth behavior of Δtsa1,2 yeast transformed with different AQPs also on growth medium containing elevated H_2O_2 levels. None of the tested AQP expressing Δ*tsa1,2* yeast transformants displayed an obvious or consistent hypersensitivity to elevated NO levels in the yeast growth medium (Figure 14.2).

In contrast, the hypersensitivity of certain AQP expressing transformants to elevated H_2O_2 levels was not only striking but also highly repeatable in several independent repetitions of the growth experiments (Bienert et al., 2007). These observations made in a 28–30°C yeast growth incubator fundamentally changed the research direction of Bienert and colleagues from investigating a potential permeability of AQP to NO to H_2O_2.

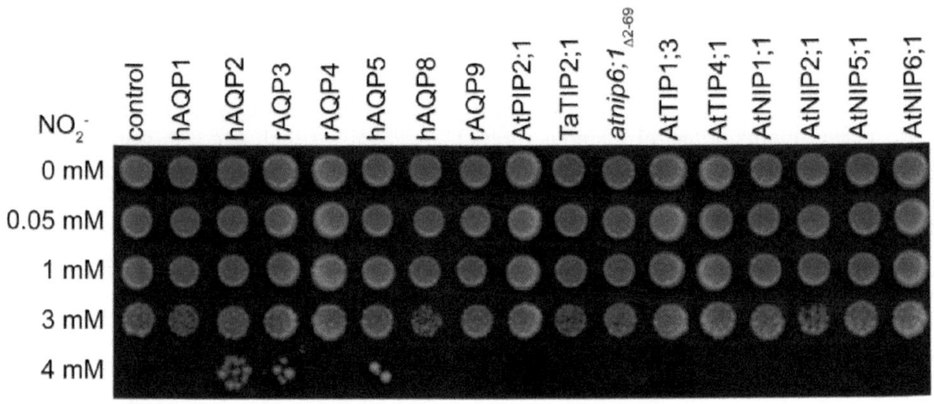

Figure 14.2 Yeast growth and survival on medium containing various concentrations of nitric oxide (NO). Δtsa1-2 yeast mutant transformed with the empty vector pYES (control) or the AQP homologs hAQP1, hAQP2, rAQP3, rAQP4, hAQP5, hAQP8, rAQP9, AtPIP2;1, TaTIP2;1, Atnip6;1$_{\Delta 2-69}$, AtTIP1;3, AtTIP4;1, AtNIP1;1, AtNIP2;1, AtNIP5;1, or AtNIP6;1. Yeast transformants were spotted at an A_{600} of 0.01 on media containing various concentrations of $NaNO_2$, as indicated. Nitrosative stress was applied via NO formation out of $NaNO_2$ under mildly acidic conditions (pH 5.2). Growth was recorded after 6 days at 30°C.

In the following, numerous AQPs were intensively and thoroughly tested for their potential H_2O_2 permeability in several different yeast strains differing in sensitivity toward oxidative stress. Expression of human AQP8 and plant *Arabidopsis* TIP1;1 and TIP1;2 in yeast decreased growth and survival in the presence of H_2O_2 in all yeast strains independent of the yeast strain intrinsic sensitivity toward H_2O_2 (Bienert et al., 2007). Further evidence for AQP-mediated H_2O_2 transport was provided by a fluorescence assay with intact yeast cells using an intracellular ROS-sensitive fluorescent dye (Bienert et al., 2007). Moreover, the application of silver ions, which block AQP-mediated water diffusion in a fast kinetics swelling assay, reversed both the H_2O_2-induced growth repression and the H_2O_2-induced fluorescence in the ROS-sensitive dye loaded yeasts expressing AQP8, TIP1;1, and TIP1;2. The assembly of experimental evidence that certain AQPs are permeable to H_2O_2 highly benefitted from a continuous critical scientific input of Ian Max Møller, who became a topically exotic frequenter in the Plant Nutrition lab of Schjoerring those days. Møller had long-term expertise and knowledge on redox biology and ensured that conditions and control assays were carefully planned and included.

The results obtained by Bienert and colleagues presented the first molecular genetic evidence for the diffusion of H_2O_2 through specific members of the AQP family (Bienert et al., 2007). A year later, in 2008, Dynowski and colleagues published supporting evidence, also using the heterologous yeast expression system that specific plant plasma membrane water channels conduct the signaling molecule H_2O_2 (Dynowski et al., 2008). These studies paved the way for many more molecular, structural, and functional studies on **peroxiporins** in the following years. Finally, groundbreaking studies were able to undoubtedly demonstrate that **peroxiporins** are physiologically highly relevant H_2O_2 signaling gates in plants (e.g., Tian et al., 2016; Rodrigues et al., 2017; Zhang et al., 2022) and vertebrates (e.g., Miller et al., 2010; Bertolotti et al., 2013; Vieceli Dalla Sega et al., 2014; Bertolotti et al., 2016; Hara-Chikuma et al., 2016; Medraño-Fernandez et al., 2016; Bestetti et al., 2018; da Silva et al., 2021; Tesse et al., 2021) all being described and referred to in various chapters of this book. The discovery that membrane diffusion of H_2O_2 is facilitated by proteinaceous **peroxiporins** and not based solely on a simple passive diffusion was therewith a long-standing outcome of many interdisciplinary research fields.

To come back to question whether AQPs are able to transport other nitrogenous molecules than ammonia and urea and whether AQPs are permeable to NO, it remains to be said that Herrera and colleagues showed that NO permeates through AQP1 in 2006 (Herrera et al., 2006).

The protein, putatively an aquaporin, which provoked all this research with utmost importance and consequences on the understanding of signaling processes governing many essential processes in plants and vertebrates, still slumbers incognito potentially in the plasma membrane of the internodal cell of the green algae *C. corallina* and is awaiting to be identified and functionally characterized.

14.4 ACKNOWLEDGMENTS

This work was supported by an Emmy Noether grant from the Deutsche Forschungsgemeinschaft to G.P.B. (1668/1-1/2). We thank the many scientists whose laborious and elegant work contributed to uncover the features and roles of aquaporins in H_2O_2 transport across the diverse membrane systems of cells. We apologize to the colleagues whose contributions to these research areas could have not been mentioned due to space restrictions.

This article is dedicated to Prof. Dr. Ernst Steudle, the ideator of the term „peroxoporins".

(† June 22, 2011)

REFERENCES

Ardon M. (1965) *Oxygen*, New York: W.A. Benjamin INC.

Bertolotti M, Bestetti S, García-Manteiga JM, Medraño-Fernandez I, Dal Mas A, Malosio ML, Sitia R. (2013) Tyrosine kinase signal modulation: A matter of H2O2 membrane permeability? *Antioxid Redox Signal.* 19(13):1447–1451.

Bertolotti M, Farinelli G, Galli M, Aiuti A, Sitia R. (2016) AQP8 transports NOX2-generated H2O2 across the plasma membrane to promote signaling in B cells. *J Leukoc Biol.* 100(5):1071–1079.

Bestetti S, Medraño-Fernandez I, Galli M, Ghitti M, Bienert GP, Musco G, Orsi A, Rubartelli A, Sitia R. (2018) A persulfidation-based mechanism controls aquaporin-8 conductance. *Sci Adv.* 4(5):eaar5770.

Bienert GP, Desguin B, Chaumont F, Hols P. (2013) Channel-mediated lactic acid transport: A novel function for aquaglyceroporins in bacteria. *Biochem J.* 454(3):559–570.

Bienert GP, Møller AL, Kristiansen KA, Schulz A, Møller IM, Schjoerring JK, Jahn TP. (2007) Specific aquaporins facilitate the diffusion of hydrogen peroxide across membranes. *J Biol Chem.* 282(2):1183–1192.

Bienert GP, Schjoerring JK, Jahn TP. (2006) Membrane transport of hydrogen peroxide. *Biochim Biophys Acta.* 1758(8):994–1003.

Bienert GP, Thorsen M, Schüssler MD, Nilsson HR, Wagner A, Tamás MJ, Jahn TP. (2008) A subgroup of plant aquaporins facilitate the bi-directional diffusion of As(OH)3 and Sb(OH)3 across membranes. *BMC Biol.* 10(6):26.

Calamita G, Ferri D, Gena P, Liquori GE, Cavalier A, Thomas D, Svelto M. (2005) The inner mitochondrial membrane has aquaporin-8 water channels and is highly permeable to water. *J Biol Chem.* 280(17):17149–17153.

Chaumont F, Tyerman SD (eds) (2017) *Plant Aquaporins: From Transport to Signaling—Signaling and Communication in Plants.* New York: Springer. https://doi.org/10.1007/978-3-319-49395-4

Choi WG, Roberts DM. (2007) Arabidopsis NIP2;1, a major intrinsic protein transporter of lactic acid induced by anoxic stress. *J Biol Chem.* 282(33):24209–24218.

Clarkson DT, Carvajal M, Henzler T, Waterhouse RN, Smyth AJ, Cooke DT, Steudle E. (2000) Root hydraulic conductance: Diurnal aquaporin expression and the effects of nutrient stress. *J Exp Bot.* 51(342):61–70.

Dainty J, Ginzburg BZ. (1963) Irreversible thermodynamics and frictional models of membrane processes, with particular reference to the cell membrane. *J Theor Biol.* 5(2):256–265.

da Silva IV, Cardoso C, Martínez-Banaclocha H, Casini A, Pelegrín P, Soveral G. (2021) Aquaporin-3 is involved in NLRP3-inflammasome activation contributing to the setting of inflammatory response. *Cell Mol Life Sci.* 78(6):3073–3085.

Dynowski M, Schaaf G, Loque D, Moran O, Ludewig U. (2008) Plant plasma membrane water channels conduct the signalling molecule H2O2. *Biochem J.* 414(1): 53–61.

Fu D, Libson A, Miercke LJ, Weitzman C, Nollert P, Krucinski J, Stroud RM. (2000) Structure of a glycerol-conducting channel and the basis for its selectivity. *Science.* 290(5491):481–486.

Gerbeau P, Güçlü J, Ripoche P, Maurel C. (1999) Aquaporin Nt-TIPa can account for the high permeability of tobacco cell vacuolar membrane to small neutral solutes. *Plant J.* 18(6):577–587.

Hara-Chikuma M, Watanabe S, Satooka H. (2016) Involvement of aquaporin-3 in epidermal growth factor receptor signaling via hydrogen peroxide transport in cancer cells. *Biochem Biophys Res Commun.* 471(4):603–609.

Henzler T, Steudle E. (1995) Reversible closing of water channels in Chara corallina provides evidence for a composite transport model of the plasma membrane. *J Exp Bot.* 46:199–209.

Henzler T, Steudle E. (2000) Transport and metabolic degradation of hydrogen peroxide in Chara corallina: Model calculations and measurements with the pressure probe suggest transport of H(2)O(2) across water channels. *J Exp Bot.* 51(353):2053–2066.

Henzler T, Ye Q, Steudle E. (2004) Oxidative gating of water channels (aquaporins) in Chara corallina by hydroxyl radicals. *PCE.* 27:1077–1195.

Herrera M, Hong NJ, Garvin JL. (2006) Aquaporin-1 transports NO across cell membranes. *Hypertension.* 48(1):157–164.

Hüsken D, Steudle E, Zimmermann U. (1978) Pressure probe technique for measuring water relations of cells in higher plants. *Plant Physiol.* 61(2):158–163.

Jahn TP, Møller AL, Zeuthen T, Holm LM, Klaerke DA, Mohsin B, Kühlbrandt W, Schjoerring JK. (2004) Aquaporin homologues in plants and mammals transport ammonia. *FEBS Lett.* 574(1–3):31–36.

Kamiya T, Fujiwara T. (2009) Arabidopsis NIP1;1 transports antimonite and determines antimonite sensitivity. *Plant Cell Physiol.* 50(11):1977–1981.

Kirscht A, Kaptan SS, Bienert GP, Chaumont F, Nissen P, de Groot BL, Kjellbom P, Gourdon P, Johanson U. (2016) Crystal structure of an ammonia-permeable aquaporin. *PLoS Biol.* 14(3):e1002411.

Liu LH, Ludewig U, Gassert B, Frommer WB, von Wirén N. (2003) Urea transport by nitrogen-regulated tonoplast intrinsic proteins in Arabidopsis. *Plant Physiol.* 133(3):1220–1228.

Loque D, Ludewig U, Yuan L, von Wirén N. (2005) Tonoplast intrinsic proteins AtTIP2;1 and AtTIP2;3 facilitate NH3 transport into the vacuole. *Plant Physiol.* 137(2):671–680.

Ma JF, Tamai K, Yamaji N, Mitani N, Konishi S, Katsuhara M, Ishiguro M, Murata Y, Yano M. (2006) A silicon transporter in rice. *Nature.* 440(7084):688–691.

Marinelli RA, Marchissio MJ. (2013) Mitochondrial aquaporin-8: A functional peroxiporin? *Antioxid Redox Signal.* 19(8):896.

Maurel C, Boursiac Y, Luu DT, Santoni V, Shahzad Z, Verdoucq L. (2015) Aquaporins in plants. *Physiol Rev.* 95(4):1321–1358.

Maurel C, Reizer J, Schroeder JI, Chrispeels MJ. (1993) The vacuolar membrane protein gamma-TIP creates water specific channels in Xenopus oocytes. *EMBO J.* 12(6):2241–2247.

Maurel C, Reizer J, Schroeder JI, Chrispeels MJ, Saier MH Jr. (1994) Functional characterization of the Escherichia coli glycerol facilitator, GlpF, in Xenopus oocytes. *J Biol Chem.* 269(16):11869–11872.

Medraño-Fernandez I, Bestetti S, Bertolotti M, Bienert GP, Bottino C, Laforenza U, Rubartelli A, Sitia R. (2016) Stress regulates aquaporin-8 permeability to impact cell growth and survival. *Antioxid Redox Signal.* 24(18):1031–1044.

Miller EW, Dickinson BC, Chang CJ. (2010) Aquaporin-3 mediates hydrogen peroxide uptake to regulate downstream intracellular signaling. *Proc Natl Acad Sci USA.* 107(36):15681–15686.

Mollapour M, Piper PW. (2007) Hog1 mitogen-activated protein kinase phosphorylation targets the yeast Fps1 aquaglyceroporin for endocytosis, thereby rendering cells resistant to acetic acid. *Mol Cell Biol.* 27(18):6446–6456.

Onsager L. (1931) Reciprocal relations in irreversible processes. *Phys Rev.* 37:405.

Petropavlovskiy AA, Tauro MG, Lajoie P, Duennwald ML. (2020) A quantitative imaging-based protocol for yeast growth and survival on agar plates. *STAR Protoc.* 25;1(3):100182.

Pommerrenig B, Diehn TA, Bienert GP. (2015) Metalloido-porins: Essentiality of Nodulin 26-like intrinsic proteins in metalloid transport. *Plant Sci.* 238:212–227.

Preston GM, Carroll TP, Guggino WB, Agre P. (1992) Appearance of water channels in Xenopus oocytes expressing red cell CHIP28 protein. *Science.* 256(5055):385–387.

Rodrigues O, Reshetnyak G, Grondin A, Saijo Y, Leonhardt N, Maurel C, Verdoucq L. (2017) Aquaporins facilitate hydrogen peroxide entry into guard cells to mediate ABA- and pathogen-triggered stomatal closure. *Proc Natl Acad Sci USA.* 114(34):9200–9205.

Sies H, Jones DP. (2020) Reactive oxygen species (ROS) as pleiotropic physiological signalling agents. *Nat Rev Mol Cell Biol.* 21(7):363–383.

Stroud RM, Miercke LJ, O'Connell J, Khademi S, Lee JK, Remis J, Harries W, Robles Y, Akhavan D. (2003) Glycerol facilitator GlpF and the associated aquaporin family of channels. *Curr Opin Struct Biol.* 13(4):424–431.

Takano J, Wada M, Ludewig U, Schaaf G, von Wirén N, Fujiwara T. (2006) The Arabidopsis major intrinsic protein NIP5;1 is essential for efficient boron uptake and plant development under boron limitation. *Plant Cell.* 18(6):1498–1509.

Tesse A, Gena P, Rützler M, Calamita G. (2021) Ablation of aquaporin-9 ameliorates the systemic inflammatory response of LPS-induced endotoxic shock in mouse. *Cells*. 10(2):435.

Tian S, Wang X, Li P, Wang H, Ji H, Xie J, Qiu Q, Shen D, Dong H. (2016) Plant aquaporin AtPIP1;4 links apoplastic H2O2 induction to disease immunity pathways. *Plant Physiol*. 171(3):1635–1650.

Tsukaguchi H, Shayakul C, Berger UV, Mackenzie B, Devidas S, Guggino WB, van Hoek AN, Hediger MA. (1998) Molecular characterization of a broad selectivity neutral solute channel. *J Biol Chem*. 273(38):24737–24743.

Uehlein N, Lovisolo C, Siefritz F, Kaldenhoff R. (2003) The tobacco aquaporin NtAQP1 is a membrane CO2 pore with physiological functions. *Nature*. 425(6959):734–737.

Vieceli Dalla Sega F, Zambonin L, Fiorentini D, Rizzo B, Caliceti C, Landi L, Hrelia S, Prata C. (2014) Specific aquaporins facilitate Nox-produced hydrogen peroxide transport through plasma membrane in leukaemia cells. *Biochim Biophys Acta*. 1843(4):806–814.

Vissers MCM, Hampton M, Kettle AJ (eds) (2017). *Hydrogen Peroxide Metabolism in Health and Disease*. CRC Press.

Wong CM, Zhou Y, Ng RW, Kung HF, Jin DY. (2002) Cooperation of yeast peroxiredoxins Tsa1p and Tsa2p in the cellular defense against oxidative and nitrosative stress. *J Biol Chem*. 277(7):5385–5394.

Wudick MM, Li X, Valentini V, Geldner N, Chory J, Lin J, Maurel C, Luu DT. (2015) Subcellular redistribution of root aquaporins induced by hydrogen peroxide. *Mol Plant*. 8(7):1103–1114.

Ye Q, Steudle E. (2006) Oxidative gating of water channels (aquaporins) in corn roots. *Plant Cell Environ*. 29(4):459–470.

Zhang M, Shi H, Li N, Wei N, Tian Y, Peng J, Chen X, Zhang L, Zhang M, Dong H. (2022) Aquaporin OsPIP2;2 links the H2O2 signal and a membrane-anchored transcription factor to promote plant defense. *Plant Physiol*. 188(4):2325–2341.

Zhao XQ, Mitani N, Yamaji N, Shen RF, Ma JF. (2010) Involvement of silicon influx transporter OsNIP2;1 in selenite uptake in rice. *Plant Physiol*. 153(4):1871–1877.

Zhu GL, Steudle E. (1991) Water transport across maize roots: Simultaneous measurement of flows at the cell and root level by double pressure probe technique. *Plant Physiol*. 95(1):305–315.

Zimmermann HM, Hartmann K, Schreiber L, Steudle E. (2000) Chemical composition of apoplastic transport barriers in relation to radial hydraulic conductivity of corn roots (Zea mays L.). *Planta*. 210(2):302–311.

INDEX

Note: Page numbers in *italic* indicate a figure and page numbers in **bold** indicate a table on the corresponding page.